The Electric Vehicle

The

Pleasure

—and your Rauch & Lang or Baker Electric is a car of Pleasure.

You find pleasure in the utility by which you so easily reach the out-o'-way places or make a social call.

Pleasure in the ease of control—in the roomy interior, in the genuine coach work, and in the knowledge that your car *is* a Rauch & Lang or Baker Electric.

Rauch & Lang Electric

"The Social Necessity"

Baker Electrics

THE BAKER R & L COMPANY
Cleveland, Ohio

ELECTRIC VEHICLE

Technology and Expectations
in the Automobile Age

GIJS MOM

The Johns Hopkins University Press
Baltimore and London

The translation of this book was made possible by a grant from the
Netherlands Organization for Scientific Research (NWO).

An earlier version of this work was published as *Geschiedenis van de
Auto van Morgen,* © Kluwer Bedrijfsinformatie BV, 1997

The Johns Hopkins University Press
2715 North Charles Street
Baltimore, Maryland 21218-4363
www.press.jhu.edu

Translation by Jenny Wormer

Library of Congress Cataloging-in-Publication Data

Mom, Gijs, 1949–
    [Geschiedenis van de auto van morgen. English]
    The electric vehicle : technology and expectations in the automobile
age / Gijs Mom ; translation, Jenny Wormer.
        p. cm.
    Includes bibliographical references and index.
    ISBN 0-8018-7138-7 (hardcover : alk. paper)
    1. Automobiles, Electric—History.   I. Title.
    TL220 .M6613 2003
    629.22′93′09—dc21   2002151369

A catalog record for this book is available from the British Library.

Title page illustration: Rhetorical jugglery in 1916: a Rauch & Lang Company ad for its Baker Electric car rhapsodized on the Baker's ability to deliver a mother and her children to any field of clover. The main players in the American electric-vehicle game, the electric utilities, stubbornly referred to the electric passenger car as a pleasure vehicle, implying that the *true* electric was a truck. This image and the following two adhered to leading features of the dominant gasoline-car culture and reinforced the classical picture of the electric as an innocent and environment-friendly plaything for well-to-do women, a fantasy that haunted automotive historiography ever after. (Author's private collection)

To Charley

Contents

Preface

ON THE COMPLETION OF THIS BOOK, I have the feeling of coming home again after a journey lasting almost twenty years. It led me—as if on an anthropological voyage of discovery—deep into the territory of the tribe of engineers, characterized by its own culture and its own language. Only those who have been there know what it is like.

I first obtained a master's degree in Dutch literature at the University of Nijmegen, the Netherlands. This study was completed by a collectively written literature-sociological thesis, after a turbulent time in the student movement there. But in 1978, I decided to start a second study at the HTS-Autotechniek (a college for automotive technology) in Apeldoorn. Thus I crossed the bridge between the Two Cultures.

The impulse to start thinking about this transfer was rather trivial. Shortly before my graduation, I had bought a Renault R4 from a friend but did not have the money to take it to a garage for maintenance. In the tradition of the Socialist Student Unions, I then started a self-help group that tried to learn the basic principles of the automotive-technological doctrines. I enjoyed the exhilarating feeling that craftsmanship can give: soon the Renault mechanism no longer held any secrets for me.

The field of automotive technology seemed a paradise full of certainties to me. After having been bogged down in a laborious theoretical thesis about a literary labor movement during the Depression, I anticipated a pleasant future full of established truths. A bolt is either tight or loose, doubt about this—or so it seemed—was impossible.

The gap between the Two Cultures really exists. After a year at the HTS, a former fellow literature student asked me whether my interest in technology in general and in that of the car in particular had meanwhile been satisfied. She did not have the faintest notion—just like me, a year earlier—what endeavors are required outside of High Culture to become familiar with that other culture, usually not written with a capital C. The culture shock was great for me, indeed. I was a poor mechanics student, because I used the textbook in the wrong way. While I studied the axioms and definitions and became entangled in a web of contradictions, my much younger buddy with only a high-school background received better grades because he did what was expected of him: he just did the exercises. Slowly I mastered the technology and the jargon of the engineering practice. Doing so, I discovered that a large part of the study (in my recollection more than half) was spent learning a trade jargon, an attitude, and a group consciousness that were hardly related to technology as such. Led by priests and high priests in automotive technology—and under a much stricter discipline than I was used to during my language studies—much time was spent (often unconsciously) acquiring an equally vehement delimiting mechanism as I had experienced in my previous life. There it had taken the shape of fear and sometimes aggression against everything that had to do with technology. As a unique icon of this technology, the automobile was a particular victim of this aggression. Thus, I no longer had much in common with my former fellow students, who had meanwhile become established teachers of Dutch, and we grew apart.

But, apparently it is hard to suppress one's inclinations. In the middle of my automotive studies, my interest in the historical and cultural aspects of my surroundings began to surface again. It arose out of a strongly felt lack of expressiveness of automotive-technological theory. During a training period in Paris (indeed, at Renault!) I had to apply a method of analysis of the friction losses of a combustion engine (the so-called Willans method) used there. In my attempts to do so, I was overcome by serious doubts about the usefulness of the method. Half of my time in Paris I spent studying the origin of the method: examination of the archives took me back to the era of the steam engine, and thus I was back at the beginning. Meanwhile my boss at Renault wondered how he could continue building cars on the basis of my hobby horse. And, believe it or not, it was possible to do so by means of a method that was clearly attuned to the steam engine. In Paris I learned that in technology a theory is "true" if it works.

Yet, history appeared to be an important tool for putting technical knowledge into perspective. This holds for at least two important areas, both of which I entered in the past fifteen years, after completing my engineering degree. First, history is eminently suited to bridging the gap between the two cultures. History, after all, cannot be practiced sensibly without integrating the social context in which technology thrives. In the hundreds of articles I wrote after 1982 for the Dutch and Belgian automotive and engineering press about advanced automotive technology, the historicizing approach to a problem appeared to be an excellent introductory way to explain complex artifacts or parts of them. It also makes technology comprehensible for nonengineers, without requiring concessions to the level of technical explanation. If one scraps the jargon and the quantitative aspect of the technical problem, one is usually left with a core that is relatively easy to convey. Conversely, a competent historical approach on the part of the engineers helps break down the misconception that the social side of technology cannot be exciting. So, history, of which the development of technology is an inherent part, may lead to a mutual enrichment of both cultures. This enrichment is eminently possible with the modern forms of technological history, which have long left behind the domain of technological narrow-mindedness and navel-gazing.

Second, history can be of excellent service in technical education, as I have seen repeatedly since my employment as a lecturer at the HTS-Autotechniek in 1989. The historical approach to a technical problem not only helps to put the solution into perspective, but by showing how contemporaries struggled toward that solution, history also helps to explain the problem itself. As a result, the problem loses its axiomatic, ideal-typical character (the form in which technical "truths" are often presented in education nowadays). It becomes alive, and it helps students to recognize that technology is a process, in which not only strictly technical, but also economical, political, and sociocultural factors play a role. As engineers, the students themselves will play a role in this process. The "truth" of technology is a historical truth, as the historicizing approach shows again and again.

In this sense, the whole story following here can be read as a long plea for an integrated approach involving both history (in which technology plays its role) and technology (in which the social aspects only can be ignored at the cost of sterility and technological narrow-mindedness). Both sides of the technological-historical model must be elucidated on a knowledgeable basis. This insight forms the basis for critical treatment of the social-historical approach to technology in the prologue and the essay on methodology: the history of the electric car presented to the reader may be read as an attempt to "save" automotive technology for these approaches. I will leave it to the reader to decide how far I have succeeded in this attempt. If so, he or she might become as convinced as I did that the automobile can be fruitfully analyzed as a mirror, as a lens for societal processes, in our case mainly focused on the emergence of a consumption society.

So, I AM HOME AGAIN, but meanwhile others live in the house than those I left behind. Especially during the many years of searching archives in five countries and gathering additional information in another seven, I have made new friends and acquaintances who alleviated the loneliness inherent to the writing process. They are too numerous to mention them all here: those who have been important for a specific section of my research are mentioned in the chapters concerned. Furthermore, I include in the Bibliography a list of people and organizations I visited during the past years for the purpose of this survey, or with whom I corresponded.[1]

Of the many librarians, archivists, and curators of technological museums, I like to thank especially Mark Patrick (National Automotive History Collection, Detroit Public Library) for his never-ending help and hospitality. I also vividly remember Douglas Tarr's professionalism (Edison National Historic Site ENHS, West Orange) and Wolfgang Zur's meticulousness (Berliner Feuerwehrmuseum), just as the generous help in finding illustrations, offered by Harry Niemann and Stanislav Peschel of the DaimlerChrysler archives in Stuttgart. No less helpful were Hannelore Stöckl and Peter Poll of the Technisches

Museum in Vienna, Ladislav Mergl of the Technisches Nationalmuseum in Prague, G. Casobon of the Musée National de la Voiture et du Tourisme in Compiègne, and Mrs. F. Henry of the Musée des Transports in Paris. A pleasant surprise was the meeting with Jean-Luc Krieger, grandson of one of the most important French builders of electric vehicles.

From an early stage Ariejan Bos (NCAD) took care of the design of all figures and tables. My new friends meanwhile also include Kurt Möser (Landesmuseum für Technik und Arbeit, Mannheim), Thomas Köppen (Museum Achse, Rad und Wagen, Wiehl), Alan Loeb (formerly Argonne National Laboratory, Washington), and Andreas Knie (Wissenschafszentrum Berlin für Sozialforschung), with whom I was able to share my interest in the early history of automotive technology. The discussion with this small group of experts has considerably honed my analysis. I also consider Michael Schiffer (University of Arizona, Tucson) as a friend. He was the first to publish an analysis of the early electric vehicle and he completely disinterestedly gave me all his documentation regarding the nonautomotive-technological journals and all his copies of many weeks' work at the ENHS. Another good friend is David Kirsch (Robert H. Smith School of Business, University of Maryland). When I called him on the phone a decade or so ago, he found out—to his initial dismay—that someone else was writing a dissertation on the same subject as he was. David and I started an exchange of documents, opinions, and eventually whole chapters of our dissertations. I found this heart-warming in an academic world, which has meanwhile become quite competitive. Moreover, David was in a disadvantageous position, because he finished his dissertation at the moment I started writing, so that I was able to read his texts, but he could not read mine. I am indebted to David for his generosity.

When the writing reached its final stage, I received important editorial support, also with regard to content, from Martijn Bakker (Nijmegen) and Johan Schot (TU Twente and TU Eindhoven). That was also due to Harry Lintsen's organizational talent, who has supervised me effectively during the whole project. Harry Lintsen also made it financially possible for me to be detached to the Technical University Eindhoven for five months. Without that dispensation of my teaching and other obligations at the HTS-Autotechniek (taken over by my colleagues Frank de Vries and Jan Voortman, for which I am very grateful) I most certainly would not have been able to write this book.

Since my dissertation appeared in print in Dutch in the fall of 1997 a lengthy revision and translation process started. I could do this because of a professorship I acquired at the Technical University of Eindhoven, where the Foundation for the History of Technology (SHT) has been active for more than a decade under the directorship of Harry Lintsen and Johan Schot, respectively. The intellectually stimulating group dynamics I encountered amidst the dozens of historians, philosophers, anthropologists, and other social scientists (and, yes, some carefully preserved engineers) within SHT are, to my knowledge, unique in the discipline. At the same time I could satisfy my urge to cooperate on an international scale by partaking in the friendly and encouraging debates organized annually by both ICOHTEC and SHOT. Both international scholarly organizations (the last, perhaps, a trifle too "American") are true cradles of professionalism, indispensable for anybody who—like me—at a later moment in his career, wishes to become familiar quickly with the main actual research problems of our field.

The Netherlands Organization for Scientific Research (NWO) provided a grant for the original translation into English (including all citations in other languages than Dutch), which was done by Jenny Wormer. She admirably steered a middle course between the pitfalls of the automotive-technological and electrotechnical terminology on the one hand, and the instructions by the American editor, in the person of Bob Brugger of Johns Hopkins University Press, on the other. My special thanks go to my friend Clay McShane, who at a crucial moment was an important intermediary between me and the publisher. I also wish to thank Maria E. denBoer, who guided me—as a non-native speaker—through the forest of the idiosyncrasies of the American language.

In the end, about a third of the original text has been omitted. The original, including appendices, comprised about 650 pages. At the same time, I used this opportunity to considerably revise the analysis of the early gasoline car as "adventure machine" in chapter 1. Moreover, I added the results of the most recent research, mostly done in Germany, on the relation between automobile and fin-de-siècle culture. Throughout the process of revision, I was supported by the critical comments of JHUP reviewers Paul Israel and Bruce Seely.

The last lines I have reserved for my soulmate and partner for life, art photographer and children's norms and values teacher Charley Werff. For the past decade, she participated physically and psychologically in my anthropological and technological-historical forays. She also took care of most of the photographic reproduction work for this edition. The extent of the appreciation and admiration I feel for her, I hope to express in another way: I dedicate this book to her.

EINDHOVEN, NOVEMBER 2002

The Electric Vehicle

Substituting for the Horse,
Choosing Propulsion

Progress happens when all the factors that make for it are ready, and then it is inevitable. —Henry Ford, 1934

A technology does not succeed, because it is technologically superior, but it is considered technologically superior, because it has sociologically succeeded. —Werner Rammert, 1994

UNTIL RECENTLY, the electric-powered automobile has not received its fair share of attention in the historiography of road transportation. No one seemed to care about the possibilities of electric propulsion until a few years ago, when it began to be considered as an alternative for the polluting, noisy combustion engine. Prior to that the electric car had been dismissed as the hobbyhorse of penniless environmentalists. But when the established automotive industry became interested in the subject and California pioneered with a zero-auto-emissions statute, the mood changed. Would a serious successor to the gasoline car emerge? Were designers about to lay the foundation for "tomorrow's car"? For automotive engineers, a new, exciting time of innovative technical development dawned. Consumers seemed receptive to a wider selection of vehicles. Small city cars, spray-painted in lively colors and adapted to their personal tastes, would become the norm. Local and regional authorities already envisioned a drastic reduction of environmental problems. The automotive industry could restore the image of "the automobile," which pollution had badly damaged.[1] As if invented on the spot, there it was: the electric car, seemingly without a past.

The story of the electric vehicle begins about the same time as the history of the gasoline-powered automobile, in the 1880s. At the time, no one knew how the separate narratives would develop or how the stories would end. A century ago the notion of the electric car as the "car of tomorrow" was commonly held by some of the important people involved in automotive development.

Especially in the United States, between 20,000 and 30,000 individual owners drove their electrics in and around cities in the East, the Midwest, and the far West. After World War II another wave of electric enthusiasm started in the United States. It rolled over the industrialized world during the 1960s and 1970s, and again two decades later. As a result of this interest, at the dawn of the twenty-first century, DaimlerChrysler and Ford have announced that by 2010 electric passenger cars based on fuel cell technology will be available to the general public.

The history of the electric vehicle offers ample evidence that the choice of electric propulsion over steam and gasoline was not based on blind dreaming or irrational partisanship. When in 1908, Maximilian Reichel, chief of the fire department in Berlin and well-educated in steam engine technology, decided to replace his horse-drawn fire engines with electric-powered equipment, he knew exactly what he was doing. Several reports and tests had convinced him that steam traction would be too expensive and internal combustion too unreliable. Within a half decade, several of his colleagues in other large German cities followed his example. Shortly before the outbreak of the Great War, 164 heavy, electrically propelled fire engines were in use all over the country.

Reichel certainly was not the only fleet owner who, during the first decades of the automobile age, opted in favor of electricity. In Berlin, for instance, 574 electric taxicabs roamed the streets at about the same time; in Amsterdam a fleet of nearly 80 electric cabs appeared to be very profitable until well after the war. In the United States more than 10,000 electric trucks were in use in several large eastern cities. Between the wars and after the Second World War, tens of thousands of British "milk floats" delivered dairy products to family homes, 5,000 electric vans and trucks were deployed in Germany and in France, especially around Lyon, dozens of electric buses were in use, and tens of thousands of industrial trucks made intrafactory deliveries in every industrialized country.

Yet during the century-long history of the automobile, the widespread conviction that electric propulsion was an inferior technology remained stubbornly in place, as if based on solid engineering knowledge. Electricity worked mainly in niche applications, so the assumption went, where the battery's low energy density, high mass, and low range on one charge were not critical. Traditional automotive historiography has made a remarkably consistent judgment about electric cars as a promising, although on closer inspection, inferior technology that failed because the battery was too heavy and the range too short. Initially, the electric car had a chance of success, because the gasoline car had not yet reached the level of perfection as we know it now: "The reputation of the electric vehicle lived off the imperfections of the gasoline engine." This assessment of the American automotive historian John Rae represents the prototype of such opinions : "In spite of intensive recent experimentation with the electric car," he concluded in the midst of its postwar revival, "there is no evidence that the limitations of the battery are about to be overcome, although this is a field where a technological breakthrough is always possible." Such opinions directly reflect only a part (but a dominant part) of the source material, where they are expressed hundredfold. Remarkably enough, they coincide with the prevailing opinion within the community of automotive engineers and, through them, the public at large.[2]

How is this possible? Why did the electric car fail to develop into the prevailing technology? How should the failure of this technology be interpreted when we know that several initiatives were quite successful, even outside the canonized electrical niches? And what do inferiority and superiority mean in this context? Are not these terms in fact intellectual shortcuts to a technocratic answer to the dilemma of technological alternatives—thus blocking a thorough insight into the roots of automobile culture? These questions provide the starting point of this work, and they underscore the truth that we cannot separate technology and culture—especially when an artifact like the gasoline car carries the label "icon of the twentieth century." In studying the rise and fall of the early electric car, one might suppose that a superior technology failed (i.e., did not penetrate the market because of nontechnical failure factors). Yet because people bring about technical change, the larger question is : Why did an increasing number of people opt in favor of the car with an internal combustion engine?

On Concepts and Vocabulary

When, a decade or so ago, I decided to analyze the failure of the electric vehicle, it soon became clear that an answer could not be found in the vehicle itself. The problem lay entangled in a web of technological and cultural interaction between the electric and the dominant gasoline vehicles. This insight opened up a vast and complex field of study. Even if, for instance, we determine why the electric taxicab lost terrain to its gasoline competitor, those conclusions do not necessarily apply to fire engines, street-sprinkling vehicles, or vans and trucks. Besides, engineers and manufacturers view their products from different perspectives than do users, whether they are individual passenger car owners (or their wives or daughters), municipalities, or commercial-fleet managers. Moreover, applications and users' views differ as to geographical context (in this book mainly confined to several European countries and the United States), and they change over time.

The project forced me to devise several analytical instruments, mostly metaphorical in nature. They do not supply a model in the sense of the social sciences, but they provide a set of tools to bring some order to the complexity. Elsewhere I have attempted to explain and justify the use of these tools to the specialist reader.[3] One of these tools is the field concept, borrowed from electromagnetic theory. The field of application comprises the areas of user applications and prototypical, real-world experiments of a vehicle type (a passenger car, a truck, a racing car) in a certain stage of its development and in a specific place. In the field of application, the choice of propulsion takes place. The field of expectation contains the historical actors' technical and social fantasies and images of future applications. These fantasies play a large role in technical change; they influence choices. But between expectation and application lie a number of intervening fields, all of which work as filters and lenses. They consist of sociocultural, technical, political, and economic factors that enable or constrain, help expand or block the field of application. We concentrate on the technical field without, however, excluding from our analysis important parts of the other fields in this cluster. In the technical field the conversion of the schemes and expectations into a concrete artifact takes place. The size of the technical field also determines the ease with which the conversion of a desired artifact into a realized artifact is possible (feedback also takes place here: fantasies are adapted and changing wishes lead to changes in artifacts). If a technical field does not support an expectation, we may refer to it as closed, at least at that point. Once opened, its boundaries are also subject to social negotiations: even though the artifact's technical properties are without dispute (i.e., even though the artifact realizes the technical process expected from it), it may still end up in a museum. Society, or part of it, may decide that the price for such functioning (i.e., the realization of the technical properties during and through the artifact's use) is too high, not only in a financial, but also in a sociocultural sense, which seemed briefly to be the case with the internal combustion engine in California.

In analyzing the history of technology, it is important to be aware of the fact that artifacts consist of components that have been combined in a meaningful way to perform certain technical properties (for example, the turning of the wheel) and thus enable users to realize a sociocultural function (driving a car, for instance).[4] Because of the meaningful grouping of the components, an artifact thus can be viewed as a system in which we can observe a hierarchy of components in which one component fulfills a general function (a gearwheel, for example) while another plays a role more tailored to the specific function of the artifact (a piston of a car engine). To avoid confusion with the usual notion of system in the history of technology, however, I reserve the term *internal structure* (or simply *structure*) for the systematic construction of the artifact.[5]

The structure of an automobile can be divided into subsystems (electrical, propulsion, or chassis), which in turn consist of functional groups that have a coherent, relatively autonomous function within the subsystem (such as the braking system, the engine, or the wheel suspension). The functional groups themselves are composed of subsystems of a lower order and of a divergent complexity (such as a brake, the valve gear of the engine, or a shock absorber). Even deeper in this constructive hierarchy are components of varying complexity (such as a brake shoe, a piston, or a shock-absorber rubber). Finally, one arrives at the basic components of bolt, nut, gear wheel, and so on. A different application (at the fire department, the post office, or a taxicab company) implies a different function (as fire-extinguishing machine, as parcel delivery van, or as taxicab) and a different vehicle type (as special vehicle, as light commercial vehicle, or as adapted passenger car). Functions are realized by users, but they have to be made possible by technical properties. For instance, a road-sprinkling function was enabled by a type of automobile characterized by a very low driving speed (either by very big gear reductions in the case of the gasoline automobile, or by applying electric propulsion in early vehicle technology). Similarly, a mail delivery function could only be realized by an auto-

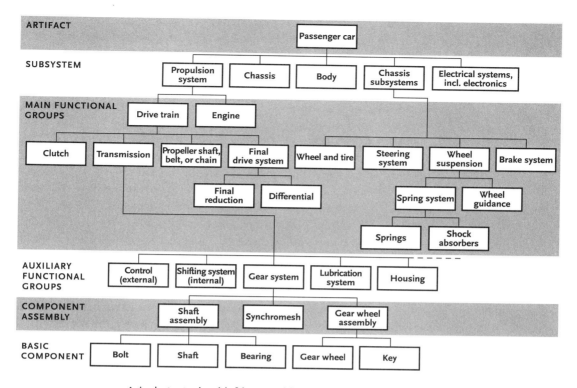

ARTIFACT							

A simple structural model of the automobile, showing its main functional groups. For the purpose of simplicity, only the transmission has been partly broken down into its constituent parts.

mobile that allowed for very frequent stops and starts. Automobile structure thus varies by place and changes over time. Components are removed, new ones are added, and in general their properties are subjected to a process of continuous, incremental change. Precisely how this process works is still an open question. This work demonstrates that in the case of the competition between electric and gasoline propulsion alternatives, which is an example of rivalry between artifacts, functional transfer between these alternatives, rather than an exchange at the level of components, played a decisive role. In other words, the rivalry between electric- and gasoline-powered vehicles throws light on the ways in which competing technologies, in trying to be more attractive in the marketplace, tend to borrow (or steal) each other's properties and functions.

In order to make this process comprehensible, I distinguish between three (or, with postwar development included, four) generations. The first generation consisted of electrified horse-drawn vehicles (however different they may have been in Europe and America). The exterior of the second generation often was not much different from that of the first, but it was characterized by a greater adequacy, a greater reliability, due to a gradual improvement of technology, especially the battery, but also, in many cases, the pneumatic tire. Some electric cars of this generation began to incorporate characteristics of the prevailing (gasoline) automotive technology, such as the elimination of central pivot steering, the use of steel chassis, or the single-motor system with differential in the United States. Also a hooded "nose" started to crop up here and there, used to accommodate the batteries. The third generation consisted of electric cars that presented themselves more and more as an electric version of the gasoline car. Often, they were consciously modeled after the gasoline rival, which in some applications already dominated the automotive field. They thus foreshadowed the real conversions from a later period: the fourth-generation electric cars derived directly from the gasoline-car structure.

This study follows sociotechnical experience over these four generations, with separate chapters on the privately owned passenger car and some type of commercial field of application, such as the taxicab or the truck, the bus or a specialized vehicle (ambulance, street-cleaning vehicle, garbage-collecting truck, firefighting truck, mail van). Apart from this temporal and functional distinction, differences between European countries and between Europe and the United States are also presented. These are mostly dealt with in every chapter, except in the case of the heavy truck. Here, the distinctive developments of the European and the American fields of application, for reasons of composition, have been treated under second and third generations, respectively. This does not mean, of course, that there were no American second-generation trucks (they figure in an introductory way in chapter 6), nor that Europe did not know third-generation trucks. Their history belongs in chapter 7, which functions as a transition to the epilogue. Put another way, the following narrative offers the reader three different triptychs: one on the passenger car for private use (chapters 1, 3, and 7), one on the taxicab (chapters 2, 4, and 7), and one on the heavy-duty version of the automobile (chapters 5, 6, and 7).

Electric Propulsion before the Automobile

The people and organizations involved in automobilization were not only part of a contemporary network full of technical, social, and cultural interactions; they were rooted in historical tradition. Especially in popular automobile literature, this tradition is usually construed by means of the many earlier proposals for the introduction of automobile vehicles for individual transportation, particularly the steam car. This, however, does not do justice to historical reality. One may call the image of the automobile for personal use a topos in preautomobile history and as such even label the desire for (auto)mobility an "anthropological constant." But the proposals were far too incidental to speak of a real tradition. Moreover, this wrongfully suggests that tradition would mainly be of a technical nature, as if an inferior, old-fashioned technology was simply replaced by a modern, technically more perfect propulsion system.

The picture thus emerges that the gasoline car was a direct descendent of the steam car. This linear idea has been repeated until it has become one of the canons of automotive history. It discards all alternatives as dead ends, inferior technology, or failures. By extending the prehistory of the automobile to nineteenth-century land transportation, however, we get a view of other, nonautomobile vehicles, such as the horse-drawn taxicab, the electric streetcar, and the bicycle. These have been at least as important as a source of inspiration to the automobile pioneers. In this perspective, the nineteenth century was an era of spectacular expansion.

From the 1830s, long-distance passenger transportation over land occurred increasingly by steam train. But even before that, "personal travel was already a growth industry." In the first half of the nineteenth century, attempts were made to motorize the heavy, rail-less vehicle by means of the steam engine. This happened in two distinctive waves, especially in Great Britain. But such incidental endeavors eventually were abandoned in favor of a further development of the steam train. This process did not directly lead to a replacement of the horse, however, even in long-distance transportation. Traffic on through roads that ran perpendicular to railroad lines often increased, because the transportation needs of the residents of small towns, not located on a railroad, had grown as well. Also, in many countries the expansion of a railroad network was accompanied by road network building and paving on a regional scale to accommodate regional transport between secondary cities and their hinterland.[6]

The train satisfied more than just a transportation need. It was the first motorized vehicle to supply a sensation of speed: the tripling of speed in comparison to that of the stagecoach was at first experienced with fear and trepidation, but gradually became part of the achievements of nineteenth-century society. In response, the dominant technology started competitive services, as in Germany, for example, where the *Eilpost* displayed an as yet unprecedented speed, or in England, where the *Mailcoach* provided a network of fast connections across a large part of the island. In this field of application a substitution process was put in motion, and the steam-propelled train would become the dominant technology for a century to come.[7]

The "industrialization of space and time" that accompanied the train not only revolutionized speed, but also the travelers' observation. Forced to look out of the windows sideways, as the speed increased, the traveler could no longer distinguish separate objects, unless they were in the distance. The view becomes blurred in both cases: the foreground changes into colorful lines, the distance consists of dots that are equally colorful and equally hard to distinguish. Because of this, a panoramic view of the landscape developed and marked a provisional end to the foreground, which had been so characteristic of preindustrial travel.[8]

Whereas the emergence of the train, against the background of general industrialization, suggests that the nineteenth century was the age of the steam engine, with regard to urban transit and at the regional level this period could be characterized as the century of the horse. For this field of application the technical field of horse traction was by no means absolutely closed. It took another century for the horse to be successfully displaced from the various application fields it had occupied for such a long time. The prevailing horse economy, however, should not so much be viewed in the sense of a continuation of an old tradition, but as a modern phenomenon, as it was, especially in the major cities, large-scale and rested on rudimentary principles of fleet management. According to a recent American study, historians "ignored this increase in animal powered vehicles, because it happened at the same time as the more spectacular arrival of the electric trolley and because it is difficult to document."[9]

Two types of vehicle dominated the passenger scene: the horse-drawn taxi-cab and the horse-drawn bus. In Berlin, for example, there are clear indications that traffic growth had started before the arrival of the train and urban public transit. Until the 1860s, the horse taxicab accounted for the largest part of passenger transit in German cities. Many American cities started a road surfacing program at the end of the eighteenth century (New York started earlier, in 1691); paving kept pace with the increase in the number of carriages. Shortly after 1800, the horse taxicab also appeared in America on the newly paved roads.[10] The horse carriages and the horse buses increased in number from the moment the train station appeared at the edge of town. Traction power and comfort of the horse bus increased considerably when it was put on rails after the example of the steam train and influenced by the poor condition of the roads. In most cities of the industrializing world the horse streetcar, as soon as it was introduced, quickly became the first urban mass means of transit.[11]

In the United States the construction boom in streetcar tracks began in the 1850s. In Europe the development of the *chemin de fer américain* (American railroad), as it was called in Paris, only got under way during the 1860s and especially in the 1870s. The rail-less horse bus was all but abandoned in most cities before the end of the nineteenth century. But in London—the biggest city in the world during the entire century—the horse bus experienced a revival in the last quarter of the century. London may well serve as a model, however, for all those other big cities in the West, where the growth of train traffic (even though, in the case of London, this penetrated unusually far into the city center) could not keep abreast with urban traffic: "Old horsepower, better organized, was in this respect performing better than the latest steam." In most Western cities the growth of transportation was higher than that of the population.[12]

It is not certain what exactly triggered the subsequent motorization of urban public transport, including taxis. It has been suggested that the horse economy ran into serious problems of urban pollution, but this seems to be a confusion of the field of expectations and the application field and, thus, to be more a reflection of later propaganda by automobile proponents than solid historical evidence. Nevertheless, it is remarkable that in some cases the growth in the horse population started to diminish in the 1870s, well before motorization gained momentum, and that criticism of horse traction only became louder when alternative types of propulsion became available. Did the horse economy reach its physical boundaries within the big cities before motorization? Perhaps so, but just as numerous as the social impulses for horse substitution were the technical impulses. For potential alternatives to horse traction were abundant, based on all energy sources imaginable at the time. The field of expectations abounded with possibilities and promises.[13]

As for the application field, Paris looked like a gigantic testing station in the beginning of the twentieth century. For besides the *automotrices Rowan* (streetcars with an integrated steam engine, 1889), "fireless streetcars" (1878), with an external source that generated the steam pressure, were also deployed there. Compressed-air traction was also applied, even with some success. A famous application was the *funiculaire de Belleville*, a cable streetcar for the steep slope in Belleville based on a central steam engine. It became especially popular in America and to a lesser extent in Great Britain and Australia. Traction by means of a coiled spring, ammonia vapor under pressure, and, of course, the internal combustion engine still in its infancy—they were all tried and applied here and there for a shorter or longer period, depending on the local and geographical circumstances. In most big and medium-sized cities in the industrializing world, some version of this competition between the alternatives took place.[14]

This variegated technological spectrum was the source of great confusion and uncertainty to the local authorities. In the middle of all this confusion, a small train was driving across the site of the Industrial Exhibition of 1879 in Berlin. Its electric motor was powered by an underground cable. Werner von Siemens had designed and built this show train, which was the first electric streetcar in the world.[15]

The attractiveness of electric traction was not determined solely by technical aspects. The field of expectations was often at least as powerful. In the city culture of the fin-de-siècle, based on the horse economy, electricity presented itself as a nervous (as it destroyed the night), but clean technology. It was simultaneously an expression of scientific and technological progress as well as a symbol of the destruction of the old class society and its values. The feeling prevailed that this was an era full of scientific and technological innovations, being swept along by ever accelerating progress, symbolized by everything that worked on electricity. This feeling spread across the entire Western world and not only caused a general euphoria, but also some concern about the future. Uneasiness about the social and cultural consequences of this development was stronger in Europe than in America. Due to a shortage of skilled labor, the belief in the "makeability" of a new society by means of a *technical fix* was much more evident in the United States.[16]

According to a recent American analysis, the "electric enthusiasm" in the United States and the entanglement of interests of the lighting and traction industries were the decisive factors behind the fast replacement of the horse and the victory over most other motorization alternatives. A central figure in this electrification movement was Frank Sprague, who in 1888 in Richmond, Virginia, developed a standard configuration with overhead wiring. In general, however, cost advantage appeared to be of secondary importance, if it played a role at all. The obscurity of cost calculation, especially with regard to depreciation, provided enough space for "interpretation": first one system was presented as the cheapest, then another one. It was actually the *expected* cost advantages that stimulated financiers—infected by the electric virus—to make investments, a mechanism that functioned as a self-fulfilling prophecy.[17]

In the United States, both the streetcar companies and the power stations were privately owned. In the extended electric streetcar network, they were hardly troubled by rules and regulations. In European eyes this resulted in a real wilderness of overhead wires and a jumble of wooden poles in the cities. That was the reason why the local authorities in Europe were not very enthusiastic about the arrival of the American streetcar manufacturers, who moved there when their own market threatened to become saturated. Entirely in line with the spirit of the times, the aesthetic argument was often used to prohibit the implementation of the streetcar with overhead wires in the inner city, with its historical buildings and characteristic, narrow streets. This hesitation by the authorities led to large variations in the starting point of and the pace at which the motorization of the streetcar was executed in the various European countries. And that was not all: the decision vacuum created room for alternative technologies without overhead wire, of an electrical as well as a nonelectrical nature.[18]

In places where the electrification of the streetcar network by means of overhead wires was impossible due to aesthetic, political, or economic reasons, another electrical alternative emerged in the 1880s: the battery streetcar. The rail-less version, the electric bus, was also tried here and there, although somewhat later. The energy source of both vehicle types was the lead battery, developed by Gustave Planté in 1859 and initially meant for stationary applications. The Planté battery was heavy, because the electrodes were formed out of solid lead, a process in which a chemical conversion occurred on the lead surface by repeated charging and discharging. The process is reversible, so that electrical energy can be derived by connecting a load (a lamp, an electric motor). In 1880, another Frenchman, Camille Faure, developed the predecessor of the grid-plate battery, which could accommodate a prepared lead paste. This led to a growing interest in battery traction, because the same amount of energy could now be obtained from a much lighter battery. The first mobile tests based on the Faure concept took place in Paris, where Nicolas-Jules Raffard started a series of experiments in 1881 on a converted horse streetcar, accompanied by a lot of publicity.[19]

Although the early experiments were not very successful, the 1890s saw a renewed interest in the battery streetcar, for two reasons. First, electric rail traction appeared to be a gold mine. Second, because of the ban in many cities on streetcars with overhead wires, the problem of the choice of propulsion system shifted to horse traction versus the electric alternatives, excluding the option with overhead wires. Although during the first year the costs of the battery streetcar were somewhat higher than those of the horse streetcar, the larger capacity of the streetcar carriages often enabled much higher revenues. This attracted— as in the case of the version with overhead wire—powerful financial groups, such as those in Boston and Manhattan, where the Whitney syndicate had managed to create a monopoly on public transportation. For such financiers the battery streetcar alternative was attractive because of the low initial costs, compared to the overhead wire version. Because of its simplicity and the flexibility of installation, the battery traction system was eminently suitable for application to the horse-streetcar carriages.[20]

Horse-drawn streetcar carriages, however, were far too light for this application. As a consequence, there were many breakdowns, stimulating the need for a lighter battery. The heavy glass battery box—usual for stationary purposes—was replaced by an expensive, but less fragile and closed hard-rubber box. A lighter type of solid battery plate became available, or Faure's pasted type was an option. Whereas in the earliest experiments the batteries accounted for half of the vehicle mass, for the second generation this was around 25 percent and sometimes even less. The British Electrical Power Storage Company (EPS) was the first battery producer to exploit the alternative Faure patent successfully. The active mass was put in holes in the lead grid. Between 1888 and 1901, the energy density of EPS's grid-plate batteries increased from 7.6 Wh/kg to 25 Wh/kg (Faure's battery had a density of 7 Wh/kg). The higher energy density made the batteries more vulnerable, however, to the short, but high discharging currents.[21]

Most commercial applications of the battery streetcar took place in Europe. According to a count in 1890, sixty-four battery streetcars, mostly converted horse streetcars, were in use that year. They often operated only a few months, but sometimes more than a year. While in the years up to 1902 the battery streetcar in America never accounted for more than 2 percent of the total vehicle fleet, in Europe that percentage fluctuated between 6 and 8. Due to the late electrification of the streetcar network in Great Britain, in 1890 and 1891 these percentages amounted to 33 and 20, respectively. After 1895 the interest in the battery alternative diminished quickly. But under special circumstances battery traction still remained attractive, as in the mixed system. These hybrid lines with an overhead wire were extended to less busy city districts by installing a battery set in the carriages. Because of the short routes, a lighter battery set, which was charged on the stretch with overhead wire, could be applied.[22]

It was by no means a coincidence that the second German battery-streetcar boom started in Hagen, Westfalen, where since 1887 the Accumulatorenfabrik AG (AFA) was located. The boom also reached Bremerhaven, Charlottenburg (near Berlin), Hannover, Dresden, Ludwigshafen, and Karlsruhe. In Berlin the battery-streetcar fleet, with 335 carriages in 1901, was the largest by far. When the opposition against the overhead wires abated, the technology had evolved to a mixed system there as well. In September 1900, conversion of the Berlin network to an entirely overhead-wire system was begun. That process was completed in 1902, but not until a fierce discussion had raged in the electrotechnical trade press about the causes of failure.[23]

That year Dr. E. Sieg, director of the Kölner Akkumulatoren-Werke (KAW), joined the chorus of German battery defenders. The KAW was the company that delivered its light grid-plate battery to Bremerhaven. Just as the AFA, the KAW initially was part of the "anti-Faure reaction in Germany," a reaction to the problems with the bonding of the active mass in the lead grids. After the expiration of the Faure patents in 1896, however, the KAW was to grow into the most important German representative of the grid-plate battery, due to its involvement in the first electric-car experiments. Meanwhile, the AFA developed its own version of the Planté concept: the large-surface plate. This plate combined the advantages of the Planté and Faure concepts by consciously sacrificing the active mass during the first discharges. But this did not happen until the thick, grooved positive plate had continued formation by means of this mass. Thus, an important part of the time-consuming and costly formation process was shifted to the customer. Heavy, but durable—that became the product philosophy of KAW competitor AFA. In the early 1890s, the latter company grew into the major (stationary) battery producer, when it started a strategic cooperation with Siemens and AEG.[24]

Due to its monopolistic position in the stationary-battery market, the AFA, for the first time in history, was able to put the problematic life span of the lead battery on the agenda. Inspired by an initiative of the British EPS, it initiated a *Revisionsorganisation*. Against an annual subscription of 5 to 10 percent of the purchase price, trained engineers periodically monitored and overhauled the battery installation, replacing defective plates with fresh ones. This passed an important part of the trials (and the costs) of new designs on to the customers, but also offered them the opportunity to calculate depreciation and maintenance accurately and in advance. So, the fundamental predictability of battery life was not brought about by an intrinsic constructive perfection, but by the monitoring of specialists, a principle that would later be adopted in electric vehicle fleets and even functioned as a model for the car system in general by delegating maintenance to a specialized garage infrastructure. In the terms of the field notion, because the technical field was conditionally closed for this application, AFA decided to expand the field of expectations by taking the maintenance into its own, specialized hands. This is a clear example of how a technical constraint can be bypassed by remedial action in the field of application. Although a *perpetuum mobile* cannot be made to work by changing the application field, an unacceptably short battery life span apparently can. At any rate it made the battery into a reliable source of energy for stationary purposes (especially in power stations). Thus, the life span of the batteries was prolonged by a continuous renewal of faulty plates, which themselves had a short life. It was a measure that was new in machine building as well as in the electrotechnical industry, where the responsibility for the product typically had ended after its sale.[25]

Despite these technical and organizational advances, in 1905 only the battery-streetcar line in Bremerhaven was still operational, whereas the mixed system was in use only in Dresden. In this period, pure battery traction, on the other hand, was increasingly used for the train, especially in the so-called *Vorortverkehr* (suburban traffic). But the batteries of such *Bahnen höherer Ordnung* (railroads of a higher order) were of the robust type with large-surface plates. After the victory of the streetcar with overhead wire, the so-called shuttle-lines—short routes with a relatively low number of passengers—also became a favorite target of the battery industry in New York. The Electric Storage Battery Company (ESB) became the major battery producer here. Founded in 1888, the ESB, with W. W. Gibbs in charge, had become the monopolist of the American battery industry by buying many hundreds of patents, including those on the grid-plate principle.[26]

The grid-plate patents owned by Brush, the "American Faure," had been in the possession of the Julien Electric Company, which changed its name to Consolidated Storage Battery Company in 1890. In 1894, the ESB took over this company out of purely defensive motives, for it based its monopoly on the technology of the lead-chloride battery, which was imported from France. The chloride cartel was founded in the early 1890s as a counterweight against the German competition, especially the AFA. It consisted of the American Electric Storage Battery Company (ESB), the British Chloride Electrical Storage Syndicate, and the French Société Anonyme pour le Travail Electrique des Métaux (TEM). In the chloride battery, cast lead-chloride "pills" were inserted in a cast lead plate. The zinc-chloride added to the pills was subsequently dissolved, resulting in a porous structure. Later, the pills in the positive plate were replaced by spiraled lead strips. At the ESB the chemist Herbert Lloyd developed this technology into a practicable battery, not only for stationary applications, but also for the battery streetcar.[27]

When in the early 1880s stories about a "horseless vehicle" started to appear in the press, many contemporaries—especially the average urbanite, but also interested individuals like Thomas Edison—might have expected this vehicle to be electric-powered. Moreover, the horse-drawn carriage seemed to be the perfect target for motorization, as it was the established individual means of transportation. This would have meant a further impetus to the substitution of horses, which had made such a promising start with the advent of the electric streetcar.

Neither of these expectations would be realized. At the start of the twentieth century, it became clear that the noisy, unhygienic, and unreliable gasoline car was going to be the dominant means of individual transport. Nor had the substitution of horses by motorized transportation become a success.

How could this have happened?

THE FIRST GENERATION

(1881–1902)

1

Separate Spheres:
Culture and Technology of the Early Car

It is certainly much fairer and wiser to study both the merits and
the defects of the different motive powers, and assign each to
its proper sphere as soon as possible. The field is the world, and
no one power can fill it. —*The Horseless Age, 1897*

In 1881, when Charles Jeantaud, the Parisian engineer and carriage
builder, in cooperation with Camille Faure, equipped a light "tilbury" with an
electric propulsion system by means of a Gramme motor and Fulmen batteries, Benz's and Daimler's "invention of the automobile" was still five years in
the future. Six years later Jeantaud made the news again when he equipped the
same carriage with a British electric motor. After another six years, in 1893, a
Swiss motor was tested and built into the same carriage. This motor was en-
ergized by a tubular plate battery, designed by Donato Thommasi, and had a
surprisingly high energy density of 27 Wh/kg.[1]

The developments took such a long time because the pioneers of the first-
generation electric cars—like those who developed their gasoline counterparts—had to go through a long and painful learning stage. This chapter in-
vestigates, first, the technical and application fields of the first-generation
electric passenger cars in private use, and then its field of expectations, against
the background of the rapid rise to dominance of the gasoline car culture. Of
course, just as its rival, the electric's culture knew its different national styles,
but in view of the nearly total lack of any analysis in this field, these differences
can be only touched upon.

Electrified Carriages and Bicycles: Technology and Manufacturers
Even though the technical structure of the electric car owed much to the electric streetcar, liberating it from the rails meant that the drive train had to be

thoroughly revised. It had to be attuned to the requirements of an independent vehicle, driven by an untrained driver along mostly unpaved roads. The electric motor, a reliable history of which is lacking up to the present, was copied from streetcar traction. It was a direct-current motor with series excitation. In order to generate an electric field, the electromagnets in the stator were connected in series to the electromagnets in the rotor, which was then put in a rotating motion by the electric field. A reduction in motor size meant a reduction in efficiency, however, so that adapted motors with special windings began to be developed. The method of variation of the rotational frequency, carried out by means of a mechanical controller, had also been borrowed from streetcar technology. But in the electric car the application of heavy preswitch resistors for low vehicle speeds was soon abandoned, also because of energy considerations.[2]

Motor speed control in early electric cars was accomplished by a technique called voltage switching. When the driver turned the controller, different groups of battery cells were successively connected in parallel and in series, so that the voltage on the electric motor—and through this, the vehicle speed—could be changed stepwise. Among the major disadvantages of this "battery switching," as it was also called, were the high compensation currents between the parallel-connected battery cells, which occurred as soon as their capacity differed and seriously threatened the life span of the battery. Such currents became disastrous when one or more cells short-circuited. This happened frequently, due to active mass becoming detached from the lead grid plates. As early as 1894, however, a survey of the technology of the American electric streetcar mentioned that the resistor method was gradually abandoned in favor of the series-parallel method. In this case this meant the successive connection in parallel and in series of the two motors, supported by a preswitch resistor only during acceleration from a dead halt. Engineers in Europe copied this "motor switching," and equipped their electric cars with two motors. For single-motor electric cars, the electromagnetic field was sometimes generated by different windings in the motor in order to avoid battery switching. Each winding could be separately excited in parallel or in series. Soon after the turn of the century, American builders of electric cars followed the European example.[3]

Copying the streetcar battery involved adaptation problems as well. After the conclusion of the battery-streetcar experiments, the further development of a reliable mobile energy source basically became the exclusive domain of the electric-car builders and their suppliers. For at least a quarter of a century after the first Jeantaud experiments, both "schools" of streetcar practice also coexisted for road vehicles: the grid-plate (Faure) concept and the large-surface plate (Planté) concept. Just like for the motor, for both battery types a reduction in size led to a disproportionate increase of inactive mass, resulting in a decrease of energy density.

It was through a multitude of incremental changes that the process of downsizing of both the motor and the battery took place. In the case of the Planté concept, this involved mainly the design of thinner plates, without losing surface area required for sufficient electrochemical reactions between the active lead and the electrolyte. In the case of the Faure concept, the design effort focused upon the bonding between the active lead mass and the supporting grid structure. In the second half of the first decade of the twentieth century, this evolutionary development was interrupted by some revolutionary changes leading to new battery structures and materials.

JEANTAUD, THE "FATHER OF THE ELECTRIC CAR," laid the foundation for early French electric-automotive engineering, including its main structural characteristics, based on his experience as a builder of heavy luxury coaches. In a way, the electric propulsion system was an engineer's dream, assuming that simplicity and ease of packaging within the automobile structure were primary parts of this dream. Thus, from an engineering point of view, electric propulsion seemed to be the perfect choice if one wanted to motorize an existing, not too small vehicle, without a lot of constructional effort. The batteries could be placed under the seat or in a separate box under the bodywork. A space-devouring gear box was in principle unnecessary, because the torque characteristics of the motor were ideally suited for road vehicle propulsion: the highest amount of energy was available at the very moments it was needed—at low vehicle speeds. These structural and performance characteristics of the early electric car gave the builder a decisive edge over his competitor, who had to accommodate the drive train of the gasoline car somewhere inside the coach, and so from the very start was compelled to look beyond the accepted coach-building technology for a solution to his structural problems. Therefore, this *constructional* advantage of the electric propulsion proponents was later to turn into a *marketing* disadvantage.

The early stage of automobile history was dominated by a hodgepodge of car builders, searching and groping in all imaginable directions, just as had been the case with the streetcar. For the Paris–Rouen race (1894), the field of expectations was stretched to its limits, and even beyond, because the 102 initially registered entrants were announced to be not only powered by electricity, steam, and gasoline, but also by compressed air, hydraulic engines, "pedal engine," and even compressed water, multiple levers, "combined liquids," and gravity. But the twenty-one cars that eventually competed in the 120 km race were all powered by gasoline or steam engines. Not a single electric car participated.[4]

In June, a year later, electric cars did take part in the 1,170 km nonstop Paris–Bordeaux–Paris race. Many believed the outcome of this race marked "year I of the future automobile civilization." Eight of the nine cars that returned within the 100-hour limit were gasoline-powered (the ninth was Bollée's steam car). Jeantaud, who was just as clever in manipulating the press as the organizers of the race, reached Bordeaux after a ride described by the press as heroic. He managed to reach the finish because in advance he had set up stations every 40 km for charging his batteries.[5]

The theme of Jeantaud's "electrified luxury coach" set the example for most French electric carmakers, who tried one drive train configuration after another, without achieving a stable standard setup. The only person who stuck to his chosen concept was the electrical engineer Louis Antoine Kriéger (1868–

For the Paris taxicab company L'Abeille, Louis Kriéger built his first electric car, an electrified horse carriage in the literal sense: the horse has been replaced by a front-end traction device. (*La Locomotion Automobile*, 18 March 1897)

1951). After he had graduated from the École Centrale des Arts et Manufactures, Kriéger in 1894 converted a victoria of the horse taxicab company l'Abeille, by placing its front side on an electric tractor. This tractor and trailer combination was used by the taxicab company from November 1894 until mid-1895. The finished vehicle weighed 1,430 kg, of which the Fulmen batteries accounted for 285 kg. The fact that the batteries only amounted to 20 percent of the total weight proved how much the structure of the battery-powered streetcar had influenced these pioneers. It also meant that initially they did not consider its range as problematic, even though this was a mere 30 km. In 1895, Kriéger built a type with central pivot steering, but a year later he switched to the Ackermann steering principle, which has each front wheel turn separately around a pivoting point.[6]

From then on Kriéger consistently opted for the highest possible battery load and fitted an electric motor as closely as possible to each of the two, large, circular front crown wheels. His electric motor was remarkable: the compound motor (which apart from the series winding also accommodated a parallel winding for the excitation of the field) was preeminently suitable for so-called regenerative braking. When using the motor for braking, the parallel winding simply converted the motor into a generator. In this way the kinetic energy of the vehicle could be transformed into electric energy, so that the batteries were recharged. Each motor shaft was provided with a small driving pinion wheel, so that a simple dual-gear transmission was created, which forced Kriéger to design a motor with an unusually low rotational frequency. Thus Kriéger broke the prevailing preference for fast-running electric motors (800 to 1,200 revolutions per minute), suggested by the need to economize on weight. By turning quickly a smaller electric motor could achieve an acceptable level of power. Kriéger, however, was able to compensate for the reduction in performance of a "slow-runner" without using a drive train configuration with chains or drive shafts and a differential between energy source and wheels. This solution avoided substantial mechanical losses as well as a lot of noise. Like Jeantaud, Kriéger opted for the type of bodywork that was reminiscent of the closed-body horse-drawn carriage and so catered to a market niche of wealthy city-dwellers.[7]

Whereas Jeantaud also supplied bodywork to gasoline-car builders, Kriéger founded a specialized company. When in 1898 the interest in electric vehicle propulsion suddenly increased in France, he reorganized his company as the Société des Voitures Électriques, Système Kriéger. That year the company delivered an electric van to the Grands Magasins du Bon Marché and took a stake in L'Électrique, a company that rented out electric cars as *voitures de grande remise* (rental vehicles) to rich clients on a subscription basis. But in February the company went bankrupt. It was taken over by the Société Française pour l'Industrie et les Mines (Indusmine), a small French-Swiss investment bank with stakes in mining and the fledgling automotive industry. In 1898, Indusmine had already established the Compagnie Parisienne des Voitures Électriques, probably an "empty" enterprise, but which was "filled" in July 1900 with Kriéger's assets and patents. The factory moved to Puteaux and under the same name, with the addition *procédés Kriéger*, forty-three electric cars were built until the end of 1901, and in the six months after that probably sixty-five. An important number of these (at any rate, twenty-five in 1901) were rented out from the factory's garage, which had its own charging station.[8]

In other European countries the initiatives in the area of electric-car manufacturing for private persons added considerably to the variety of the field of application as well as the technical field. In England, for example, Thomas Parker and Paul Bedford Elwell started their own business, Elwell-Parker Ltd., which first introduced the (large-surface plate) Planté concept in England and after that further developed Faure's (grid-plate) process. Subsequently they sold this company to the Electrical Power Storage Company (EPS). Elwell had already built an electric car in 1884. Four years later Radcliffe Ward founded the Ward Electric Car Company, after he had developed an electric bus and van for the Postal Service in London. In 1895, this company started to boost the implementation of electric buses in London. A year later its name had been appropriately changed to London Electric Omnibus Company. Walter C. Bersey, an electrical engineer, designed Ward's second electric bus. He was also responsible for building an electric mail van, which was tested for half a year in London in 1894, and covered about 1,000 miles.[9]

In Germany two important electric-car manufacturing centers emerged, one in Cologne and one in Berlin. In Cologne, Heinrich Scheele's coach factory turned into the first electric-car factory in Germany. Also in Cologne, an electrical engineering company, the Kölner Electricitätsgesellschaft vorm. Louis Welter & Co., imported a Kriéger from France in 1898 and turned to the KAW in nearby Kalk for the energy source. In 1899, KAW succeeded in developing a light grid-plate battery (type W) with an energy density of approximately

20 Wh/kg, and, even more important, a life span in the laboratory of about 300 charging/discharging cycles. The imported Kriéger with type W battery was used for business and private purposes. The results were apparently so satisfactory that the two companies established the Allgemeine Betriebs-Gesellschaft für Motorfahrzeuge (Abam) in 1900, which announced plans for starting an electric taxi service. The electric propulsion system was produced under Kriéger license, and the cars had solid rubber tires.[10]

In Berlin, Ad. Altmann & Company, founded in 1879, became a manufacturer of internal combustion engines in 1887. In 1897, it closed a licensing deal with the American Columbia Electric Vehicle Company and accordingly changed its name to Allgemeine Motor-Wagen-Gesellschaft mbH. At the same time the factory moved to a large industrial park in Berlin-Marienfelde and thus laid the foundation for the later Berlin Daimler factory. In the meantime it was renamed again and became the Motor-Fahrzeug-und Motorenfabrik Berlin-Marienfelde. Production in the new factory started in February 1899. Its production range was wide, from "motor-locomobiles" (internal combustion engines on a wheeled undercarriage), to gasoline cars, to electric-powered boats and cars. Up to 1901, when production was stopped, the company built electric passenger cars, buses, and trucks according to the Pope license, at a royalty of 5 percent. On 16 August 1902, the company merged with Daimler-Motoren-Gesellschaft (DMG) in Stuttgart.[11]

On an experimental scale the European electric-car pioneers started earlier than their gasoline competitors, but from "year I of automobile civilization" they were faced with a numeric dominance of gasoline-car builders. There was a peak in the number of established companies per year in both Great Britain and France (figure 1.1). Apparently, as soon as the market was occupied by a large number of trendsetters, the problem of propulsion type for the followers assumed a different character. This seems to be true regardless of the propulsion type first chosen. This phenomenon, known as the founder effect in genetics, can, at the local level where the propulsion choice took on a very real character, be illustrated by means of the Lohner Company in Vienna, which was one of the oldest and largest coach manufacturers in Europe.[12]

Ludwig Lohner began to be interested in automobile manufacture in 1896. After a study tour of Europe, he decided to start the development of a car in order to compensate for the decrease in coach sales. Due to his late start, however, he found it hard to interest associates among the automotive pioneers: Benz was allied to an automobile factory in Nesselsdorf, and Daimler wanted too much influence in exchange. The French Mors company was out, too, because of a lengthy delivery time. In the seller's market of just before the turn of the century, Lohner was forced to come up with his own design. He started with an engine of the French company Le Fèbre Fessard, which cracked after a quarter of an hour during the first experiment. "Our position in the [gasoline] automobile trade is hopeless," Lohner wrote to his friend, Georges Kellner, the renowned Paris coach builder, with whom he was a trainee during his study tour. At a loss what to do, he even considered the application of a "petroleum engine," for which he contacted Rudolf Diesel. But his first

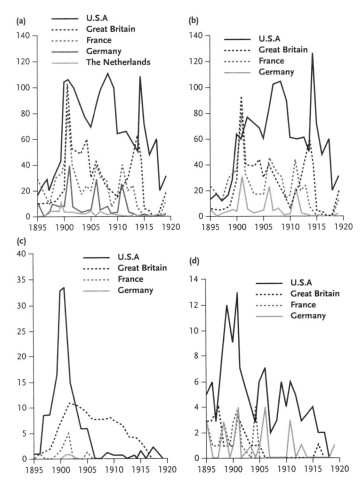

FIGURE 1.1. Automobile manufacturers according to year of foundation, 1895–1919, in the United States, Great Britain, France, Germany, and the Netherlands. (Adapted from Foreman-Peck and Hayafuji, "Lock-in and Pan-glossian Selection in Technological Choice"): a. total; b. gasoline-car manufacturers; c. steam-car manufacturers; d. electric-car manufacturers

gasoline car turned into a "problem child": "selling would be easy now—but manufacturing it . . . ," Lohner sighed. In cooperation with Belá Egger's electricity company, he then started the development of an electric-powered version. According to him, such a propulsion system had "the complete sympathy of the public, that adores electricity above all."[13]

The extant letters of Ludwig Lohner show how a beginning car builder at the periphery of European developments in this area struggled with technology as well as culture. "Eventually I have to repeat," he wrote in March 1898 to an unknown correspondent, "that I considered the whole gasoline technology only as a transition—otherwise I would long ago have lost the pleasure in it." It is remarkable how in his correspondence the ecological argument suddenly pops up as an alibi for an economic interest: "Leave us 'world villagers' the last

A Lohner-Porsche electric vehicle with wheel-hub motors, one of the paradigms of early electric vehicle construction, featured at the Paris World Exhibition of 1900. (Technisches Museum, Vienna)

remnants of oxygen and clean air that our wonderful society has bestowed on us. This air is polluted ruthlessly by the combustion products of the growing number of gasoline engines," Lohner wrote by the end of 1898, at a time he was still busy working on the gasoline car.[14]

Shortly after that, he broke up with Egger due to frequent problems with the electric motors. The model he subsequently built in cooperation with Ferdinand Porsche, an electrical engineer from Egger, drew a lot of attention at the Paris World Fair in 1900. Like Kriéger, Porsche found a way to circumvent the losses in the drive train, but his solution was even more radical. He constructed two big hub motors that directly powered the front wheels without any gearwheel transmission. The idea of the electric hub motor was not new, but, consciously developed as an alternative to Kriéger's system, the combination with the controllable front-wheel drive was an innovation that, after the independent (single and double) motor design and Kriéger's designs, offered the electric car a third structural possibility. Thus, Kriéger and Lohner-Porsche may be regarded as the trailblazers of the front-wheel drive concept, which was not seriously tackled by gasoline competitors until two decades later.

IN THE UNITED STATES the early history of the passenger car, including the electric version, deviated considerably from that in Europe. Development started later and was initially strongly influenced by European experiences. John Rae, the American automotive historian, describes this early period as "wrestling with mechanical problems already solved in Europe."[15] As we will see later, Rae's conclusion is biased toward gasoline propulsion because American electric vehicle pioneers soon started to export considerable portions of their production to Europe, where they offered solutions to problems not yet solved.

The American pioneers were not faced with a technical dilemma, but by a technical trilemma: the steam car played an important role in the United

States, due to the appearance of cheap petroleum on the market and the American preference for low initial costs to expensive labor. Gasoline-car manufacturers were the "followers": while the number of newcomers to steam and electric propulsion shows a peak, for gasoline traction this is not the case (see figure 1.1).[16]

When the first news trickled through about Benz's three-wheeler, and the electrification of the urban streetcar system was gaining momentum, an "American system of manufactures" began to develop simultaneously. The bicycle industry, the coach industry, and the machine tool industry were the most important proponents. At the World's Columbian Exposition, held in 1893 in Chicago, coaches were still dominant, but they were manufactured more and more by just a few major brands, such as Studebaker and Durant-Dort. The coach, especially the light buggy with its high wheels, suitable for driving along poor city and country roads, was to a high degree "democratized." The only car of American origin at the exposition was Morrison's electric surrey, nowadays regarded as the first American electric vehicle.[17]

As in France, motoring in the United States gained momentum due to a contest, organized by "Judge" Herman H. Kohlsaat, director of the *Chicago Times-Herald* and inspired by the Paris–Bordeaux race. Unlike in France, however, the emphasis was not on speed. "When we come to consider the motor vehicle in its present stage of development," Kohlsaat explained, "the element of speed sinks into relative unimportance. Other considerations, such as simplicity, economy of operation, ease of control, etc., are entitled to much greater weight. A motor vehicle may attain great speed, and yet be practically worthless for the ordinary purpose of everyday use." This early utilitarian approach to the car is typical of the American development. It helps explain the initially modest role of the less reliable gasoline car and is in sharp contrast with the European, especially the French, development in the same period.[18]

Kohlsaat ordered his staff to organize the contest for 4 July 1895, but due to a lack of competitors the event was postponed. On 28 November 1895, there was five inches of snow on the ground. There were eleven registrations for the run to Evanston and back, of which three withdrew and two others did not manage to get to the start in time. So, six cars stood at the mark: four gasoline cars (three of them of the Benz brand) and two electric cars. The number of spectators was small, and only about fifty people watched the finish. The only car not to receive a cash prize was Morris and Salom's "Electrobat." Because Morris and Salom did not have time to take care of battery exchange stations along the route, "[they] announced that they would simply make a short run to show that the electric wagon could make satisfactory headway through deep snow," exactly the type of load electric propulsion could master. They therefore received a gold medal "for best showing made in the official tests, for safety, ease of control, absence of noise, vibration, heat or odor, cleanliness, and general excellence of design and workmanship." This award caused a French commentator to sigh that the jury was "undoubtedly very electric-minded," but the decision was less controversial than was often assumed afterward. In France some people apparently could not imagine such contests to involve anything but speed racing. But even Charles Duryea, Frank's brother, who finished first, said afterward that "the crying need of to-day is not speed, but better construction and better operation, and possibly, to suit the desires of the masses, less cost." The Duryeas were soon to become the most fanatic pursuers of speed.[19]

The Electrobat had an important influence on the further American development of the electric car. Pedro G. Salom (1856–1945), the chemical and electrical engineer, and the inventor Henry Morris had started their developmental work in Philadelphia in June 1894. They adapted a General Electric ship's motor, while the chloride battery came from the Electric Storage Battery Company (ESB), where Salom worked as a battery specialist. Although the first Electrobat carried a battery pack of some 400 kg, the spring of 1895 saw the birth of the Electrobat 2 with bodywork designed by a coach builder. Of its total weight of about 825 kg, the batteries accounted for a mere 10 percent—half of Kriéger's Abeille taxicab. It was this model that took part in the Chicago contest and formed the source of inspiration for the first American-manufactured automobiles, including the electric-powered variety, which had made a hesitant start in 1897 and was supported by an important contribution from the bicycle industry. Typically, all these early designs reflected the simplicity enabled by the structural flexibility of the electric propulsion system. In its most extreme version this can be seen in the American "piano box" design, a buggy reduced to the bone.[20]

The most stirring success story was that of the Pope Manufacturing Company, founded in 1877 by "Colonel" Albert Augustus Pope (1843–1909). By 1890, it had become the biggest bicycle manufacturer in the United States, annually producing 60,000 "safeties" of the Columbia brand. Located in Hartford, Connecticut, the Pope Company was surrounded by a large number of precision manufacturing factories and workshops. Pope also owned the Hartford Rubber Works Company, established in 1881, which was to produce the single-tube tires for Pope in 1891. But Albert Pope and his cousin George became especially known for their production of the cold-drawn seamless tube. Converted to a tubular frame, this was to become an important structural characteristic of the first American electric cars.[21]

Harold Hayden Eames, a former lieutenant commander, played a crucial role in this development. He made the tubing factory into "one of the industrial wonders of New England," in part by managing to attract competent engineers. One of these was Henry Souther, who in 1892 established a scientific metallurgic laboratory and testing facilities to improve the bicycle design. Eames not only was a pioneer of modern management (based on the three principles, "Standardize, Deputize, and Supervise"), but also knew how to make the steel tubes extra strong by subjecting them to a special heat treatment.[22]

Pope's "chief lieutenant" was George H. Day (1851–1907), former head of the Weed Sewing Machine Company, who initially assembled the bicycles for Pope and had given up his sewing machine production to do so. Hermann F. Cuntz, a mechanical engineer, became Pope's patent expert. But for our story the most important individual was Hiram Percy Maxim (1869–1936), the son of the inventor of the machine-gun and a graduate of the Massachusetts Insti-

S. Blimline of Sinking Spring, Pennsylvania, built this electric runabout in 1897. It typified American electric vehicles of the first generation—piano box design, tiller steering, and a lot of bicycle technology. (Smithsonian Institution, Washington, D.C.)

tute of Technology. After his graduation Maxim became involved as an engineer in grenade production, where he met Eames as navy ordnance inspector. As of 1895, he was in charge of the Motor Carriage Department at Pope that had been established the year before. In November 1895, he had been an official in Morris and Salom's car during the *Times-Herald* race.[23]

Maxim had already converted one of Pope's three-wheelers to gasoline propulsion in his spare time and was intent on developing a gasoline car. He relates in his memoirs how by the end of 1896 his supervisors asked him to construct a delivery van with gasoline engine, "something simple, light, not an elegant carriage—just the place for a gasoline-engine . . . it should be a cycle and not a wagon or a carriage." The results confirmed the Pope management's view that electric propulsion had the future. For when Day saw Maxim's first gasoline creation, he said with a shake of the head, "Well then, Maxim, let me tell you something. We are on the wrong track. No one will buy a carriage that has to have all that greasy machinery in it. It might be that young fellows like you and Souther would buy a few of them as interesting toys, but that would be only a drop in the bucket. No, Maxim, we are on the wrong track!" Albert Pope was of the same opinion, according to his one-liner that keeps cropping up throughout automotive historiography: "You can't get people to sit over an explosion."[24]

During an open day in May 1897, the first electric Columbias were presented to the press and invited guests. In accordance with an agreement with ESB, chloride batteries were going to be used from then on. Three years later the factory in Hartford was incorporated in the EVC conglomerate (see chapter 2). At the time it was at the peak of its development, and assembly of the various function groups took place under the supervision of twenty-one qualified engineers. This assembly was carried out in the form of a kind of inverted assembly line: groups of workers successively finished a number of chassis, placed in a row, and then started again at the beginning, with a new row. This made the Pope factory the largest automobile producer in America: in the fall of 1900 the Pope Company had produced 1,000 cars. Several of them were shipped to Paris (80 vehicles), Berlin, London, Madrid, Mexico, and Canada. Of the 540 cars that left the factories in 1898 and 1899, 500 were electric and 200 of these were exported to Europe.[25]

After the Hartford factory had been taken over by the EVC and was continued as the Columbia Electric Vehicle Company, the number of employees increased from 400 to 1,700 in 1900. That same year a second producer of electric vehicles was taken over, the Riker Motor Vehicle Company in Elizabethport, New Jersey. Andrew Lawrence Riker (1868–1930) was one of the few American pioneers who had tackled all three forms of propulsion. He dropped out of Columbia University after one year of study and committed himself totally to his inventions, for which he established the Riker Electric Motor Company in Brooklyn, New York, in 1888. Riker's pursuit of the light electric car was unparalleled. In 1895, he built his first experimental lightweight, four-wheeled car. He constructed this car by connecting two bicycles with steel tubes; each rear wheel was powered by an electric motor. In 1897, an electric victoria (an open city carriage) appeared on the scene, provided with electric motors after Riker's own design. The wire spoke wheels were provided with pneumatic bicycle tires, just produced by Pope for the first time.[26]

New England became the cradle of the American car industry, which, during those earliest days, was to a large extent electric in scope. Also, Cleveland (Ohio) was a pivotal center of electric vehicle activity. One of the factories, which also became known in Europe, was the Cleveland Machine Screw Company, a subsidiary of the White Company that would later become a renowned steam-car manufacturer. Elmer Ambrose Sperry contacted the company in 1898. Freshly graduated from Cornell University, he had come to Cleveland in the early 1890s to supervise a streetcar project. With his friends Thomas Edison and Walter Baker, the electric car builder, Sperry was among the many drivers who did not meet the deadline of the *Times-Herald* run. Sperry is one of the examples of well-educated engineers with ample experience in electrical engineering—from control engineering to the electric streetcar—who got stuck on the seemingly simple problem of the "horseless carriage." He obtained patents on all possible function groups of the automobile, including an electric motor with air cooling, a transmission, a clutch, and a braking system. His first car, with a two-cylinder gasoline engine, was destroyed by fire in his workshop in Cleveland in the spring of 1896. "That's what you get for deserting electricity," he later wrote to one of his sons. In 1898, Sperry had finished six electric cars, complete with a battery after his own design, which drew the attention of the Machine Screw Company. That same year the company was taken over by a French outfit, and it sent Sperry to Europe, where his six cars were shown at the Paris automobile exhibition. Just as Pope had experienced, the interest in

The car as a double bicycle. An experimental electric vehicle, at the time called a "motorcycle," built in 1896 by Andrew L. Riker in Brooklyn, New York. Riker later became president of the Locomobile Company of Bridgeport, Connecticut, and also of the Society of Automotive Engineers. (Smithsonian Institution, Washington, D.C.)

American cars was considerable there, so that Sperry went home with an order for a hundred cars and a licensing contract.[27]

But in the further fortunes of Sperry's patents, Pope again played a role. After the takeover of his factory in Hartford by the EVC and with the looming slump in the bicycle market, Pope managed to forge a mammoth merger of forty-five bicycle manufacturers with fifty-six factories under the umbrella of the American Bicycle Company (ABC) in 1899. That same year ABC also bought two automobile manufacturers: H. A. Lozier & Company in Toledo, Ohio, manufacturers of steam trucks after a British design, and the Waverley Company, by origin a bicycle factory. The latter was the result of the diversification policy of the Indiana Bicycle Company in Indianapolis, which employed a thousand workers in 1898, and that year started a close cooperation with the American Electric Vehicle Company, located in Chicago.[28]

To make the story even more complicated, the latter company was the result of an association between C. E. Woods Company and a group of capitalists from Chicago and New York. Clinton Edgar Woods was the son of a coach builder and a graduate of the Massachusetts Institute of Technology in mechanical and electrical engineering. He was one of the few American electric car pioneers to focus attention on the utilitarian application of electric traction right from its beginning in 1897. Woods set up the American Electric Vehicle Company with the support of none other than Samuel Insull, the "utility czar" of Chicago, and a number of "Standard Oil magnates."[29]

As if this were not enough, an ESB competitor also got involved in this tangle of mutual interests. The Fisher Equipment Company, also located in Chicago and set up by Woods as a car factory, closed a deal with Sipe & Sigler in January 1898. These were former jewelers from Cleveland and partners of T. A. Willard from the same city. Willard had worked on further development of Planté's large-surface plate by making the solid lead plates extremely thin. When Woods and Insull, in April 1899, established the Woods Motor Vehicle Company, they subsequently took over the Fisher Company. But in the fall of that year the company went bankrupt, "and Woods became an automobile dealer." Waverley became ensnared in the Pope syndicate as well, but this brand remained in existence as Pope-Waverley until the First World War.[30]

In 1901, Lozier and Waverley were reorganized under the flag of the International Automobile Company, one year before the collapse of the ABC. At the same time the patents and electric car supplies of the Cleveland Machine Screw Company were taken over, including Sperry's patents. From 1901 on, the Sperry battery was produced by the National Battery Company and marketed with a special separator, made of fibers. As manager, Sperry remained involved with the production of his battery for several years.

Back in Cleveland, in his friend Sperry's workshop, Walter C. Baker (1867–1955) was experimenting with electric propulsion as well. Baker was the son of an inventor of sewing machine parts (and of roller skates). After his engineering studies at the Case School of Applied Science (later the Case Institute of Technology), he got a job at the White Sewing Machine Company. This was the continuation of the above-mentioned White Manufacturing Company and the parent company of the Cleveland Machine Screw Company. There he became a specialist in the field of ball bearings. In 1897, Baker built an electric car, the first one in Cleveland. A year later the newly founded Baker Motor Vehicle Company launched an electric buggy weighing only 250 kg. Baker was one of the first automobile manufacturers to institute a regular production process, characterized by assembly, as with many early American carmakers. For the electric motors Baker turned to the Elwell-Parker Company, located in Cleveland. His first customer was Thomas Edison, who at that moment had just abandoned his initial preference for the gasoline car.[31]

WE HAVE LOOKED at only a selection of the many initiatives taken, in order to provide a background for our study. At this stage in car production, sales to private persons were generally very modest. This is true for the United States as well as Europe. In the latter case sales in this market segment may have amounted to 250 cars at most around the turn of the twentieth century, but probably fewer. It seems, therefore, that the initial increase in the number of carmakers was more a reflection of a blown-up field of *expectations* of the producers than of a real demand. If the number of exhibits entered in the Paris trade shows is any indication of these expectations, it initially did increase. At the Salon du Cycle of 1897, one electric car was present; at the first International Automobile Exhibition in 1898, 29 vehicles by 10 exhibitors were on display; and the 1899 exhibition had 19 participants with 63 electric vehicles. But in 1901 enthusiasm for the electric car among French automobile builders had fallen off. And the Salon Automobile of 1905 in Paris left little doubt about the preference of most wealthy Parisians: whereas there were 12 companies showing their electric vehicles, 98 exhibitors showed gasoline cars. Only Serpollet (in the meantime associated with the American Frank Gardner) still had a steam car on display.[32]

At first sight this was altogether different in the United States. It is remarkable, for instance, that at the 1900 World Fair in Paris the majority of the 237 automobiles exhibited were gasoline cars (176 vehicles); most of the electric cars (40 vehicles) and steam cars (21 vehicles) were from the United States, which had sent few gasoline cars. It thus seems that at the end phase of first-generation automobile development, technology transfer for the gasoline propulsion system originated in Europe, whereas for both other types of propulsion the direction of transfer was the other way around.[33]

According to the American census of 1900, 109 manufacturers produced 4,192 vehicles that year: 1,681 steam cars, 1,575 electric cars, and only 936 gasoline cars. Almost half of these were produced in New England. The total value of the electric cars surpassed that of the two other alternatives together and amounted to twice that of the steam cars and thrice that of the gasoline cars. The steam car was the least expensive and was the most popular among private customers. In 1900, Locomobile, for instance, series-produced 1,600 steam cars a year, a number that decreased rapidly after that. Foreman-Peck, the British economic historian, explains the initial dominance of the steam cars by the low fuel costs (lamp petroleum) and the low purchase price. According to his calculations, the American steam car was 0.75 cents per mile cheaper than the gasoline car. This advantage only disappeared after a distance of almost 53,000 km, when the gasoline car became more economical due to its lower fuel consumption. In Europe the advantage in operating expenses of the steam car was less substantial (0.67 cents per mile), and the purchase price was considerably higher.[34]

Although Foreman-Peck's calculations may reflect an economist's bias toward costs in a phase and a market segment where costs were the least of all problems faced by early automobile users, the picture of the early, absolute dominance of the steam car in the American street scene is supported by anecdotal proof of the largest automobile market, New York. For example, a private traffic count was held in 1902 between Forty-second Street and Fifth Avenue, where from 11 to 12 o'clock in the morning 20 steam cars, 12 electric cars, and 8 gasoline cars were counted. Another example was Locomobile's announcement in May 1903 that of the 4,000 automobiles registered in New York State, 53 percent were powered by a steam engine, 27 percent by a gasoline engine, and 20 percent by an electric motor. The picture is entirely different in Chicago, however, where the electric passenger car was predominant. At the beginning of 1901, a total of 378 licenses had been issued, 226 (60 percent) for electric cars, 92 (24 percent) for steam cars, and only 60 (16 percent) for gasoline cars. The picture also deviates considerably from what the earliest American pleasure rides showed. For example, during the first run the Automobile Club of America organized on 4 November 1899, in New York City, the majority (22) of the participating 35 cars were electric, whereas 9 were equipped with gasoline engines and 4 with steam engines. But the American production numbers of the first eight months of 1902, published in the *New York Sun*, clearly show that in the meantime the gasoline car had taken the lead in the American market. Thus, one should beware of confusing average national market distribution figures with local and regional distribution statistics. It is at the local level that explanations should be sought for the dominance of one of the three propulsion alternatives. Chicago, for instance, was known for its well-paved, flat roads and the presence of influential actors in favor of electric propulsion. New York, with its hilly suburbs and bad roads, lent itself better to the gasoline car. Differences at the local level were enormous and call into question the usefulness of national averages at this early stage: in the small New England city of Portland (Maine) 88 percent of the fifty-two vehicles registered up to 1902 were steam-powered (8 percent electric and 4 percent gasoline); in Los Angeles in 1902 half of the hundred cars were electric.[35]

One of the accepted observations in automobile history is that the initial dominance of the electric car over the gasoline car in America had radically changed three years later. As detailed figures are not available, historians often refer to the reduced numbers of electric models at the automobile exhibition of 1903 in New York and the diminished interest in electric propulsion in automobile magazines.[36] However, a calculation of the Pope production puts this "reversal" in perspective. It shows that at least 850 vehicles of this production were delivered to the EVC as taxicabs. So, the suspicion is justified that in 1900 the number of electric cars sold to private customers was already lower than that of gasoline cars. Nevertheless, this means that the number of privately owned electric cars was double or triple that in Europe.[37]

The question, then, arises of how to explain the phenomenon that the first-generation electric car was primarily used as business vehicle. While this field of application will be the exclusive subject of the following chapter, the question remains why the reaction of private persons initially was surprisingly aloof. Pointing to an inadequate technology or insufficient infrastructure does not suffice. After all, gasoline-car pioneers had similar problems. Foreman-Peck's argument of cost difference seems appealing at first sight, but it cannot be upheld at closer look. To arrive at this conclusion, we have to dig deep into early car culture, especially into that of the gasoline version and its cultural predecessor, the bicycle.

Racing, Touring, and Dirty Hands: Automobile Sport and Fin-de-Siècle Culture

Long before the car came into its own as the mainstay of individual transportation, the bicycle had done its preparatory work. Invented in the 1810s as a toy for adults in the form of a "hobbyhorse," the *vélocipède* initially was received with interest in mainly fashionable, aristocratic circles in France. After 1830, the hobbyhorse lost its appeal, until Pierre Michaux (1813–83) and his fellow-worker Pierre Lallement equipped the vehicle with pedals and a crank mechanism in the early 1860s.[38]

The "ordinary," with a front wheel that grew larger and larger, quickly became the favorite vehicle between 1875 and 1885. Mainly young, adventurous men dared mount and ride the monsters, so that other models were developed for ladies and older gentlemen, such as three-and four-wheelers. In 1885, the British J. K. Starley (1854–1901) introduced the "safety," which would soon become the standard model with two wheels of equal height and a tubular framework, mainly the work of Rover, the bicycle manufacturer (and later carmaker). This "democratic" bicycle was to become the vehicle of the bicycle craze during the 1890s.

Yet, the ordinary cannot be called an aberration in the development of bicycle technology. On the contrary, it fashioned the social-cultural line along which later bicycle models developed. With the emergence of the safety, one did not simply fall back on the use of the old hobbyhorse, because the aspect of speed remained embodied in the construction. This embodiment of a cultural value took place *literally:* in light tubular frame and friction-reducing ball bearings in the light wire spoke wheels, but most of all in the pneumatic tire, reinvented in 1888 by John Boyd Dunlop. And yet this is only half of the explanation for the safety's success. For this very same light construction and these very same pneumatic tires also enabled different uses of the bicycle, uses that so far had been restricted to three- and four-wheelers: touring safely outside the city and performing figure riding in the arena. The American Chris Sinsabaugh, for example, recounts in his memoirs how as a boy he fell under the spell of the bicycle craze: "I had wanted one of those ordinaries, but I was too timid at first, so I bought the safety." And he explained elsewhere: "Those who couldn't ride fast enough to win races took up touring." Later Sinsabaugh was to become a sports reporter, a job he referred to as "speed merchant."[39]

In this sense, the bicycle concept as we use it today and which first materialized in the safety should not be described as a "socially constructed speed machine," as Wiebe Bijker in his famous analysis has done, but as an *individual adventure machine* that through its *universality* could take part in a variety of application fields. For the bicycle as adventure machine not only provided the individual cyclist with a sensation of speed, but also the glow of an unbridled, individual mobility for ladies and older gentlemen as well as for timid young men.[40]

If the aspects of speed, touring, and figure riding were still in the tradition of horseback riding, a third aspect of the bicycle as adventure machine was new: its mechanical character. Experiencing this "progress of technology" during touring had its price. It was necessary to have a certain minimum basic knowledge of the device in order to be able to maintain it. And one had to be prepared to interrupt a ride at unexpected moments for repairs, especially repairs of the pneumatic tires. The "adventure of technology" could be faced by sound preparation, know-how, and skill. Thus emerged the "strangely ambivalent phenomenon" of the bicycle tourist who adored nature, but at the same time believed in the advance of technology. "Flight" seems to be the keyword here, in a double sense. First, the bicycle allowed an escapism with modern technical means. Second, bicycling seemed to enable a euphoric experience of the "poetry of motion," a prefiguration of later ballooning and airplane experience, quite remarkable in view of the road conditions of the day.[41]

The fin de siècle, the "pivoting period" between the nineteenth and twentieth centuries, is inconceivable without this ambivalent belief in progress. France, and particularly Paris, was the shining example of this ambiguity. There, Schivelbusch's anxiety of the train traveler had reached larger sections of the population, expressing itself in poetry and novel writing (Maurice Rollinats's poetry collection *Les névroses* [The nervous], for example). Moreover, after the military defeat against Germany in 1871, a feeling of decadence and boredom had spread in the upper classes, together with a greater interest in the self. The licentiousness in literary and fashionable circles may have started earlier, but the general public became acquainted with it through a new type of media, the illustrated popular press. This licentiousness also seemed to pass on to other segments of the population, through the bicycle craze, for instance, where women from the emerging middle class rode their bicycles in "masculine" clothes.[42]

From the early 1890s the bicycle culture was divided into two distinctive domains that influenced each other culturally: that of the races and that of the touring clubs. Some of the heroism of the first domain rubbed off on the second. There were races in different varieties, each with its own specific function and each a reflection of an element of the fin-de-siècle culture. Yet the ultimate expression of the burgeoning self-consciousness in the fin de siècle was the race against the clock, and so against oneself. In 1899, the American Charley Murphy briefly reached a speed of over 100 km/h. Mile-a-minute-Murphy had achieved this by "staying" behind a specially prepared train.[43]

The growing need for records can be considered a typical phenomenon of the late-nineteenth-century period. The "gigantism" of the Eiffel tower (built between 1885 and 1889); the erection between 1884 and 1900 of the Ferris wheels with heights up to 110 m in London, Chicago, Vienna, and Paris; the building of the skyscrapers in the United States in the 1880s—all these were expressions of a restless age, when breaking records was associated with the emancipation from a period of stagnation, as *Scientific American* wrote in 1899. Against this background, the record was nothing less than quantified progress, translated for the benefit of a mass public.

The bicycle boom cannot be explained solely as a consequence of the artifact's universal usefulness and constantly decreasing price. The best study of this phenomenon mentions six additional factors that shaped this "madness" (for that is what it was). A bourgeois belief in freedom, social expectations of higher prestige and class equality, belief in progress, sensation of speed, new youthfulness, and idolization of nature—all these factors led to a mostly emotionally determined enthusiasm for cycling that can best be typified as euphoric behavior. In the bicycle euphoria the mental tendencies of the era were united, and there was a strange mixture of a rational perception of technology and the romantic feeling of the reform movement. The bicycle industry recognized this irrational, mythic element in the new sport, judging from the choice of brand names (Hercules, Diana, Hermes, Apollo). For the first time since the introduction of the steam engine, millions of people experienced the phenomenon of the *Technik zum Anfassen* (hands-on technology) during their cycling trips. Through this, they could feel part of a new, progressive movement.

Women's interest in the sport disturbed many contemporaries, and the commentary on this new phenomenon reflects the late-nineteenth-century blend of euphoria for the new technology and fear for its consequences. At the same time the gradual embedding of the constantly growing bicycle movement in society also increased its sociopolitical influence. Because of this a first step could be taken toward traffic regulation (maximum speed, code of behavior in the city, sign-posting). Moreover, the quality of the roads was brought to the attention of the authorities. For instance, Albert Pope, the largest bicycle manufacturer, vigorously supported the "Good Roads" movement in the United States.[44]

The army gradually recognized the potential of the new vehicle as well, and not just as a simple mechanical replacement of the horse. Apart from the euphoric experience of nature and physical hygiene, the sport also functioned to internalize social standards and to express and channel individual aggression by means of group discipline. The increasing interest of the German army in the bicycle sport, for instance, ran parallel to a shift of emphasis within the bicycle movement from an internationalist-cosmopolitan to a patriotic-nationalist attitude.[45]

The bicycle as adventure machine bridged the contrasts that were perceived in the fin de siècle between town and suburb, culture and nature, man and woman, and technology and culture. With the improvement of road quality and bicycle construction, the expansion of bicycle club power, and the reduction of bicycle prices, the adventurous character of cycling gradually diminished and the utilitarian character began to dominate. This process had ended in most Western countries before the First World War and led to specialist bicycle constructions for racing. A comparable evolution toward a specialized touring bicycle (e.g., by offering more comfort at the cost of speed) did not take place. The universal bicycle kept carrying its adventurous function in its construction: the touring bicycle had become a "fast" touring bicycle and was to stay that way to the present day.

THE CULTURAL AND PSYCHOLOGICAL MOTIVES of the participants in the new automobile sport closely resembled those of the bicycle sport. The cyclist, and especially the racing cyclist, had considerable knowledge of the new sport, for instance, about the state of the roads, sense of direction, and keeping one's cool during all kinds of unexpected events. Long before the car was chosen to realize *individual mobility,* one had practiced these driving skills on the bicycle. It is the only way to explain the strange phenomenon that "the car . . . already (seems) indispensable at its emergence, it arrives on the scene as a long-expected guest."[46]

The car races partly determined the structure of the first automobiles, and thus the field of application forced the technical field to open as wide as possible. What was more natural with a dynamic device like the car than to take part in the general enthusiasm for speed and adventurous wanderlust, now that the problems of congestion and hygiene in the cities seemed to have been solved by electric public transportation? Would a trip outside the city not have the same cleansing effect on city culture as the electric streetcar? And was the long-distance race not the perfect means for testing the reliability and endurance of the car, considering the fact that the vibrations caused by poor road conditions became worse at increasing speed? And was the attention of the popular press not a wholesome and supportive phenomenon in this development?

Speed races were not the only public events organized during these first years. But very often the "endurance contests" degenerated into a race about who was the fastest. This happened, for example, during the "Emancipation run" from London to Brighton, organized in November 1896 to celebrate the abolition of an automobile-unfriendly law (the notorious Red Flag Act). The London Motor Car Club had invited car drivers and manufacturers from Europe and America to participate in this run. It was controversial from beginning to end, partly because of the inadequate organization, as has later often been explained, but also because testing endurance without racing encountered a cultural barrier. More than half a million spectators had gathered at the start and along the route, "probably the largest crowd the road to Brighton had ever known." During the run, intended as a parade along the London streets, the gaps between the cars (of Bollée, Daimler, and Duryea) became wider and wider. Later Frank Duryea was to write that "the British public, contrary to our habit, looked upon the event not as a race, but as an endurance contest with equal honors for all who finished. All the drivers, of course, were out to win if they could. As usual, I raced instead of merely touring." Thus Duryea arrived first in Reigate, where they were going to stop for lunch, but Bollée's tricycles ignored this rule and "buzzed through without stopping." That is why Duryea finished third after the two Bollées. One Daimler reached a speed of almost 30 km/h and a Panhard was said to have gone even faster than 50 km/h. The Motor Club's point of view was, however, "that the contest was not a race, and therefore the order of arrival was irrelevant." So, every contestant received a medal with the inscription "for punctual arrival at Brighton."[47]

As competition increased, however, and the average speed continued to rise due to special adjustments to the construction, wealthy manufacturers tended to deploy cars that were especially built for such races. The costs of these racing cars became so high that Karl Benz and Gottlieb Daimler, for instance, felt confirmed in their skepticism about the increasingly higher speeds, which they had expressed from the very beginning. Louis Renault was against higher vehicle speeds as well, because it shortened the life span of the engine due to higher wear and tear. Henry Ford initially also opposed racing. According to Ford, it even slowed down the technical development of the automobile, because manufacturers were tempted to make fast cars instead of good ones. The notion, derived from bicycle culture, that winning a race was a recom-

The anxious scene at one of the early competitive events, the Gordon–Bennett races, this one of 1903 held in Ireland. The Belgian "Red Devil," Camille Jenatzy, won the race in a 60-hp Mercedes race car (gasoline powered), at times reaching speeds of 135 km/h. (DaimlerChrysler Archives)

mendation for quality did not apply to the car, according to Ford: "I can hardly imagine any test that would tell less [than racing]."[48]

While the bicycle became cheaper and "more democratic," the car appeared to become a new means for the leisure class to "practice hygiene." Entirely in the euphoric style of the bicycle sport, Frédéric Régamey sang the praise of the car as a vehicle of health at the end of his *Vélocipédie et automobilisme* : "The roads are open to everyone; the fields offer everyone the intoxication of clean air and wide horizons, the magic of changing scenes that dispel sadness and boredom, the genial and invigorating breezes that give the short-winded city-dwellers back their health. The wide world is for everyone." Later this experience received scientific support, for instance, from the Dutchman B. ten Have: "Breathing fresh air, clean, dust-free air, that can be very thin because of the fast propulsion; the change of climate and everything related to this, is considered as very wholesome for the lungs, as it is accompanied by the unconscious performance of strengthening lung exercises."[49]

At first sight, it seems confusing that at this early stage the car was pushed to the fore as a remedy against the *Zeitgeist,* while it can also be seen as its sublime expression. To explain this paradox, we have to delve deeper into the fin de siècle culture than simply signaling a paradox, as with the bicycle. Key here is the German term *Autotherapeutik* (car driving as self-therapy). If nervousness—from 1880 diagnosed more and more frequently by American and European neurologists—denotes a "physical expression of civilizational impatience about progress," emerging automobilism provided a way out of the "ambivalence . . . between pleasure and suffering." Thus it enabled identification with the modern pace of life, with the "feverish haste, the nervous hurrying," in the words of the economic historian Werner Sombart.[50]

In 1902, well-known French automobilism promoter Baudry de Saunier devoted an entire chapter of a book to the "automobile as remedy." In that chapter he referred to the much-discussed "vibration machines" of the Swede Zander and even to Abbé Saint-Pierre, who had already invented a vibrating rocking chair at the beginning of the eighteenth century. In his time, Louis XIV benefited greatly from his travels by coach, one of the most effective means "against many diseases caused by melancholy, hysterical fits, the gall, the liver, the spleen, and other organs in the lower part of the body." Baudelaire's Parisian *flâneur*—on foot—had to use drugs to become high, but the motorist reached "flight" from the "soft vibrations." "It is by no means a matter of jolting, shaking, rocking, but a soft almost unnoticeable quiver. When the car is stationary, it is strongest; the faster it runs, the weaker it becomes," the German automobile pioneer Otto Bierman wrote in his account of a long journey by car. Motoring, a French physician summarized, was "a different form of drunkenness." Especially for men, whose hysteria at the end of the nineteenth century was identified as a "neurosis of the brain," the car as a vibrator on wheels formed the "most perfect way of passive motion." Particularly in Germany, it was concluded that neurasthenia was caused by incest, masturbation, and comparable "sensualism and passions" and threatened to lead to "complete sexual negligence." In this perspective, the car ride imposed itself as a substitute for the (emancipating) woman, who, in the course of the nineteenth century, had become increasingly unattainable for the bourgeois man. In an English epistolary novel from 1902, written by the very productive Williamson couple, the heroine (some women were also susceptible to the car "drug") could not be more explicit when she wrote to her father about her car trip: "'This is life,' said I to myself. It seemed to me that I'd never known the height of physical pleasure until I'd driven in a motor-car. It was better than dancing on a perfect floor with a perfect partner to *plu*perfect music; better than eating when you're awfully hungry; better than holding out your hands to a fire when they're numb with cold; better than a bath after a hot, dusty railway journey."[51]

The best remedy against the decadence of the time, against the "crisis of abundance" of the fin de siècle, which could only lead to "female softness," was "self-control." The "sublimated aggression" resulting from it "could collaborate with Eros to build cities, speed travel, enhance comfort, improve communications, lengthen life," Peter Gay ends his diagnosis of "The Bourgeois Experience." When such a bourgeois man drove a car, he could compensate for his feeling of powerlessness against "rebellious objects" by his "power of the wheel." For, had the French psychiatrist Charles Féré not argued "that active and challenged minds became more resistant to nervous breakdown"?[52]

In the varied spectrum of neurasthenia, as depicted in the growing body of handbooks written by medical doctors, motorists seemed to belong to a separate category: that of the "vagabond," the "aggressive-nervous" person who often is his own therapist because he vents his unrest on the outside world. Such a "nervous person does not suffer much; for, at the slightest bodily disorder he usually makes such a lot of noise, that the people gather round, which does the nervous person's sickly heightened subjectivism so much good, that he enjoys this as a triumph and quite soon forgets the insignificant pain." The euphoric feeling that remained as a result of this aggressive venting of unrest we have observed earlier in the bicycle culture. It is not surprising, then, that the "erotic tickle" of the bicycle ride was prescribed as an effective means against onanism. The German historian Joachim Radkau identifies the discovery of "relaxation in movement" as an "anthropological novelty" in his brilliant analysis of this "nervous era." Whereas half a century earlier one still pointed at the train as the source of nervousness, the car appeared on the scene when people increasingly started to look for "nerve strengthening" by means of motion. In this period the hygienic movement merged with the neurasthenia discourse: the urban nervousness could be balanced by living in quiet suburbs, by traveling, weekend trips, and garden colonies. The "sitting" culture had to be compensated by an "exploring" culture. A life style developed, in which the film and the cigarette became comparable forms of "brief pleasures of the culture of eternal dissatisfaction."[53]

THE PERSPECTIVE ON the motorized recapturing of nature by the city-dweller appealed to many rich cyclists. Baudry de Saunier, for example, well-known because of his publications on cycling, decided to change to motoring in 1897. *Automobilwelt,* the German magazine, pictured individual motorization as follows in 1905: "Now the car has arrived and it has delivered the traveling nature lover from the dominating power of space, so that he, with his freedom of movement, can enjoy the speed of the railroad and the comfort of the compartment. No-one tells him road and purpose, time and departure. He can buzz along from place to place, he can relax at a beautiful, shaded spot with his fellow-travelers, and taste the delicacies from his basket; he can, if he so wishes, change his goal at the spur of the moment and does not have to pass the beauty of regions that are situated off the road as he is forced to do by the insensitive railroad." The added value of the car when experiencing this adventure becomes clear from the idolization of another German motorist, who dreamed that he was "softly carried along, as if there was no contact with the ground, suspended, and without a worry [enjoying] the power of the human mind over matter."[54]

The state of the country roads in those days, the technical condition of the suspension, and the general construction of the automobile suggest that something is "suppressed" here. Yet something realistic and new was experienced, which historical-psychological research describes as "gliding: an unsupported movement, simultaneously as a source of fear and lust." In this connection the breakdown, the small defect that is relatively easy to mend, takes on special meaning. Already in 1877 Ernst Kapp postulated the unity of man and machine in his organ projection thesis. Since in the neurasthenia debate the car had been denoted as a prosthesis of the male body, the treatment of nervousness was not only possible for his own body, but also for his "automobile prosthesis": "not the nervous body should receive more treatment, but the still imperfect technology. In other words: the illness is 'shifted' from the human body onto the impersonal technology." Thus, the German neurologist Willy Hellspach is able to conclude in 1902: "The better half of the neurologist nowadays is—the engineer." It confirms the thesis, advanced earlier with regard to the bicycle sport, that the willingness to encounter a technical risk was an integral part of the experience of the car as adventure machine. In the plain words of a British historian: "the bad roads merely served to heighten the challenge." R. Mecredy, editor-in-chief of the British *Motor News,* who initiated his sporting friends in the secrets of the internal combustion engine, insisted on the importance of technical knowledge. In his writing, still caught up in the world of the horse economy (a garage in his terms is a "car stable"), he gave as a major advantage of such knowledge that the driver could "emancipate himself from the tyranny of the skilled mechanic." Moreover, he would also "materially increase his pleasure in the pastime, for the study of the engine affords almost as keen enjoyment as the actual driving." "Without breakdown, no good driver," two Swiss motorists exclaimed at the tenth anniversary of their automobile club. And, as late as 1907, a German critical observer of the automobile movement explained its success not only from "the feeling of eternal danger . . . in which the Horsepower is worshipped like a God," but also from the *"tickling sensation* of gloomy breakdowns in the dense fir tree forest" ("prickelnde Reiz dunkler Pannen im dichten Tannenwald").[55]

Indeed, the dirty hands of the early motorist expressed the fin-de-siècle state of mind. These motorists kept a finger on the pulse of the technology that was hidden under the hood, a remedy against the decadence of comfort and against the ever increasing distance from the natural state of mankind. Emile Jelinek (a representative of the German Daimler in Nice) let slip in an interview in 1906 that he had become a motorist "out of boredom." Folklore has it that there was a café near Paris where renowned motorists were allowed to make an impression of their black hands on a whitewashed wall. Boredom, much feared by the ideologists of this era, was not an item in the motorist's handbook, although one should not exaggerate this. Many early motorists, certainly in Europe, had a chauffeur to drive them or at any rate they had a subordinate to start the engine, fix flat tires, and clean and maintain the car after the ride was over.[56]

Even more than the bicycle sport, automobile sport was experienced as "active traveling," in contrast with passive traveling by train or streetcar, in contrast even to traveling by electric car that allowed little activity. The Englishman, T. Chambers, writing in 1907 about the technique of driving an electric-powered car, observed: "Apart altogether from its limitations of range and

A gasoline car pays a call on a Dutch village in about 1910. Early automobile culture was not only about racing and touring but also about conspicuous consumption. Two technologies converged—the automobile and the camera. (Author's private collection)

speed, it is certain that there is not much sport in driving an electric carriage. It is far too simple and too unexciting to be attractive. The fascination of the petrol engine to the man who is born with an engineering instinct is largely due to its imperfections and its eccentricities. In these respects, it possesses a soul that has much in common with the human, and one may safely prophesy that when the day arrives that every motor-car shall run with monotonous certainty, the main attraction of driving will have departed, and the amateur will turn his attention to balloons or airships, seeking for further difficulties to overcome."[57]

The "insensitive train" versus the "inspired car," being driven versus driving oneself, made one experience the inspiration of the car in one's guts. This aptly typifies the new experience the car offered. That is why the German motorist, Otto Bierbaum, entitled his report of a trip to Italy that appeared in 1903: *Eine empfindsame Reise im Automobil* (A sentimental journey by automobile).[58]

Installing an engine in the car eliminated not only the horse, but with it the second "driver" with a sense of direction and biological rhythm of its own was also lost. Only the driver himself could bring to life the dead horsepower, stored as potential energy in the fuel. This gave a new, sensational feeling of power. Sir Francis Jeune wrote: "Many persons did, and, I am afraid, some persons do still, accuse us of a love of too rapid progression" (by using the word *us,* he indicated that he considered himself to be part of a "movement"), but "there is a glorious exhilaration in the mere motion of a motor-car, strong, unweary, unresting, with no drawback of regret for strain of exertion on man or beast. The mere sense of motion is a delightful thing; the gallop of a horse over elastic turf, the rush of a bicycle down-hill with a suspicion of favouring wind, the rhythmical swing of an eight-oar, the trampling progress of a four-in-hand, the striding swoop of skates across the frozen fens—all these things of which the reminiscence and the echo come back to us with the dash and pulsation of the motor-car." In the spirit of the times, this sense of power was expressed in a military analogy: "To many of us come all the pleasures and excitement of exploration. . . . I believe that the Duke of Wellington used to say that the best general was the man who knew what was on the other side of a hill. We are all of us in that sense qualifying to be generals now."[59]

Holding the steering wheel had a peculiar effect on people. It is apparently hard to express and sometimes it is better perceived when we catch someone making a slip of the tongue or pen. Baron Henri de Rothschild wrote about his first car ride in a friend's Peugeot. When, to his surprise, he was allowed to take the steering wheel, his attitude toward the vehicle changed; from then on he talked of "my car." Another writer said about a car ride how he "enjoyed the trip as he always enjoyed danger. Danger made him feel wholly alive and pushed all his senses to their extremes. And he always enjoyed seeing someone do anything really well, utilizing skill and concentration and achieving success with what seemed absolute ease. He hated the slipshod, the unprofessional, the indifferent."[60]

The way of observing things also changed. Had Proust not described the car as a new instrument of human observation in À la recherche du temps perdu (published in 1918, but already began in 1871)? According to one of Proust's British contemporaries, the observation from a train was "wrong." For we live on roads and not on railroads, and it is only logical that this notion returns after the interruption of the train travel era: "The road is always with us. The motor-car and the bicycle have restored to us a full remembrance of the fact." From the train we only see the wrong side of the scenery and the houses. In the car, however, "we cut across roads, not wind down them." Then follows the description of a hypothetical, idyllic trip of a "British householder living in the middle of Kent" with his family to the seaside. At first the family went by the train, but later changed to a car, or cars rather, for all in all three cars were necessary for the transportation of family and provisions.[61]

The panoramic view of the train trip was adapted to the new vehicle. The perception of the motorist was no longer determined by the vast landscape, but by "the thousand sights of beauty and interest under his eyes." The car gave back the foreground that had been lost by the train: "A railway has no foreground, unless telegraph posts on an embankment half-clothed, and not at all ashamed, can be said to constitute such a feature. To a road and the traveller on it the foreground is everything. . . . We revive in these later days much of the spirit of the old coaches." It is remarkable that this regressive aspect of the automobile culture returns in much of the "confessional literature" about the early experiences in automobile sport: the car restored an old experience by modern means. Thus, the bourgeoisie recaptured the country road that it had lost to the common folk in the beginning of industrialization.[62]

For the motorists described here, it was not so much the new vehicle's speed as its motion, or rather, its *change* of motion. Speed itself is not felt, unless one is exposed to the elements, which was certainly mostly the case. But speed was not new (trains, bicycles). What was new was self-controlled change of motion. By the laws of mechanics this change of motion has two aspects: the change of speed (exerting a force on the driver when accelerating or decelerating) and the change of direction (producing a centrifugal force when turning corners, or generating a gravity force when riding over bumps in the road). This change of motion in space and in time, controlled by the driver, determined the new automobile sensation, in addition to the tension that developed during the ride between the "power over the wheel" and the uncertainty, the lack of power, with regard to the unreliability of the new technology. And although one could diminish this uncertainty by studying automotive engineering, not even the best driver-mechanic could overcome some of the technical uncertainties. The pneumatic tire was the biggest specter. Alfred Harmsworth, the English automobile pioneer and newspaper magnate and one of the richest men in England, mentioned in 1901 that his tires bill amounted to 500 pounds that year, more than the price of a light car.[63]

In a later section we will discuss the relation between these experiences and the choice of automotive propulsion. But at this point we can already determine that whatever *technical* arguments we may encounter against the electric car, even a perfect electric car, yes, *exactly* such a perfect car, will hit upon an important psychological barrier with its potential customers. No wonder, then, that Edison, when he started a publicity campaign for his new battery, in a Freudian opposition against the flight forward as enabled by the gasoline car, praised its rival as a "dangerless electric auto." No wonder, either, that when Alfred Harmsworth granted his first car (a steam-driven Locomobile) its silence, he felt obliged to add: "but so is a corpse."[64]

CERTAINLY, THE field of expectations regarding the car was not monolithic: the early automobile sport undoubtedly was an elite pastime. Charles Rolls, Lord Llangattock's third son, was regarded as a dangerous revolutionary by his acquaintances, but his revolutionary behavior was apparently not very threatening to society, for the Prince of Wales (the later King Edward VII) declared in 1900: "I shall make the motor car a necessity for every English Gentleman." The attitude of the ruling monarchs or other aristocrats of influence served as an example in most European automobile countries. For Emperor Wilhelm II and the German nobility, the car was not much more than a "fashionable equivalent for horseback riding and hunting, for example." In Austria many of the early motorists belonged to the nobility as well. The German emperor and his brother, Prinz Heinrich (who patented a windshield wiper and wrote under pseudonym in an automotive magazine), lent their names to races and contests.[65]

Whereas the bicycle was praised for its democratic influence, the car seemed to emphasize old class distinctions. Speed became a means of social distinction: "[The automobile] invokes quiet jealousy with the excluded proletarians, anger with the rural population suffering damage, and with the elite it feeds the old privileged instincts of arrogance and callousness," wrote *März*, a German magazine. The later American President Woodrow Wilson, at the time still chancellor of Princeton University, in 1906 declared himself an opponent of motoring. "Nothing has spread Socialistic feeling in this country more than the use of automobiles. To the countryman they are a picture of arrogance of wealth with all its *independence* and *carelessness*." In France the division into "two Frances," that of the "Dreyfusards" and that of the "anti-Dreyfusards," initially seemed to reflect the possession of a bicycle or a car. Count Albert de Dion (1851–1946), "playboy-aristocrat with a passion for machines and publicity," gambler, duelist, lady-killer, anti-Dreyfusard, and chauvinist, in 1899 spent two weeks in prison because of a demonstration against President Loubet that got completely out of hand. The Automobile-Club de France (ACF), of which de Dion was one of the founders, was shortly after closed down by the authorities as a "den of conspiracy against the Republic." In 1902, de Dion became a Member of Parliament as a right-wing Bonapartist, which helps explain some of the measures against the automobile, issued by left-wing city councils in the provinces. To these authorities the car was a "right-wing machine" with strong chauvinistic traits. In an interview with a British journalist de Dion exclaimed: "We control this business and we want to keep it that way. It is of no use for you to try it. You may try it, but you will never beat us. We will destroy you—destroy you!" When Félix Faure, president of the Republic, during a visit to the automobile show of 1899, added a technical touch to his dislike of the man ("Your cars are quite ugly and they are pretty smelly"), he may have spoken on behalf of the Republican part of the nation. But it did not influence sales.[66]

The ACF, founded amid the publicity wave surrounding the Paris–Bordeaux race in 1895 (which made it the oldest automobile club), was an elitist society indeed. The list of founders reads like a Who's Who of Paris aristocracy and the upper middle class. The first president, Baron Etienne van Zuylen de Nyevelt, descendent of a Dutch family, represented the enormous family capital of the Société Générale de Belgique, which had stakes in the Compagnie Financière Belge des Pétroles of the Rothschilds' Paris branch of the family. Van Zuylen as well as the Rothschilds had financial stakes in the oil industry. Van Zuylen was also financially involved in Count de Dion's automobile company and furnished 10,000 francs as prize money for the race to Bordeaux. The first ACF executive committee was a fair representation of the most important parties involved: aristocracy, rich businessmen, automobile manufacturers, and the automobile press.

The German nobility was less interested than the French in the new means of transportation. In Germany the phenomenon of the *Herrenfahrer* (gentleman driver) is an indication that the automobile sport had expanded toward the *Besitzbürgertum* (upper middle class). Was this development the reason why in British aristocratic circles the car was dubbed "the poor man's yacht"?

Also, the emphatic presence of military men and the consciousness of serving a national interest by participating in the automobile sport was remarkable in Europe. For instance, the proceeds of the famous 1,000-mile race in England in 1900, organized by newspaper magnate Alfred Harmsworth, was used to help finance the Boer War.[67]

In most other European countries automobile clubs were also founded before 1900—as in Belgium (1896), Italy and England (1897), and Austria and the Netherlands (1898). In the United States the founding of the Automobile Club of America took place in 1899. In Germany the process of club formation has been researched in detail, and the clubs there cannot be said to have propagated a political view in the sense of an existing party ideology. It is noticeable, however, that the German labor press paid little attention to the automobile, and that the Social Democrats and the progressive, leftist liberal parts of the bourgeoisie in Parliament were less outspoken against the automobile than the conservatives, the national liberals, and the center parties. The latter parties voted in favor of a luxury tax on cars in 1905.[68]

The propaganda of the national automobile clubs seems paradoxical at first sight and contrary to the need to create "their own territory." This paradox was concretized by the excessive ticket prices for the Gordon–Bennet race of 1904 in Berlin, where start and finish were an exclusive affair for members only, but the race itself could probably be enjoyed by the general public. The paradox appears to be false: status only makes sense socially if it is based on a certain acceptance. So motoring had to attract interest in order to continue this "territory of its own," even though it was unthinkable that at the current prices the car would ever become a democratic machine. As vehicles for this propaganda, entirely in the tradition of other spectator sports, the choice fell on exhibitions and races. The glamour of these spectacles would automatically rub off on touring, for only "by the demonstrative consumption of valuable goods the distinguished gentleman acquires prestige," Thorstein Veblen observed as early as 1899.[69]

The car not only emphasized class distinction, it also highlighted the distinction between town and country. The electrification of city lighting made the countryside seem even darker, and although the first motorists romanticized this natural state, for the rural population cars represented city life on wheels. In Germany, for example, there were more cars in the southern and western parts of the country than in the agricultural east. In contrast to what one would expect, the first motorists were not welcome in the American countryside either. The local authorities often supported this hostile attitude of the population, as in the case of Vermont, where each car had to be preceded by an adult man with a red flag at a distance of one-eighth of a mile.[70]

In nostalgic automobile literature, there is the tendency to dismiss the resistance of the rural population to the automobile as anecdotal folklore. Yet the first cars were actually seen as a serious infringement on their lives and their reaction cannot be rejected as merely backward resistance to progress. The population's attitude toward the new sport was largely hostile. They had reason for concern, because the car exported the hectic city life to the country and

The bucolic hazards of
early automobile touring.
(ARM Archives; repro-
duction Charley Werff)

scared the horses. So, resistance to the car was caused by much more than mere criticism of the car as status symbol. In the traditional view of the population, the road belonged to everyone. It was the place where an important part of public life took place and where children could play safely, occasionally interrupted by a squeaking wagon passing by or a group of noisy cyclists. Whereas in the city the process of disciplining traffic had already started, in the country this was not yet the case. This concerned not only children, but also dogs, chickens, and sheep, which frequently became the victims of motorists speeding by.[71]

The motorist experienced this resistance as a war. Frequently armed with a gun or a whip, he indulged in the "romance of the countryside." The French automotive magazine *L'auto* even included resistance to the car under the general vulgarization of society and increase in petty crime. It recommended fighting the "vandals" by means of self-defense, following a course in combat sports or the use of guns. Also a German motorist, who wrote a brochure on this subject with the telling title *Der Krieg gegen das Auto* (The war against the car), had little understanding for this "anger . . . against the automobile monster. The poor rich automobile owners can resist as much as they want, but it won't help them. The desire of the masses for excitement always needs a lightning rod, and the automobile serves well here, just as ten years ago the bicycles and 75 years ago the trains. The defensive battle of the car is rarely one against intelligence, more often automobile owners and drivers have to take into account the prejudices of that class of people, against whom even the gods fight in vain." The author of these lines showed a great deal of insight into the psychology of the motorist, for the "temptation of driving" was also felt in the villages. If a hostile crowd surrounded him, he would invite the biggest spectator for a ride. A Bavarian, who was thus invited, stated afterward that if someone made him choose between his wife and the car, he would pick the latter. Thus "from the Saul a Paul was made," and as readers we have caught a glimpse of "propaganda in practice," complete with a choice of words that reminds one of the conversion zeal of a religious movement.[72]

It seems that resistance to the car was less violent than that against the bicycle, but generally such experiences reminded motorists of what had happened ten to twenty years before. In 1899, the American *Automobile Magazine* explicitly referred to the "tremendous hostility" in the early bicycle period. It encouraged motorists to follow the bicycle movement and get organized, to be in a better position to fight social resistance. Often a warning for motorists to stop careless driving was added because this cast a slur on the "movement," but hit-and-run accidents kept happening, if only out of fear of the population's reaction. Although accidents involving automobiles were mainly an urban phenomenon, the magazine thought it necessary to compile a comprehensive file of newspaper clippings. It cited from it extensively to show that especially the drivers of coaches and wagons, together with their shy horses, were responsible for the inconvenience of both parties.[73]

Mutual misunderstanding was common. This touching incident is cited by a British author from his early diary, who drove down the main street of Hertford: when he approached an old woman, she prostrated herself face down on the road, convinced that she was facing an "inevitable, immediate and painful death." The tongue-in-cheek style of such anecdotes makes one fear the worst about other confrontations with endings less happy that have not been recorded.[74]

In the Shadow of the Gasoline Adventure: The Subculture of the Electric Car
While the prevailing culture tended to favor danger and excitement as therapy, what field of application lay ahead for the quiet and "dangerless" electric? Surprisingly, for not a few vehicle manufacturers, this was the wrong question to ask: their field of expectations was littered with images of a flight from the city, like that of their rivals. Proponents of the electric vehicle realized their limitations in the media hype around long-distance races, but they initially found an arena on their own terms: that of the short burst of energy, at the level of the technical field supported by an unsurpassed high torque at low motor speeds and by the fact that the electric motor could be briefly overloaded. This was shown, for example, during the Concours de Chanteloup in France, an uphill race of 1,782 m, in which forty-eight cars participated, all with gasoline propulsion systems, except for one: the Belgian Camille Jenatzy's. At a speed of almost 35 km/h, Jenatzy, whose electric car of 1,800 kg was the heaviest, beat all of them. Unfortunately, this win was not shown in the official results, which made a distinction between vehicle classes, so that Jenatzy raced only against himself. Surprisingly, this did not arouse any protests. On the contrary, the electrotechnical journal *L'Industrie Électrique* sneeringly figured out that during the climb the Fulmen batteries, which accounted for 44 percent of the total vehicle mass, had to deliver four times, and on the steepest parts maybe even seven times, their normal amperage.[75]

But most famous was the contest that took place from December 1898 to

The first automobile
to exceed a speed of 100
km/h, the electric La
Jamais Contente, and its
driver, the famed Camille
Jenatzy. (Author's private
collection)

May 1899 between Count Gaston Chasseloup-Laubat and Jenatzy for the speed
record on the "flying kilometer." Organized by the auto magazine *La France
automobile,* it was held on a specially prepared stretch of macadam road in the
Parc Agricole d'Achères near Paris.[76]

The race began with twenty-one participants, among which there was only
one electric car, that of Chasseloup-Laubat. Soon after, Jenatzy challenged the
Count—who had left the other drivers behind at a speed of 63 km/h—to a
duel, according to the traditions of the Automobile Club. In four months' time
the record was broken time and again, until, on 29 April 1899, Jenatzy appeared on the course with his secret weapon. Being hauled by a gasoline car to
the Agricultural Park, his La Jamais Contente (Never satisfied) that day broke
the magic boundary of 100 km/h at a speed of "105.882 km/h," as the press
in the whole Western world reported with exaggerated accuracy (figure 1.2).
Shaped like a grenade, Jenatzy's electric racing car was a visual curiosity. This
streamlined form, made possible by the application of rolled light metal,
served mostly a symbolic function, for the driver rose from the waist up from
the bodywork and thus considerably enlarged the frontal surface (decisive for
air resistance).[77]

A similar development occurred in the United States. Whereas during the
Times-Herald contest in Chicago speed was listed below general reliability on
the array of assessment criteria, that was altogether different at the race held
on 7 and 8 September 1896, organized during the annual State Fair in Narra-
gansett Park in Providence, Rhode Island. Due to the bad condition of the
American roads, it was the first circuit race and was hailed by *The Horseless Age,*
because "conditions for high speed are more favorable, and spectators will enjoy a constant view of the competing vehicles." Of the twelve registered cars,
five withdrew because of the required minimum speed of 24 km/h. The races
were held in heats of 5 miles on a good road surface, so that Morris and Salom's
Electrobat and Riker's electric car proved to be the fastest during all five heats.
The maximum speed was almost 43 km/h. John Rae, the American automotive historian, reports how the 5,000 bored spectators responded by calling,

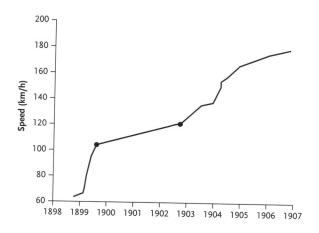

"Get a horse!" suggesting that the speed was low. But this was not the case at all, and was even contradicted by the race itself! But apparently a fast, but quieter electric car was perceived to drive more slowly than a slower, noisy gasoline car. The speed of a car, as confirmed by this incident, is not purely a measurable category, but must be felt, heard, experienced. This not only goes for the crowds that came pouring in at the time, but also for some automobile historians.[78]

Meanwhile, other American electric carmakers also had acquired a liking for racing. At the end of 1901, Walter Baker appeared in the arena with his electric Torpedo (valued at $10,000). Like Jenatzy's car, it was aerodynamic, covered with oilskin, and the "cockpit" was covered with leather in which small mica panes were mounted. The Torpedo weighed 1,250 kg and reached an incredible speed of 192 km/h during a test drive. But when Baker wanted to establish a world record with it on Memorial Day 1902 on Staten Island, a rear wheel hit a streetcar track, and at a speed of 120 km/h he was catapulted into the crowd, causing two deaths and five seriously injured. Baker was arrested and accused of murder, but was released shortly after.[79]

Such races were held just when the first feelings of doubt started to appear about the direction in which automotive engineering seemed to be developing. This doubt was strengthened during the recession of 1900–1901, when the first sales crisis began to loom. Opposition had already started before this among the rivals of gasoline propulsion. "No more contests, no more contests," exclaimed Jeantaud, when La France automobile asked for his opinion about this tricky subject. "In all honesty, I am against it," steam car builder Léon Bollée joined in. But Count de Dion found the races "absolutely necessary for a couple of years more." "A contest is of more use to us than twelve months of labor in the workshop or at the drawing board," he still gave as his opinion two years later.[80]

But in 1900, when the French automobile press gloomily began speculating about la crise de marasme (the stagnation crisis), the opposition to races became stronger. La locomotion automobile was the most outspoken skeptic. One of its editors, L. Béguin, said in an editorial that he had changed his opinion.

The Riker brothers' electric racer, a low-riding version of the two-bicycle structure. (*Horseless Age* 8 [20 November 1901]: 737; DaimlerChrysler Archives)

Paris–Bordeaux, Paris–Marseilles, it had all been spectacular: "But now the people have seen it, they have even seen too much of it, they have understood, that is enough! Let us turn to more serious exercises." Especially when during the Paris–Madrid race a large number of fatal accidents (including that of Louis Renault's brother) had occurred, the aversion against the "delusion of speed" reached its temporary culmination. De Dion felt obliged to declare that he had always been against high-speed races, "that in his opinion offer no practical features, and as a result have no usefulness whatsoever." Even Baudry de Saunier publicly declared his "dislike of modern contests" by exclaiming, "The racing machine is our cancer." Some people now began to realize that only a more utilitarian approach to the car offered any prospects for the automotive industry.[81]

But the supremacy of the electric racing car was over by then, especially because races were now organized over longer distances, a choice entirely geared to the culture of the gasoline car. And although in automobile historiography the "victory" of the racing car with gasoline engine is, after 1902, usually construed as the "loss" of the electric car, a different interpretation is also possible. Generally, among electric carmakers the tendency to a utilitarian deployment of their products was initially stronger. The increasing criticism of the "speed madness" may have strengthened this element in their field of expectations even further. There is, however, no reason to assume that Jenatzy's performance in 1899 could not have been continued with specially prepared racing cars. In this interpretation the choice for racing over large distances is an expression of a *cultural victory* of the gasoline car rather than a *technical defeat* of the electric car.

APART FROM THE AREA of the adventure of time there was yet another uncultivated area some electric car builders felt challenged to enter by their gasoline rivals: the adventure in space, or, touring. The ACF took up the challenge in 1898: in cooperation with the Syndicat professionnel des Industries Électriques it decided to produce a map of France "indicating all the places where

The ultimate electric adventure machine. Walter C. Baker's Electric Torpedo, which set a record by covering a mile in 58 seconds (reaching 99.87 km/h) on Staten Island, New York, in late May 1902. (Smithsonian Institution, Washington, D.C.)

automobiles could find electric current and under what conditions it could be furnished." The idea was Georges Prade's, editor of *L'Auto-Vélo* and fervent promoter of *électro-tourisme*, as he called it. The idea was that "the drivers would find charging stations, about every 20 to 50 kilometers, as easily as gasoline cars nowadays find fuel."[82]

Like racing, touring outside the city was initiated by gasoline car manufacturers and then adopted by electric car proponents. At the beginning of 1899, a company was established with a capital of 5 million francs and the sonorous name "La poste électrique internationale." It announced the plan "to establish on the auto routes of France and abroad electric stations for traction, illumination, and other electric applications" with a distance of about 20 km between stations. This combination of automobile-technical and general electrotechnical products is remarkable, particularly because in the general field of expectations the picture of what exactly constituted "an automobile" had apparently not yet been fixed. If gasoline car manufacturers and oil producers could build up an infrastructure, the electric car manufacturers and electricity producers should be able to do the same, as those in the electric camp apparently expected. The French plan included fourteen routes from Paris to as far as Italy, Spain, Switzerland, and Belgium and five routes from Brussels to Germany and the Netherlands, among other destinations.[83]

In theory the first-generation electric car could be used for controlled electric tourism along the invisible "rails" of the routes, although the driver had to accept a considerably longer charging time than his gasoline-car colleague. But as we have seen in the previous section, the essence of the space variety of the automotive adventure was the glow of independence, of the individual construction of a foreground of one's own. At the start of 1899, Kriéger entered this supposed niche in the electric-car market. Under the meaningful title

"Electricité contre pétrole" (Electricity versus gasoline), he made his bet public: the vehicle that would come to a halt first, after having started from the same point and without stopping while on its way, had lost. Remarkably enough, nothing was heard from the gasoline camp: perhaps there was a fear that, being severed from its fuel distribution infrastructure, the gas guzzling petrol cars would not have traveled very far on one fill of the gasoline tank. On 2 July 1899, however, Chasseloup-Laubat, who had lost his race with Jenatzy for the fastest kilometer, drove to Rouen, a distance of 110 km, in a Jeantaud, using just one charge. A gasoline car drove along to keep an eye on things.[84]

The appetite for long distance also became apparent late in the summer of 1899, when the electric vehicle maker BGS tried to promote electric tourism to the Normandy coast. Articles appeared in the automotive magazines that were clearly intended as advertisements, in which the organizers expressed their surprise at the unexpectedly large number of charging stations along the route, but also about the discrepancies in kWh prices. Moreover, these rides were not on pneumatic tires, but on so-called compounds (soft, solid rubber tires), and among the "tourists" were the two inventors of this tire type. Such pursuits of the large range, however, were not so much a reflection of the density of the network of charging stations, but of the need to surpass the gasoline rival. They reached their culmination in 1899, when Georges Prade suggested a "Tour de France électrique." This idea was still alive two years later in the automotive magazines, as a trip of 2,400 km with charging stations along the route.[85]

BGS established its name as producer of electric touring cars especially by its participation in the Deuxième Critérion des Électriques, organized by *Sport Universel Illustré* in June 1900. The company covered a distance of 262 km on one charge with a specially prepared car of 2,300 kg (with a battery share of as much as 50 percent). On 16 October 1901, Kriéger, who took up the challenge by its arch-rival, broke BGS's record by covering 307 km between Paris and Chatellerault on one battery charge. Along the river Loire the car reached a speed of 26 km/h. Kriéger's car weighed 2,400 kg, including load and crew. It contained 1,230 kg of Fulmen batteries, 51 percent of the total weight, and thus much more if the weight of the load would have been subtracted. "Electric acrobatics, comparable to those performed by Mr. Jenatzy at Achères," according to electric-car expert É. Hospitalier, could only have "a disastrous influence on the spirit of the general public." He also suspected that the batteries had been specially prepared by inserting extra thin lead plates and applying nearly undiluted sulfuric acid as electrolyte. Kriéger and Fulmen engineer Brault denied this most emphatically. In their opinion, the experiment was "a first stage toward the electric tourist vehicle, capable of covering practically 200 km a day." Hospitalier rejected this claim, as long as he did not know "the *number* of discharges that the batteries could withstand at this debilitating rate." And one might wonder indeed about the use of such an exercise with a lead battery that generated an improbably high energy density of 41 Wh/kg, caused by (indeed) the higher density of the electrolyte.[86]

Unlike its parvenu counterpart, conventional electrical engineering, with its strong tradition of urban streetcar transportation, sneered at such frivolities. Records of distance "baffle[d] the bourgeoisie," but proved absolutely nothing with their battery-vehicle mass proportion of 45 percent, according to Hospitalier's harsh judgment. Motoring was a sport, not a circus, he opined (clearly indicating, by the way, that touring, as much as racing, in those early days generally were both seen as "sport"). Nevertheless, hopes for national electric tourism were kept alive. In the summer of 1901, Abel Ballif, president of the Touring-Club de France, announced that the number of charging stations on the "electric map" had increased to 700 and that he hoped this number would soon be 1,100. Publication of this map was set to be at the beginning of 1902.[87]

In England a modest campaign for long-distance runs also started, initiated by the Electrical Undertakings Company. Their car of 1,000 kg and a battery share of 45 percent covered the distance of 83 km from London to Brighton at an average speed of 18 km/h on one charge. Shortly after the turn of the century, Kriéger took the lead under the meaningful name "Powerful," an electric touring car manufactured under license by the British and Foreign Electrical Vehicle Company. In November 1900, the Powerful won the electric-car contest sponsored by the British automobile club, covering 95 km on one charge.[88]

Such expectations of electric tourism along the strands of a tightly woven web between the big cities also existed in America. They were less intense, however, due to the poor condition of the roads and the concentration of early automotive engineering on urban vehicles and utilitarian fields of application. Nonetheless, in 1897 Hiram Maxim compiled a list of charging stations on behalf of touring around Hartford. It appeared that even then there was a sufficient number of charging stations between Boston and New York to make such touring possible. Inspired by this initiative, Insull's Woods Electric Vehicle Company, in electric-car city Chicago, launched a plan for establishing "a system of cross-country runs from Chicago to Milwaukee, St. Louis, Cleveland and Indianapolis." But in this case, in contrast to Europe, a timetable would be used. The idea of an electric touring-bus network that would start with fifty cars and would mean an extension of initiatives with bus lines in the big cities was rejected by *The Horseless Age* as "visionary to the point of lunacy," as it meant a futile attack on the railroad network. "The car," so much was clear in the gasoline camp, should not be pushed as part of a system, because as such it would be doomed.[89]

But even Hiram Maxim, proponent of a utilitarian deployment of electric propulsion, could not resist the temptation to undertake a long-distance trial run. Together with ESB engineer Justus Entz, he covered 160 km on a single charge in a Columbia, and was accompanied by controllers on bicycles. Thus, for a while the record went from French hands into American hands. This can only be explained by the lower (gross) vehicle mass of 1,112 kg, as the share of the battery mass (41 percent) was not extremely high. This result was remarkable, because it shows that the same field of application could be realized through two different technical fields. The success of the French record pursuers was based on a superior battery technology, whereas the low energy density of the American chloride battery was compensated by the lower vehicle mass and the lower rolling resistance this caused.[90]

ALTHOUGH SOME ELECTRIC-CAR BUILDERS, in Europe as well as in the United States, kept following the trail of the electric adventure, others realized that this was nothing but the pursuit of the gasoline adventure by electric means and that it was doomed to fail. Yet this controversy has remained an inseparable part of electric-car history until the present day. But then as now there is no escape from the undeniable fact that the lead battery allowed either high speed or a large range, but not both at the same time. This was due to the reciprocal relationship between battery capacity (measured in ampere-hour) and power density (measured in kW/kg). If one is looking for a nearly absolute technical constraint in the history of electric vehicle technology, it is to be found here.[91]

After the first years of experience with electric cars, optimization calculations began to appear in scientific publications. These showed that at a given energy requirement of 100 Wh per ton-kilometer and an energy density of the lead battery of 20 Wh/kg, the battery mass should not be allowed to be lower than 25 percent nor higher than 50 percent of the gross vehicle mass. Above this range, the required reinforcement—and so increase in weight—of the vehicle more than canceled any energy gain. Below this range traction problems occurred. Such calculations agreed well with common practice, which in the meantime had converged to an average battery-vehicle ratio of 30 to 40 percent. From this it was not difficult to calculate that, at a speed of 18 km/h, a theoretical action range of 60 to 80 km was possible.[92]

Such calculations formed the scientific foundation for the division of the field of application between the three propulsion alternatives. It popped up in any number of varieties and in all magazines that paid attention to the "problem of choice," usually influenced by the national transportation culture. In Great Britain, for instance, Sir David Salomonson, president of the Self-propelled Traffic Association, formulated this division of tasks as follows: gasoline propulsion was "presumably" suitable for motorcycles, steam propulsion for all other cases "when real work is called for, and where a return upon capital expenditure is required . . . Electric energy, if the necessary adjuncts exist, has a great field open in towns, as a luxury, where the question of upkeep is not a vital item."[93]

In France the problem of choice was understandably seen in a different way. In 1898, *L'Industrie électrique* considered steam propulsion suitable for heavy vehicles for the transportation of goods and passengers, gasoline propulsion for high speeds, *grand tourisme,* and "excursions," but also in the city for those who were not repelled by a complex mechanism, oil, and grease. Electric propulsion on the other hand would power taxicabs and city cars, "waiting for newly realized progress once the batteries will have extended their range." This last addition is particularly meaningful, suggesting the emergence within the field of expectations of a "miracle battery" that, next to the image of the electric car as a would-be petrol car, would become a second theme of electric-car history and was to leave an important mark on it.[94]

In America, in the early part of the twentieth century, *The Horseless Age* deemed electric propulsion suitable for "boulevard and park carriages, especially for ladies and elderly persons . . . ; for physicians and business men, where streets are in good condition, and as a complement to the 'stable' of those automobilists who are anxious and can afford to use the best suited motive power for any trip they may want to make." Despite the different views of the division of tasks between the three propulsion alternatives, judgment about the electric car was virtually unanimous: the electric car was a *city car*.[95]

THE QUESTION NOW ARISES: What trajectories were open to the potential private electric-car driver, and what form was this version of motoring to assume, in technical as well as sociocultural respects? Those who did not wish to be influenced by the racing results in their choice of propulsion type were overwhelmed by a confusing flow of information from the automobile magazines, the automobile shows, and the much publicized contests. Two of these contests were of great importance to those who felt attracted to electric propulsion.[96]

The Paris Concours d'accumulateurs, organized by the French Automobile Club as of June 1899 and proposed by Jeantaud, received the most publicity. In organization and approach, it betrayed a preference for a utilitarian use of electric propulsion. For six months the competing batteries were subjected to a series of tests for durability, efficiency, and energy density, carried out in a charging station set up and operated by experts. The aspect of costs was reserved for a subsequent contest. A special shaking machine, built by Jeantaud, simulated the motion of a car, but due to constructional problems the machine was in operation only 20 percent of the time. G. Forestier, grand old man of French electrical engineering, chaired the organization committee. Among its members were not only renowned electric car pioneers like Kriéger, Jeantaud, and Jenatzy, but also the director of the biggest Paris taxicab company, Bixio, and his engineer De Clausonne. The jury report, written by Hospitalier, was finished in January 1900 and received a lot of publicity.

On the whole, the results of the severe test were disappointing, especially as far as the lighter batteries with grid plates were concerned, which in France comprised the majority, judging by the number of contest registrations. Only six of the sixteen batteries tested lasted longer than sixty charging/discharging cycles. All light grid-plate batteries broke down prematurely. All six "survivors" were either provided with large-surface plates (such as Tudor's) or with quite thick grid plates (Fulmen). Also the TEM battery (with a positive large-surface plate and a negative grid plate, both based on lead chloride)—related to the American ESB battery—belonged to the last six. According to the most thorough contemporary analysis, based on the jury report, which may have been colored a little too much by the experiences of battery streetcar practice, the tests proved the superiority of the heavy (large-surface plate) battery. But it cannot be denied that the Fulmen battery, which (together with Phénix's tubular-plate battery) came out best with an energy density of 15 Wh/kg and a life span of a hundred cycles, had plates that were almost as heavy as the large-surface batteries.[97]

To many the results must have been so disappointing that the following year no battery contest was organized.[98] One of the results was, however, that the Mitteleuropäischer Motorwagen-Verein, the German automobile club, at the last moment added a test of its own to the Berlin auto show of 1900, which it delegated to Dr. M. Kallmann, *Stadtelektriker* (municipal electrical engineer) of Berlin. Between 23 and 28 April 1900, fourteen electric vehicles on show that year were investigated. Half of them were passenger cars, built by Scheele, Henschel, Kruse, Fiedler & Company and the Fahrzeugfabrik Eisenach. Among the members of the jury were chief superintendent Vogel, head of the Berlin fire brigade Giersberg, representatives of the Postal Service and the automobile club, and a number of independent engineers, including Franz Wilking and Max Zechlin. For his assessment Kallmann had invented a complex system of premium points, in which aspects of construction, business organization, and business economics had been incorporated as well as maintenance and purchasing costs. The maximum speed constituted only one of the 60 points. Some of these points were established by calculation and measurement, others were determined by a jury, so that it is not surprising that the result of the trials did not provide a clear winner. In the thirty-six-page report, the winner was not mentioned by name, although he did receive a gold medal and the motivation of this choice took up many pages. This breach with the tradition of races and contests that expressed technological "progress" in a single figure, preferably higher than the previous one, could not prevent Kallmann's conclusions from playing a role in the further developments of German electric propulsion.[99]

Kallmann recorded a battery share in the total car weight of 19 to 37 percent, while the average range was 50 km, with a maximum of 66 km. The most economical vehicle consumed 57 Wh/(ton-km), a value that increased to 84 Wh/(ton-km) in bad weather. The price of the automobiles was between 4,000 and 9,000 German marks, the batteries accounting for a share of 17 to 33 percent. During tests, it was determined that the batteries with large-surface plates could be overloaded up to three times their nominal amperage, the grid-plate batteries only up to two times. The most important conclusion concerned the costs of operation, however, calculated on the basis of a maintenance subscription of a third of the battery purchase price of 1,100 German marks and an annual mileage of 15,000 km. These were 3.3 pfennigs (pf)/km, 2.5 pf/km of which were reserved for battery maintenance. Apparently, the life span of the batteries was estimated to be unreasonably high. At least that can be deduced from Kallmann's final comment: "If a life span of the battery was guaranteed for 250 discharges only [2.5 times the life span of the best battery of the Paris contest!], the costs would triple and would probably exclude the profitability of an electric transportation company." So this test eventually led to a plea for a businesslike utilization of electric propulsion. With the field of expectations defined as such, what did the application field at the level of individual users look like?

It will not come as a surprise that those confronted with the dilemma (in Europe) or trilemma (in America) were private owners, who even at an early stage and in the middle of a quite different prevailing car culture, approached their property as a utilitarian vehicle. In automobile circles, the medical doctor's car was touted as the showpiece of the utility car. Medical circles already supported this notion in 1898. In the midst of certainty about the use of the automobile in general and uncertainty about its propulsion system, the British *Lancet* wrote: "the motor-carriage in *some of its forms* will prove admirably suited for the requirements of medical men, and it will not be long in coming into extensive use."[100]

But the question was: powered by what type of energy source? Information about this struggle at the level of the individual customer is hard to find. Therefore, the two opinion polls organized by *The Horseless Age* in 1901 and 1903 are of vital importance. In dozens of letters American physicians and a few businessmen described their experiences with the three propulsion systems. In 1901, the letters mostly came from the major cities on the East Coast, but in 1903 they arrived from all over the country. The magazine also held interviews, for instance, with physicians in New York to ask about their experiences. One physician recounted that he initially ordered an (electric) rental car with driver, but returned to the familiar rental horse carriage due to high costs, poor service, and drunken drivers. Typical of the American situation is the story of another physician, who first bought an electric car, but changed to a steam car because the electric's range was much less than the manufacturer had promised. The steam car proved unsuitable (unfortunately, the interviewer did not mention the reason why), so that eventually he preferred "a strong, simple gasoline vehicle with slow-running, single-cylinder motor." Another physician bought two vehicles at the same time: a steam car for his regular visits and an electric car for emergencies. The majority in the poll of 1901 had a gasoline car (ten physicians), immediately followed by steam (nine physicians), while only two of the respondents owned an electric car. The remainder of the twenty-five letter writers consisted of special cases, such as the motorcycle owner, or the physician who returned to horse traction after trying steam traction, or the person who rented a mechanic with his steam car. One of them had a gasoline car and insisted on keeping a spare vehicle on the side.[101]

In the second poll of 1903, which had a more nationwide character, the willingness among these pioneers to undertake do-it-yourself maintenance was remarkable. Most of them had a workshop of their own and some claimed to use the gasoline car also for recreation, to tinker with it in their leisure time, and to understand the mechanics. The preference for the gasoline car with some was inspired by the wish to go touring out of town during the weekends or the holidays. One physician mentioned that he had already driven 1,000 miles in the summer of 1900.

With regard to the role of *The Horseless Age* in the campaign against the "Lead Cab" (see chapter 2), the polls are probably biased toward the gasoline car. They do document, however, the country physician as the ideal user: used to tinkering and improvising with his medical equipment, he could combine adventurous and utilitarian deployment of an artifact that at the time could by far not function on its own. But the magazine only printed one letter that mentioned the renting of electric vehicles, whereas in Newport, for instance, such cars were, according to an electric propulsion proponent, "in great demand by physicians for the removal of convalescents." Apparently, the subculture of the electric passenger car was not very well documented in the trade press of the day. Nevertheless, it would be wrong to suppose that maintenance of the electric car would be less problematic than that of the gasoline car. For instance, W. Hutchinson, a medical doctor, concluded after one-and-a-half years that his electric car was "a piece of apparatus that will require constant, skilled and intelligent care and attention if it will be of service . . . And the manufacturers themselves now say the same thing, although they did not make it so clear a year and a half ago." Hutchinson explained how he covered 25 to 50 km a day and subjected his battery to a "painstaking inspection" once a week. This meant that each week the battery pack had to be lifted from the car "by the aid of two muscular stable hands." He then started the maintenance job in the kitchen, "much to the disgust of the tidy housemaid." If a defective lead plate had to be washed or exchanged, the lead connections were sawed off and then soldered together again by means of an open flame, a job that was not particularly pleasant due to the presence of hydrogen gas that was released during charging. Hutchinson concluded that such maintenance was a definite obstacle to any further success of the electric automobile, that one could not trust a single manufacturer in this respect, and that the only solution was a public charging station.[102]

The few nonprofessional users who took the time to respond to the polls presented the same general picture: as far as they made use of an electric car, it was just as reliable as an electric streetcar. This statement, by the way, came from the pen of an electrical engineer, who was also an electric-car owner. These users preferred not to get their hands soiled by the adventure machine. No, to quote a French representative of this subculture, for them "a *real* car . . . always had to be ready for use." And this was exactly the bottleneck, not only for the early electric-car user, but also for the historian who tries to get a grip on this subcultural field of application.[103]

Despite the scarcity of documentation about the very early use of the electric car by private persons, the conclusion seems justified that medical doctors, due to their practical inclination, were more prepared to take the risk of purchasing one than others. Because of their education, they were also able to write about it in an articulate way. The doctor and the businessman (who was far less heard of in the automobile magazines) can be seen as pioneers, who combined a utilitarian use with adventure; a vehicle was to be enjoyed on the weekends and the holidays. So, it is not surprising that they preferred the gasoline car and the steam car.

It is much more difficult to track down the early users of the electric counterpart of the pleasure car, as it was called in contrast to the car for utilitarian purposes. In order to identify the different market niches of the electric pleasure car, we follow the upper rungs of the social ladder. To start at the top: like (and even before) the gasoline car, the electric car appealed to a royal clientele. As early as 1896 the London company Thrupp & Maberly built an electric car for the queen of Spain. Five years later the British Queen Alexandra bought an American Columbia from the City and Suburban Electric Carriage Company. Jeantaud mentioned in 1898 that he was building three large electric breaks— on order, as was customary at the time—for Prince Strozzi in Milan, Prince Galitzin in St. Petersburg, and the duchess of Alva in Madrid. It seems that especially royal women preferred the electric. In this respect King Leopold of Belgium was not exceptional when he mentioned to a journalist of the *New York Herald,* on the occasion of the Paris automobile show of 1902, that he wanted a car "faster than all others. If I can't drive 130 km/h, it is no good."[104]

The market segment below this is even harder to trace. About the *beau monde* of Paris, it was rumored that it indulged in *"concours d'élégance* of electric vehicles," as in June 1901 on the occasion of a *polo de Bagatelle.* It was not any different in the United States, where *The Automobile Magazine,* in its first issue of October 1899, observed a "present craze for automobiles among the leaders of American society." The magazine spent many pages on a *concours d'élégance* in Newport, where during the summer the women had joined the auto fad, "insisting on running their own automobiles." Mrs. Stuyvesant Fish, Miss Greta Pomeroy, Mrs. Whitney, Mrs. Drexel, and Mrs. John Jacob Astor drove electric and gasoline cars that were garlanded with flowers from tire to hood. The magazine paid detailed attention to these decorations, even more than to the technology. And at an "automobile procession . . . the largest ever taking place outside of Paris," organized on 24 May 1899 in New York, eleven of the fifty-one participating electric cars were driven by women.[105]

If we descend one more step on the social ladder, the picture of the users becomes anonymous and vague. In 1900 in Chicago, for example, no licenses were given out to women for steam and gasoline cars, because these were considered "unsuited for use in feminine hands." Women did not seem to be bothered by this, because in November of that year the city engineer of Chicago, member of the local automobile inspection, mentioned that he received many reports about lady drivers without licenses. He said he would "arrest on sight any lady found driving her machine without the proper authority."[106]

Many potential electric-car users were as much bothered by the trilemma as the American medical doctors. For instance, in 1896 Paul Meyan, editor-in-chief of *La France automobile,* mentioned the many letters from readers that he summarized as follows: "I would buy an automobile, but for the bad smell of the gasoline, the vibrations and the engines that are still too complicated; steam requires carrying along lots of fuel." Then he posed the question: What about the electric car? Two years later the sister magazine *La locomotion automobile* mentioned that without any doubt the public preferred the electric car. But the electric car was seen in the meantime as the *"Vehicle of Tomorrow,"* whereas the gasoline car was considered the *"Vehicle of Today,* ridden with all kinds of awful faults: bad odor, complicated, etc.etc."[107]

For such potential users it was important that the charging grid in the city was as dense as possible. An initiative in this direction took place in Paris in

As a racer and touring vehicle, the auto carried strong male connotations. Yet early automobile sport also knew a "female" form, the flower contest. If women comprised a small minority of early automobile drivers, they were well represented at these events. Here in 1895 a gasoline-powered Daimler Riemenwagen parades in Eastbourne, East Sussex, England, a camera at the ready. (DaimlerChrysler Archives)

1898. A cooperation of electrotechnical professional societies and the French automobile club organized a contest for the design of a charging pole, but the plan never advanced beyond the prototype stage. A year later, when the *Annuaire générale de l'automobile* was published, it mentioned the number of "factories and charging stations." For all of France there were 265, fifty of these in Paris. In America such initiatives can be seen too, such as that of the General Electric Company that had developed an "electrant" around the turn of the century. A contraction of "electric hydrant" (inspired by the fire hydrants), it could be operated by inserting coins.[108]

The Parisian electric vehicle makers Charles Mildé and Robert Mondos, however, launched the most daring urban plan. They seem to have been inspired by Edison, who introduced his light bulb as a tiny part of a comprehensive system of electricity production and distribution. Shortly before 1900, they proposed a newly designed luxury car that they accompanied with a detailed plan of an electric automobile service covering the rich 8th, 16th, and 17th districts of Paris. In this area around the Arc de Triomphe, they had projected eleven charging stations. The two declared that the plan was no harder to realize than many other plans that seemed impossible at first, such as the telephone system and the pneumatic tube system for the distribution of mail. An "Electric House," part of the plan, consisted not only of a garage for the electric car, but also of electric lighting, an electric elevator, and an electric heating and ventilation system. The house, according to the designers, was "pre-eminently . . . a House of Luxury . . . at the same time healthy and clean." In this plan the electric car was meant for all those Parisians who owned a horse and carriage worth about 3,000 to 5,000 francs. Those with a horse worth less than 1,200 francs were politely but decidedly referred to the "gasoline horse."[109]

Although many of these plans never left the drawing board, the charging grid in some of the big cities was undoubtedly quite extensive, so that a potential motorist could have cut the Gordian knot of the dilemma or trilemma in favor of electric propulsion. Why did this not happen on a larger scale? The British electric-car builder Walter Bersey gives part of the answer. "I have no belief whatever," he wrote in a booklet, published in 1898, "in the idea that persons will be able to buy an electric carriage and keep it themselves, and charge it with electricity from the ordinary electric light arrangements that they may have fitted up to their house or stables. It might happen in a few cases that persons could do this, but they would have to have considerable expensive electrical apparatus, and they would also have to have a competent electrician to deal with the necessary charging, &c." Moreover, it was not likely that battery manufacturers would give a guarantee to individuals, "as there would be considerable doubt to their being properly looked after." Although Bersey thus presented the electric vehicle as a contraption for specialists, at the same time he saw the gasoline car as only a transitional phenomenon: "I quite admit that at present there is an enormous demand for the petroleum autocar, and the manufacturers are, I am told, asking from 15 to 18 months for the execution of an order. But it is mostly the wealthy classes and amateurs who are using it. They obtain amusement from it just as a child is amused by a new toy . . . [But a] wealthy man will never be so proud of his automobile as he is of his well-groomed thoroughbreds." Besides the belief in the future of horse traction in this rich segment of the market, it is remarkable that Bersey's focus was not on the problem of the density of the electricity grid, but on the guarantee of a reliable battery.[110]

Many of the earliest motorists owned several cars. And even if these users chose the electric car from their automobile stock, "contact with nature" was not entirely impossible. For them there was always the civilized drive in the city park, that is, if authorities allowed it. American cities had sacrificed a much greater proportion of their surface to the laying out of such "urban lungs" than was customary in Europe. Many American parks were laid out in the second half of the nineteenth century as unemployment relief work during the recession. In 1890, the total length of the "park drives" in fifteen big cities amounted to more than 500 km. The parks were inspired by British landscape architecture and appealed to the taste of the new middle classes that perceived nature primarily as scenery. A second wave of parks followed at the end of the 1890s. Although after the Great Depression the incentive behind the "park movement" gradually disappeared, exactly at the moment of the advent of the car, in several cities a heated controversy developed about the accessibility of such parks to cars. In particular, electric-car owners claimed the use of the parks, as they were used to going there with their horse-drawn carriages. The local authorities often supported their claim.[111]

Initially all cars, whatever their motive power, were banned with the argument that they scared the horses. This was the case in "conservative Boston," for instance, which *The Horseless Age* called "foolish, reactionary and worthy of a Dogberry." In Colorado Springs cars were kept out "until such times as they are as odorless and quiet as a horse conveyance." But on 13 November 1899, the New York City Park Board published a list of twenty-six licenses for owners of open electric cars. The only exception concerned a motorcycle ("an experiment"), whereas three separate licenses were given out for closed electric cars. The majority of the licensees were male (among them the millionaire J. J. Astor), but the list also contained two women. The seventeen-year-old "Miss Florence E. Woods," daughter of the electric-car builder, made the headlines because she was the first woman to obtain such a license, after she had "demonstrated her ability to operate her automobile skilfully." In the summer of 1900, four electric buses, for twelve passengers each, were admitted to the park for tourist rides.[112]

In 1900 in Druid Hill Park in Baltimore, electric cars were admitted, while gasoline and steam cars were not. But in December "a party of socialites," supported by the automobile press, including *The Horseless Age,* provoked a lawsuit by leaving their electric cars at home and driving into the park with their gasoline cars. The state authorities then overruled the local authorities by issuing a law that allowed automobiles to drive in the parks at a maximum speed of a little under 10 km/h. In Philadelphia, the largest park area of all the big cities, Fairmount Park, was opened to motorists in the fall of 1900. Around the same time San Francisco also opened its city park to automobiles, "irrespective of the motive power of the machine."[113]

These examples show that the electric car as *exclusive* city car did not stand a chance, because early automobile owners were all-rounders. They did not have to face a technical trilemma, simply because they owned a variety of cars that they could use for their "specialist tasks" according to their type of propulsion system. And touring, even in the not very adventurous surroundings of the city park, apparently belonged to the application field of the gasoline car for many of these pioneers. For them, Foreman-Peck's retrospective comparative cost-benefit analysis, had they heard of it, must have appeared quite meaningless.

THE ALLEGED "INFERIORITY" of the electric car in automotive historiography is largely based on the first generation of individually owned passenger cars. It is also based, as we have seen, on a false opposition between the initial dominance and subsequent decline of electric enthusiasm. In reality, the electric car's field of application, as far as private use is concerned, has always been modest in comparison to the gasoline rival.

It seems nonetheless enigmatic that the field of expectations among early electric vehicle proponents was full of convictions about a bright future for their product. Generally, the electric engineering community supported them in this belief. Apart from the fact that many of them regarded the gasoline car as only a temporary fad, they had a very good reason to expect more. This was due to the fact that, because of its intrinsic structural flexibility, the technical field of the electric was much wider than one would expect, thus enabling a broader application field as well. To analyze this, we have to go back into the final five years of this period, and study some well-documented experiments of first-generation electric taxicab deployment.

2 Failed Experiments:
The First-Generation Electric Taxicab

The time of the electric taxicab will arrive faster than the
race from Paris to Marseille and back would lead us to believe.
—*L'Industrie électrique*, 1896

THAT THE ELECTRIC TAXI EMERGED before the turn of the century in relation to the horse taxi business, which it was intended to supplant, may seem obvious, but one other impetus may be at least as important to its creation. For, is it a coincidence that when, around 1895, the battery streetcar business decreased, the interest of engineers and companies in that other (semi)-public transportation vehicle, the taxicab, gained momentum? The relation between the battery streetcar and the taxicab has not been thoroughly researched so far, but there surely are strong indications of its existence.

The taxicab motorization experiments aroused great public interest and were accompanied by vehement commotion in the automobile industry, the taxicab business, and the trade press. Their consequences for the battle between electric and gasoline propulsion were far-reaching. The unfortunate outcome of all early experiments must have left a deep impression on those directly involved.

The test-tube character of the three most important first-generation taxicab experiments, in London, Paris, and a number of cities in the United States, enables us to answer the question of the degree of "closedness" of the technical field. Surprisingly, there are two possible causes for its failure: the battery and the pneumatic tire, which had evolved from streetcar and bicycle technology, respectively. Were these factors decisive, or did other factors (both sociocultural and managerial) at the level of the application field play a comparable role? This question is also important because its answer will shed more light on the comparable choice problems within the application field of the much less documented private passenger car.

The Burden of the Battery Streetcar: The London Electrical Cab Company

On 12 November 1896, two days before the Red Flag Act was going to be replaced by new legislation, the London Electrical Cab Company Ltd. was founded. Everything—from the composition of the board of directors to the speeches at the opening to the technical structure of the cab itself—culminated in just one wish: replacing the horse as the propulsion system of the London taxi. But the company itself had no roots in this business. The key technical figure of the company was "electrical manager" Walter C. Bersey, who brought in his patent of 1894 on an electric car.[1]

On the cab company's board of directors, all the social forces needed to make the plan succeed were represented. The chairman was the renowned coach builder H. H. Mulliner, and among the other members of the board were the directors of the London Electric Supply Company, the Great Horseless Carriage Company, and the Daimler Motor Company. The latter was also director of the Northampton Street Tramways Company. The Daimler Company was part of the British Motor Syndicate, established by monopolist Harry J. Lawson. He also managed to acquire Bersey's patents, so that he could collect a licensing fee of 4 pounds per taxi per year. Lawson was also in charge of the Great Horseless Carriage Company, which obtained the exclusive right to build the cabs.

According to the prospectus, the choice of electric propulsion was obvious: "Whilst petroleum may become the motive power in country districts, and steam will probably be used for very heavy vehicles, there is no doubt that electricity will be the most advantageous where the traffic can be located within a radius." The management expected that charging stations would be installed in different quarters of London, so that the cab drivers could exchange the batteries without having to return to their own station. According to the founders, electric propulsion had nothing but advantages compared to the gasoline car: "There is no smell, no noise, no heat, no vibration, no possible danger, and it has been found that vehicles built on this Company's system do not frighten passing horses."[2]

The taxicab was designed by an engineering consultancy company that had a reputation for building electric streetcar lines. The company carried out tests on batteries "from every make of repute" by means of a special shaking machine. In anticipation of the results of these tests, the first vehicles were equipped with batteries of the Faure-King type. These grid-plate batteries, developed by Frank King and Camille Faure for the battery streetcar and put on the market in 1896, were supplied by the Electrical Power Storage Company (EPS). This company was under contract to take care of maintenance at 10 percent of the purchase price.[3]

The vehicle that emerged was, according to the most thorough contemporary analysis, "not integrated"; all components were connected to each other as separate parts, so that no "one part formed part of other parts." A closed coupe was placed on a heavy steel frame, while the rotating parts, placed on separate subframes or plates, were connected to the main frame by means of leaf springs. The entire construction weighed almost two tons and was powered by a 2.2 kW Lundell motor. The electric taxicab had relatively light wheels with thin wooden spokes; the steel wheel rim was provided with a solid rubber tire. The tire was very narrow because this was expected to give better grip "on greasy pavements."[4]

This much was clear: the cab was not conceived as a system; it was not the work of an automotive engineer à la Jeantaud or Daimler, but of a cooperation between renowned coach builders and electrical engineers with experience in streetcar construction. The composite character of the vehicle construction could also be seen in its external appearance: it gave the impression of a carriage for horse traction that was being hauled by the driver's cab of a battery streetcar. Developing a car from a streetcar was no mean feat, considering that the electric vehicle, removed from the rails and with a battery mass of 15 to 20 percent of the total vehicle mass, had to overcome a rolling resistance on a level road of about 2.5 times as high. This required an energy consumption in watt-hour per ton-kilometer of a factor 3 to 4 higher than on rails, to be delivered by a much smaller battery pack.[5]

A charging station with adjoining garage was built in Lambeth, in the neighborhood of Westminster Bridge. In the garage a hydraulic lifting device made battery exchange possible within two to three minutes. The principle of this exchange charging, in which *mass* instead of *energy* was charged, was not new. Although not common, it had been applied in the battery streetcar days by the Belgian Edmond Julien. In comparison to the gasoline car practice, however, it was of strategic importance: at one stroke it undid the disadvantage of long charging time.[6]

The company's electric charging energy was going to be supplied by the London Electric Supply Company from its power plant in Deptford. In the charging station, two motor generators of 75 kW each would convert the alternating voltage of 2.4 kV. But the price of the electricity supply was so low that the kWh price would be virtually the same as when electricity production had taken place in the station itself.

ON 19 AUGUST 1897, "the first 12 Electrical Cabs in Europe," and, according to Bersey "for the first time in the history of the world," were shown to the press and invited guests. The following day it appeared that fifteen taxicabs were finished, but ten of them were immediately rented out to private customers on a daily basis. Just as with the taxicabs, the rates were the same as for horse-drawn vehicles.[7]

The first year the company suffered a loss of more than 6,200 pounds. Then, in the spring of 1898, when about twenty-four cabs were on the street, the company decided to replace the original design by a growler type of cab, to be built by the Gloucester Railway Wagon Company. Fifty were ordered. Some months later the fleet had grown to seventy-one electric cars. Forty of them were used as taxicabs and together had covered almost 320,000 km.

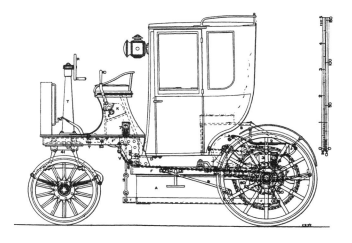

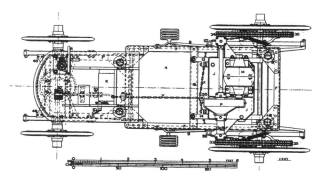

Technical structure of the Bersey taxicab, typical of the first generation, side and underside views. Note the central-pivot steering, reminiscent of horse carriages. (Beaumont, *Motor Vehicles and Motors*, p. 398)

Management not only had to deal with financial problems, but also with complaining drivers, who were often taken over by private persons. An internal survey brought to light five different reasons for the drivers' complaints. One of them was a drive chain that proved too weak. It was replaced by a strong double roller chain, but this chain produced more noise. Furthermore, both the car body and the electric motor with the batteries were now separately sprung. Moreover, although the construction was lighter the battery pack had become bigger, now weighing 750 kg, almost 40 percent of the total vehicle mass.[8]

But to the surprise of those involved, the most insistent complaint concerned the rubber tire, which showed a markedly different pattern of wear and tear than the one used on the horse-drawn carriages. In hindsight, with our present knowledge of automotive technology, it is not difficult to find the reasons for this. First, the mass of the electric taxicab was much higher, and second, instead of a single function (i.e., carrying the axle load), the wheels now performed a triple function. Besides this carrying function, the taxi wheels now also performed a propelling (rear) function and a steering (front) function. Also, the rubber compound had to be made extra hard to counter a lateral slip. In horse traction, lateral slip could be corrected by the horse pulling the carriage out of the slip, but in automobiles lateral slip initiated the beginning of a skidding movement that made the car revolve around its vertical axis.

The struggle of this taxicab pioneer with automotive technology suggests that the technical field was absolutely closed for the desired application. The pneumatic tire was not a viable substitute: despite the weight reduction of the improved models, the Bersey taxi was far too heavy for this tire. In response to the drivers' complaints, an extensive survey was conducted to determine the most suitable solid rubber tire. Six or seven tires were tested. Eventually a manufacturer was found who was willing to risk a nine-month guarantee. Director Mulliner had to confess to his shareholders "that a large part of our loss has been in connection with rubber tyres." As a justification he referred to a big automobile manufacturer on the continent who had assured him "that he was convinced the success of moto-carriages in the future depends not on the accumulators nor the motors, but on the possibility of obtaining a rubber tyre which will stand the severe strain."[9]

But the cab company's loss was not only caused by technical deficiencies. Mulliner also felt that the electricity company was to blame, because the electricity costs appeared to be much higher after rectification than it had promised. The cab company then decided to generate its own power. This costly operation led to a decrease in electricity costs to a quarter of the original amount, but the installation of a steam engine and a generator caused a further increase in expenses, even though the engineering company managed to get rid of the two old converters without loss.

Battery supplier EPS caused problems as well. According to Mulliner, it was true that electricity and batteries "are not amenable to contracts—the unforeseen always happens." Nevertheless, he managed to convince EPS that the batteries no longer had to be purchased, but could be rented at a price that was comparable to that of maintenance costs. But the financial situation did not improve. Problems with the springs, typically encountered by automobile pioneers in the days before the pneumatic tire, is an indication of this: the vehicle mass appeared to be too high, even for solid rubber tires. So, the battery plates shook and cracked due to the rough road surface, despite the tests with the shaking machine that had been carried out before production started.

As a result, the second series of taxicabs was taken off the roads. Then, on 24 May 1899, the cabs appeared again. After another couple of weeks, however, they were gone again. This was the result of a fierce press campaign against electric propulsion, initiated by cab drivers who, as a historian of the London taxicab later concluded, "feared for their livelihood, and their conspirators, [who] won the support of the capital's newspapers which carried stories, often contrived, of breakdowns, accidents and injuries to passengers appertaining to the electric cabs. So brutal was the campaign . . . that public opinion of the vehicles was irreparably damaged." In March the bond holders exercised their rights: supported by a court order, they put up for sale the charging station, the taxicabs, the land, and the tools. In August of that year—two years after the first taxicabs had started to operate—the goods were sold, including thirty-six complete and forty-one incomplete vehicles. All in all the taxicabs had been operative for about a year.[10]

The electrotechnical trade journal *The Electrician* in a short commentary named a lack of revenue and excessively high expenses for electricity, battery, and tire maintenance as reasons for the debacle. Yet it concluded: "Be that as it may, [the reasons] relate to difficulties of an engineering character that must inevitably be surmounted before very long." The most severe criticism came from automobile expert W. Worby Beaumont, who wrote in the handbook he had just finished, "They were practical cabs, but not commercial cabs. They were very much too heavy in the frames . . . They were too heavy as a result of the separate conception in parts instead of as a whole, and they were heavy in every part. They were under-powered, and were too slow for people accustomed to London hansoms. After the novelty had worn off, public patronage fell off, and adverse criticism as to the rumble of the machinery and the jerking when stopping, starting, and slowing was very common. It cannot be doubted that improvements could be made on these cabs; but even the mystic power of the name 'Electrical' will not haul cabs, without insufficient foot-pounds [sufficient torque], though it may attract the precious metals from what are usually considered safe places."[11]

The Electrician suggested yet another reason for the demise of the cab company: the influence that the problems with the Lawson monopoly might have had on its fortunes. This was not unimaginable, because recent scholarship has described the foundation of the taxicab company as Harry Lawson's "last coup." Until then Lawson had been able to deceive his shareholders, but in the spring of 1897 he had to admit that he still had not built a car, despite the purchase of many patents. Part of Lawson's paper empire was the Great Horseless Carriage Company that he set up in May 1896. He put all the patents he had gathered into this company, and promised that mass production would begin at the end of the year. The boards of directors of the Great Horseless Carriage Company and the cab company partly overlapped. In the spring of 1897 the shares of his companies started to fall, and at the end of that year his empire began to break up, shortly after Bersey's first electric taxicabs appeared in the streets. This did not yet mean the end of his interest in the taxi market, for a little later he would pop up again behind the scenes of the EVC in New York. It would be a couple more years before Lawson, in 1904, would be found guilty of "an ingenious system of fraud carried out over a long period" and was sentenced to prison for a year, with hard labor.[12]

Although failing management and an unfortunate involvement with a speculator may have played a role in the failure of the London electric taxicab initiative, it cannot be denied that constructional factors formed the basis of the disaster. It is clear that the technical field at the time was full of closely related problems. In this case, battery technology (lead paste being shaken off the grids), tire technology (spring function), and general vehicle technology (high vehicle mass) formed a Gordian knot that somehow, by someone, had to be disentangled. Whatever the outcome of this disentanglement, the Bersey case made quite clear that the lack of a resilient function of the tires represented a formidable bottleneck for heavy-duty electric vehicle applications. To put this into the perspective of our overall story, however, the Bersey case does not allow irrefutable conclusions with regard to the batteries as a possible failure factor, even though there were indications that these were not functioning satisfactorily. No documents are available for forming an unambiguous conclusion, such as an accurately kept logbook of the expenses per taxi. However this may be, at the dawn of the new century not a single electric taxicab was left in London, or any other British town for that matter.[13]

Hummingbirds and Horses: The Compagnie Générale des Voitures à Paris

The Electrical Power Storage Company thought it could play down its share of the responsibility for the London disaster by referring to a Paris initiative of taxi motorization that had just started and for which a number of Faure-King batteries had been ordered. The initiative in Paris was started by Maurice Bixio, member of the Automobil-Club de France and chairman of the board of directors of the Compagnie générale des Voitures à Paris (CGV).

Unlike the Bersey experiment, the Paris initiative was firmly rooted in horse taxicab practice. The CGV was the biggest horse-drawn taxicab company in Paris. It had been forced to reorganize shortly before the turn of the century. Whereas the number of horse taxis had reached record highs, the proceeds started to fall, especially due to the competition of the streetcar. In 1898, the first vehement restructuring of the horse-drawn taxicab market began. That year ("the start of the horse taxi crisis") a record number of taxi companies went bankrupt. Moreover, personnel at the big companies began to become restless. Since the 1850s the cabman in Paris had been a subentrepreneur, because he rented his cab from the company at a fixed daily fee, the so-called *moyenne*. In 1896, Bixio was forced to avoid a strike by increasing the *moyenne* by 1.26 French francs, which that year resulted in a decrease in revenue for the CGV of 1.3 million francs. Also, the Paris taxes claimed an increasingly larger share of the revenues. And due to a horse disease (glanders) in the previous two years, 2,392 horses had died, which cost an additional 1.7 million francs.[14]

And so, for the first time in automotive history, the field of expectations of a big fleet manager was charged with the choice of propulsion technology. The pressure to motorize became even more intense when all kinds of Bersey-like companies were set up around the CGV that were involved exclusively in the motorized taxicab. Most of these enterprises did not come from the world of the horse taxi. After initial hesitation at determining his strategy, Bixio considered all his 8,000 carriages fit for motorization in a report for the shareholders meeting of 26 April 1897. And he did not leave it at that: whereas the CGV had presented a coupe and a milord with an internal combustion engine at the Exposition du Cycle of December 1896, Bixio now pronounced his preference for electric propulsion, "without absolutely rejecting gasoline." The company would first carry out tests with its own material, so that there would be sufficient time for "the search for a definitive, completely new type of vehicle."[15]

With more than 50 million francs on the balance sheet, Bixio set his standards extremely high. In an interview by the *Pall Mall Gazette,* Bixio declared that the London electric taxicab did not meet his specifications. He wanted a vehicle that would not weigh more than 1,000 kg, including passengers—less than half the weight of the Bersey taxicab. He said that at the moment he was waiting for the dispatch of an automobile from Berlin, London, and New York. But "the accumulators are troubling us . . . An ordinary [horse-drawn] cab covers about 60 miles a day, and in order to be ready for emergencies the accumulators ought to have a driving power of, say, 80 miles." Bixio's high standards can only be explained by the fact that he could not picture the new vehicle other than as a motorized horse taxicab: "Of course, there is the alternative of re-charging in the middle of the day, and in this case 40-mile accumulators would answer the purpose. But this would lose the cabman's time, and would also entail extra expense at the charging stations." Bixio initially wanted only the benefits and not the trouble of the new vehicle. At the time, an average horse taxi in Paris covered no more than 60 km in sixteen hours (and not 60 *miles,* as the British magazine wrote); one-third of that distance it was without passengers.[16]

The Paris correspondent of the *Gazette* then asked the obvious question: why he did not take his chances with the gasoline car industry. But, as if he had orchestrated his answer with Bersey, Bixio responded that "the petroleum car has no chance at all. As soon as we get the electric car ready, the petroleum automobile will be seen no more: it is not strong enough—that is to say, it cannot be relied upon to ascend an incline, such as the Rue des Martyrs or the Rue Lafayette. . . . I am of the opinion . . . that neither the electric nor the petroleum car will ever come into universal use. Automobilism is only likely to replace horses in public conveyances and in the case of the delivery carts and wagons. The wealthy classes will always keep to their horses."

Soon, however, Bixio had to accept that automobilization also involved a reorganization of everyday business, and consequently a different ratio between revenues and expenses. In September 1897, a "hummingbird" (as the Bersey cab was called in London) arrived in Paris that shortly afterward covered a distance of 80 km on one charge. Meanwhile, the required range had decreased to 100 km, but not a single battery manufacturer wanted to guarantee such an energy content. Mid-1897 Bixio once again had to scale down the required limit, this time to 60 km. That same year the CGV established the Compagnie française des voitures électromobiles (CFVE), with a corporate capital of one million francs. The CGV took a stake in it of 230,000 francs. The extent of Bixio's motorizing plans was clearly demonstrated by the order of 120 electric Lundell-Johnson motors from The Electric Vehicle Syndicate Ltd., in which Bersey and Mulliner had an interest.[17]

While preparations for taxicab motorization were in full swing, Bixio, in the presence of his shareholders, announced that the Automobil-Club de France would organize a "general contest of automobile taxicabs" on 4 April 1898. According to the regulations of this contest, the participating taxicabs would be judged on the following items: the daily costs at a minimum run of 60 km during sixteen hours (exactly the daily average of the horse taxi), comfort and ease of handling of the vehicle, and finally "the frequency of feeding, importance and ease of making repairs." Over a period of fifteen days, fifteen different runs would be completed at a maximum speed of 20 km/h. The jury report also had to pay attention to "elegance of appearance, noise of the vehicle, and the comfort of the passengers."[18]

The taxicab contest was delayed by two months; it finally took place from 1 to 12 June 1898. It triggered enormous international publicity and, according to some contemporary observers, led to a renewed interest in the electric car. The number of runs was reduced to three, and each of these runs had to be covered three times from 2 to 10 June, so that a total distance of 540 km had to be driven. This simulation of taxicab practice was to be preceded by an examination of the vehicle and measurements of the battery, and was to be interrupted by a parade to Versailles. On the eve of the tests, it appeared that twenty-six vehicles had registered, two with an unknown propulsion system, eight with gasoline engines, and the other sixteen with electric propulsion. The latter group included all the well-known French (and Belgian) electric-car builders, notably Kriéger, Jenatzy, and Jeantaud, and of course the CGV and the CFVE associated with it.[19]

On the day of the contest, it appeared that most of the gasoline cars had been withdrawn, because it was feared that the electric cars would be given preferential treatment. Only Peugeot, with a closed-body coupe with gasoline engine, withstood the electric predominance. But the CGV and its associate company also withdrew their automobiles, which were all electric-powered. This was not surprising, as Bixio was a member of the jury. In fact, the arena had now been reduced to the established electric-car trade (Kriéger with four, Jeantaud with six, and Jenatzy with one vehicle). Jeantaud's hansom (with its driver at the back) was by far the lightest in weight (1,090 kg including passengers, 2 kW). If we disregard this exception, gross vehicle mass varied between 1,270 and 1,802 kg and motor power between 3 and 4.5 kW. Battery mass accounted for 28 to 32 percent of vehicle mass—remarkably low and yet another indication that electric taxicab builders were inspired by the battery streetcar during the first-generation initiatives.[20]

In two respects the participants showed a remarkable unanimity, as if they had drawn their conclusion from the impending Bersey debacle. All cars were provided with Michelin pneumatic tires (only two of Kriéger's cars had solid tires on the rear wheels, also Michelins). Furthermore, all electric-car manufacturers appeared to have chosen the Fulmen battery with (quite thick) grid plates, mounted in a celluloid battery box. When measured on the first day, this appeared to deliver 26 Wh/kg of energy at a consumption of 5 W per kg, a remarkably high energy content. Measurements further confirmed the generally held notion that battery capacity decreased as power delivered by the battery increased. So, a higher speed was achieved at the cost of range, and vice versa. Jenatzy drove the longest distance on one charge: he passed the 60 km boundary and managed to cover a total distance of 105 km (table 2.1).

The consumption measurements during the nine runs led to a controversy among the jury members. Chairman Forestier, grand old man of French electrical engineering, published his full report in *La Génie Civil* as early as

September, complete with a cost comparison of the alternatives: horse, gasoline, and electric. The jury had not approved yet. É. Hospitalier, "general inspector of Bridges and Roads," especially found Jeantaud's consumption results incredibly low. Hospitalier was the only jury member who voted against the report when it was being discussed.[21]

Later commentators pointed at Forestier's "too great optimism" in the assumptions he made in comparing the costs of horse, electric, and gasoline propulsion. But even so, this comparison, calculated on a daily basis (as was usual with the large horse taxi companies), turned out to be in favor of horse traction. The gasoline car did very poorly, although the results were negatively influenced by specially provided cheap electricity at a special "contest discount." Eventually, the two French participants, Kriéger and Jeantaud, were pronounced the winners, but according to *La Nature* the contest was particularly a "triumph for the Fulmen battery."[22]

Later, after a ten-month private experiment, Jeantaud was to arrive at real battery maintenance costs of 7 to 8 francs a day, which was double that of the taxicab contest. The commentators concluded that despite such large-scale experiments, for the time being it was impossible to get a realistic picture of the actual costs of an electric taxicab company. The cause of this was the battery.[23]

Despite these results, and clearly forced to motorize because of the horse cab crisis, the Parisian company then began an ambitious project. In Aubervilliers, on a cheap plot of land north of Paris and at about a distance of 5 km from the Opéra, it built a huge hall that could accommodate 1,000 automobiles. The company had obtained a loan of 10 million francs to realize its plans. In the building two steam engines were installed that via two generators could supply 159 charging outlets with electricity. Because of this the kWh price came to 12 to 14 centimes, more or less the amount Forestier had assumed for the taxicab contest. The exchange-charging system was also applied, based on the hydraulic lift. Electrical engineer De Clausonne derived the design of the power station from the battery streetcar line between St. Denis and the Madeleine. Next to the building a test course was built with various slopes; city scenes were simulated using piles of bricks and human forms cut out of cardboard. The test course was meant for the training of drivers before they had to take an exam at the Préfecture de Police.[24]

Undoubtedly the manufacture of the vehicles must have started well before the end of 1898, and the plan had certainly been determined by then. Initially, the CGV intended to have 50 cars on the road by 1 November of that year, followed by another 50 on 1 January 1899. Those cars (110 of them, including 10 spare vehicles) were indeed ready to go on the specified date, but, unfortunately, the EPS batteries had not been delivered. The battery manufacturer was subsequently subpoenaed, while in the meantime another manufacturer was found, who promised to deliver 20 batteries (for 10 cars) at the beginning of April 1899.[25]

Thus, on Easter, 2 April 1899 (at the same time that the London property was up for sale), two electric taxicabs appeared in the streets of Paris. One day later, there were six more. To the surprise of many, there were no Jeantauds or

TABLE 2.1 Participants of the Concours de Fiacres and Their Test Results, Paris, 1898

	Kriéger 1	Kriéger 2	Kriéger 3	Kriéger 16	Jenatzy 13
Empty mass, including driver (kg)	1,360	1,130	1,310	1,370	1,662
Curb mass (kg)	1,640	1,270	1,590	1,770	1,802
Battery-vehicle mass ratio	27.9	27.7	28.8	25.8	31.3
Number of gears forward	6	4	6	6	resistance
Engine mass (kg)	2 × 65	2 × 50	2 × 65	2 × 65	
Nominal power (W)	3,000		3,000	3,000	

Tests on flat, clean, and dry macadam road on June 11, 1898

	Kriéger 1				Kriéger 2		Kriéger 3			Kriéger 16			Jenatzy 13
Average speed (km/h)	10.7	15.45	21.9	26.4	4.7	13.7	5.9	11.8	20.4	8.8	19.9	25.7	8.5
Battery voltage (V)	88	88	87	86	47	90	46	88	88	48	90	87	45
Current (A)	15	23	36	44	9.5	15.5	14	16.5	29	20.7	24.4	3.38.15	25.4
Power (W)	1,320	2,024	3,132	3,784	447	1,395	664	1,450	2,552	1,000	2,190	3,430	1,143
Specific power (W/ton)	805	1,233	1,919	2,306	352	1,100	405	912	1,605	565	1,237	1,937	636
Specific energy (Wh/km)	123.4	131	143	143.2	94.7	101.7	109	122.8	125.4	114	115.5	133.4	134.5
Specific energy (Wh/[ton.km])	75.3	79.9	87	87.2	74.7	80.3	68.6	77.5	78.6	64.4	65.3	75.4	74.7

Tests on slopes of 82.8 promille (Mont-Valérien) on June 11, 1898

	Kriéger 1		Kriéger 2	Kriéger 3	Kriéger 16	Jenatzy 13
Average speed (km/h)	6.55	7.1	5.77	7.4	6	7.9
Battery voltage (V)	85	85	84.5	85	87	80
Current (A)	52.85	59.8	64.7	53.6	60.7	74.2
Power (W)	4,490	5,080	5,470	4,560	5,280	5,940
Specific power (W/ton)	2,770	3,100	4,300	2,865	2,980	3,300
Specific energy (Wh/km)	688	716	950	615	880	752
Specific energy (Wh/[ton.km])	419.5	436.5	747	387	497	418

Energy consumption in the charging station from June 2 to June 10, 1898, determined by the terminal voltage of the battery

	Kriéger 1	Kriéger 2	Kriéger 3	Kriéger 16	Jenatzy 13
Total (kWh)	98.53	80	103.54	100.81	119.29
Per day (kWh)	10.95		11.5	11.2	13.25
Specific consumption (Wh/km)	182		191	185	220
Specific consumption (Wh/[ton.km])	110		120	100	120

Calculation according to Hospitalier

	Kriéger 1	Kriéger 2	Kriéger 3	Kriéger 16	Jenatzy 13
Specific consumption (Wh/[ton.km])	176		188	186	221
Specific consumption (Wh/km)	107		118	105	122.6
Charge-discharge factor	1.33		1.52	1.32	1.63

Source: Industrie Électrique, June 10, 1898, pp. 281–85.

	Jeantaud 21	Jeantaud 22	Jeantaud 23	Jeantaud 24	Jeantaud 25	Jeantaud 26
	1,590	1,476	1,520	1,340	1,270	950
	1,800	1,616	1,660	1,480	1,410	1,090
	28.9	28.3	27.6	27.3	28.7	32.3
	4	4	4	4	4	
	3,500	4,500	4,500	4,500	3,000	2,000

Jeantaud 21			Jeantaud 22			Jeantaud 23			Jeantaud 24		Jeantaud 25		Jeantaud 26			
8	16	20	6.05	12.25	15.8	11.75	15.15	16.45	11.5	16.15	7.82	16.95	6.9	12.95	13.8	17.5
52	104	103	45	88	88	88	87	87	89.8	89.8	44	88	46	86	83	80
21	22.5	31	19.8	20.8	25	21	24	26	16.7	24.2	12.56	16.58	14.35	13	16.2	20
1,092	2,345	3,187	900	1,830	2,200	1,850	2,090	2,260	1,500	2,178	544	1,460	660	1,118	1,345	1,600
	1,305	1,774	562	1,131	1,360	1,115	1,260	1,360	1,014	1,470	386	1,035	605	1,025	1,235	1,467
137.8	147	160	148.5	148.1	138	157	138	137	130.4	135	69.7	86.2	95.5	86.4	97.5	91.5
78.6	81.6	88.5	92	91.7	85.5	94.5	83.2	82.5	88.2	91.2	49.3	61.1	87.5	79.2	89.5	84

	Jeantaud 21	Jeantaud 22	Jeantaud 23	Jeantaud 24	Jeantaud 25	Jeantaud 26
	10.85	8.1	8	7.66	9.9	8.5
	95	85	85	80.6	82	72
	74	67	68.7	55	57.2	48.7
	7,025	5,695	5,840	4,440	4,680	3,500
	3,900	3,520	3,520	3,000	3,321	3,210
	647	703	730	580	473	412
	359.8	435	140	392	335.5	378

	Jeantaud 21	Jeantaud 22	Jeantaud 23	Jeantaud 24	Jeantaud 25	Jeantaud 26
	102.25	58.35	110.3	57.89	93.16	57.89
			12.25		10.35	
			204		172	
			122		120	

	Jeantaud 21	Jeantaud 22	Jeantaud 23	Jeantaud 24	Jeantaud 25	Jeantaud 26
			202.8		167.8	146.5
			121		119	134
			1.45		1.91	1.5

Illustrating the professionalization of the taxicab business, the Compagnie Générale des Voitures à Paris (CGV) constructed a 700 m track with simulated street scenes for instructing their chauffeurs. The cabs themselves closely resembled the London Bersey cab. (Baudry de Saunier, Dolfuss, and de Geoffroy, *Histoire de la locomotion terrestre*, p. 319; reproduction Charley Werff)

Kriégers with CGV logos (after all, they had been the winners of the taxicab contest), but adapted hummingbirds, that is, Berseys. That same month thirty cabs were on the road, because a new shipment of batteries had arrived. After the experimental series of 100, the aim was to have a fleet of 1,000 vehicles in operation at the opening of the World Exhibition of 1900.[26]

This third version of the Bersey, assembled by the CFVE and complete with central pivot steering and heavy frame, had a gross weight of 2,300 kg and so did not have pneumatic tires on the front wheels. The CGV still maintained that no definite decision had been made with regard to the batteries. But after experiments with EPS and Julien batteries, the first series appeared to have been equipped with TEM batteries. The construction of the vehicles had cost a million francs.[27]

The results of the large-scale experiment were disastrous. In October 1899, *The Horseless Age* reported that only one taxicab was still operative in Paris. This was the electric Jeantaud with number 16060, and according to the American magazine, it was "a manufacturer's hobby." According to a French contemporary, CGV's taxicab misery lasted only a few months. After that, the CGV "denumbered" a total of sixty taxicabs, that is, it removed them from the register of stationing taxicabs. The end of the CGV experiment is controversial, but it is certain that the behavior of the drivers (who, as we saw, were not directly salaried by CGV) was an important incentive to the denumbering of cabs: because of the high *moyenne,* they refused to accept customers for short trips and only accepted orders by the day or half-day.[28]

Although for 1899 Bixio had set a target of 40,150 vehicle days, due to the late start and the lower number of cars (100), these only added up to 5,522. The following year, the year of the World Exhibition, a total of 12,446 vehicle days was reached, but the vehicles were no longer in operation as taxicabs; they were now "rental vehicles." The CGV management eventually capitalized on a drivers' strike to terminate the service. All in all, 149 cabs had been built.[29]

The debacle plunged the CGV into a deep crisis, so severe that Bixio openly wondered in a report to the shareholders in 1901 "whether . . . the Compagnie générale should disappear." In his commentary, Bixio noted the very good response by the public. The electric motor and the electric installation had also functioned properly. It was, Bixio said, the battery that had failed: because it did not allow 100 km per charge, two per taxi were needed and on top of that no manufacturer could be found who was willing to take care of maintenance at a fixed rate. Therefore, he had decided to stop the experiment prematurely. As long as the battery did not allow a large enough range, further efforts were senseless. This view seemed to be confirmed by the maintenance costs: these were exactly double the amount (50 francs per day) charged to the drivers and a factor 2.5 higher than the costs Forestier had calculated on the basis of the taxicab contest.[30]

In his report to the shareholders, Bixio, answering his own question, concluded that it made sense to continue, but that dividends would become considerably lower. Although net profits initially (over 1900) were doubled due to the Exhibition, the following year the CGV suffered a loss of about 2 million francs, for the first time in almost half a century. It took until 1904 before the company was profitable again and the shareholders received a dividend for the first time in four years. A year later Bixio stepped down as director.[31]

At first sight Bixio's criticism of the battery is supported by the "battery contest," held in June 1899, shortly after the first CGV taxicab appeared on the road. As we have seen in chapter 1, the results of the contest proved disappointing for most batteries. In this severe trial, the TEM battery was among the average batteries, with an energy density of only 9.7 Wh/kg. It lasted only eighty-two charging/discharging cycles. So, on the assumption of one cycle a day, CGV's TEM battery would not last three months. Even a less intensive use of one cycle every two days (in the rental vehicles, for instance) would yield a life span of barely six months. Some criticism was expressed about the batteries not being treated in the same way during the tests. Some manufacturers withdrew their batteries if the voltage at discharging dropped too much. A week later, after maintenance, they reentered their batteries. This convincingly proved that even if they were scrupulously maintained, batteries had life spans that were too short for a cost-effective operation: the technical field was absolutely closed for this application. It is also important to note that EPS's failure to deliver its light grid-plate batteries is not surprising. The battery contest showed irrefutably that all light grid-plate batteries (including the Faure-King batteries) could not cope with the heavy-duty requirements of the taxicab business.[32]

The CGV initially tried to hide the fiasco by deploying the taxicabs as rental cars, but eventually withdrew these from circulation in the spring of 1901. This move, however, was in vain. Although the CFVE had attracted patents of electric carmakers like Doré, had concentrated on the production and operation of rental cars, and had enjoyed good revenues during the Exhibition, it had to admit: "it is impossible for us, considering the present state of the batteries, to reduce the price [of a ride] in such a way as to be able to fight successfully against animal traction." At the moment, the company had thirty rental electric cars and two charging stations in operation, one in Versailles and one in the rue Lord Byron.[33]

According to a technical survey of 1901, most of the small companies set up in the preceding years had an almost exclusively financial character. The management often knew little about the technical aspects of the company. Although the suggestion of speculation practices surely does not apply to the CGV, the second part of the criticism does seem to apply to Maurice Bixio. For even though the battery was rightly blamed as the most important factor of failure, Bixio kept searching for the cause of CGV's failure in the battery producers' inability to supply a "miracle battery"—a battery that would make exchange charging during the day superfluous. His field of expectations, in other words, stood in the way of finding practical solutions in the application field. Although the costs of maintenance eventually proved to be exactly double the amount of the *moyennes* paid by the drivers, nothing indicates that Bixio blamed the limited life span for this, which was *far below* the ideal he pursued. His explanation that the CGV would only start thinking about electric propulsion again when this ideal had been reached supports this conclusion. As such, Bixio's behavior is the earliest example of an expectational field deformed by the expectational field of the gasoline automobile.[34]

Bixio's odd choice of the Bersey can only be explained by the fact that this car was easier to build and convert by a coach builder with little experience in automotive engineering. The Jeantaud or the Kriéger was structurally speaking much more unified. But most important, both were, just as all the participants in the first taxicab contest, considerably lighter than the Bersey. Bixio, as well as those in charge of the London Electrical Cab Company, seemed to have conceived the taxicab as a truck or a battery streetcar. This was suggested by their conviction that, after a short, fashionable period of success, the light touring car would soon be a thing of the past and the rich would return to their horses.

In London, a futile attempt was made to mechanize the horse taxicab, that is, to literally replace the horse while preserving all other characteristics of the horse economy. It was the horse economy without horse. It was, in other words, nothing but economy. In Paris the same goal was pursued, but there was the added complication of an impossible range. Thus far, those who blamed the catastrophe mainly on poor management were right. But even at a horse-cab company with a tradition of almost half a century of scientific maintenance, motorization did not pay. The technical field, the boundaries of which were especially determined by the battery, was too narrow for this. The question then arises whether the same applies to the third, American, electric taxicab experiment of the first generation, which developed in a quite different set of fields. How far, within this different context, was fleet management hampered by constraints in battery and tire technology, and how did the American management try to circumvent the trap, so revealingly set by the results of the Parisian battery contest?

Network Building and Speculation: The Electric Vehicle Company

The third electric cab experiment was organized by the Electric Vehicle Company (EVC). The roots of this company can be found in Philadelphia, with its well-paved roads. EVC had no dealings with horse taxicab practice. At the beginning of 1896, Henry Morris and Pedro Salom founded the Electric Carriage and Wagon Company, with a capital of $10,000. After ten years of previous experiments, they had used their Electrobats 1 and 2 as a basis for a further development on behalf of the rental and taxicab business. The initiative of the two men must have appealed to the Electric Storage Battery Company (ESB), located in the same city, for after a few months the start-up company appeared to be registered in New York State, now with a capital of $300,000 and an office on Broadway. The board of directors then included, besides Morris and Salom, Isaac L. Rice and W. W. Gibbs, president and vice-president of the ESB. Later EVC director Herbert Lloyd noted that Morris and Salom approached the ESB board with the suggestion that the electric car could be a promising alternative for the battery streetcar, since this did not sell batteries at a rate as was originally assumed.[35]

Initially, there was much confusion about the exact definition of the field of application. Due to the ambiguous use of the English word *cab* in those days (in contrast to the French distinction between *voiture de place* [taxicab] and *voiture de remise* [rental vehicle]), it is not always clear whether a taxi or a rental practice was pursued. Usually it was a combination of the two, as in the case of Morris and Salom, for according to their announcement the cars could "be hired or ordered by telephone." But initially there was no question of stationing. The fledgling company's plans were in the first place targeted at sales on the private market. Apparently, the original plans were changed because of the takeover by the ESB. For in the fall of 1896 it was explicitly announced that the company did not intend "to fill individual orders," but that it would restrict itself for the time being to "building a number of electrical hansoms and coupes for public service either in New York City or Philadelphia." To this end charging stations would be installed "in every large city in the Union where a private individual who desires a carriage can have it properly charged and cared for." Shortly before the first cars appeared on the road, a representative of the company announced that there would be "trips between certain points," adding "that the route had not been fixed." Thus, the fledgling company was clearly put in the tradition of battery streetcar practice, but soon developments would take a different turn altogether.[36]

In 1897, the headquarters of the New York company were established on West Thirty-ninth Street, around the corner from Broadway. In January the first four electric hansoms appeared on the road and on 27 March, five months before Bersey and Lawson's "world premiere" in London, a modest fleet of twelve taxicabs was available to the public. At first sight, it seemed that battery streetcar tradition had been abandoned and that the company had joined the world of the horse taxicab. But a few weeks later the London *Electrical Review* mentioned that only seven or eight were being used, "and these are not located at cab stands, but are kept in a stable ready to be brought out when ordered."[37]

The first cars were all built after Morris and Salom's original design. The bodywork of the hansom was done by coach builder Chas. S. Caffrey Company, and the car was equipped with two fast-running (1350 rpm) Lundell motors of 1.1 kW each, with single windings that powered the front wheels via a leather pinion wheel and an internally geared gearwheel, mounted on the wheel. Around this gearwheel a band brake was mounted. The indebtedness of the designers to streetcar technology could be seen in the suspension of the motor. Although it is true that two electric motors that deliver the same power as a single motor (such as Bersey's) have a lower efficiency, this effect was compensated for by a much simpler drive train, without the differential and large gearwheel reductions of the London design.[38]

The wheels, running on ball bearings and with tangentially placed spokes, were derived from bicycle technology. They were provided with large (3-inch wide) pneumatic tires with extra thick walls and a tire pressure as high as 6.9 bar. Steering was done by the smaller rear wheels, a characteristic that was "from a scientific point of view . . . faulty," according to a British commentator. Nevertheless, axial pivot steering had been applied, and not central pivot steering as in the Bersey and CGV designs. Also, unlike these designs, there was no steering wheel. Instead, a lever could be moved back and forth to make a right or a left turn.[39]

The forty-four battery cells of the ESB (with a total weight of about 450 kg, comprising 36 percent of the net vehicle mass of 1,250 kg, and placed under the driver's seat) were divided into four groups, so that battery switching was possible. This old-fashioned concept had been abandoned in the Bersey and CGV taxicabs in favor of motor switching, due to energy considerations. The four battery boxes could be taken from the car separately and exchanged for charged ones. The range was 40 km at an average speed of almost 10 km/h. The maximum speed was twice as high.[40]

In the many commentaries in the international electrotechnical and automotive trade press, the chances of the company being successful were rated very high, partly because of the extremely high horse-taxicab rates in New York, which were also adopted by the new enterprise. The initial results were encouraging. After eighteen weeks Morris and Salom published an overview, from which can be deduced that their initiative—"the first and only in the world"—was a success. The ten hansoms in operation had transported 4,765

The Electric Vehicle Company of New York City built cabs that followed the model of the London hansom, the driver sitting behind the passenger compartment, and above the steering wheels. In this version of 1902, earlier bicycle-type spoked wheels, higher at the passenger end, have been replaced by uniform wheels, this innovation among many EVC made without ever solving the "tire problem." It took a mere 75 seconds to change the batteries. (Smithsonian Institution, Washington, D.C.)

passengers over a distance of more than 23,134 km. Only forty breakdowns occurred—one per 578 km. The management thought these were good results in comparison to the horse taxi, especially because of the fact that the drivers were still inexperienced. In brief, the results were a proof of the *"future reliability of the service,"* so that the manufacture of 100 more vehicles was announced to begin within a month. Fifty of these would be of a new type, the coupe.[41]

On the basis of the figures provided by Morris and Salom, we can calculate an average daily distance per hansom of almost 18 km. Assuming an average occupation of 1.5 passengers, we get 1.5 rides per day. Even with the most favorable occupation of just one person per ride, we get an average of a little over 2 rides per day. This shows that the cars did not station, but were used as "rental vehicles." "A number of vehicles are usually sent out to cab stations," an electrotechnical trade journal was told, "but ordinarily the station is kept quite busy supplying calls from hotels, clubs and even private residence for the electric-cab service." Thus, a third form of application was introduced, after the stationing and rental cab practice in London and Paris: that of the "cab on call." In such circumstances, exchange charging of the batteries was not deemed necessary. The chloride batteries could even cope with peaks of double the daily average of 18 km on one charge a day, without any problems. At least, this was true for the first months, as the Paris battery contest would demonstrate shortly afterward. Nevertheless, on the basis of these figures, it is hard to imagine how the company could be profitable. The figures are misleading, however, to the extent that they also incorporate the preoperating period, for just two weeks later it was announced that the charging station near Broadway was run "already on a paying basis." So, it seemed reasonable to give the company a larger financial base for the fleet expansion announced. This was absolutely necessary, because the demand for electric transportation "far exceeded the supply." To this end Rice founded the Electric Vehicle Company (EVC), which took over the company in Philadelphia on 24 September 1897, with a capital of $10 million. This took place half a year after the first taxicabs appeared in the streets of New York. Morris and Salom had faded into the background by then.[42]

The announcement of the EVC foundation must have made a deep impression on the burgeoning automotive world. In the middle of the following year, only 150 automobiles had been manufactured in all of the United States since the beginning of the automobile age. The EVC would at once become the greatest purchaser of automobiles when it placed its order. And that was not all: with a view to the American automobile production in 1897 of about fifty vehicles (of which the largest producer, the Pope Manufacturing Company, built a few dozen electric cars) it was by no means clear who was going to manufacture the vehicles ordered.[43]

It does not come as a surprise, then, that the expansion of the EVC fleet happened in a much more modest way than had been originally announced. In September 1897, fourteen vehicles of an improved design replaced the old taxicabs. Of the planned fifty coupes only one was operative. In one year's time the fleet had grown to twenty cars, but demand for transportation still was many times larger than the supply. "We are now renting cabs by the months," EVC's financial man, Frederick Vieweg, said in an interview at the end of 1898, "and number among our patrons ladies who maintain their own carriage [with horse traction], but prefer to use our cabs at night. Most of our drivers have a regular list of patrons." Vieweg also announced that the EVC, in cooperation with the Edison company, would install charging stations "all over town," "where not only their [EVC's] cabs, but all other electric vehicles can get power." Taxicab rates in the meantime had been cut to 30 dollar cents per mile (20 cents lower than the horse-taxicab rates), but the "rate for shopping, visiting, etc." remained $1 per hour.[44]

In the fall of 1898 the charging station on Broadway was finished, near the business, shopping, and entertainment district, as well as the trains and ferries. It was an impressive building, designed by chief engineer G. Herbert Condict in cooperation with mechanical engineer A. W. Gilbert. *The Horseless Age*, in an editorial under the heading "Our Electric Cab Station," exulted that Paris might have its automobile salon, but New York now boasted "the most extensive plant in the world for the care of electric vehicles." The scheme to generate electricity there had apparently been abandoned, because the energy was obtained from the power station of the Edison Electric Illuminating Company, at a distance of a few hundred meters. The station was equipped to process 150 battery boxes (1.5 per car at a planned fleet size of 100) and so was "manipulated by a minimum of manual labor."[45]

Exchanging the batteries took exactly seventy-five seconds. Each car was

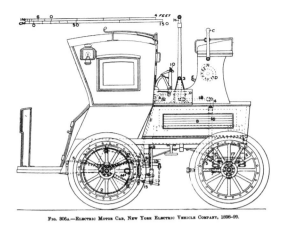

Fig. 806a.—Electric Motor Cab, New York Electric Vehicle Company, 1898-99.

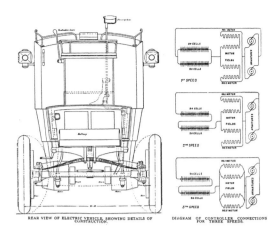

REAR VIEW OF ELECTRIC VEHICLE SHOWING DETAILS OF CONSTRUCTION.   DIAGRAM OF CONTROLLER CONNECTIONS FOR THREE SPEEDS.

Technical structure of the EVC electric cab, with pneumatic tires and rear-wheel steering by a lever. The battery set lay under the driver's seat. (Beaumont, *Motor Vehicles and Motors*, p. 425)

"carefully inspected by competent men and thoroughly cleaned" on entry. It was then driven onto one of the two platforms, lifted by a hydraulic hoist, and moved to the right spot by hydraulically operated buffers. Next, the battery box was hydraulically pulled from the opened battery compartment and put on a conveyor belt system that was electrically powered. It was operated by one man, who was positioned in the cabin of an eight-ton crane and who subsequently transported the batteries to the charging board by means of automatic grippers. If the car was not going out into the street again, it was placed on an upper floor by an electric elevator. A second elevator would bring it down again.[46]

The new hansoms, first introduced in September 1897, were considerably heavier, because, among other things, they were equipped with forty-eight instead of forty-four ESB batteries. Their total weight was 624 kg, 43 percent of the net vehicle mass of 1,450 kg. This now enabled a speed of 24 km/h, while the range was estimated at between 40 and 56 km. All batteries were placed on one tray, so that putting them in and taking them out was easy, while they made direct electrical contact with the propulsion system when the tray was pushed in. Condict later obtained a patent on this arrangement. The two motors were more powerful as well. From then on they were delivered by the Westinghouse Electric & Manufacturing Company. They were constructed with four poles, and each motor supplied 1.5 kW at 800 rpm. An emergency button, to be operated by the driver's heel, could render the whole system powerless. The series-parallel connection had now been extended to the motors: by connecting the two groups of batteries as well as the two motors in parallel and in series, three speeds were realized. The bicycle wheels had been abandoned; four big steel disk wheels of equal size were being used now. They were provided with pneumatic tires with a width of as much as 5 inches, walls almost 1.5 cm thick, and a pressure of more than 4 bar. The Diamond Rubber Company had especially designed them. If larger tires had been available, the management announced, even larger wheels would have been mounted. Both measures—the larger air chambers and the possibly larger wheel diameter—would increase comfort and reduce jolting of the battery and the mechanical parts.[47]

The fleet of 100 vehicles was complete in January 1899, three years after the modest start in Philadelphia. Most of the vehicles were rented out on long-term contracts. "Many aristocratic people . . . are so enthusiastic that they declare they will not bring their horses to the city another winter, but will leave them at their country places." Of the forty-five taxicabs, not even one stationed: "No cabs can be spared to the stands now," *The Horseless Age* wrote. According to American historian David Kirsch, who managed to unearth the only two complete daily records left (for 30 and 31 January 1899), the average daily distance of these cabs had increased to 45 km (about 40 percent of which they were empty) during the previous two years. But three cars covered more than 80 km on one of these days. The maximum amounted to 90 km. Seventeen cars remained below 34 km a day. Unfortunately, the number of rides is not known. It now appeared that the investment of 1.5 batteries had made sense, because (assuming a range of 40 km on one charge) just over 50 percent of the forty-five cabs needed a second battery on the same day. So, Kirsch concludes that "the New York electric cab service had expanded and was probably covering enough paid miles to generate an operating surplus."[48]

But the pneumatic tires and not the batteries initially caused the main problems. Whereas the latter hardly needed maintenance and performed excellently, the company had tried more than twenty types of tires made by eight different manufacturers, ranging from solid to pneumatic tires, with all varieties in between. In a presentation to the New York Electrical Society in February 1899, chief-engineer Condict defended himself against the advice of "authorities" in tire technology (who wanted him to choose a solid tire). He used the argument that the public, once used to the comfort of pneumatic tires, would not accept any other type. Pneumatic tires were necessary not only to withstand the bumps and jolts of the "inadequate and despicable cobble" of the Manhattan roads, but also to help cross the streetcar tracks that in the United States protruded from the road surface. In an interview with a rubber trade journal, Condict declared that the EVC had fifteen different types of pneumatic tire in use at the time, but none of them was satisfactory.[49]

Shortly after Condict's presentation, the announcement of an order for another 200 vehicles followed: seventy-five hansoms, and the remainder coupes and broughams of different types. They would have to be delivered by 1 June 1899. In Philadelphia a similar company was established, with W. Gibbs as director and a capital of $6 million. In New York the opening of a second charging station on Third Avenue and a factory with adjoining workshop was planned. In Philadelphia shares went up.[50]

The execution (and adjustment) of these plans was no longer the responsibility of the EVC in its old form, however. For by the end of February, reports appeared in the press (the *New York Times*, among others) about rumors that a giant trust was in the making. On 21 February 1899, a new company, the New York Electric Vehicle Transportation Company, apparently had been founded with a capital of as much as $25 million. And although EVC President Rice declared that it concerned "nothing more than an enlargement of the present concern" here, *The Electrical Engineer* added: "Harmonious relations with Metropolitan Street Railway interests are broadly hinted at." Although it is hard to reconstruct what exactly happened during the few weeks that followed, the sources available do allow construction of the following scenario.[51]

The apparent success of the EVC had opened the eyes of other electric-car manufacturers to the lucrative possibilities of the taxicab and rental-car market. In Chicago, for instance, taxicab plans were being devised by a group consisting of the Fischer Equipment Company (manufacturer of the well-known Woods cars) and Sipe & Sigler (owners of the Willard battery). Samuel Insull was in charge. Initially the initiative was explicitly set up "in opposition to the Electric Vehicle Company" and with the aim of making Willard's large-surface plate battery into "a formidable rival for the 'Chloride' battery" of the ESB. But just a month after this announcement, the EVC had persuaded Insull to consider a merger.[52]

But there were many more contenders waiting for their chance. New York must have buzzed with rumors in 1899. In May of that year, for instance, the New York Central & Hudson River Railroad announced the foundation of its own taxicab service with a hundred electric cars that would serve the Grand Central Station. Also, in April the General Carriage Company (capital of $20 million) was established, which according to rumors had managed to obtain an exclusive license for nonrail-bound transportation for the entire state of New York. It started a taxicab service with eight electric hansoms at the very low rate of 25 dollar cents a mile and took over two stables of a horse streetcar line. The financiers of the New York Auto Truck Company were supposed to be behind this. This enterprise, which together with many other companies was part of the International Power Company, was one of the many examples of stock jobbing movements that afflicted the transportation business at this stage. Even Henry Lawson was said to have departed from England to the United States in order to set up a "new automobile trust" with a capital of $75 million, under cover of the Anglo-American Rapid Vehicle Company and in cooperation with Studebaker. This syndicate, set up by Lawson in cooperation with his American counterpart, the flamboyant Edward J. Pennington, was organized to bring about a merger between the major British and American automobile manufacturers, on the basis of the more than 200 patents they controlled. W. Gibbs (no longer working for the EVC at the time) was also involved. According to the American amateur-historian James Doolittle, the EVC was part of these plans. The syndicate also owned the British rights to the Kriéger taxicab and managed to get the New York Auto Truck Company interested in it. The latter company had already launched plans for a bus service based on compressed-air traction. It was announced that the Kriéger car would be produced in "large numbers." Later it apparently only concerned imports from France. But the most alarming news for the EVC must have been that the taxicabs were going to be powered by Fulmen batteries. These constituted a direct threat to the ESB's "monopoly batteries," because the Fulmen patents had priority over those of the ESB.[53]

Whether the EVC, for fear of a takeover, wanted to raise its capital in these turbulent months can no longer be ascertained. But it does seem likely, judging from the date of the foundation of the New York Electric Vehicle Company

(21 February 1899) and the almost simultaneous establishment of the Philadelphia branch. Another fact that supports this likelihood was the agreement
with the ESB for the supply of batteries at a considerable discount—as the new
owners later noticed, to their dismay. However this may be, it did not help, because the rumors in the press proved to be correct. According to engineer Hermann Cuntz, who noted in his memoirs half a century afterward, the negotiations had already started in March 1899. Besides those directly involved (the
ESB and the Pope Manufacturing Company in Hartford, Connecticut), William Collins Whitney also appeared at the negotiating table. He apparently held
the financial trump cards.[54]

What Samuel Insull was for Chicago, "Colonel Whitney" was for New
York. He was a former secretary of the navy under President Grover Cleveland
and as leader of the Whitney–Philadelphia syndicate he had acquired the exclusive rights to the horse streetcar in Manhattan in the 1880s. In the following decade he had the streetcars electrified. Gradually the financial interests
of the group also extended to the electricity companies, specifically New York
Edison, while it dominated virtually the entire New York streetcar network
with its Metropolitan Street Railway Company. This made Manhattan the
"battery-streetcar city" of the United States for a while.[55]

Since the beginning of 1896, the group had gradually acquired a majority
share of $1 million in the ESB. Rice, who had replaced Gibbs at the ESB in
1897, strongly opposed the looming takeover. But in June 1899, he gave in,
after the shares had been boosted from $20 to $120 and he could sell his for
$141 to Whitney. He was then made president of the EVC, which the Whitney
group had also acquired via the ESB. At the ESB the banker George H. Day, the
man who had carried on the negotiations with Whitney, became president. At
Pope, its founder with the same name had faded somewhat into the background to make room for the retired director "captain" Hayden Eames. During the negotiations in Hartford, the order was increased to 1,600 cars and
Whitney made an offer of $1 million in cash for the expansion of the Pope fac-
tory. The contract was eventually sealed mid-April 1899. The Whitney group
first wanted to be sure that monopolies on both the electric-powered and
gasoline-powered cars had been obtained. This came about by the acquisition
of the Selden patent, which would rock the foundations of the fledgling American automotive industry for the next ten years.[56]

In the following months the field of expectations was blown up to gigantic proportions: the outlines of a real empire emerged, expressed in constantly
changing plans that were eagerly published by the press. The intention was
clear: "a worldwide network of branch EVCs" was being developed. First, the
patents of Pope and the ESB were joined, each with an estimated value of a $1
million. Added to Whitney's $1 million, the capital was used to set up the Co-
lumbia Automobile Company on 19 April 1899. The two companies each took
half of the shares. Pope, Day, and Eames formed the board of directors. The
new company merged with the EVC (read: Whitney) on 3 May and became
the Columbia and Electric Vehicle Company. The latter would take care of the
manufacture of the automobiles in the Pope factory in Hartford. It purchased

the batteries at almost 20 percent above cost from the ESB (read: also Whitney). The EVC remained in existence as a holding.[57]

Furthermore, in Hartford the New Haven Carriage Company was taken over because of its experience in bodywork, while the group also became the owner of Siemens & Halske Electric Company in Chicago, where the electric installation would be built. The vehicles were exclusively sold to the New York Electric Vehicle Transportation Company, which would be responsible for operating them. This company would make an initial down payment in shares and hand over 2.5 percent of the gross proceeds to the manufacturer. The New England Electric Vehicle Transportation Company (in Boston) and the Pennsylvania Electric Vehicle Company (in Philadelphia) also obtained this status. The Illinois Electric Vehicle Transportation Company, established later that year in Chicago, also became part of this arrangement. Except for Philadelphia ($6 million), all operating companies had been capitalized with $25 million. Furthermore, a few smaller "agencies" (with a capital of $100,000) were planned in the remaining states and would—on paper at least—eventually amount to seventeen. Without this last group, the total capital of the conglomerate was $100 million in September 1899, of which $7.4 million was cashed "for the purchase and operation of the product controlled," as the new EVC President Rice expressed it. If the other plans were realized as well, a $200,000,000 enterprise would have emerged. In comparison, the value of all 4,192 cars manufactured in the United States in 1900 amounted to hardly $4.9 million, whereas the average capitalization of automobile companies amounted to somewhat more than $100,000.[58]

Due to the mega-order, the factory in Hartford became the undisputed leader in the American automotive industry, a fact largely neglected in automotive historiography. There even was some talk of raising the order to 12,000, but on 12 July 1899, Rice announced that besides the agreed-on 2,000 landaus, hansoms, and broughams for the taxicab and rental-car company, an equal number of passenger cars (runabouts, golf traps, victorias) were ordered as well. Studebaker would produce most of the bodies, while a separate battery factory would be set up in Hartford. According to Rice, the automobile production was going to reach a capacity of 100 a day within a year, a prediction that he adjusted to "at least 8,000" for the coming year only a few months later. In 1900, a second electric-car manufacturer, the Riker Motor Vehicle Company in Elizabethport, New Jersey, was taken over. This involved $2 million and was apparently meant to get hold of Riker's patents, for his company was subsequently dismantled.[59]

The press could hardly keep up with all the news about the expansions. In New York, for instance, in March 1899, two new charging stations would be set up at first, each with 100 cars. A month later the old idea of the electric car as a "complement" to public transportation came up again, without doubt under the influence of the new leaders within the EVC. According to this idea, 1,500 cars would be deployed in New York, which would serve the streetcar network of the Metropolitan Street Railway (of Whitney and Co.) and would also obtain their energy from this company. At a rate of about 15 cents for a transfer right the passengers could be transported to the streetcar from their houses on the sidestreets. That plan apparently failed, for after one year some 200 cars were in use, in as many as twelve different versions. Part of them were accommodated in a new charging station on the corner of Fiftieth Street and Eighth Avenue. Another year later, in May 1901, this branch had moved for the fourth time, to a building of half a block wide and 150 m deep, a little farther down Eighth Avenue, with washing and reading rooms for the drivers. In the meantime, 350 cars were present there. The charging station, built to accommodate 640 cars but with a capacity of 1,000, was connected to the Edison power station. Cars from which the batteries had been removed could be driven to their parking space by a trolley system via cables running along the ceiling. There were six platforms for charging and discharging the battery boxes, and the charging room contained 750 tons of batteries when at full capacity. There was no doubt about it: the new charging station again was "the largest as well as the most completely equipped automobile charging station in the world."[60]

The subsidiaries in other cities displayed taxicab initiatives too. The most noticeable initiative took place in Chicago in 1898, where a charging station was opened for the growing number of electric cars in that city. A year later the establishment of a taxi service followed, led by none other than Samuel Insull. "Soon," Insull announced, 1,000 taxicabs and rental cars would be in operation there.[61]

In the spring of 1900, the newly founded Washington Electric Vehicle Transportation Company started a taxicab service at the terminal of the Baltimore & Ohio Railroad on Capitol Hill. It announced it would set up such services in New York, Philadelphia, Baltimore, and Chicago as well. Cab drivers not only had to wear uniforms, but also had to line up for inspection.[62]

By mid-1899, the "New England" division had provided 5 "cabs" and 12 open-bodied cars for summer rental in residential Newport, a chic seaside resort. "The cabs now in service are in constant use," it was announced. After the summer of that year it was decided not to expand the Boston fleet, but to direct all efforts at Newport, where the cars were rented out at $150 a month and where demand far surpassed supply. Even the 10 cars ordered, but not yet delivered, had already been reserved. In the beginning of 1900, there were 20 hansoms, 20 broughams, and 1 or 2 delivery vans in Boston, while 7 broughams were rented out on a monthly basis. At that moment the plan was announced to purchase 25 delivery vans that would be rented out to shopkeepers. In Boston the drivers were mostly recruited from among the streetcar operators—not surprising if one bears in mind that the EVC branch in that city also controlled the Boston Transit Company.[63]

Before the beginning of the summer of 1900, it was announced that in Newport a large building would be opened on Downing Street, where a "practice ring" and a repair shop were prepared and where "100 automobiles for general use" would be garaged. This shift to the business of renting out vehicles to tourists was further encouraged by the installation of charging stations along the Atlantic coast of New Jersey (in Sea Bright, West End [Long Branch], Allenhurst, Spring Lake, and Atlantic City). These stations were meant for "charging and caring for Columbia automobiles owned by private parties." Yet, in the fall of 1900, there were 100 "cabs" and 50 delivery vans operative in Boston that could utilize a charging station of their own. Besides, another 75 open-bodied cars were rented out on a monthly basis, probably mainly in Newport.[64]

In the beginning of 1901, an independent survey by R. Fliess of the EVC's vehicle stock resulted in an estimate of the total number of "vehicles" (probably including the heavy-duty commercial vehicles) in the three major cities. He came up with 650 cars: 300 in New York, 250 in Boston, and 100 in Washington. This excluded the stock of 109 vehicles in Chicago and the fleets in a few other big cities, such as Philadelphia. Probably no more than a few dozen cars were present there, so that—with a correction of 50 for the stock in New York a few months later—a total fleet size of 850 seems a safe estimate at the beginning of the twentieth century. That was a large fleet and up until then unparalleled, but nonetheless only a shadow of the number that had been promised.[65]

Yet this picture is somewhat misleading: the agencies were independent units and the 850 cars were not located in one place. Nevertheless, the fleets in both New York and Boston were considerably larger than those in London and Paris. This picture becomes different, however, when we narrow the comparison with Europe to stationing taxicabs. In that case the figures presented so far may justify the assumption that the fleets in Paris and New York were in principle not that much different in size.

Internal documents allow us to get a glimpse of some details of the application field. When engineer William F. Kennedy inspected the installation in Newport on 8 November 1899, he found eleven of the nineteen cars "out of commission." "Nothing on the vehicles is right," chief-engineer Condict wrote in November 1899 to W. Johnson, general manager of the EVC. The controller had such a faulty construction that it had to be designed anew. At the Hartford factory, the roller bearings had apparently been replaced by cheaper friction bearings (causing more friction losses). Moreover, the wheels were not mutually exchangeable and so badly aligned that the range of the cars was reduced to a fraction of what was normal. Obviously, the management was more intent on the fast buck that could be earned by the sale of automobiles and batteries to the operating companies. Many constructional changes were implemented, such as the change to front-wheel steering and rear-wheel drive in the hansoms. Moreover, sturdy wooden wheels replaced the steel disk wheels. Due to all these changes, the net vehicle mass had risen to about 2,000 kg, of which the batteries accounted for 40 percent. It will come as no surprise that the pneumatic tires were sacrificed as well.[66]

From some rare internal documents of the subdivision in Atlantic City (set up in the summer of 1900 by the New England division) it can be concluded that in the meantime the situation had hardly improved. Rapporteurs came up with the same shopping list of complaints as Kennedy a year earlier in Newport. "No records of any kind have ever been kept," one of them wrote, who further commented that the instructions, drawn up as a result of the Newport experience, simply were not consulted. Moreover, ten of the fourteen batteries delivered were completely out of order. One of the rapporteurs "was inclined to lay most of the fault on the negative battery plates themselves— they are a summer output which have always proved to be inferior to winter ones." Furthermore, the energy supply apparently did not start until midnight (the electricity producer apparently had completely sold its evening peak). Because of this the automobiles drove off only partly charged, and in between they were recharged at a high amperage, which considerably shortened the life span of the batteries. Finally, the electric energy, if it was supplied, was unreliable because the voltage fluctuated too much. All this made manager Johnson sigh that "all the money we made and a few hundred per cent more will be spent in repairs," and he summoned his rapporteur to New York to discuss closing the station.[67]

As if this were not enough, the EVC also began to confront a hostile environment. In the middle of 1899 (shortly after the takeover of the EVC by the Whitney group), Editor-in-Chief E. P. Ingersoll of *The Horseless Age* started a publicity campaign that gradually turned into a kind of personal crusade. He had been alarmed by reports in the *New York Times* about Whitney's giant trust. But he probably was mostly upset by the rumors about the Selden patent, which threatened to block the development of the gasoline automobile. In May of that year, he expressed the hope that the syndicate behind the General Carriage Company would continue its plans for producing the Kriéger taxicab and thus instigate a "determined onslaught upon the fortress of the Electric Vehicle Co.'s boasted monopoly." When the reports about the failed London and Paris taxicab experiments began to arrive, and especially when the results of the battery contest in Paris became known, Ingersoll's attack gradually shifted toward the battery. When Hiram Maxim in a lecture said that he was in favor of electric propulsion, Ingersoll extended his repertory of verbal onslaught to personal attacks on the man. This "blind and infatuated partisan of the lead wagon," who defended "this sick baby, the storage battery, a heavy, inefficient, delicate and destructible apparatus wholly unfit for general locomotion," deserved a reply, because his lecture "smacks so strongly of the Wall Street curb . . . Let them [the financiers behind the EVC] consult competent and disinterested engineers, and learn the difference between generating electricity direct from a steam engine and boxing it up in a ton of lead and trundling it over cobble stones and car tracks. . . . The outlook for the lead cab is utterly hopeless."[68]

It was this last qualification—and the "Lead Cab Trust" derived from it— that became common wisdom in the discussion about the pros and cons of electric propulsion, even in the newspapers. The shares that had a value of $20 in 1897 had risen to $97 in the fall of 1899, but started to fall at the end of October, although a dividend of 8 percent was still being paid. On 18 December, the crash followed—according to Ingersoll, due to the desperation of the "chief conspirator, the Electric Storage Battery Co." And he lost himself in biblical (or should one say, cowboy) rhetoric : "bluff, bravado and misrepresentation will postpone but cannot prevent the final day of reckoning."[69]

The direct impulse for the collapse of the EVC fund was an investigation by the *New York Herald* into a loan that the EVC had managed to get by illegal practices. The paper called it "one of the greatest banking scandals of a decade" and concluded with satisfaction: "The financiering methods of the Lead Cab promoters are beginning to leak out at last." The scandal severely damaged the reputation of the EVC in financial circles, and the value of the shares of the operating company in New York suddenly dropped from $30 to 75 dollar cents.[70]

In February 1900, Ingersoll decided the limit had been reached. As a result of the rumors about Lawson and Pennington's Anglo-American Company and after an extensive and expressive explanation of stock jobbing, he declared that from then on he would refuse the trusts' advertisements. He printed a long list of names of the "floater(s) of watered stock companies." Mid-1900, when the plans of the Whitney group with the Selden patent became publicly known, he called on the automotive industry to oppose them by forming an alliance. Ingersoll's campaign continued well into 1901. Each maneuver of the EVC he accompanied with cutting criticism on the lead battery. When the EVC, in an attempt to find funds to cover its losses and attract working capital, sold a large number of shares far below their value to the ESB, a small shareholder revealed that dividends had been paid out to the Whitney group while the EVC was in debt.[71]

In light of these details, it is amazing that in the electrotechnical trade press hardly any mention was made of "one of the greatest banking scandals of a decade." It continued to print uncensored reports of initiatives in the field of the electric car as well as of the EVC. Also, some automobile magazines began to revolt against the "auto-electrophobe" Ingersoll. The rival *Motor Age,* which coined the term, declared it had no personal or financial interest in the matter and warned its readers against the "untruthful and vicious diatribes" of its sister magazine. Although from then on Ingersoll started to sing a milder tune, the tone was set for the coming years: the "real" automobile was a gasoline car and until a lighter battery was invented, any attempt at developing an electric propulsion system was doomed to fail. Thus, for the second time in automotive history (after Bixio and his CGV) the field of expectations of the electric vehicle became thoroughly distorted by its neighboring field for the gasoline automobile. And again the core of this distortion was the evocation of a "miracle battery."[72]

By the spring of 1901, two years after the Whitney group appeared on the scene but more than five years after Morris and Salom's original initiative in Philadelphia, the EVC's house of cards began to crumble. Chicago was the first to go under. President Insull capitalized on a drivers' strike on 2 March 1901 to liquidate the company, after consultation with New York. At the moment the annual turnover was $137,000, but the loss was almost as much. Two months later the end came for Boston. The year 1901, up to March, showed a loss of $212,000.[73]

One can only guess at how much the stock jobbing had yielded the Whitney syndicate. The explosive growth in profits of the ESB (also owned by the Whitney group) speaks for itself, although one should bear in mind that during this period the popularity of the battery for stationary purposes began to spread in America. From the ESB's annual reports it is evident, however, that the delivery of battery plates to the EVC in the year 1897–98 formed a significant share of 14.4 percent of ESB production. So, the battery giant's 1897 profits of $180,000 tripled the following year, the year after it even increased ninefold, and in 1900 it amounted to no less than $6.7 million. Then the profits sharply decreased again, and did not exceed $3.75 million during the whole first decade of the twentieth century. It may be concluded that the ESB managed to reinforce its monopolist position by its involvement in the first American taxicab experiment. To round off this position, they took over the patents and further assets of Sipe & Sigler in March 1902, and established a branch under the name Willard Storage Battery Company. This increased its share in the American battery market to 95 percent.[74]

However, in the historiography of the EVC, the commotion about its financial collapse drowns out the clearly recognizable continuity in the business. For while in financial circles and in the press the death-knell of the "Lead Cab Trust" was sounded, in New York production and business were conducted as usual, but not without an important shift of emphasis. In the first place, the sale of electric cars to private customers became more and more important; the operating companies could take advantage of this too by offering maintenance and battery-charging contracts. In the second place, production gradually shifted to the delivery van and truck and the electric-powered bus. In January 1901, the EVC opened a regular service with "electric stages" on Fifth Avenue. This plan—like a similar plan in July 1901—increases the likelihood of a relationship between the Whitney syndicate and the shadowy Auto Truck Company, just as James Doolittle suggested. For it appears that the EVC had an exclusive license on establishing such lines "over the streets of New York not at present used by street cars." For this service the EVC only selected the "well-paved streets." Due to these initiatives, characterized by a fixed, calculable route, the taxicab business was further marginalized within the operation. But as can be deduced from the history of the EVC and its forerunners, as reconstructed here, the *stationing* taxicab never formed the backbone of the business.[75]

In this context it is remarkable that the ESB did not utilize the Exide batteries that it had presented in 1900. According to American historian Richard Schallenberg, the batteries were a "stopgap design, developed quickly to do something to replace the ESB's original electric cab cells which were driving the EVC to financial ruin." In an environment where making a fast buck was all-important, this move would seem to confirm the thesis that the American battery giant was less interested in the costly maintenance of its battery than in the guaranteed sale of its products to the EVC. We can only guess at the truth of this thesis, but we do know that several years later in Germany the development of a marketable version of the same battery type encountered huge technical problems and took about two and a half years. This must have been the case here too, for even in 1902 two-thirds of the EVC taxicabs were equipped with the chloride version. Thus, it seems that the new, lighter battery, which would play such a decisive role in the electric vehicle's technical field of the second generation, could not play a role of any significance at a time when it was important: it appeared on the market in an experimental form when the operating companies outside New York had already been closed.[76]

A last shift in emphasis (and this is an understatement in the light of what was to follow) was the increasing interest in including the gasoline car in the product range of the Hartford factory. This interest was inspired by the importance of the Selden patent for an enterprise that saw its electric-car activities diminish, and for its survival was more and more dependent on licensing revenues from a patent that it regarded as a monopoly. Meanwhile, Condict had replaced Herbert Lloyd as director of the New York operating company. Eames and Maxim had already left the EVC because they could not work with the "arrogant" Riker. The latter left shortly afterward for Locomobile, but Maxim later returned to the Hartford factory. Half a year later, in the beginning of 1902, the company reduced its capital from $25 million to $5 million, whereas in November of a year earlier it had still made a profit of $16,000 with a turnover of $52,000. It simultaneously changed its name to New York Transportation Company. The electric car had disappeared from the name.[77]

The turbulent history of the Selden patent and the involvement of the EVC and the Whitney syndicate is outside the scope of this study. It is sufficient to remark that Henry Ford was one of the few who refused to comply with the EVC's wishes. But before he was proved right, after a legal battle of many years, the EVC—as one of the first of many victims—had already tripped over the economic "Panic of 1907." On 10 December 1907, the EVC declared bankruptcy. But until 1911, under different management, it would continue to play an important role in the second wave of motorizing the taxicab business in New York.[78]

The Battery as a Failure Factor: An International Comparison

The extremely varied picture that emerges from the reconstruction of the EVC activities given here makes the general question about the failure of the EVC as such by and large misleading. For the EVC was not a monolithic unity and can only be adequately described if one makes a distinction between the various departments and the holding. Moreover, one should bear in mind that the company's activities involved sale and rental of passenger cars, delivery vans, and trucks, and the operation of taxicabs-on-call and stationing taxicabs. The large variety of vehicle models is remarkable and seems contrary to the concept of a rational operation of stationing taxicabs. Besides, cars of third parties were also maintained and charged. The mixture of these activities not only differed by department, but also changed over the course of time, and developed in such a way that the emphasis shifted more and more to sales and maintenance for third-party vehicles.

In what respect, then, can we speak of a "failure"? What failed at any rate was the Whitney plan for forming a monopoly and the ambition to set up a gigantic national system with 12,000 vehicles. What also failed was the supervision of the operating companies by the holding company, thus creating an abominable operation of the taxicab business, aggravated by poor product quality for this heavy-duty application. This last aspect may also be related to the ideas of those in charge at Hartford about the future of the car. Greenleaf at least suggests that they assumed that the car would be the subject of an equally short-lived fashionable craze as the bicycle at an earlier date, remarkably in accordance with the fields of expectation of other fleet managers (Bersey in London and Bixio in Paris).[79] According to this analysis, the EVC management wanted to profit from this situation, but the conviction that it would soon be over would not have encouraged solid, costly quality control.

From this perspective, it really is amazing that the EVC New York remained in existence. It did not "fail" until 1907, but even after that it still played an important role in the taxicab motorization of New York. This makes the answer to the cardinal question of the role of the battery even more complicated. Nevertheless, determination of whether the EVC taxicab operation failed cannot be based upon the initially satisfactory functioning of the fleet. After all, the Paris automobile contest had demonstrated that the problems with the batteries only started after a couple of months, when the active mass started to fall from the negative grids. The rapporteurs in Atlantic City had also observed this. The fact that, in the only document related to the battery, these systematic flaws were blamed on "summer quality" and on the complexities within the application field of electric road vehicles is another proof of the strength of the battery-streetcar tradition and in general of the complete novelty of the electric-car technology under very heavy conditions.

Schallenberg suggests in his analysis that company managers knew about the failure of the batteries, because they had already had two years of experience with the same battery type in the streetcar: "we can only wonder at the careless abandon with which the syndicate's creators poured millions of dollars into a technology which they had taken no steps to evaluate." It seems, however, a bit too far-fetched that the EVC creators were so stupid. And indeed, neither a theory about villainous speculation nor a conspiracy theory is necessary to explain the failure of first-generation taxicabs in the United States.[80]

At first sight the continued existence of the New York department seems to contradict the conclusion that the battery was in fact the most important failure factor. But there is a catch in this. Whereas London decided on two batteries per car, and Paris initially restricted itself to one due to cost considerations, New York adopted a middle course. But with 1.5 batteries per car it would be impossible to run a stationing taxicab service in a big city, unless one served the client from a garage. And that was exactly what happened: from the beginning the business evolved toward a service with taxis-on-call and a rental car system. The lower revenues resulting from this must have contributed to the demise of the subsidiaries. Moreover, the shift toward maintenance for third parties (without renting out the batteries) points at the tendency to dismiss the financial consequences of a faulty concept. Establishing electric-bus lines (with a fixed, computable route) fits in with this view. So, if "New York" failed because of the battery, this could not be blamed on low energy density (for this problem had been solved by a change in the field of application), but on the life span. After the collapse of the conglomerate, the emphasis in New York turned even more to sales, rental business, maintenance for third parties, and trucks.

In the explanation given here, this change of emphasis was imposed by the limited energy density of the chloride battery.

But was the battery the main cause of failure? The answer to this question can only be found by an explicit comparison of the three cases. When comparing different applications of the same vehicle type it is important to indicate first the major similarities and differences between the *fields of application*, so that one does not compare apples and oranges. Put in another way: in the end the (technical) *properties* may have been largely the same in the three artifacts under consideration, but their *functions* may have varied in disparate geographical settings.

As far as the similarities go, first, in all three cases the battery streetcar tradition was an important determining factor. Second, the tradition of the horse taxicab also played a role. In New York this is seen in the concept of the taxicab (hansom); in London, and especially in Paris, in business management and the salary structure of the drivers; and in all three cases in the selection of the drivers. In the third place, the taxicabs appealed to the public, at least initially (London, New York). In all three cases, this absolutely convinced those involved that the electric car would definitely replace the gasoline car after the automobile sport fashion had blown over. Furthermore, in all cases the cars were used intensively, at any rate when compared with privately owned passenger cars. That is why in all three cases the tire appeared to be a restrictive factor: the life span of both the solid tire and the pneumatic tire (New York, initially) appeared to be very short, so maintenance was extremely costly. Although one can dismiss the European hesitation to apply the pneumatic tire and opt for the solid rubber tire as regrettable constructional conservatism, New York showed that the pneumatic tire was not suitable for such heavy vehicles, not even—and this is surprising—within the city.

At closer consideration it appears, however, that the American field of application and the technical field were of a different character than the European ones, differences that we can summarize under three headings: vehicle structure, organization of business management, and battery.

As far as vehicle structure is concerned, Morris and Salom's hansom was much more modern, despite its "antique" exterior, although there was some doubt among contemporaries about the decision to use the rear wheels for steering. The two inventors had converted an existing, originally two-wheeled taxicab model, in which they had avoided the energy-wasting differential by the application of two electric motors and the central pivot steering system by implementing axial pivot steering. They had also put the vehicle on pneumatic tires. But most important, they had made a considerably lighter construction. In the course of the EVC development this construction had to be adapted, and a remarkable process of convergence of the American and European models emerged; the American taxicab became heavier and lost its pneumatic tires as a result.

As far as business management is concerned, first of all the ambitiousness of the American initiative is noticeable, at least in terms of production. This led to all kinds of quality problems at assembly. As far as the operation of the taxicab fleets by city is concerned, the American and European experiments are comparable. Although evidence for the United States is lacking, there are strong indications that even in New York the stationing fleet did not comprise more than a hundred taxicabs.

In contrast with Europe, electricity in America was not generated independently, but was purchased from the Edison power station. Furthermore, the relatively (compared to the horse taxicab) low rates are noticeable. In Paris this was not the case, as the operation of both types of taxi was carried out by one company. The EVC further differed from Europe in that the drivers were salaried employees and had no interest in profitable operation.

And finally, the direct involvement of the battery manufacturer is unique to the American situation; those responsible for the operation of the taxicab had no choice as to battery type. Because of this the emphasis in management shifted to sales and car rental and moved farther and farther away from the "classical" taxicab business.

With regard to the battery, its low range is noticeable. This was not so much the result of a different battery concept (which was the same for New York and Paris; London had the grid-plate battery). It was brought about by the "half-and-half" solution to the maintenance problem: only half of the New York fleet could utilize the principle of exchange charging. If exchange charging could not be used, one was forced to turn to quick charging in emergencies, which affected the life span of the battery even further. In all cases the battery-vehicle ratio eventually amounted to about 40 percent.

The reconstruction of the London and Paris experiments shows that technology was the most important factor of failure. For Paris this conclusion can even be specified by reference to the battery. For London the battery is a serious potential factor of failure as well, because all light grid-plate batteries failed at the Paris battery contest, and the EPS failed in the delivery to the CGV. The fact that, at closer consideration, the EVC as a whole was *not* a taxicab company in the European sense (with a very high deployment and high daily average distances) made the battery problem less acute. Thus, a battery that functioned poorly in Europe (compare the battery contest and Bixio's criticism) could do better in the United States, because its use was different. One can, as has been mentioned before, turn this reasoning around and state that due to this "monopolistic battery type" the EVC was forced to neglect the taxicab service and to evolve toward the rental and sales business. In this sense, the technical field induced a change in the field of application.

So, in the end, the question of the role of the battery in the EVC case remains. Was it the main failure factor, or were social and managerial factors to blame? According to some historians of technology, the question of whether an artifact functions depends on the assessment of its contemporaries. This is only true, however, in a restricted sense: if the EVC had decided to use a brandnew battery every six months, the rental-car service could have functioned well (and the ESB would even have had a higher turnover). In this sense, maintenance is a social category, just as is life span. The latter may become clear from the use of the pneumatic tire on privately owned gasoline cars. It did not function "better" there in a technical sense, but initially wealthy individuals paid with pleasure for the high costs of the "repair adventure." So, while the technical properties in both cases were exactly the same, their function was quite different: a production function in the case of the taxicab versus a technical challenge important to a largely male user group. This has nothing to do with the well-known "flexibility of meaning" from constructivist theory. On the contrary: function comes before meaning and is realized during use. Conscious reflection upon this use may not coincide with function, which is why this function is not always traceable in the written sources. In this respect there is a remarkable parallel with personal computers today: the programs are user-unfriendly and often annoying, but who wants to return to the days of pen and paper? For its use in a "rational business organization," however, the extra time such disadvantages require compared to the existing technology to be replaced play a decisive role. For the first-generation electric cars, the costs formed the decisive difference between the fields of application of the taxicab and the privately owned automobile. With regard to these costs, the environment of the horse taxi set a limit to the fares the EVC could charge, and when competitors appeared, it even had to lower its fares. Thus, this life span is expressed in the operating costs, and those costs eventually led to the demise of the taxi service. In this sense, the life span of an artifact generally can serve as a criterion to check its adequacy.

It is certainly true that better maintenance at the subsidiaries, which was neglected by the holding company, could have extended the life span of the battery. A more accurate automobile manufacturer could also have taken care of this by removing the energy wasting friction points in the drive train. But because of the battery, the EVC was either forced to utilize a depot service that did not yield enough and accelerated the road to bankruptcy, or run a taxicab service that refused to turn to exchange charging for the entire fleet. If the EVC had pursued deployment similar to that of the regular taxicab business in Paris, there is no reason to assume that what failed in Paris might have succeeded in New York with the same battery type. Only if the purchase and maintenance costs (as the Parisian example shows) had been more than halved in America would there have been a chance of success.

The comparison with Europe demonstrates that the EVC showed how it *should not be done*. Maintenance is a social category, but not in the sense that it can be placed opposite technical arguments. Battery maintenance was of a totally different order than maintenance of machines, indeed, even than maintenance of the gasoline car. The life span of the battery and of the pneumatic tires, defective maintenance, and high costs made up the vicious circle of failure factors of the first-generation taxicab, but the battery was the most important. Even exemplary maintenance, as in Paris, could not make an inadequate concept viable for taxicab application. In the United States a regular taxicab service based on a chloride battery was speculative at best, even if the EVC would not have been involved in stock jobbing.

The argument that the EVC was top-heavy and had little technical content does not hold, at least if this argument is intended to create a contrast with gasoline propulsion: both the gasoline car and the electric car were unreliable. But what for the one initially was a merit, was for the other an insurmountable barrier. It is, therefore, not a coincidence that in this first phase of automobile history no partially successful taxicab experiments with gasoline cars can be found. From all examples given in this and the previous chapter, it becomes clear that electric propulsion was the only possible choice for this heavy-duty application. As far as this is concerned, the EVC founders did not bet on the wrong horse, but on the right one.[81]

But what remained completely beyond the horizon of gasoline propulsion appeared to be too high an aim for electric propulsion as well. This conclusion sheds new light on the technology to be replaced: it commands respect for the deployment of the horse.

The Second Generation

(1902–1925)

3

Horse Power:

The City Car, the Touring Car, and the Crisis of 1907

The principal defect in a storage battery is its modesty. It does not spark, creak, groan, nor slow down under overload. It does not rotate. It stays where it is put, and will silently work up to the point of destruction without making any audible or visible signs of distress. If it does not cry for attention none will be given it. —*Electrical Review, 1902*

AT THE BEGINNING OF THE TWENTIETH CENTURY, proponents and opponents of electric propulsion alike agreed that what was wrong with the electric car was its poorly functioning battery. This opinion was brought about by the failure of the taxicab experiments, but there is room for doubt whether the harsh judgment also applied to privately owned cars that were not operating in a fleet. After all, at first sight the experiences of the taxicab business cannot simply be transferred to the dilemma of choice of the potential private electric-car drivers. Private owners used their cars less intensively; thus, the narrow technical field of electric propulsion seemed to be suitable for the field of application of the private car, but not for that of the taxi. Michael Schiffer, for instance, pursues this reasoning in his study of early electric-car culture: "It is clear that a privately owned electric, whether used for light business or pleasure, did not exhaust batteries at the same prodigious rate as New York taxi cabs."[1]

In this chapter, we will question the apparent logic of such assessments against the background of an emerging new stage in the development of battery and tire technology. As in chapter 1 we will, in three separate sections, first continue and deepen our analysis of the user culture of the urban electric vehicle, stressing the importance of maintenance and the crucial role of a garage culture. Then we will direct our attention to a shift in the user culture of its gasoline counterpart, brought about by the crisis of 1907. In the third section, we will return to the electric's response to this applicational shift and witness the emergence of a second-generation technology, introduced in part by a new type of manufacturer with well-established roots in horse-drawn carriage production.

Car Rental and Automobile Culture: The Electric Passenger Car in the City

Although Schiffer's argument is appealing, at closer consideration the factor that determined the boundaries of the technical field during the first-generation taxicab experiments—the life span of the battery—was not the essence of the private owner's trilemma of choice between gasoline, steam, or electric propulsion. For, as has been argued in chapter 1, for the first automobile owners costs were not decisive. On the other hand, for private owners an accurate, weekly checkup of the battery was just as necessary to prevent a breakdown as it was for the fleet manager. Here, a technological difference was at issue that made the lead battery essentially different from a machine, such as the combustion engine in the gasoline automobile. A broken-down machine could be put aside for a while and be repaired at a convenient moment. A defective lead battery had to be repaired immediately, otherwise its life span would be shortened. Even a partly discharged battery of the grid-plate type had to be recharged immediately to prevent "sulfating" of the lead paste in the grids. This phenomenon, by which the plate surface becomes impermeable by the formation of coarse-crystalline lead sulfate, occurred especially when the battery was idle.

Moreover, the nature of the car breakdown was different. In 1902, *Horseless Age* contributor Albert Clough formulated it as follows: "If one runs short of gasoline on the road one may walk to the next grocery store and bring back a can full, but one cannot bring back a can full of electricity for his stalled electromobile, but must surrender to the despised 'hay motor', who can take supplies wherever the grass grows." This not only imposed a different automobile culture, in which the element of planning played an important part, but also saw to it that the technical field was as narrowly defined for the private-car owner as for the taxicab business: without an adequately functioning battery, no electric motoring was possible. So, just as for the taxicab, the development of a reliable, low-maintenance battery was crucial for spreading private-car ownership.[2]

It was here that the cry for a "miracle battery" found its roots, but the hope for a miracle was connected with other elements in the field of expectations of the electric vehicle camp, especially with the conviction that such a battery would reverse the balance of the battle between the propulsion alternatives and that it, in the end, would dethrone the gasoline car. As late as 1907 (some months before the crisis of 1907 and well before the coming-to-market of the Ford Model T), a German observer opined that "the problem of the People's Car will never be solved by the gasoline engine . . . The gasoline car will, until the great turning point of the small accumulator, remain in existence as a luxury car *par excellence,* and as a rental car and a commercial vehicle, but it will vanish as a car for the people with an annual income of only 50,000 marks or less."[3]

Against the background of this perceived stalemate between both rivals (here a lack of a light battery, there a lack of general reliability), the concept of the electric as city car, its high purchase price, its technical properties of easy operation, and the absence of vibrations, noise, grease, and fuel made it eminently suitable as a replacement for the horse for the affluent city dwellers. Usually such potential users had personnel that could maintain the car, but considering the type of maintenance required it was more likely that they had this done by a public garage. Hence, some drawbacks of driving electrically, such as the awkward battery checkup and maintenance, were effectively shifted into the hands of specialists. Hence, the necessity of proper maintenance may well have been recognized earlier in the electric camp than in the gasoline camp. In any case, given the commercial application field of the electric taxicab, maintenance was professionalized and systematized earlier, compared to a gasoline car culture, where initially repair and overhaul were considered to be a part of the sport of car use itself.

The setup of such garages had been derived from the traditional stables for the *voitures de remise* (rental carriages). The first years of the twentieth century showed a gradual expansion of this garage culture, pioneered by electric vehicle proponents. At the same time, this gradual growth of a garage culture suggests that the gap between the first and the second generations—as far as the use of the electric car is concerned—was less wide than is suggested by automotive historiography. These garages were often located in or near wealthy districts, such as the avenue Montaigne in Paris, where l'Électromotion (a branch of the American Pope concern) had set up such a garage in 1900. A little later, when all sixty parking spaces were occupied (all by private customers), Pope representative Hart O. Berg announced the opening of a second garage with a capacity of 150 cars in the neighborhood of the Arc de Triomphe. Another enterprise, the Société Française d'Automobiles Électriques (usually shortened to l'Électrique) had made a modest start in 1898 with two cars in a garage in Levallois. In 1902, when it was planning to move to the center of Paris, it was in charge of eighty vehicles. Here the garage formed an extension of the manufacturer: the customers bought the cars without batteries, and l'Électrique rented out the batteries and the drivers.[4]

In 1903, Kriéger's parent company, Indusmine, had bought a large plot of land at the expensive rue de la Boëtie in the 8th district of Paris, where it built a garage and a charging station. As the annual repayment turned out to be too high, on 2 April 1905, a subsidiary, the Société des Garages Kriéger et Brasier, was founded with a capital stock of 2.5 million francs. Here Indusmine colleague Brasier's gasoline cars were accommodated as well, and the buildings that were ready for use in August were rented from the parent company. The garage could house 300 automobiles; 200 of these belonged to private customers and 100 were in use as rental cars. The sale of the Kriéger and Richard Brasier brands also took place there. At least until 1906 this system was a success.[5]

The City and Suburban Electric Carriage Company, established in London in 1901, two years later had "230 carriages of different types" in its garage,

Signs of second-generation technology appear in this electric, rear-driven Columbia, which clearly owed a debt to horse-carriage body construction but came equipped with a front hood borrowed from gasoline-automobile design. (*The Horseless Age* 11 [12 January 1903]:131; DaimlerChrysler Archives)

which were "examined and repaired, and their accumulators kept in an efficient condition." In 1905, the number of garages in London had increased to six or seven, where about 400 to 500 electrics were housed. Once left in the care of specialists, these vehicles were very reliable: in a period of six months one of these garages only had to deal with sixty-one "involuntary stops" on a total distance of 255,000 km (at an effective range on one charge of about 40 to 55 km). "It would be a most profitable business if their battery loss was not so enormous," a representative of Thomas Edison wrote to America in 1903.[6]

In the United States, such garages dated as far back as 1897, when in the summer Pope opened a charging station in the chic resort of Newport, Rhode Island, "to introduce society to the automobile." In New York, the New York Electric Vehicle Transportation Company—heir of the infrastructure and furnishings of the EVC—was in charge of "the Largest Service in the World" in 1902—a fleet of no less than 616 vehicles. Besides about forty trucks and touring buses, this number largely consisted of hansoms and broughams. In the meantime rechristened as the New York Transportation Company, the outfit was led by President Richard W. Meade, former manager of the Metropolitan Railway Company, associated with the Whitney syndicate. George H. Day was one of the directors.[7]

The company had apparently learned its lesson from the unfortunate past. All vehicles were equipped with solid rubber tires and the new ESB Exide battery. During the summer, almost half the fleet was rented out, including drivers, at the considerable rate of $250 to $350 a month, depending on vehicle type. In 1904, the company set up a garage in the old headquarters of the Metropolitan

Railway Company, specially for the district east of Central Park, where the rental of electrics by month was very popular. Moreover, the "boarder," the private car that was maintained and charged in the company's garages, was a lucrative part of the business. In 1905, the company started with gasoline cars of the Columbia brand, of which it also held the agency for New York. In May 1906, it took over the garage of the bankrupt Vehicle Equipment Company, so that it could add over sixty electric vans and trucks to its fleet. The company also owned two subsidiaries. The Fifth Avenue Coach Company operated horse-drawn buses, but also experimented with hybrid (combined electric and gasoline) propulsion. The Park Carriage Company operated two electric "sightseeing breaks" and two comparable Mack gasoline vehicles. In June 1906, the company ordered fifty electric landaulets of the latest type ("Mark LXVII") and fifteen gasoline touring buses from the EVC factory in Hartford. A fully equipped factory took care of the repair and construction of bodies, the revision of the batteries, and all other necessary work.

The company had found a creative solution to the problem of the low energy density of the American battery: the 350 to 400 drivers (paid by the hour actually driven) were sent out after an order by telephone, whatever the length of the ride. At a speed of about 20 km/h and a range of 50 to 65 km, the driver sometimes would have to go to a charging station during the ride. In such a case, the customers would be parked in a waiting room and after exchange charging for a couple of minutes the ride was continued.

THE EARLY ELECTRIC automobile culture followed its own course, separate from the gasoline adventure machine. Hiram Maxim suggested in this respect that driving about in an electric had replaced driving in a horse carriage, as steering a carriage in city traffic no longer offered the pleasure of earlier days.[8]

One of the standard themes of automotive historiography is that women at this early stage were preeminent electric-car drivers, but this is not supported by the few figures available. If it were true at all, it only applied to the very earliest period. According to the women's column in *The Car* of 1902, thirty-five women in Chicago had a driving license that year. Half of them were single and, virtually without exception, they were young. According to Clay McShane, who investigated two American states for which the early registration lists have been preserved, in 1905 only a quarter of the registered female car owners in Washington, D.C., owned an electric car. But Washington was an exception. In 1911, only three of the 214 female car owners in New Hampshire owned an electric.[9]

It is more remarkable to see the small share of female car owners, *independent* of the question of propulsion type. It is true that 21 percent of the car owners in the District of Columbia were female, but four years later this share had decreased to 15 percent. McShane explains this by a wider distribution of the car among the upper middle classes, where female car ownership was less accepted than in the upper classes. As has been mentioned above, Washington, D.C., was an exception, which McShane does not explain, but it may be related to the high share of diplomats in that city. Virginia Scharff, who examined the registration lists for Tucson (Arizona) and Houston (Texas), confirms this picture.[10]

In Europe the situation was not much different. This can be deduced, for instance, from a list of buyers (unfortunately undated) of the German Mercedes Electrique; about 10 percent of all electric passenger cars were sold to women. And a buyers' list of the Lohner company in Vienna with a total of 354 electric and hybrid cars sold between 1900 and 1915 contains an even lower percentage of women. It is true: owning and driving a car are not the same, and the earliest American registrations most likely underestimate the real number of privately owned vehicles (but, on the other hand, they also contain duplications, because registration was necessary in each state one wanted to drive through). But the tendency is clear enough: even if many male owners let their wives and daughters drive, it cannot be denied that women's share in early automobile use and its culture, even in that of the electric car, was small.[11]

This state of affairs leads us to our conclusion that women and electrics had one thing in common: they were both outsiders in the prevailing automobile culture. The cultural difference between electric car and gasoline car use also led to misconceptions about the popularity of the electric car in general (irrespective of gendered use), as an American electrotechnical trade journal reports: "somebody wrote recently in a newspaper that very few electromobiles were to be seen at race meetings, meaning horse-race meetings. From this he argued that the popularity of that type of vehicle is on the wane . . . The fact is that electric vehicles are more popular than they ever were before, but the class of people who own private vehicles of this sort are not particularly interested in race meetings; they are to be found in great numbers in such places as Newport and Bar Harbor, but the betting ring of the Coney Island track knows them not."[12] If this assessment is true, its consequences for a history of the private electric vehicle are far-reaching indeed: whereas the early gasoline car culture was visibly and loudly connected with racing (and was celebrated thousandfold in a multitude of magazines), the electric car thrived in a quiet, elitist atmosphere, effectively resisting the historian's grip on its subculture.

Despite this obscurity, we can get a glimpse of at least one part of the field of automotive application in general: that of the American medical doctor, although here the same bias toward the gasoline car may be operative. Shortly before the crisis of 1907, in April 1906, an American medical journal published the results of an extensive poll it had taken among physicians from all over the country. It is instructive to compare the results of this investigation with those of the polls by *The Horseless Age* of 1901 and 1903, discussed in chapter 1.

Sixty-seven of the approximately 50,000 American physicians had taken the time to give an extensive account of their experiences. Most of them had bought their vehicle in the previous three or four years, and an annual mileage of 10,000 km was not unusual. Many of them had started with a one- or two-cylinder runabout, and had later changed to a heavier version with a three- or four-cylinder engine. The pioneers among these physicians mostly recounted bad experiences with the first-generation cars. Apart from a single dissenter, who called the car "practically useless," those polled unanimously focused on the liberation from the horse and on the "simplicity" of the car. No one wanted to return to the days of the "doctor's buggy" (with horse traction), but the ideal doctor's vehicle did not yet exist, as many of them concluded.[13]

The results largely confirmed the picture of the earlier polls: the car was praised because of its surplus value compared to the horse, manifesting itself in the saving of time (the almost unanimous conclusion of a time-saving factor of 2 or 3 was noticeable) as well as in "fun." The time saved was not spent on more professional work, but on maintenance and on touring in the evenings and on the weekends. Some of them spent a few minutes a day on maintenance, others an hour, and most of them enjoyed doing it. "Many small repairs are constantly necessary." Among the sixty-seven physicians, only one (Mr. Kahlo from Indianapolis) had bought an electric, which he explained by pointing at the fact that he did not have a coachman. Another physician owned a steam car, whereas two had changed from the electric to the gasoline car. One of them had done so because the electric car was not suitable for long runs. He praised his new acquisition for its adventurous character, causing what he called "auto-intoxication" that made his pulse increase by six beats a minute, which he had measured. Also unusual was the physician from St. Paul, Minnesota, who had replaced his gasoline car with an electric vehicle "for comfort and pleasure"; but, in contrast to the physician in the earlier poll, he kept a gasoline car on the side for emergencies.

The views about the costs were divided, but the largest expense item clearly was the pneumatic tire, which many replaced with the solid tire. The latter type of tire was better suited to the cart tracks in the country than the pneumatic tire. From this poll the same picture emerges as we saw before: the desire for "something mechanic" for individual transportation was very strong and the choice of vehicle was often based on the fact that many physicians had their own driver.[14]

For our purposes, however, the conclusion from this poll must be that medical doctors did not choose the electric alternative, which was more reliable, at least for taxicab purposes. Why?

Reliability and the Crisis of 1907: The Taming of the Gasoline Car

While electric proponents boasted about the greater reliability of their propulsion alternative, the gasoline car seemed to escape all engineering rationality by embedding itself in a male sports culture of technical challenge and adventure. But in the midst of the race to dominance of the gasoline car as an adventure machine, the mainstream car culture seemed to run up against boundaries that threatened to jeopardize its further expansion. This became quite clear during the crisis of 1907, which represented much more than an economic disturbance. This crisis can be typified as a catalyst of the developments that would lead to a radical change in the technical, the applicational, and the expectational fields of the automobile. As a consequence, the crisis would have an important impact upon the interaction between the electric car and the gasoline car.

At closer analysis, the 1907 crisis appears to display not only economic, but also important sociocultural and technical aspects. In the first place, it was a sales crisis that started in the United States, which could soon be observed in the European automobile nations as well. The crisis also manifested itself as a safety crisis of urban automobile traffic, and as a crisis in the image of the car as adventure machine. A technical crisis in tire technology was directly related to this. All these aspects had been visible before 1907, but the recession translated them into a financial alarm signal for manufacturers.[15]

As far as the sales crisis is concerned, even according to historian James Laux, who calls the recession "mild," the years 1907–9 formed "a turning point in the European automobile industry." The same was true in a very literal sense for the United States: due to the financial problems that threatened the existence of the EVC and the Pope concern, the center of automobile manufacture shifted to the Midwest, led by the Ford factories in Detroit.[16]

When the crisis reached Europe, the Italian automotive industry was especially hit hard, leading to fierce pricing competition. In Germany, Daimler (which had only built 800 of the planned 1,200 cars in 1907) started to lay off workers: of the 3,030 workers in 1906, only half were left a year later. It was not until 1910 that the workforce would grow again, to 2,300. In France, exports collapsed and everything pointed to a saturation of the market. Minister Méline appeared to have predicted this crisis as early as 1904. In his *Retour à la terre* (Back to earth), he wrote that France had an automobile market of 30,000 buyers at most. Once this market was saturated, production of 5,000 to 6,000 cars a year would be sufficient, and could be covered by three or four manufacturers. Méline's opinion corresponded with that of the manufacturers. Initially they did not believe in a mass market, because the average income in France was lower than in the United States and because they thought that manufacturing costs could not be lowered any further. So, the crisis of 1907 for some producers was a signal to diversify—to build aircraft engines (Renault, Clément) or trucks (Renault, Berliet, Delahaye, Peugeot).[17]

In the United States this crisis led to an expansion of the batch production of the so-called runabout, the less expensive car for driving on bad roads. Whereas before 1907 a real touring car could not be produced for less than $2,000, even in America, the success of the automotive industry in Michigan was due to runabouts that were sold for $650 to $750. As early as 1904, such cars accounted for 56 percent of total production, as opposed to 33 percent for the more expensive cars, a clear indication that the American consumer switched to the automobile as means of transportation earlier than the European consumer. Thus, the original broad variety of car types, including the relatively cheap *voiturettes* after 1900, was narrowed to luxury cars. And although the recession of 1907 led to the production of "cyclecars" (cheap, light cars based on bicycle technology), their success remained mainly limited to Great Britain.[18]

The second aspect of the crisis of 1907, the safety issue, has hardly been researched by historians. Although the increasing number of accidents was often blamed on the high speed of the automobiles, the first accident statistics show that many of the culprits were urban coachmen. Because people do not get killed by percentages, but by vehicles, it is important to establish that the absolute number of accidents caused by cars also increased. In Germany the major culprit in the cities was the taxicab, and the majority of the victims were laborers, merchants, and servant girls.[19]

The number of road casualties in New York City doubled between 1902 and 1907. This increase was attributed to the higher speed of the electric street-cars, but also to the lack of a modus vivendi for five different traffic modalities (pedestrians, cyclists, streetcars, horses, and automobiles). Horse traction caused most of the accidents. Three-quarters of the victims were probably pedestrians, "mostly children playing in the street, inattentive to the new ve-hicles." After 1907 there was a sudden reversal in the number of casualties, caused by the application of air-brake systems on the streetcars and the virtual disappearance of cyclists from the streets. As of 1909, the number of deaths started to rise again, this time caused by the increase in automobile traffic. By 1911, the automobile had become the dominant factor in New York traffic.[20]

The third aspect of the crisis of 1907 involved a speed crisis and was very concretely expressed in the technical crisis of the pneumatic tire. As the utili-tarian aspect of the automobile gradually became stronger, its technical unre-liability—manifest in tire trouble—caused automobile users to perceive the nuisance of tire trouble in a different light. Besides a large expense item, the tire now became a source of irritation that caused one to question the useful-ness of the automobile for daily business, as became clear in the case of Amer-ican medical doctors. Thus, the technical field of the gasoline car threatened to become too narrow for the changing field of application. To analyze this piv-otal situation, a deeper look into the technical field of the tire is called for.

What exactly was the weak spot of the pneumatic tire that made the tire problem expand into a full-blown crisis? The crucial problem was the transla-tion of bicycle-tire technology into that of the automobile. Here not only the higher axle load and the higher speed played a role, but especially the greater lateral load on the tire, at the spot where the tire bead grips the rim edge. The first pneumatic tires were made of natural rubber, enriched with zinc-oxide powder and at first strengthened by linen. Soon, under the influence of Amer-ican tire manufacturers, linen was replaced by the so-called Sea Island cotton. Such tires were so soft that they ripped open very easily; yet this softness made them easy to remove from the rim for repair. In 1904, the removable tire, the so-called clincher tire, was part of the standard structure of the American automobile.[21]

The crossed textile fabric for strengthening the tire (the "canvas" as it was called in Britain) had an unpleasant property. As the pneumatic tire constantly flattened out and subsequently resumed its original form as it rolled over the road surface, a scissorslike action developed between the cotton threads that led to the heating of the rubber and so to a seriously higher chance of damage. This effect became a problem when the structure—borrowed from the bicycle tire—had to carry a much higher mass. According to Michelin, the wear and tear on the pneumatic tire increased with the third power of the axle load. For a dynamic function group like the wheel, this meant that an increase in speed led to a serious increase in wear on the tire.[22]

For the lighter gasoline car, use of the hollow solid rubber tire seemed possible in this period, and as a matter of fact, as we have seen, some medical doctors from the 1906 survey replaced their pneumatic tires with solid tires. For these cars the switch was not so radical, because the pressure of the pneumatic tires on the first automobiles was at least twice as high as it is nowadays, in order to limit distortion on the road surface. So the tire crisis mainly played a role in the heavier gasoline cars that cruised the country roads at high speed. This may well have been an added incentive for gasoline car manufacturers to focus on the *city car* for the application of their products. However that may be, the tire crisis—which was as much cultural as technical—started in France, because of a fatal car accident involving the Marquis d'Audiffret-Pasquier in the summer of 1904. The accident stimulated an overwhelming number of letters to the editor of the leading car magazines with complaints about the "killer tires," a "little panic," as the writer Baudry de Saunier commented soothingly. The examples of near-fatal blow-out tires filled the pages of the automobile magazines for weeks on end. At the time Michelin started to increase its publicity activities by writing weekly columns for these magazines. Its defense of the pneumatic tire conveyed confidence in modern automotive technology and a harsh criticism of first-generation technology. It was the beginning of a phase characterized by an unparalleled vitality in automotive technological development.[23]

Between 1906 and 1909, at least 140 proposals were published calling for the replacement of the "Achilles' heel of the automobile" (as a German magazine called the pneumatic tire) with a mechanical suspension system, either in the hub, or for the replacement of the spokes or the tire around the wheel rim. In automotive historiography, especially its nostalgic variety, such "elastic wheels" are often dismissed as curiosities. But from the relief the rubber trade journals expressed after the apparent failure of such innovations, it can be deduced that the tire business felt threatened by what essentially was a technical crisis. At the beginning of this crisis "De Cadignan's elastic wheel"—one of the best known innovations of this period—was introduced, equipped with coil springs instead of spokes. A rubber trade journal suggested that this design should trouble the pneumatic tire industry: "we have here a serious competitor for the pneumatic tire and our manufacturers had better not neglect this." It is noticeable that such proposals were made in relation to those for the installation of shock absorbers: for the first time wheel suspension was approached as a system. The compromise between comfort and safety required complex solutions like spring-absorber combinations. The shock absorbers stayed, but the elastic wheel did not. Apart from the fact that it broke down easily, it had a major disadvantage that the pneumatic tire did not have: small, high-frequency vibrations in the elastic wheels provided an upward acceleration force of the compartment where the driver was sitting. The pneumatic tire, on the other hand, "absorbs the obstacle," as a relieved rubber trade journal formulated it, referring to the well-known Michelin slogan that had been known for several years by then. For heavy trucks with a speed lower than 30 km/h, elastic wheels may have been useful, but for the passenger car they were not.[24]

Technically there was a glimmer of hope when the American Goodyear Company introduced a pneumatic tire in 1910, built from rubber that was made about ten times more wearproof by the addition of soot. Thus, at the beginning of the second decade of the twentieth century, one of the most important functional obstacles to the adoption of the gasoline car as a universal vehicle seemed to have been removed. This was supported by a simultaneous economic tendency, characterized by strongly fluctuating, but up to 1905 constantly rising, prices of raw rubber on the world market. Despite this volatile raw-material market, a price increase of the tires was impossible. Because of the price competition between the numerous manufacturers, the German Continental Company, for instance, was forced to lower its tire prices first by 4, subsequently by 6 and then once more by 16 percent at the outbreak of the automobile crisis of 1907. In September of that year, Dunlop also lowered its prices significantly, while Michelin announced its fourth price cut of 10 percent in April 1908. In 1910, rubber prices suddenly soared to an unparalleled peak, which again led to a crisis mood in the automobile press: "Will the pneumatic tire kill the automobile?" But that same year prices returned to a normal level as a result of a crash of raw rubber due to overproduction. During the next sales crisis of 1911, a fierce price competition developed again among the tire manufacturers. Michelin, for example, lowered its prices by 20 percent in 1911 and shortly after by another 15 percent.[25]

One should, however, not draw the conclusion from these developments that the crisis of 1907 heralded the end of the automobile as adventure machine. This was certainly not the case—just like the safety bicycle, which did not lose its adventurous characteristics as a "fast bicycle." Instead, the automobile's adventurous character gradually disappeared underground, or rather, it descended from the brains to the guts, where it found a safe haven for decades to come. Illustrative examples of this process of internalization are the Prinz-Heinrich contests in Germany that were organized from 1908 on and were really meant as reliability tests for touring cars. Initially (in 1908) the engine power of participating cars was limited, and subsequently (in 1910) the cylinder volume. The only way to reach a higher vehicle speed was via an increase of the engine speed (combined with a lighter, aerodynamic body). This development was stimulated by German tax legislation, the basic principles of which were adopted by most European countries. The tax was based on a formula obtained from the automotive industry, in which the stroke and the diameter of the piston as well as the number of cylinders were taken into account. The engine speed, which greatly contributes to the engine power, was left out of the tax formula. So, even with a higher engine speed, one nevertheless stayed within the same tax category.[26]

In yet another respect, the Prinz-Heinrich runs were an expression of the dilemma the automotive industry was confronted with after 1907, for the organizers of the contest had not prescribed the shape of the body of the participating automobiles. Whereas the automotive industry wanted to concentrate on the manufacture of touring cars, in these and other such races it let itself be lured by the competition into building increasingly lower, lighter, and aero-dynamically sound special bodies—the opposite of what was intended. Because of this, not only the tall electric cars (in which one could sit straight even wearing a top hat) with their short wheelbase, but also the cheaper gasoline versions now looked like representatives of bygone times. For example, the Brush Runabout, a one-cylinder buggy from the United States, was described as "wooden body, wooden axles, wooden wheels, wooden run." When the recession of 1907 heralded the saturation of the luxury-car market and the automotive industry began to pay more attention to the cheaper market segment, the new construction principles also penetrated this segment. The private, cheaper "utility car" thus retained the characteristics of a vehicle type that originally had been designed for a different purpose.[27]

"Dark Ages"? The Revival of the Electric Passenger Car

Shortly before the First World War, the American electric-car manufacturer McAllister Lloyd called the period between 1900 and 1908 the "Dark Ages." Yet, one should put this in perspective to some extent, because this metaphor also seems to have been inspired by the spectacular expansion of the gasoline car, and so implicitly documents the disappointment of an unrealized universality claim by the first-generation electric-car proponents. However, if one rejects the image of the electric car as a kind of failed touring vehicle, the limited figures available to us do not show that there was a reduction in its *manufacture*. Earlier in this chapter we have seen that there were no indications of an important reduction in its *use* either. On the contrary, the number of public garages for electric-car owners and renters increased after the turn of the century. On top of that, from about 1905 on, a revival can be observed in the manufacture as well as in the use of the electric car.[28]

The historiography of the automobile too often fails to recognize that the expanded use of the car started in the *cities* and initially was the concern of the *local* authorities.[29] It is also not sufficiently recognized that these authorities had different interests than the regional and national authorities, who looked upon the automobile on the state highways primarily as touring vehicle (and gasoline car). While national car ownership was increasing, local authorities often appeared to prefer the electric car. Given the evidence at hand, this may not be convincingly established for the American case, although the opening up of the American city parks to electric cars exclusively (sometimes revoked by state authorities, as we saw) is an indication that this might have been the case. For Europe much more evidence exists of the preference given to the electric taxicab at the cost of the gasoline car by local authorities, as the following chapter will confirm. Despite this preference of the local authorities, the gasoline car penetrated their domain. On the Champs Elysées in Paris, for example, a German observer composed his own *Gelegenheitsstatistik* (occasional statistics), by tallying eighty passing cars in a quarter of an hour on 25 March 1905. Sixty of these cars had a gasoline engine and he only counted nineteen electrics and one steam car. Nevertheless, a year earlier, another German ob-

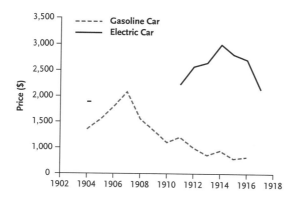

FIGURE 3.1. Average price of the American gasoline car, 1904–16 (adapted from Barber, *Story of the Automobile*, 175). The average prices of electric cars are also given for a few years (1904: Kirsch, "The Electric Car and the Burden of History," note 32; remaining years: my own calculations based on various volumes of *EV* [U.S.]; see also figure 7.2).

server mentioned a stock of at least 600 "luxury cars with battery" in Paris, without doubt the largest number outside America.[30]

Thus, it is not surprising that the efforts to establish the gasoline car as a reliable city vehicle raised hopes among electric vehicle proponents: they could justifiably expect to beat the new intruder in a field of application where they considered their product as "superior."

In 1904, American gasoline-car production amounted to 18,699 vehicles, versus 1,425 electric cars and 1,568 steam cars. As we have seen earlier, at the Census of 1900 the average product value of these types of vehicle was $938, $1,822, and $682, respectively, but now these figures had changed to $1,279, $1,804, and $1,466. Due to effects of scale at its manufacture, the gasoline car took over the first position from the steam car (in 1900 still the cheapest of the three). The electric was the only alternative that became a little cheaper, but was still 40 percent more expensive than the gasoline car. Just as in Europe, the latter started to become more expensive, until the Model T would disrupt this proportion drastically (figure 3.1).[31]

In both Europe and the United States the revival of the electric led to the creation of new brands, as the auto shows of 1905 and later demonstrate. This means that this revival had already started before the 1907 crisis and seems to be a result of a final general upswing in the sale of the automobile as an exclusive luxury article, which formed the overture to the subsequent collapse of this market segment. In France, Dinin, Gallia, Electros-Cardinet, and Védrine entered the leagues of the manufacturers. But the old-timers had not yet given up: in 1901 Mildé, for example, took over the insolvent estate of EVC representative l'Électromotion, with the help of the Rothschilds. He continued building electric cars under the same company name until 1907. Even gasoline-car builders such as Georges Richard and de Dion-Bouton had a go at the electric car, the latter even with a production of his own until at least 1903. Kriéger still led the camp of the "electric tourists," however.[32]

The Kriéger automobiles were the best-selling cars of Europe: according to a later source, the brand sold about 800 cars in the ten years up to 1908 in France and some 400 via licensees abroad. Mainly because he set up a modest network of branches abroad, Kriéger was unique among the European electric-car builders. Except for the British Electromobile Company in Britain (as of

1900), a short-lived expansion to Italy between 1906 and 1912, and the attempts, described in the previous chapter, to penetrate the American taxicab market, the branch in Germany was particularly important. There, in 1905, the French company, together with the Abam, established the Kriéger Automobil-Aktiengesellschaft in Berlin, where it set up a garage with charging station on the Wilhelmstrasse. But on 17 May 1906, there was a new initiative in Bremen: on the instigation of and with financial support from Indusmine, the Norddeutsche Automobil-und Motoren A.G. (Namag) was founded, which took over the Kriéger patents from the Abam. The Namag was a joint project of the Norddeutsche Maschinen- und Armaturenfabrik (set up in 1902 as a subsidiary of the Norddeutsche Lloyd), a banking consortium led by the Deutsche Nationalbank in Bremen, and the Compagnie Parisienne des Voitures Électriques (Procédés Kriéger) S.A. The new company was founded with a regional-political objective (making Bremen less dependent on shipping and trade in times of crisis), and it obtained the licensing rights of the Kriéger system for the entire German Empire. It also took over "the Berlin branch of the parent company in Paris," which was subsequently liquidated in 1907. Georg Heinrich Plate became chairman of the supervisory board; he held the same position at the Lloyd. After long negotiations, the French company managed to obtain three-fifths of the profits on each car sold.[33]

According to the German historian Gerhard Horras, the Namag's start-up problems were hushed up in the official company history: the basic capital of 2.25 million marks had already been invested in buildings and machines before production started. Only after a 3.5 million mark loan could production of vehicles begin in a new factory that was probably not finished until April 1907. Shortly before this, in January, the Namag had taken over a coach-building company, renaming it Bremer Wagen- und Karosserie-Werke vorm. Louis Gärtner m.b.H. That year the Namag introduced its first electric car, the "Lloyd."[34]

The first half of the financial year ended with a surplus of a little less than 16,000 marks, but the first complete financial year (that ended 30 September 1907) showed a loss of 56,000 marks. In 1907, twenty-two cars were produced, and in 1908 the Namag's total electric-car output had risen to 130; among these cars was also a vehicle for the German emperor. The Namag became especially known for its production of taxicabs, fire engines, ambulances, and mail vans.[35]

It is tempting, but misleading, to compare these production numbers to those of the largest German gasoline-car brands, because it would mean comparing automobile types that had a *different function*. Nevertheless, the Kriéger brand was much esteemed, especially in Germany. Kriéger was officer of the French Légion d'honneur and supplier of the king of Italy (three Kriégers), the king of Great Britain, the shah of Persia, and the khedive of Egypt. According to the German automobile press, Kriéger had built an automobile with "a performance that surpasses the average touring car with gasoline combustion engine" in his "biggest electric-vehicle factory of the world." The electric components were produced under the supervision of technical director Sigismund Meyer, a mechanical and electrical engineer, who had worked ten years in Britain and the United States at British Thompson Houston and General Electric.[36]

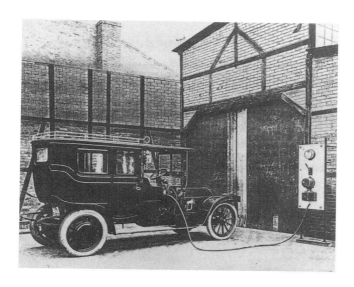

The internationalization of Kriéger was not an expansion, but a relocation of production, a development that ran parallel to both a geographic relocation and a shift of emphasis within the European application field of the electric, as can be deduced from the automobile shows of Paris and Berlin. For the Namag initiative was not an isolated case. It was the culmination of a development in Germany that had made a modest start before the turn of the century and since then had continued without notable interruption on the same scale. In other words, there seems to have been no German "Dark Ages." On 24 December 1901, for example, the AEG founded the Neue Automobil-Gesellschaft (NAG) by taking over the Allgemeine Automobil-Gesellschaft Berlin GmbH that since 1899 had functioned as sales company of the Elektrizitätsgesellschaft vorm. Schuckert & Company in Nürnberg. After that, NAG took over the automobile department of Kühlstein Wagenbau, located in Charlottenburg near Berlin, which had built electric cars since 1898 under the supervision of Joseph Vollmer. After a surprising start with gasoline taxicabs, during the "revival" of 1905 NAG also put an electric on the market.

Competitor Siemens also started at the same time with a program of electric passenger cars, trucks, and buses, but without the "detour" via the gasoline car. Not until the end of 1909 did it also start to produce gasoline cars, after it had taken over the engine factory Protos. Moreover, battery manufacturer KAW decided in 1905 to start its own automobile manufacture. To this end, the company not only started to produce pneumatic tires in a factory of its own, but it also utilized its own newly developed battery (type W extra) with an energy density of 30 Wh/kg (50 percent higher than the old type W). At the Berlin automobile show of 1905, the company introduced the Urbanus, a luxury city car that—judging from its characteristic front-wheel drive—had been inspired by the Kriéger design and was also meant to serve as a taxicab. The automobile department of KAW built 1,500 electric cars until 1909, according to its own later specification. If this number is correct, the Cologne company equaled the Namag in automobile production.[37]

Another important factor influencing the German electric-car market was the Lohner-Porsche concept that, via a complex, not yet satisfactorily explained play of interests, eventually ended up in the hands of the Daimler-Motorengesellschaft (DMG). It was ironic that Emil Jellinek, whose daughter had been the inspiration for the Mercedes name, provided the decisive impetus. Jellinek, who owned 40 percent of the DMG shares and had a French sales monopoly that became increasingly troublesome for the company, tried to get hold of the Lohner-Porsche patents in the spring of 1906. To this end, he established two new companies, located in Paris. The Société des Automobiles Commerciales bought the factory from the Österreichische Daimler-Motoren-Kommanditgesellschaft, whereas the Société Mercedes Electrique would take care of the sale of electric and hybrid cars that were going to be built in Vienna. Jellinek was the largest shareholder of the latter company. Ferdinand Porsche became the technical director of the Austrian factory, but only after Jellinek had forced a switch to rear-wheel drive. Supposedly, 1,200 orders for electric cars were waiting. A British sales company was set up as well, but it did not last long and went bankrupt in June 1908. Lohner then retired from car manufacturing and from then on concentrated on the building of car bodies and aircraft. That year the association between Jellinek and the DMG also came to an end, according to the Austrian automotive historian Hans Seper, because Jellinek was disillusioned by the fact that mass production of the electric car failed to take place. After this he "was prepared to give his share away." After its sale to DMG, production was partly moved to Germany, where from 1909 on the truck versions were sold under the name of "Elektro-Daimler."[38]

It is not quite clear how many cars were built according to the Lohner-Porsche concept. The German historian Thomas Köppen mentions that until the end of production in 1915, 354 units were sold, the majority (272 electrics and 7 hybrids) in the period from 1906 on. Of these, 91 were delivered to the Société Mercedes Electrique. According to the already mentioned undated sales list of the DMG in the DaimlerChrysler archives that must have been drawn up well before the end of production and probably around 1908, 333 "Mercedes-Electrique (Lohner-Porsche)" had already been sold. This number included 105 heavy-duty commercial vehicles (67 fire engines and 29 buses, including 23 trolley buses), and 75 taxicabs. The remainder were sold to private customers. The combination of both sources makes it likely that the peak of production was in the second half of the first decade (figure 3.2). If we deduct all commercial vehicles from the total, we have 153 passenger cars left, that is, 46 percent. More than a third of these were sold to the nobility, high officials, and the like.[39]

In the United States the situation changed as well. Here the turning point came two years later, in 1907, but, as in Europe, before the crisis.[40] In December 1906, the Association of Electric Vehicle Manufacturers in New York (established the year before) held a special meeting for nonmembers in an attempt to boost membership and to encourage activities for the association. To this end the board had invited a special guest, Hayden Eames, who even then was known as "the apostle of the juice wagon." He presented a long speech, in-

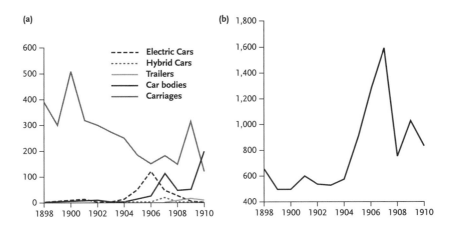

(a)

600 —
500 —
400 —
300 —
200 —
100 —
0 —

- - - - Electric Cars
· · · · · · Hybrid Cars
——— Trailers
——— Car bodies
——— Carriages

1898 1900 1902 1904 1906 1908 1910

(b) 1,800 —
1,600 —
1,400 —
1,200 —
1,000 —
800 —
600 —
400 —

1898 1900 1902 1904 1906 1908 1910

FIGURE 3.2. Sales and turnover of Lohner, Vienna. (Steinböck, *Lohner*, 29): a. number of vehicles sold; b. turnover (in thousands of crowns)

tended as a pep talk, and at the end was made a member with an ostentatious ceremony.[41]

The turning point of 1907 was preceded by a gradually crumbling EVC. From internal EVC documents, it appears that the complaints of the New York Transportation Company had not yet subsided. They were mainly about the lack of support from the parent company with the sales and maintenance of cars and batteries. A year later engineer Frank Armstrong mentioned a fast erosion of the company's reputation as the "absolute [leader] in the manufacture of electric vehicles."[42]

And so the second wave in the United States was dominated by players that deviated considerably from the EVC—even in their production-technical origin. All newcomers of any significance were coach builders who had succeeded in adapting to altered circumstances at a time when the market of luxury horse traction was shrinking. Studebaker, for instance, had been a renowned coach manufacturer when it started with electric-car experiments in 1898, after the company had signed a contract with the EVC for the delivery of 100 bodies. According to a recent American study, the company shortly afterward concluded yet another deal for 1,000 vehicles (excluding the electrical installation). In 1902, it launched its first electric car. That year twenty automobiles left the factory, but in 1904 the range was extended to gasoline models. At that moment Hayden Eames had become director of the Studebaker Automobile Company, while engineer William Kennedy, another EVC veteran, also became involved in the company.[43]

Another newcomer, the Rauch & Lang Carriage Company in Cleveland, was a coach factory as well. In 1903, according to local historiography, the company was impressed by the flourishing electric-car market and decided to start manufacturing cars. The first car was launched two years later. That year 50 vehicles were built. In 1908, after the takeover the year before of an electric-motor factory, production increased to 500 vehicles. A year later production capacity was doubled with the manufacture of 1,000 automobiles annually. The Rauch & Lang Company was to merge with Baker during the First World War. This last brand managed to sell 400 electric cars in 1905 and a year later 800. In

June 1907, Baker became known by covering more than 160 km on one charge and by beating 18 gasoline competitors in an uphill race that same month. According to *The Automobile*, this was "the first time in a long while that an electric has come to the front in competition with gasoline cars." Like Rauch & Lang, Baker was a manufacturer of decidedly luxury electric automobiles.[44]

But by far the most important contribution to the second boom of the American electric was the Detroit Electric, launched by the Anderson Carriage Manufacturing Company in June 1907, initially manufactured at a pace of 10 to 12 vehicles per month. Founded in Detroit in 1884, William Anderson's company had produced about 160,000 coaches until the turn of the century. In 1907, it manufactured 125 cars, a year later 400, another year later 650, and the year after, in 1910, production had increased to 1,600. In 1911, the company name was changed to Anderson Electric Car Company. The Detroit had been designed as a typical city car: a high and spacious interior, a bonded aluminum body, and the characteristic "tiller steering," consisting of a bar that the driver could lower as soon as he sat down. The cars sold for $1,400 to $2,400, depending on the model.[45]

The second wave of electrics led to a reinforcement of two tendencies that had already become visible during the first generation: the "miracle battery" and hybrid drive. They developed into an increasingly concrete feeling of uneasiness about the most vulnerable aspect of the electric car: its limited range in comparison to that of the gasoline adventure machine.

With regard to the first tendency, the mood among electric proponents can best be summarized by one of the many quotations that made the future of electric propulsion dependent on the search for a "miracle battery." "The development of the electric automobile," an American electrotechnical trade journal wrote in an editorial at the end of 1901, "has proceeded to that point where it is safe to say that the machine is practically perfect, if—and the if is a large one—the proper accumulator can be discovered." For all involved, it was clear that such a battery could only be realized by a different combination of electrodes, and it was Thomas Edison who seemed to be able to provide it. In 1901, Edison announced in a press campaign that he was about to put a battery on the market with an energy density that was twice as high as before. A similar initiative to perform a miracle took place in France, where Jeantaud in 1903 proposed a lead battery based on a highly controversial new material which he called "allotropic lead," of which the name ("Le Messie," the Messiah) spoke for itself. During the following two years, Jeantaud further developed his battery into a model, of which he claimed an energy density of no less than 41.2 Wh/kg and which he dubbed "E.I.t," the electrotechnical product of voltage, current, and time that together deliver "energy."[46]

Whereas Jeantaud's hopes did not materialize, Edison in 1904 put his alkaline battery on the market, the result of three years of very thorough developmental work preceded by an extensive study of patent literature. The idea for this battery type was not new (Kriéger, for instance, had taken out a patent on a similar concept in 1901). It was based on the insolubility of the electrodes in an alkaline electrolyte and the fact that this electrolyte did not participate in

One of Our Nine Rare Creations
For 1912

WHEN you buy a Detroit Electric you *anticipate* the future, because of this car's many new, exclusive and patented features.

Remember, the Detroit Electric has the *exclusive* right to use in electric pleasure cars, Thomas A. Edison's greatest invention—the Edison nickel and steel battery.

The simple operation of the Detroit Electric makes it the safest and most easily controlled car made. With one hand on the controller lever, you are absolutely master not only of all speeds, but in addition you can apply powerful brakes to the rear wheels with one instinctive, backward pull on this lever, without even touching the foot pedals.

This controller lever practically *thinks* for you. It is impossible for you to make a mistake as every movement is safeguarded. It has another advantage. It is horizontal and allows *full seat* room.

Aluminum body panels add to the strength of the body, beauty of finish and life of the c r. They do not check, warp or crack.

For those who do not care to make the expenditure necessary for the Edison battery at first, we furnish the Detroit Electric Guaranteed Lead Battery with our own warranty—the first lead battery made by and *guaranteed* by the manufacturer of the car. This makes it possible for us to guarantee the car in its entirety.

Beautiful illustrated catalog sent upon request. It tells you about the many other exclusive features of the Detroit Electric, made in the largest factory in the world, devoted exclusively to the manufacture of electric automobiles.

THE *Detroit* ELECTRIC
"Chainless"
Shaft
Drive

Buffalo
Brooklyn
Cleveland

Anderson Electric Car Co.
409 Clay Avenue   Detroit, U. S. A.
Branches
New York, Broadway, at 80th Street
Chicago, 2416 Michigan Avenue
(Also Branch at Evanston, Ill.)

Selling representatives in all leading cities.

Kansas City
Minneapolis
St. Louis

Placed on the defensive by gasoline car producers, Detroit Electric in 1912 advertised one of its "Nine Rare Creations" for the year as especially appealing to women drivers, leading later students of automotive history to the controversial view that women preferred electric over gasoline vehicles. (Private collection Hans van Groningen, The Netherlands)

the electrochemical reaction between the electrodes. In principle this required a much smaller amount of electrolyte than for the lead battery, of which the electrolyte accounted for a quarter of the total battery mass. In combination with the application of electrode materials with a lower density, the hope for a much higher energy density of the battery seemed justified.[47]

The only known commercial predecessor of the Edison battery was the copper-zinc battery of the Waddell-Entz Electric Company, founded by Montgomery Waddell and Justus B. Entz in 1889. This battery had been tested in the battery streetcars of the Whitney group in Manhattan in the first half of the 1890s. Like several others, the Swede Waldemar Jungner also began developing an alkaline battery in the early 1890s, and he took out a patent in 1899 on iron and cadmium as possible anode materials. Jungner was probably the first to replace copper as cathode material with nickel; the most important application of this he patented on 31 March 1899. Mid-1900 Edison reached this stage too: he became convinced of the importance of insoluble electrodes and focused his attention on nickel and cobalt oxide as positive electrode.[48]

When the Swedish inventor founded the Jungner Accumulator Company in the spring of 1900 and shortly after that made a deal with AFA, Edison began to feel threatened. Moreover, in the meantime a world cartel in the field of the lead battery had been formed, led by AFA and ESB. Edison, whose most important patent dated from 6 February 1901, established the Edison Storage Battery Company in May of that year. He then began a publicity campaign, which he knew how to control as no other. In scientific circles the discussion was opened by Dr. A. E. Kennelly's lecture at the annual meeting of the American Institute of Electrical Engineers in May 1901. The following decade the discussion swept with great intensity through the electrotechnical and automobile trade press of the Western world, but newspapers also carried articles about it. Some reporters praised Edison as the "Wizard of Menlo Park"; others mocked the new stunt of the "Old Man." "If the new type of battery is all that is claimed," wrote the *Electrical Review*, "it may indeed be said that the automobile problem is practically solved."[49]

Edison did all he could to keep this expectation alive. At the automobile exhibition of New York in 1901, a prototype of the alkaline battery was one of the main attractions. The following years the trade press never stopped writing about the long-distance runs with the "steel battery" that held the promise of a new future for the electric. During one of those runs, in October 1903, a distance of almost 400 km was covered between Boston and New York. Unfortunately, the charging facilities along the route were adapted to the lead battery. Only one charging station appeared to be suitable for quick charging with a high current, a property that appeared to make the Edison battery especially suited for electric tourism.[50]

In early 1903, the legendary patent fight between Jungner and Edison began. In an interview for the *New York Sun* Edison promised an energy density twice as high as that of the lead battery and a charging time half as long. A year later the *North American Review* quoted Edison as saying that a Baker of hardly 500 kg with a battery mass of not even 31 percent had covered almost 100 km on one charge during a run along "country roads, containing many grades." According to Edison, an electric car costing about $700 was now "within the reach of the man of moderate means." That year the steel battery was presented at the New York automobile show as being ready for the market.[51]

"You are not going in for high-speed machines, are you?" *Electrical Review* asked in 1903, when in the meantime the rate of production had increased to a battery a day, intended for Edison's own experiments. "'No,' said Mr. Edison, 'I have had enough of that. These machines will be geared for twenty-five miles an hour, and they will make it right along. I believe that with one of these machines I will be able to beat, or, at any rate, keep up with any gasoline machine on a long run. If they run faster than my machine on a level, I will be able to go down hill just as fast as they will dare to, and for hill climbing, the electric motor is just the thing, so I will beat them there. On rough roads they will not dare to go any faster than I will; and when it comes to sandy places, I am going to put in a gear of four to one which I can throw in under such circumstances, and which will give me 120 horse-power of torque [sic], and I will go right through that sand and leave them away behind! If the gasoline machine is stopped for any reason at all I will beat it easily. If they have no trouble whatever they may possibly beat me, though I doubt it. My machine will not break down at all.'" This set the tone for years to come in the area of the traction battery. Needless to say, expectations ran high when Edison put "type E" on the market in 1904. It was almost identical to the Jungner battery: the nickel, processed to tiny "flakes," was mixed with graphite and pounded into special, elongated "pockets" in the battery plate. The new battery especially fed the universality fantasies of the most outspoken electric proponents, such as some of the power plants. Thus, the in-house magazine of the New York Edison Company thought it likely "that the day is not far distant when electricity will also replace gasolene [sic] and steam in the long distance field."[52]

Initially the American Edison batteries seemed to perform well. Baker, National, and Studebaker adapted some of their models to accommodate the new batteries. The transportation department of the Adams Express Company was especially enthusiastic. At the beginning of 1905, 144 automobiles with Edison batteries were on the road. A year later Edison sold a couple of hundred of them to the Adams Express Company.[53]

This feverish battle for the electric adventure machine also branched out to Germany, when KAW, the most important German manufacturer of light grid-plate batteries, sent Paul Schoop, the well-known engineer and author of battery books, to Sweden. KAW let itself be persuaded by an enthusiastic Schoop to buy the patent rights for Germany, but it did not succeed in producing a lighter alkaline battery. KAW then focused on a durable version that would have a long life span, even when treated poorly, but the energy density of the new design was hardly half that of Edison's. When those closely involved began to become familiar with the real properties of the Edison concept, KAW chief-engineer Sieg calculated in 1905 that it did not even come close to the lead battery as far as costs were concerned. Whereas a lead-battery pack for a taxi-cab cost 1,500 marks, the alkaline alternative was sold for 4,080 marks. Sieg calculated 2 pf/km for maintenance costs, versus 0.75 pf/km for his lead battery. Added to a much lower charging/discharging efficiency of 50 percent (compared to 80 percent for the lead battery), Sieg arrived at a total cost of 3.5 pf/km, whereas KAW charged only 3 pf/km for complete car and battery maintenance. In Sieg's opinion even a 100 percent wearproof Edison battery could not rationally be utilized, unless at very fast discharges and very low electricity prices, so that the difference in performance was of less consequence. These were exactly the two conditions that appeared to be fulfilled more and more in the United States.[54]

The fight even became more complicated when Sigmund Bergmann, Edison's "old friend and ally," for whom the American felt an "amused admiration," also got involved. In the 1880s, Bergmann had made small parts for Edison's lighting system in his factory in New York. He returned to Germany in 1890, where he started a company for the manufacture of electrical devices, the Bergmann-Elektrizitätswerke A.G., which also sold electric-car motors. In October 1904, he established the Deutsche Edison-Akkumulatoren Company G.m.b.H. (DEAC). On the board of directors was "the famous inventor

Thomas A. Edison, with whom director Bergmann has had friendly personal relations for many years." In 1905, Bergmann began building Edison batteries in the factory of the Deutsche Garvin-Maschinenfabrik in Berlin, while constantly urging his American friend to work more quickly, especially after the Jungner patents in Germany had been legally defeated. "I think we are in a position to put lead cells out of business," he wrote, expressing the same expansionist field of expectations as some of his fellow electric vehicle proponents. Although Edison admonished him to be patient, Bergmann started to build electric cars for the Edison battery (the first were shown at the Berlin exhibition of 1907). His relations with the American inventor were somewhat disrupted when Edison heard rumors that Bergmann had closed a secret deal with AFA. Internal AFA documents show that Edison had reliable sources of information, for not only were the AFA engineers intensively busy with the "miracle battery" (and not impressed with its performance), but Bergmann indeed had approached the German lead battery colossus with a request for market segmentation. But his proposal, according to which the AFA would have to retire from the car-battery market and Bergmann would not put any stationary lead batteries on the market, was laughed off as an April Fool's joke by the parties involved.[55]

In the United States in the meantime major technical problems had cropped up: the steel battery boxes began to develop hairline cracks along the hard-brazed welding joints, while battery capacity collapsed after only a few discharges. Eventually Edison closed his factory and retired to the laboratory, only to emerge again five years later. Nevertheless, it is not surprising that the lead-battery manufacturers looked upon these developments with anxiety. The ESB, for example, was one of the first battery manufacturers that learned a lesson from the poor results of the chloride concept during the first battery contest in Paris. When ESB in 1900 released the Exide battery, the positive plate no longer contained any lead chloride, while wooden separators between the lead plates made dense packing without risking a short-circuit as well as higher energy density possible. It is likely that this important development in the area of the lead battery was prompted by Edison's simultaneous attempts to achieve a higher energy density via the same solution of a shorter distance between plates in his alkaline battery.[56]

In June 1901, the Exide batteries were installed in the electric buses in New York, after they had first been tested in the EVC taxicabs that had driven a total of 4000 km with them. Naturally, it was said "that excellent results have been obtained," but in 1906 the life span of the Exide did not appear to be significantly higher than the batteries of the Paris battery contest. In the meantime (in 1903, the year that the Brush lead battery patents expired), the negative chloride plate had disappeared and was replaced by a "box negative," copied from the AFA, which had turned the concept of the double grid-plate construction into a world standard. According to Richard Schallenberg, Edison's predictions about the range were based on the *old* chloride technology, but in view of the thoroughness with which Edison pursued his goal, conscious misleading of the over-eager public cannot be ruled out.[57]

When ESB's German cartel partner AFA at the Berlin automobile exhibition in 1905 saw that a new wave of interest in electric cars had arisen and that KAW surfed along on this wave with its lighter grid-plate batteries, the company sought advice from its American colleague. In the meantime the ESB had managed to improve its design, so that the short-circuit problem in the old lead battery caused by the active mass falling from the plates had largely been solved. Moreover, some compounds in the wooden separators appeared to have a chemically beneficial effect on the negative lead electrode. Of the two electrodes, this one had the shortest life span, because it slowly oxidized to a hard, nonporous mass under the influence of the battery acid. This process was counteracted by the lignin (much later identified as the active compound in the wood) and under certain conditions even restored. By 1904, virtually all American battery manufacturers had adopted the wooden separator.[58]

Alarmed by the German second-generation competition and encouraged by the establishment of large Berlin taxicab centers, the AFA then developed a relatively light battery in little over two-and-a-half years. The company had taken great care that the plates were not so thin that the life span would be adversely affected. Nevertheless, Adolph Müller, founder and director of the AFA, had to use all his powers of persuasion to make his engineers give up the robust large-surface concept. In 1905, the first AFA vehicle batteries were delivered. It soon became clear that AFA would have to maintain the principle of *Revision,* which it had developed into a smoothly running organization for stationary batteries. This principle was based on continually monitoring the charging state and the technical condition of the battery plates by company mechanics—a principle that had made the AFA large and financially strong during the fifteen preceding years. Without continuous maintenance, charged to the automobile owner as an amount per kilometer driven, the life span of the battery could not be kept within reasonable financial boundaries. This was caused by an annoying side effect of the wood compounds: they affected the structure of the positive electrode, a phenomenon called *Verrottung* (rot, decay). It was not until 1908 that the AFA would overcome its initial hesitance to sell its batteries without monitoring them. Until then, the KAW with its concept of very thin lead plates was the sole player on the private market (table 3.1).

The other tendency that characterized second-generation electric car technology was the hybrid propulsion system based on a combination of two different energy sources, usually a thermal one and an electric one. In a certain sense one can argue that the second generation was thematically dominated by this fourth type of propulsion, next to gasoline, steam, and purely electric. Surely in sports-loving France the interest in the *pétroléo-électrique* (gasoline-electric) was an expression of the disappointment in the limited range of the electric car. Hybrid propulsion has a prehistory that is almost as long as electric propulsion, but it became especially known due to a design by a Belgian company, Pieper. This concept garnered a following during the first decade of the twentieth century, to such an extent that the brand name "Auto-Mixte" (mixed car) became a generic name. Pieper had been known since 1867 as a gun factory in Liège. In 1889, it founded the Compagnie Internationale d'électricité and

TABLE 3.1 Application of Traction Batteries, Supplied by AFA, According to Vehicle Type, 1905–11

Year	Taxicabs	Luxury Cars	Trucks and Delivery Vans	Ambulances	Fire Engines	Total
1905	263	23	5	—	—	291
1906	71	21	4	—	9	105
1907	141	54	5	4	10	214
1908	92	63	10	9	14	188
1909	80	122	14	2	25	243
1910	292	48	53	7	51	451
1911	167	40	68	4	36	315
Total	1,106	371	159	26	145	1,807

Source: AK, January 20, 1912, 11, Varta archives.

further diversified into bicycle manufacturing in 1896. Henri Pieper, son of the founder, that same year developed a hybrid, which he perfected to the point that the combustion engine was even equipped with an electromagnetic control of the carburetor. His patents were acquired by the German company Siemens-Schuckert and by the British Daimler Company in Coventry, whereas in France the Société Générale d'Automobiles Electro-Mécaniques (GEM) put it on the market in 1908 as "Synthesis of the Automobile." AFA also occupied itself with the Pieper automobile in the middle of the first decade. It tried, for instance, to develop a special buffer battery (*Pufferbatterie*) for the hybrid propulsion system, based on the positive large-surface plate.[59]

It will not come as a surprise that Kriéger proved to be an outspoken proponent of hybrid propulsion. In France this type of propulsion assumed the character of a transition toward the gasoline car, the adventurous nature of which many desperately wanted to project onto the electric car. To increase the range of the electric car, Kriéger first added a one-cylinder combustion engine (3kW, de Dion), running on alcohol, which at the time was attracting a lot of attention as a fuel. He began to sell his hybrids in 1902, and a year later he developed a new type of "series hybrid," which had an electric system that, due to the reduction of the battery pack size, had the character of an "electric transmission"; an 18kW gasoline engine of Richard-Brasier powered a dynamo that provided both motors with energy, just as in a power station.[60]

Technically speaking, the electric car had now become the functional equivalent of the gasoline car and the emancipation from its streetcar past seemed complete. The power station, which had been the symbol of mass transportation due to the electric streetcar and an impediment to electric tourism because of its limited distribution in the country, was now hidden in the car in a miniature size. By borrowing from its deadly enemy, the electric car seemed to have escaped its limitations and had become a touring car among touring cars.[61]

In 1903, as a token of his hopes for a new future for the electric, Kriéger drove one of his hybrids to Florence, as a participant in the Paris–Florence rally. A year later the Kriéger hybrid was the sensation of the Paris motor show.

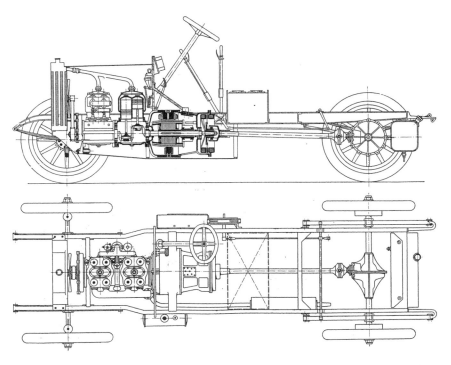

Technical structure of the Belgian Pieper hybrid. This drive train configuration allowed for the installation of a smaller, lighter set of batteries. Repolarized, the electric generator could become an electric motor for starting a gasoline engine—a feature gasoline car manufacturers later adopted. (*Zeitschrift des Vereins Deutscher Ingenieure*, 6 July 1907, p. 1064)

Kriéger would make an ultimate attempt in 1906 to introduce hybrid propulsion into the taxicab business, but especially attracted attention with a patent, in which the thermal component of the hybrid structure had been replaced with a gas turbine. The hopes of the parent company Indusmine were running high: in 1907 it hoped to achieve the production of 450 automobiles (300 electric and 150 hybrids) and an annual production of 600 for 1908. It was expected that this increase in production would be entirely due to the hybrids. A commercial success in this area did not come true, however, as the annual report of 1906 had to conclude.[62]

Lohner-Porsche had an early hybrid version as well, which was put on the market by Jellinek via a separate sales company, the Société Mercedes Mixte. The Viennese company built fifty-two hybrids between 1898 and 1910, including two buses. The German fire brigade also briefly bet on the hybrid horse shortly before the First World War. But it looks like this concept did not sell more than 100 vehicles in Europe up to the war, due to its high price and its complicated structure.[63]

Yet the disproportionate attention in the trade press on this type of propulsion was an important signal. The hybrid as transitional form, as a technical compromise, was attractive from two vantage points. For, on the one hand, the system could be considered an electric car with an enlarged range by virtue of the addition of a combustion engine; on the other hand, the system could be considered a gasoline car, in which the mechanical transmission (in this phase still presenting a lot of problems) had been replaced by an electric one. As early as 1903, Jenatzy proposed a "progressive transmission"—actually an electromagnetic clutch-cum-transmission—that was an early prototype of an electric continuously variable transmission. In an extreme version, especially promoted by Kriéger, the battery had been reduced to a small starter battery. This structure could no longer be called a hybrid; it was a gasoline car with a purely electric transmission. This version was the most extreme example of a general distrust of the lead battery, also among the gasoline-car manufacturers who around this time largely changed to magneto ignition. The most popular version, however, was the "electricity-generating automobile" (*voiture automobile électrogène*), a concept that today is known as a series hybrid. In this structure the battery was still a full-fledged second energy source—although smaller and lighter—that was kept charged by the combustion engine, connected to a separate generator. By reversing the action of the generator it was possible to start the car electrically as well. Kriéger's hybrid, for example, had a gross weight of 1,585 kg, of which the battery pack accounted for only 25 percent. It could go 45 km/h on a level road. The hybrid, in other words, enabled electric tourism with a much more acceptable battery-vehicle ratio. Thus, these pioneers experimented with concepts that would partly be adopted many years later by the gasoline car, including the electric starter motor.[64]

AT THIS STAGE the first attempts appeared in the trade journals, often on the occasion of an automobile exhibition, to categorize cars—not an easy task, as the structural flexibility of the electric drive train offered so many possibilities. This was especially true for the first generation, when cars were so different that any attempted categorization would have been futile. For the second generation, however, categorization started to make sense. Categories were generally based on the number of motors and the place of the driven wheels.[65]

After the EVC debacle, systems with one central electric motor were especially popular in the United States. Just as with the gasoline car, these systems always powered the rear wheels, due to the problematic combination of steering system and drive shafts. In America this solution was chosen not only because of its lower production costs, but also because of its lower weight. Consequently, most electric carmakers soon abandoned the two-motor system with its higher mass and its lower total efficiency. According to *The Horseless Age*, the European electric cars "as a rule" not only had "a very clumsy appearance," but they also had a much more complex structure than the American electrics.[66]

The two-motor system dominated the European market, especially due to the special concepts of Kriéger and Lohner-Porsche, but the German brands Scheele and Stoewer also applied it. The French brands Electros-Cardinet and l'Electrique applied this concept to the front wheels. KAW, the Abam, and, after Jellinek's interference, Daimler (Lohner-Porsche) applied the two-motor system on the rear wheels.

The Lohner-Porsche concept garnered a significant following during this period. By raising the number of electromagnetic poles of the armature, Porsche had in fact created a slow-running electric motor that showed a satisfactory efficiency. Because the Berlin taxicab company Bedag and the Berlin fire brigade opted for this system, the competition also started to work on the wheel-hub motor. Dr. E. Sieg, chief-engineer at KAW, tried to dampen the enthusiasm somewhat in 1908, although he praised the concept because of its "brilliant construction." Sieg had noticed a strange phenomenon in such motors, especially in the Lohner-Porsche concept with its so-called internal poles. As soon as these motors had to function at lower speeds, not only did energy consumption increase strongly, but they also could no longer be overloaded, a property that up until then had been one of the distinct advantages of electric propulsion systems. Because of this, such motors might suddenly come to a halt on even the slightest slope, "like an overloaded gasoline engine." If a wheel happened to be lodged in a hole, there was no way to get the car moving again. Thus, the electric motor in its technical properties began to resemble the engine with internal combustion. In Sieg's opinion, the electric car constructed according to the Lohner-Porsche concept was only suitable for a "smooth ride on level ground." For a hilly city it was better to choose electric cars with fast-running motors.

The success of the wheel-hub motor was so overwhelming, however, that in the Berlin automobile show of 1905 almost all electric-car manufacturers introduced some version of the wheel-hub concept. Even KAW started to manufacture such a system, but it used the external-pole concept, for Sieg had demonstrated that for an acceptable efficiency the minimum speed of the wheel (and so of the motor) had to be 150 rpm. This led, because of the wheel size, to a maximum motor diameter of about 45 cm. This had forced Porsche to make his system wider in order to accommodate it in the wheel. By now opting for the concept with external poles, Sieg could lower the vulnerability of the motor considerably.[67]

Shortly after 1905 most of the people involved concluded that electric-car technology had grown into a full-fledged, mature branch. It differed from gasoline-car technology in a number of important respects. It was also plagued by some technical weaknesses. Most electric vehicles by this time were equipped with a separate chassis, just as their gasoline counterparts. But in contrast with the gasoline car, the body did not need as many adaptations, because far fewer "essential parts [were] fitted to the body." Better batteries enabled the construction of lighter vehicles, because more energy could be extracted from the same battery mass. The new KAW type W extra battery, for instance, which appeared on the market in 1905, enabled a decrease in vehicle mass of 350 kg due to its higher energy density, at the same range. The culmination of this development was the light electric car by Siemens, introduced at the Berlin automobile exhibition of 1907. In a way the Siemens concept (according to KAW director Sieg, with a chassis that was too light) was ahead of its time: its designers had taken over so many characteristics of gasoline car structure that it already foreshadowed third-generation technology. But for most of the second-generation electrics, vehicle speed had increased to the point that in the city there was hardly any difference compared to the gasoline car.[68]

Yet the electric was considered by many as a "decadent car," which confirms our earlier assessment of the gasoline car as the best culturally adapted to the fin-de-siècle. The French *La Vie Automobile* published a review of it in 1909, in which the magazine pointed out the "psychological argument" that "for the masses, the gasoline car is reassuring, despite its complexity and despite what one has dared to call its barbarism." The magazine explained this by the "essentially material" character of this vehicle type, of which the small defects could be seen, felt, heard, and even smelled. Only the magneto-ignition was not included in this judgment, because it was electric "and it either runs or it doesn't." The electric car was an enlargement of the "cruel enigma" of electricity: "one needs an intelligence and an education of a degree above that required for a gasoline vehicle." So, it was not a coincidence that the review finished with an analysis and a recommendation of the hybrid: the combustion engine counterbalanced to some extent the decadent comfort of electric propulsion. As proof of this automobile-cultural gap, embedded in the technology, the magazine published the results of a poll of its readers, complete with a list of 3,437 automobiles. The results showed that Renault, with 385 respondents, was the most popular vehicle among the readers, followed by Peugeot (331), de Dion-Bouton (282), and Panhard-Levassor (233). Other brands often mentioned were Bayard-Clément, Mors, Berliet, Darracq, and Lorraine-de Dietrich; Kriéger, Mildé, and Jeantaud are not found on the list. In the prevailing automobile culture, there was no place for the decadent electric propulsion system.[69] Here again we are confronted with the contradiction, the paradox even, between the manufacturers' and the engineering community's enthusiasm for the reliable electric on the one hand, and the potential users on the other, who modeled their field of expectation after that of the "barbaric" gasoline car culture.

IN EUROPE, the center of the electric-car culture shifted to Germany, leading to the end of the electric passenger car in France. An important factor in its demise was the fragmentation of French battery manufacture and the lack of cooperation between these manufacturers and the electric carmakers. Whereas Jenatzy had already abandoned electric propulsion in 1903 to pursue a career in racing with gasoline cars, and Jeantaud had closed his company in 1906 (and committed suicide shortly after), the direct victims of the crisis were Brillié, the supplier of the Paris bus company (see chapter 5), and Kriéger. In 1910, Védrine was closed down, as well as Charles Mildé fils et Cie.[70]

The fall of Kriéger displays striking parallels with the American EVC disaster. In his analysis James Laux speaks of "manipulations," because the founding of the new Kriéger companies in 1905 and 1906 was not only suggested by "insufficient sales of electric cars and overly optimistic expansion," but also by "the troubles of the Indusmine bank." As far as the first failure factor is concerned, this must have been less important than Laux assumes, judging from the licensing revenues from Italy and particularly Germany—an aspect he ignores. As far as the second factor is concerned, the Indusmine bank, which also had complete control of Kriéger, was the first to stumble over the crisis. A Spanish mine, a British manufacturer of textile machines, a Russian cement factory, and especially a South African soap producer caused heavy losses, which the bank tried to offset by borrowing from Kriéger, "one of its more solvent creations." The Indusmine companies were so closely intertwined that it was impossible to establish who pushed whom over the edge. The *Globe,* a financial newspaper, blamed the catastrophe on disorderly management. At any rate Indusmine was the first to go, in March 1907, followed by Kriéger on 11 January 1908. At that moment the Kriéger shares of nominally 100 francs were quoted for only 8.5 francs. The parallel with the EVC does not stop here, for Kriéger did not completely disappear from the scene: in 1915, Kriéger and Mildé merged and showed up again in the period between the wars. How many Kriéger automobiles were produced until the declaration of bankruptcy is not known. A later French source mentions 400 in France and 2,000 under license abroad. This seems an underestimation of the French share and an overestimation of the German share. But there is no doubt that the Kriéger and its clones were among the best-selling electric cars in Europe.[71]

In Germany the crisis expressed itself in a different way. Important electric-car builders extended their production to gasoline propulsion systems, such as Siemens (Protos) and Namag. For fiscal year 1909–10 the Namag again suffered a loss, this time of almost 312,000 marks, six times as much as the loss of two years earlier. By the end of 1913, the company—constantly in the red due to a continual overcapacity—approached automobile manufacturer Hansa-Automobilwerke, located in Varel. On 9 June 1914, the merger took place and resulted in Hansa-Lloyd, which continued to manufacture electric cars until well into the 1920s.[72]

A comparable development occurred in the United States. The EVC instituted insolvency proceedings on 10 December 1907, but two years later the administrators were subpoenaed to demonstrate "why an offer of a reorganization committee to take over the entire assets except the cash in the hands of the receivers, should not be accepted." One of the members of this reorganization committee was Herbert Lloyd, representative of the Columbia Motor Car Company in Hartford, Connecticut. He took over the EVC furnishings and paid $430,000 in cash for it.[73]

In 1909, some 4,000 electrics were purchased in the United States. According to a different source, 3,639 of these cars were produced that same year, versus about 82,000 gasoline cars. The entire stock of electric vehicles numbered 20,000. Three years later, in 1912, Studebaker stopped the manufacture of electric cars. It had produced a total of 1,800. That year Studebaker had become the third gasoline-car manufacturer of the United States, producing 28,523 units.[74]

THE final phase of the period described here as well as the early phases of the following period was a time of intensive automotive-technological change.

So far, this development has been so poorly reconstructed in automotive historiography that it is hard to draw a general conclusion about the *adequacy* of electric propulsion. In any case, however, it is clear that the revival of electric vehicle production and use was the result of a final upswing in the luxury car segment, and that the crisis of 1907 forced the gasoline car manufacturers into diversification and the development of more economical and reliable cars that were also meant to function within the city. Although the crisis brought some electric car manufacturers to bankruptcy, it enabled newcomers (especially some well-known carriage makers in the United States) to enter the market with second-generation technology. Generally characterized by much borrowing from gasoline car structure, this second-generation technology also benefited from major improvements in battery technology, which seemed to turn the electric into a well-equipped adversary of the gasoline "intruder" in the city. For the most ardent electric vehicle proponents, however, this new battery was just a stepping-stone to a general competition with the gasoline car, also outside the city. Together with the other alternative of hybrid propulsion, the "better battery" for them was just another means by which to conquer the whole automotive field.

How did this second generation perform on this battlefield? Like the first generation, this is very difficult to establish as far as the application field of the privately owned passenger car is concerned. But just like this earlier generation, we can look at other, much better-documented application fields for an answer to this question. Therefore, in the following two chapters we will try to give a detailed account of the development for two distinctive and well-documented applications, the taxicab and the freight and delivery van. Although these two studies are independent and specifically contribute to the historiography of the vehicle types concerned, they can also be used as *case studies* to get a grip on the problem of the "failure" of the privately owned electric passenger car and to help document the development of the gasoline car into a more or less reliable city car without sacrificing its already well-established touring characteristics. To understand this development toward multifunctionality and universality, we are especially interested in the role of automobile maintenance to counteract the lower reliability of the gasoline propulsion system, compared to its electric alternative.

The Trojan Horse:
The Competition for the Taxicab Market

And thus the moment cannot be far off when the urban cab
horse also disappears, just as the horse streetcar has
disappeared or has been replaced by the electric version.
—*Chief-engineer Ansbert Vorreiter, 1904*

THE TAXICAB DISASTERS in London, Paris, and some big American cities had
discouraged the automobile industry, the engineers, and the financiers from
taking new initiatives in this area. It took years before such large-scale experiments got a new chance. Therefore, it was not a coincidence that these experiments were carried out in cities where such debacles had not taken place: in
a few large German cities and in Amsterdam, The Netherlands. In the meantime, the gasoline-car industry seized the opportunity to enter this market
niche. Largely neglected by automotive historiography, it was under the guise
of the taxicab that gasoline engine propulsion entered into the urban application field and it was in this field that the gasoline vehicle learned to become
"civilized" and throw off its image of unreliability. To do so, it had to take over
the maintenance concept already developed by the electric taxicab community.

To those involved it had already been clear before the crisis of 1907 that the
taxicab market was crucial for the further propagation of motoring. "As long
as there is no 'inexpensive automobile,'" a German automobile magazine
wrote in 1906, "the motorized taxicab is the type of automobile that makes
available the advantages of motorized traffic also to the less affluent classes."
The taxicab was thus a new means of propaganda for motoring, as the successor of—or at any rate in addition to—the sports car from the first period.
Moreover, the deployment of the taxicab in the tough city environment was regarded as a realistic test of the usefulness of the automobile as a means of
transportation.[1]

The new initiatives were not simply a repetition of the earlier attempts. They took place under considerably altered circumstances of an external as well as an internal nature. Externally, the gasoline car was in the process of liberating itself from its exclusive sports function, and there were signs of a saturation of the luxury-car market. Furthermore, the huge horse-cab fleets in the various capital cities offered a welcome opportunity to compensate for the cyclic, seasonal character of automobile sales by creating a guaranteed sale.

On the constructional level, important changes had occurred as well. As far as the electric vehicle was concerned, this mostly had to do with the battery and not so much the vehicle construction as such. It is remarkable that the two most important electric cabs in this period (Kriéger's and Lohner-Porsche's) had been borrowed from the first period, without major conceptual changes. But especially in Europe, the concept of the light lead battery was further developed, partly independently (France, KAW in Germany) and partly under the influence of the Exide design (AFA). Shortly before the crisis of 1907, such batteries were launched at the same time as—and stimulated by—the emergence of new electric-taxicab initiatives. Thus, the term *second generation* with regard to the taxicab can be almost entirely attributed to the improved battery technology and tire concept. This resulted in an electric cab of a new quality. The improved battery and tire stretched the boundaries of the technical field of electric propulsion so far that now (compared to the failures of the first generation) a realistic field of application of the electric taxicab became feasible.

A Market Niche: The Emergence of the Gasoline Taxicab

In an atmosphere in which the proponents of electric propulsion had a deep-rooted fear of new failures, the gasoline car now had a chance in the city. The role of the French automobile industry was decisive here. In 1907, Renault became the largest French manufacturer because of the sales of its taxicabs. Renault's success was particularly due to the Compagnie française des Automobiles de Place (CFAP). After a year of thorough research of the taxicab market, this company was founded on 4 March 1905, with a capital of 5 million francs. Its founding was the initiative of the Société française d'Études et d'Entreprises of the Mirabaud Bank in Paris. The Société d'Études commissioned a Geneva bank to analyze earlier taxicab experiments, the results of which appeared in a detailed report. Particularly the decision of the German Allgemeine Electricitäts Gesellschaft (AEG) to introduce a *gasoline* taxicab fleet gave the rapporteurs food for thought.[2]

The CFAP's comparative experiments lasted a year, and eventually the choice was a Renault coupe. Renault's car then was subjected to a test of several months, which also enabled a calculation of the revenues to be expected. The experiments also led to a considerable structural change in the car, which would later be known as type AG. All parts of the chassis were characterized by *une interchangeabilité absolue* (absolute interchangeability). Elements that might distract the driver from his task of driving his cab as economically as possible had all been removed. The driver had only two pedals at his disposal: one for controlling the mixture quantity (via a gas valve in the inlet manifold of the engine) and one for declutching and (by pressing the pedal down farther) braking. There was no battery. The spark-plug ignition operated on the basis of a high-voltage magneto system that Bosch had put on the market a few years earlier. The AG was provided with a two-cylinder engine with a cylinder volume of 1.06 L, a power of almost 6 kW, and a maximum speed of 30 to 40 km/h. In 1914, the AGs were to comprise the majority of the well-known *taxis de la Marne*.[3]

Gone were the days when a gasoline car was associated exclusively with big engines, lots of power, and high speed, and was meant to function as much as possible at a constant speed during touring and racing. The seemingly inherent unreliability of the gasoline car's engine and drive train did not seem compatible with the increasingly motorized urban traffic with its stop-and-go-character. Yet the organization of the Parisian taxicab business facilitated the gasoline car's acceptance. The drivers (all former coachmen of horse cabs, as ordained by the police) had to pay a 300 franc deposit and always drove the same vehicle. Their income, apart from tips, came mainly from a percentage that progressively increased with the proceeds. In Levallois-Perret, west of Paris, a 650-car garage was built. There the drivers were provided with gasoline for 150 km, which they had to buy from the company out of an extra allowance from the gross daily proceeds. The drivers were not allowed to do any repair work themselves, except on the ignition and the pneumatic tires. In case of a breakdown, they had to drive back to Levallois. If this was impossible, they had to call the garage for a tow truck, so that the specialists in Levallois could take care of the technical problems. Every day the driver had to hand in a status report on his car. The pneumatic tires were checked daily for flaws. Every forty days the chassis was inspected thoroughly and the engine completely taken apart. Apart from this remarkable maintenance, the whole setup of the business resembled that of the competing electric cab business in every detail.[4]

On 9 December 1905, the first Renault taxicabs became available from the Compagnie des Autos Rouges, as the CFAP was sometimes called because of the reddish-brown color of its taxicabs. The fare initially was as much as 33 percent higher than that of the horse cab, but eventually it was only slightly higher. The following years the fleet expanded quickly (figure 4.1). In 1914, the CFAP merged with the Compagnie Autos-Fiacres (CAF), which contributed another 820 vehicles, mostly Renaults.[5]

In 1910—a year before the number of motorized taxicabs in Paris was to double again and would surpass the number of horse cabs (figure 4.2)—the French journalist F. Girardault gave an overview and an analysis of the taxicab market. The total number of vehicles in the nonpublic transportation sector in the city of Paris amounted to more than 100,000 at the time. It included 4,243 taxicabs and rental cars, nearly half of the 9,350 private cars that were also present. The remainder still had horse traction: more than 12,000 horse cabs and rental carriages, 9,500 private carriages, 1,800 special vehicles, such as ambulances, mail coaches, prisoners' wagons, and street-cleaning wagons, and almost 63,000 commercial vehicles, mostly two-wheeled. Of the 3,377 *stationing* taxicabs he counted, 1,800 units, or 53 percent, were of the Renault brand. Then followed Bayard-Clément (11.8 percent) and Unic (6.5 percent). In the

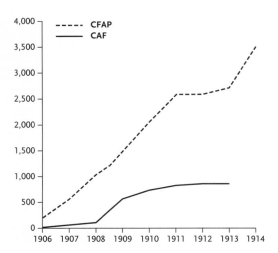

4,000
3,500
3,000
2,500
2,000
1,500
1,000
500
0
1906 1907 1908 1909 1910 1911 1912 1913 1914

- - - - - CFAP
———— CAF

FIGURE 4.1. Growth of
the gasoline-taxicab fleet
of the CFAP and the
CAF in Paris, 1906–13;
after the merger in 1914,
the firm counted 3,550
vehicles. (Boudou, "Les
taxis parisiens," 293)

middle bracket, with shares of around 2 to 3 percent, were the brands of Dela-haye, Chenard & Walcker, Duhanot, de Dion-Bouton, and Kriéger.[6]

Bayard-Clément was especially in use at the CGV. The largest horse-cab company in Paris waited until 1907 before it dared to get involved in motori-zation again, as it had hardly recovered from a financial deadlock after the World Exhibition of 1900. Without doing much research, the company this time opted for gasoline propulsion. In July 1907, the first dark-blue CGV gaso-line taxicab appeared in the streets of Paris. A year later the CGV operated about 100 of them and 400 in 1910. A year later the company had 1,200 gasoline cabs, with 400 more on order, distributed among 4 garages. At the same time it still owned 16 stables that accommodated 3,500 horse cabs.[7]

The Kriéger taxicab, with 100 units comprising 3 percent of the Paris stock, was the only one with an electric transmission, according to Girardault's list. A four-cylinder Brasier engine took care of the Kriéger's hybrid drive. The en-gine was coupled to a generator that provided the two electric motors at the front wheels with electric energy. Kriéger manufactured the car for the Com-pagnie Parisienne de Taxautos Electriques, founded on 28 January 1907. This company had placed an order for 150 cars and announced plans for the opera-tion of 1,000 units. The price of the car was 11,000 francs, almost triple that of the run-of-the-mill Renault. A specially established company, the Société des Garages Kriéger et Brasier, took care of the maintenance of the cars. It was located in the center of Paris on the expensive rue La Boëtie. For easy mainte-nance the entire power source was placed on a subframe. If necessary, it could be replaced out on the street. Unfortunately, business data for Kriéger's hybrid fleet are not available, but it is certain that the move to the larger series did not succeed. Soon after, the Kriéger company went bankrupt, which sealed the fate of this unique enterprise in early taxicab history.[8]

Between 1905 and 1910, nineteen other large cab companies were estab-lished, each operating from a few dozen to more than a hundred taxicabs; at least three of them had electric cars. Many of these companies, as well as quite

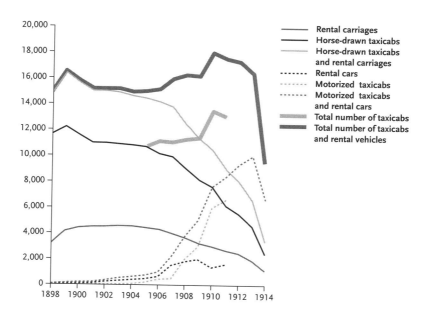

Rental carriages
Horse-drawn taxicabs
Horse-drawn taxicabs
and rental carriages
Rental cars
Motorized taxicabs
Motorized taxicabs
and rental cars
Total number of taxicabs
Total number of taxicabs
and rental vehicles

FIGURE 4.2. Horse
substitution in the
Parisian taxicab fleet,
1898–1914. (*Annuaire
statistique de la ville
de Paris*, 1898–1914)

a few of the horse-cab businesses established in an earlier period, failed in the crisis of 1907. The "spring cleaning" among the motorized cab companies was a direct consequence of the gradual decrease in the net profit per car. As far as the automotive-technological side of the business is concerned, the costs of the tires largely caused the low proceeds.[9]

The first motorized fleets of taxicabs and rental cars in Paris appeared to be the elixir of life for the early automotive industry. In 1907, the taxicabs accounted for almost 10 percent of the automobile fleet in the capital city. In 1911, this share appeared to have gradually risen to as much as 44 percent. By concentrating on the utilitarian vehicle, Renault had been able to cope with three strikes in the spring of 1906 and the crisis of the following year without too many problems. In 1909, two-thirds of the 3,000 Paris cabs were of the Renault marque.[10]

The French automotive industry received yet another boost when the same group behind the CFAP started a similar company in London. That took place in 1907 with the foundation of the General Motor Cab Company, which immediately ordered 500 Renault taxicabs of an improved type. An American observer of the London taxicab business estimated the Renault's costs of operation at $ 408.81 (excluding interest and depreciation) for half a year. As much as 33 percent of this amount was meant for gasoline, 27 percent for the (solid) rubber tires, and 18 percent for repair of the chassis. In 1909, London had 2,400 motorized taxicabs, half of them Renaults. A year later the turning point in the dominance of horse traction to motorized propulsion was reached, a year earlier than in Paris (figure 4.3).[11]

Attempts to revive the electric taxicab failed. Not until the middle of the war did the electric cab surface again, as a "crisis car." In the beginning of 1917, London had 191 electric taxicabs, distributed among four companies. Little is

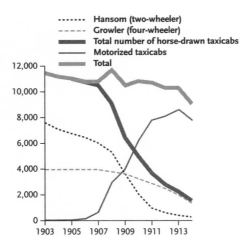

Hansom (two-wheeler)
Growler (four-wheeler)
Total number of horse-drawn taxicabs
Motorized taxicabs
Total

FIGURE 4.3. Horse substitution in the London
taxicab fleet, 1903–14.
(Georgano, *A History of
the London Taxicab*,
59–60)

known about these fleets, but the largest, the Heilford Street Motor Company
Ltd., operated 120 of them, followed by the Krieger Electric Carriage Syndicate
Ltd., with thirty-six Kriégers.[12]

IN CONTRAST TO GREAT BRITAIN AND FRANCE, in the United States the
electric taxicab did not entirely disappear from the streets. For it was the New
York Transportation Company, heir of infrastructure and furnishings of the
EVC, which introduced the phenomenon of the taximeter in New York. Ironically, the company was also one of the first to introduce a gasoline cab fleet in
the United States. That may partly be due to a fire in the charging station in
January 1907, which destroyed 200 of the more than 650 cars. The company
ran into a wall of incompetence and unwillingness when it approached the
American automotive industry in order to replenish its shortages with a reliable and sturdy landaulet. It then imported a trial Renault, but eventually decided in favor of the French Delahaye, ordering fifty. On 1 June, the company
equipped the first twenty-five electric cabs with Cosmos taximeters. Because
none were produced in the United States, they were imported from Germany.[13]

In 1904, Richard W. Meade had assumed the leadership of the company.
He would serve as general manager and president for almost fifteen years; after
that he became president of the Detroit Motorbus Company and of the People's
Motorbus Company of St. Louis. Until 1907, electric vehicles had been the
only part of the company that regularly produced an operating surplus. When
Meade, in 1910, finally committed to standardize his cab fleet by using de Dion-
Bouton gasoline vehicles and prepared to liquidate the remaining electric rolling stock, he explained his strategy in a special taxicab issue of *The Horseless
Age:* "with the exception of good management, there is probably no factor of
greater importance in contributing to successful taxicab operation than a standardized equipment." Until then, "every form of equipment" had been used;
"the embarrassment of riches was complete and its own undoing . . . Half the
equipment was always out of season." For a fleet owner, apparently, the structural flexibility of electric propulsion was no advantage at all; instead, gasoline

propulsion forced cab operations to standardize around a single vehicle design. The last EVC electric-cab model Mark 67 Landaulet was "the most serviceable electric carriage ever built" and was "unexcelled for motor cab purposes, but for the unquestioned superiority of the gasoline cab on account of its unrestricted radius of action and ability to use pneumatic tires."

Meade was comparing apples and oranges here (at the same time the Amsterdam Atax company was operating electric cabs with pneumatics without any problems), but his assessment of the promises of gasoline propulsion in an internal company memo were telling nevertheless: "There is no possible doubt that, so far as accidents are concerned, the gasoline cab is a far better risk than the electric cab, and for this reason the amount of our accident and damage claims is decreasing as the electric cabs are eliminated. This is directly contrary to the belief of every accident insurance company representative with whom I have ever discussed the matter, as they believe that the higher speed of the gasoline cab makes it a more potent engine of destruction than the electric cab with its lower power and speed. The secret lies, however, in the lighter weight and better control of the gasoline cab and the bad skidding propensities of the electric cab on solid tires."[14]

Not only the American automotive industry, but the horse-cab companies as well were notably absent at the start of the second wave of motorization of the taxicab. New operating companies filled this vacuum in the market. In New York the New York Taxicab Company was especially active. It was a subsidiary of the New York Motor Cab Company, the shares of which had been equally distributed among the London and the Paris taxicab group on the one hand and a group of rich New Yorkers on the other. Harry N. and Walter Allen, importers of Mercedes and De Dietrich, had at about the same time established the Motor Carriage Company. In the fall of 1907, they merged their company with that of the French-English group. They did not choose the Renault, but the competing marque Darracq. In May 1908, 300 Darracqs with taximeters were operative in New York, while another 500 were on order. In 1907, one-third of the 1,800 cabs in New York were Darracqs.[15]

In the beginning of the winter of 1907, not a single American automobile manufacturer had launched a taxicab model. Unlike Paris, where Renault's initiative dated from 1905, but like London, it was the "Panic of 1907" that forced these actors to move with the times. The E. R. Thomas Motor Company in Buffalo was one of the first (it was behind the foundation of the Federal Taxicab Company in Washington). Soon it was followed by the Ford Motor Company, the Cadillac Motor Car Company, and a dozen other brands. Now American taximeters began to appear on the market, some after a "study tour" to Europe by the potential manufacturer. A flood of operating companies followed in their footsteps, but of the fifty that were counted in New York City in 1911, only eight were left in 1913.[16]

THE PROMINENCE OF THE EUROPEAN AUTOMOBILE in the American taxicab fleet is an indication of the unreliability of the early American gasoline car, despite the greater emphasis on a utilitarian use of the early passenger car as compared to Europe. In 1911, 3,000 of the 5,000 American taxicabs were still European, mostly supplied by French brands, such as Renault, Darracq, de Dion, and Delahaye. The American Taxicab Company operated almost 4,000 of them, distributed among ninety locations.[17]

Had the gasoline car become reliable enough shortly after the crisis of 1907 to replace the electric cab at the New York Transportation Company and cause a wave of new gasoline cab fleets in New York and other cities? The New York Transportation Company's bankruptcy belies this impression. In 1912, it sold its "stations, rights, machine shop and other property," including more than 600 taxicabs, to the Connecticut Taxicab Company. However, this bankruptcy—and others—can also be attributed to the daily practice of big fleet management. Due to the sheer size of the American big cities, distances of 160 km per twenty-four-hour business day were quite common (very close to the distance an electric car might have covered by exchange charging once a day). As the tradition of "stationing" was poorly developed, however, a re- markably large part of the distance was driven as "empty kilometers." The five large gasoline-cab companies in Chicago did not have garages in the suburbs (in contrast to New York), and they reported a "zero occupancy" for 60 percent of the shorter rides. It was remarkable—and in contrast to European prac- tice—that this figure decreased to 40 percent on longer rides, probably be- cause suburbs in the United States were established earlier than in Europe. Longer rides thus increased the chance of finding a client who wanted a ride to the city center. The Pittsburgh Taxicab Company had an average zero occu- pancy of 50 percent. This percentage decreased by a quarter when the Ameri- can Telephone & Telegraph Company installed telephones in all quarters of the city, which the cab drivers could operate by means of special keys. Never- theless, many companies did not dare to operate a twenty-four-hour service. Most restricted themselves to an eleven- or twelve-hour service, the effective use of which was only about 65 percent (including waiting at the cab stands). Some companies that opted for a twenty-hour service not only changed the driver at each ten-hour shift, but also the cab.[18]

Cost analyses of the American taxicab companies are not available, but the wave of bankruptcies in New York before the war seems to indicate the same tendency that we observed in Paris. There the proceeds had begun to drop dras- tically right from the beginning, due to the increasing mutual competition, the new competition by the CGO bus company, and a few big strikes and skyrock- eting rubber prices. It became clear quite soon that such large fleets could sur- vive only if they were strictly run and maintained by specialists. Apart from the drivers' salaries, the rubber tire was a critical factor in the operating costs. The technical field of the gasoline cab was perilously narrow in this phase.

The question now is, to what extent the electric cab, too, was confronted with this *fleet-size problem,* and whether—under different circumstances—it would succeed. The second-generation taxicab in Germany is a good source for studying this question. There, both propulsion alternatives were present in sufficient numbers and were already thriving before the First World War, al- lowing us to compare both alternatives in detail.

Rivals in a Growth Market: The Second-Generation Electric Taxicab in Germany

In European countries where the electric cab had not proved to be such a spectacular failure, small-scale experiments were set up at the end of the first generation. All these experiments failed on technical grounds. Apparently, these small-scale failures did not have such severe after-effects, because when the infant German automotive industry (earlier than in France) discovered the taxicab as an anticyclic instrument, Berlin in particular seemed to be an attractive market also for electric propulsion. This was related to the unique monopolist character of its taxicab business, which was strictly regulated by the authorities. In 1898, the Berlin police announced a *Nummernsperre* (stop on the issue of licenses), which was not lifted until 1904. Moreover, the introduction of taxicabs with combustion engines was initially frustrated, because the fuel had to be methylated spirit. This byproduct of the agricultural industry was more expensive and harder to use as an alternative to gasoline, especially because of its deviant vaporization properties.[19]

It was all the more remarkable that the electrotechnical concern AEG was the first to discover the potential gasoline-cab market. Although the American Pope group had approached the company earlier about the manufacture of electrics, the AEG opted for the gasoline version by taking over the Kühlstein Wagenbau Charlottenburg and the Allgemeine Automobil-Gesellschaft in 1901. The latter company was rechristened as Neue Automobil-Gesellschaft (NAG), an analogy of AEG. On 18 October 1903, the NAG launched its first taxicab in Berlin. A year later the NAG established a separate operating company, the Automobil-Betriebs-Gesellschaft m.b.H. (Abag). It started with twelve NAG cars and extended its fleet the following two years to 48 and 73 units, respectively. The introduction of this fleet had become possible because of the abolishment of the *Nummernsperre* as of 1 October 1904. This led to an increase in the number of motorized taxicabs as well as horse cabs. Whereas Berlin had only 57 motorized taxicabs in 1904, in 1906 the number had increased to 245.[20]

The opening up of the taxicab market by the Berlin authorities gave the electric cab a boost. Several factors played a role here that also had consequences for the situation in other German cities. In contrast to London, Paris, and the American EVC cities, in Germany no impact in the area of electric propulsion had taken place. Despite the much-discussed failure of the Berlin battery streetcar, there was no question of an absolute breach between first- and second-generation electric cars. This was especially due to the activities in and around Cologne—the cradle of the German electric car—and initially occurred completely without the involvement of the largest battery producer AFA. In Cologne a modest taxi fleet was kept in operation, which was mainly supported by the Kölner Accumulatoren-Werke Gottfr. Hagen (KAW). This initiative led to the foundation of the Allgemeine Betriebs-Gesellschaft für Motorfahrzeuge (Abam). Under the Kriéger license it began to build taxicabs and applied the KAW's new W battery. After half a year the operating costs (calculated on an annual basis) amounted to 5,200 marks versus 3,200 marks for horse traction. Yet each car—with a daily route of 60 km and 340 effective days of operation—seemed to promise a much higher yield from the 20,400 kilometers covered per year. Soon Abam cabs were driving around Cologne, Düsseldorf, Dresden, and Frankfurt am Main.[21]

By the end of 1904, the Abam and KAW managed to interest the Berliner Fuhr-und Automobil-Wesen Thien GmbH in a new venture. To this end Thien, which had experimented with all sorts of propulsion types since before the turn of the century, set up the Elektromobil-GmbH with a capital of 125,000 marks. It would establish a charging station on the Schiffbauerdamm and purchase the cars from the Abam and the batteries from KAW. In the following months a "trial operation" was started with six Kriégers, built in Cologne. There was no exchange charging, but battery maintenance cost 2.5 pf/km. Whereas the gasoline NAGs were in use day and night in a two-shift system, the Kriégers were especially deployed at night because of their higher allowed fare. From the beginning of November 1904 until the end of January 1905, 27,766 kilometers were driven during 511 effective days of fleet operation. The average ride length was 5.7 km, 30 percent of which was empty. At an average daily distance of 60 km (with a peak of 117 km), gross proceeds of 30.8 pfennigs per vehicle kilometer were feasible, yielding a net profit of 1.8 pf/km. Next to the drivers' wages, the rubber tires formed the largest expense item (19 percent of the net proceeds), followed by cost of electricity (17.5 percent), depreciation (15.5 percent), and the battery-maintenance subscription (12.5 percent).[22]

Ansbert Vorreiter from Cologne, who published these figures, also managed to collect some data from two larger competing gasoline-cab companies. There the number of effective days of operation appeared to be only 260 per year, but the average daily route (with 2 drivers) was 159 km. Excluding the drivers' share, the tires now accounted for 19.4 percent of expenses and the fuel for 28 percent (methylated spirit) or 24 percent (gasoline). Depreciation accounted for 21 percent of expenses. Although the operating expenses were 1.3 pf/km lower than for electric propulsion, the distance-related costs were 0.25 pf/km higher. Vorreiter explained this from the fact that due to its higher speed the gasoline car was selected for longer distances, which resulted in a higher number of "empty kilometers." From these data Vorreiter concluded that the more compact electric cab company Abam was superior to the two gasoline-cab companies. This superiority was further enhanced by the decision to introduce a two-shift system for the electric cabs as well, which required an investment of an extra battery set per car; on balance, Vorreiter figured that the fixed costs per kilometer would decrease by a third. Over 1904, the Abam made a profit of 5,440 marks.[23]

On the other hand, Vorreiter also had to acknowledge that the Abag of the NAG (with thirty-five gasoline cars) had also become profitable. His conclusion could not be misinterpreted: "At present every motor cab pays, those with gasoline engines as well as electric cars, yet profitability can only be attained at a company with several vehicles." This set the tone for the following years: the electric-cab proponents recognized the usefulness of both propulsion types. Vorreiter summarized the conclusions of his analysis in two simple rules for a healthy electric cab business. First, because profitability became higher as

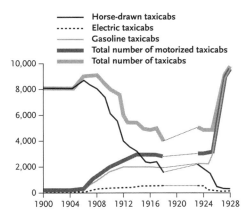

FIGURE 4.4. Horse sub-
stitution in the Berlin
taxicab fleet, 1900–1928;
from 1919 through 1922,
no statistics were kept.
(Knaths, *Die Entwicklung
des Berliner Droschken-
fuhrwesens*, 42–45)

the share of *paid* kilometers grew, electric cab companies had to increase their "load factor" as much as possible in order to lower average costs. He suggested a bonus system to motivate drivers to keep the number of "empty kilometers" as low as possible. Second, the daily route per car should be as long as possible in order to lower the fixed costs per kilometer. The same effect could be achieved by deploying a larger number of cars. Especially Berlin, "with its excellent pavement and virtually level roads, unlike any other big city in the world," seemed ready for such a future, at least, if the power plants provided sufficient charging facilities. In Vorreiter's opinion, the German automotive industry could grow into a world power in the area of the electric utility vehicle, just as the French industry had done in the area of the luxury car.[24]

The liberalization of the issue of licenses in October 1904 sparked a strong, chaotic growth of the Berlin motorized taxicab fleet (figure 4.4). The prospect of a net daily profit of 20 to 40 marks per cab resulted in a flood of speculative tendencies. Not a few individuals attended the drivers' school and subsequently bought a gasoline car for 10,000 marks on credit. From the third year of operation, however, repair and maintenance costs began to increase. The car manufacturers were also to blame for this situation, as they often promised a life span of ten years. Just as in Paris, a "taxicab crisis" developed. About 400 taxicabs were chained up and lawsuits between cab companies and manufacturers were frequent. The press painted a picture of the automobile as generally unreliable and published calculations that predicted the almost unavoidable loss smaller companies in particular would suffer.[25]

These calculations supported Vorreiter's conclusion that only large fleets could be profitable. But Vorreiter's calculations were biased in at least one respect. As we will see later, in the case of the EVAA (chapter 6) electric vehicle proponents tend to focus on the "serious side of mobility" and tend to neglect the more frivolous (or less official) forms of automobile use. According to another observer, profitability in the gasoline cab's case was exactly the opposite of the electric cab: "the smaller the company, the more profitable! The most profitable gasoline-cab company consists of one vehicle, which, for example, would be driven during the day by the father, and at night by the son." In such a case not only the overhead costs largely disappeared, the salary of a cab driver was saved as well. Nevertheless, Vorreiter's prophetic words had made an impact, for in the fall of 1905 the NAG started an electric-cab experiment with an AFA battery. A year later the Siemens-Schuckert-Werke launched electric-cab experiments as well. In 1912, the NAG-Abag fleet eventually contained not only 140 gasoline cabs, but also 100 electric cabs. This made the Abag the largest fleet manager in Germany.[26]

Technical changes in the vehicle concept allowed second-generation electric taxicabs to flourish. While the changes were not spectacular, they proved to be of decisive importance for the width of the electric cab's technical field. In December 1907, KAW director Sieg mentioned the most important departures from the first generation in his lecture for the Elektrotechnische Verein in Berlin. The first-generation taxicabs could go 15 km/h and, therefore, could use solid rubber tires without any problems, although the costs amounted to about 4 to 5 pf/km. The competition of the gasoline cab (with its speeds of 40 to 50 km/h that could really be reached in the city at night) forced its electric rival to adapt. Thus, a new element had entered the development of electric propulsion. When the gasoline car appeared in the city as utilitarian vehicle, Vorreiter had predicted a "peaceful co-existence" of the two types of propulsion. But now the two technologies began to influence each other, because a direct comparison had become possible. Thus, the large taxicab companies began to demand a maximum speed of at least 25 km/h. This meant that electric vehicle manufacturers were forced to provide their cabs with pneumatic tires as well. Initially, this made tire costs soar to irresponsible heights (15 to 20 pf/km) "and only after years of effort, we managed to lower the costs to about 8 to 12 pfennigs, depending on vehicle weight, speed, and state of the pavement." It could, however, not be denied (as Sieg, to his surprise, established after measurements at a taxi company) that energy consumption was higher with pneumatic tires than with solid tires due to the greater elastic deformation in the pneumatic tire. According to Sieg, this led to the conclusion "that one had to apply very special care to the energy consumption of the tires, especially with regard to the electric vehicle with its limited performance because of the size of its battery."[27]

The required higher speed also had consequences for the electric motor. As energy consumption had become a critical factor, a high motor efficiency was essential. This, however, required a large-size motor that would lead to an intolerable increase in weight of the chassis. The only option left was an increase in the rotational frequency, but then the replacement costs of the leather pinions in the Kriéger concept would rise. Use of a metal pinion instead would cause too much noise, so that a *technical limit* had been reached. Given Kriéger's drive train concept, which apparently was not debatable, the technical field was thus absolutely closed, and KAW was simply forced to accept the higher maintenance costs of the gearing.[28]

As far as the battery was concerned, a development similar to that involving the tires took place. The first taxicab fleet in Cologne KAW charged 2.5 pf/km for a maintenance subscription. But the demand for a higher maximum speed led to an increase in vehicle mass, because the heavier battery set also necessitated reinforcing the chassis. The old Cologne first-generation taxicabs, with their gross mass of 1,400 kg and their speed of 15 km/h, consumed a charging energy of 220 to 250 Wh/km. For the modern taxicab, consumption was between 350 and 425 Wh/km. The considerable increase of about 60 to 70 percent was only partly caused by the higher vehicle mass of about 2,000 kg (remarkably close to the first-generation taxicab masses in London and Paris, and eventually also at the EVC). The main cause was decelerating and accelerating these new cabs, which required about five times as much energy as for the old cabs. All this led battery-maintenance costs to more than double, increasing to 5 to 8 pf/km. Therefore, the total operating costs amounted to about 23 pf/km. At gross proceeds of 40 pf/km, this yielded a net profit (after subtraction of the drivers' wages) of 6 to 10 pf/km. Due to this delicate energy balance, Sieg emphasized the importance of a speed limit in city traffic. He further concluded that especially for electric propulsion the fixed infrastructural costs could only be recovered by a large fleet, whereas such costs "for smaller [electric taxicab] enterprises . . . exclude a profitability of the company."[29]

The relative flourishing of the second-generation electric taxicab was, however, not only reached on technical grounds. In fact, the narrow technical field was largely compensated by an opening up of the application field. For, on 1 April 1905, a new Berlin taxicab regulation went into effect that put the basic rate for the gasoline cab at the same level as that of the horse cab (50 pfennigs), but set the basic rate of 80 pfennigs for the electric cab. This enabled the electric cab to emerge from the darkness of the night (with its higher rates) into daylight. Thus, by working two shifts during the day, a profitable operation became possible. In this way, the Berlin taxicab fleet expanded in one year from 110 (1 April 1905) to 478 cabs (July 1906). But due to the undeniably preferential treatment of electric propulsion, the share of electric cabs increased more than 20 percent (97 units, versus 381 taxicabs with combustion engine). In 1908, a peak was reached at 23.5 percent (figure 4.5).[30]

The Berliner Elektromobil-Droschken A.-G. (Bedag), founded on 15 May 1905 after the example of the Abam in Cologne, accounted for a considerable share of this growth. It was the only Berlin taxicab company that had consciously been set up as a large company before the First World War. The initial capital of 1.5 million marks was doubled a year later. Initially, the Bedag's foundation received the full support of both AFA and AEG. The AFA was in favor because it created a welcome "laboratory" for the development of a light battery of its own, the AEG because it expected to sell more of its NAG electric cabs. The AFA offered active support, not only by giving advice with regard to the establishment of the charging station with exchange charging (which became operative in the beginning of 1906), but also by artificially keeping the subscription fee on a kilometer basis at a low level. For this company the foundation was a *glücklicher Zufall* (happy coincidence), as AFA director Adolph Müller put it, because it had nothing else to put opposite the increasingly popular light KAW battery than its renowned but heavy concept based on large-surface plates. Now, an excellent opportunity seemed to present itself to develop such a battery (the design of which was borrowed from the ESB's Exide)

FIGURE 4.5. Share of
the electric taxi in the
automobile-taxicab fleet
of Berlin, 1905–18.
(Adapted from Knaths,
*Die Entwicklung des
Berliner Droschkenfuhr-
wesens*, 42)

under the supervision of its own specialists. In fact, the NAG's decision to implement the electric cab was the decisive signal for the AFA to start developing a light lead battery with grid plates. The AFA received an order for approximately 100 battery sets, for which it had to produce about 1,000 grid plates. The planned energy density was 25 Wh/kg.[31]

But the AFA had hardly promised its cooperation, or the Bedag founders changed banks at the last moment. Besides the director of the new bank, Carl Neuburger, other financiers stepped in that the AFA and AEG did not trust. AFA and AEG then withdrew their promise of financial participation. Later, when the problems mounted, AFA also withdrew as technical adviser. But it did not want to give up battery maintenance, "because the AFA without doubt wanted to ensure the flawless functioning of the batteries." Yet the development of the AFA grid-plate battery (that became known as the Ky battery, after its designer Dr. Kieseritzy) took more than two years and was at the time certainly not flawless, even in the hands of the AFA specialists.[32]

Mid-September 1905, the first taxicabs became operative, and by the beginning of 1906, 20 of them were out in the streets. The Bedag started with a mixture of all well-known European electric cars. After a first experimental period, it placed an order for 25 electric cabs at each of four manufacturers (Lohner-Porsche, Scheele, KAW, and Kriéger). In the fall of 1906, 130 electric cars were in use, of which 100 were taxicabs. That same year another 180 electric cars were ordered, so that the Bedag fleet expanded to 235 electric vehicles, of which 186 were taxicabs.[33]

The investment of as much as 2 million marks for the first 200 taxicabs meant a unique boost for the German electric-car industry. Because of the heavy demands (a maximum speed of at least 25 km/h, among others), the industry was also encouraged to carry out extra development work. In 1907, the Lohner-Porsche concept was chosen as the best—a heavy blow to KAW in particular. It is very likely that the Bedag fleet consisted of 80 Kriégers of the Namag in 1908, whereas Lohner-Porsche in Vienna brought in at least 48 cars. By far the largest part of the remainder was probably also made according to the Lohner-Porsche concept, but supplied by Daimler. Until the end in 1911, the fleet size remained at the level of 1906, although the number of taxicabs further increased to 205 in 1908 at the cost of the number of rental cars. This development was exactly the opposite of the first-generation taxicab experiments.

Also, the high battery-vehicle ratio is noticeable here, which varied from 2.26 to 2.35 between 1906 and 1909. This high value may be partly due to the maintenance of batteries for third parties. It is remarkably close, however, to the proportion that applied to the Amsterdam Atax, as we will see.[34]

As a token of the improved technology of second-generation electric taxicabs, their annual mileage became increasingly higher. Initially, in 1906, each taxicab only did 143 day shifts and 151 night shifts on average, of a total of 2,465 service hours; 253 cab drivers together drove 2.2 million kilometers (almost 9,000 per driver) on daily trips of about 58 kilometers. The following year, the average number of effective operating days had increased to 213 per car, with the same fleet size. Only 165 drivers managed to cover more than 3.6 million kilometers, whereas the average number of rides per day was 11. From this, an annual mileage per driver can be calculated of a little less than 22,000 km. The total number of effective operating days for that year was 40,000, averaging out to a daily route of 91 km. In 1909, 205 taxicabs (19 more than in 1906) reached an annual mileage of more than 4.4 million km, approximately as much per taxi as in 1908. But even in 1906, peaks per car of 250 km a day occurred (in two shifts). The technical field of the German electric cab in the meantime had been stretched so far that nothing seemed to stand in the way of a functioning taxicab fleet, at least not on technical grounds. In other words, the widening application field for the electric cab had generated enough financial resources for the technical field to be expanded accordingly.[35]

Yet while the Bedag made a modest profit of 3,436 marks over 1905, the following year it was in the red. A reorganization followed, while squabbles among the directors led to the dismissal of two of them. Meanwhile, the cars (after a delayed delivery) could not be fully deployed. This was caused by a lack of skilled drivers, although the company had set up a drivers' school in the meantime. The year 1907 finished with a loss again, aggravated by a strike that caused the company to be idle for at least two-and-a-half months. A year later the losses had risen to half a million marks. Interestingly enough, because of the strike the batteries were not used for a long time, so that the phenomenon of sulfating manifested itself.[36]

There were plenty of explanations for the looming debacle. The Berlin paper *Tägliche Rundschau* blamed the disaster on the "irresponsibility and . . . laxness of the management." AFA director Müller became increasingly vocal in ventilating his judgment internally. In his opinion the expansion of the fleet had gone too far too fast. Bedag director Altmann had bought "initially any old vehicle junk he could lay his hands on," only to decide subsequently to manufacture in-house. While the car battery had become an important part of the AFA turnover, Müller began to fear that because of a possible disaster at the Bedag his battery would be subjected to "an unmerited discredit." He worried more and more about the unprofessional management of the company. Particularly the limited checking of the battery bothered him, which he thought was "horrendous." This checking as well as the maintenance by specialists were crucial in his opinion: "All companies [for taxicabs] that have been set up so far, have failed because of the battery."[37]

The influence of the financiers irritated Müller even more, as they shifted the technology into the background. The Bedag, in his opinion, because of how it was founded, had become "financially and personally dependent on a bank." The bank went bankrupt in 1910, and this heralded the end of the Bedag. Müller also suggested that the Bedag had merely been set up as a "jobber" company, a curious expression for a German, which can only be explained by his many contacts with the ESB and his knowledge of the fate of the EVC.[38]

Yet Müller could take the cab company's gloomy fate to heart as well. It is true that the first cars ordered had not been delivered on time, as Bedag's directors had neglected to fix a delivery date. But the AFA battery had not been finished on time either. From the very beginning the development of the new battery at AFA and the tests in the Bedag charging station were plagued by problems. Not until the middle of 1908 did the AFA engineers in charge consider the battery sufficiently "market ready" to consider sale (albeit cautiously) to third parties. Müller's irritation becomes understandable, however, if one considers that the Bedag opened a second charging station in Halensee (west of Berlin) and ordered all 200 to 300 batteries from KAW, "under such extremely favorable conditions, that we would not by far have accepted."[39]

Müller's criticism was not only of an organizational but also of a technical nature. For example, the charging voltage, supplied by the Berliner Elektrizitätswerke (BEW), fluctuated so much that business operation was endangered. Moreover, enormous current surges accompanied the charging of the batteries (despite the series resistors). This so much burdened the electricity grid that the AFA resorted to incorporating (illegally) fuses that were far too heavy. Müller in general criticized the lack of support from the electricity companies, which in his opinion, had contributed considerably to the success of the electric car in America.[40]

Even more harmful to the batteries were the high discharging currents (up to double the amount that was permitted), caused by the drivers' rough handling of the controller shifting. They would switch the controller into the drive position while the mechanical brake was still on. In a gasoline car such behavior would immediately stall the engine; in an electric vehicle this caused the battery's electrolyte to boil, a process that was not detectable from the driver's seat. On top of that, the drivers continued driving until the batteries were completely drained, so that the Bedag had to deploy electric tow trucks. But they often broke down too, so that tow trucks with gasoline engines were purchased. Müller further criticized the defective technology of German electric-car manufacture, which was not geared toward the requirements of the batteries. The batteries were separated into two units and mounted in a casing that was barely accessible. The specialist battery trade journal *Centralblatt* expressed a similar opinion. But it added an important extra factor. The Bedag was not profitable "because their material is more vulnerable than that of the gasoline cab, and because their development into a luxury-cab enterprise was forbidden." So, the Bedag did not have the chance to turn to the rental-car sector, as the American EVC had done in an earlier period.[41]

Eventually, the company filed for bankruptcy in 1911. The share capital had been completely lost, while the creditors waited in vain for a joint payment of 2 million marks. The Bedag had not paid out a dividend in any single year. The AFA was among its creditors. This company had to write off 15,000 marks in subscription fees on battery maintenance. The adventure had left it, however, with a promising light battery, with which the Bedag reached an average range of 80 to 90 km in as early as 1908. Moreover, the company had collected about 675,000 marks in the process. The AFA had supplied the Bedag with a total of 244 batteries. So, with 205 taxicabs with an exchange-charging system and about thirty private cars with a single battery set, competitor KAW must have supplied almost 200.[42]

The remarkable thing about this story is that KAW's director Sieg—after he had learned his lessons from the looming Bedag calamity—gave a considerably milder and more balanced judgment than Müller. According to Sieg, the pneumatic tire formed the largest ticket item. A return to the solid rubber tire was no longer possible, however, as this led to axle and steering-knuckle breaks. The solid tire was entirely impossible on the front wheels, as a considerable part of the battery mass rested on them. Therefore, the Bedag decided to convert part of the cabs to rear-wheel drive. They also did tests "with a new, American solid rubber tire and with elastic wheels." Other manufacturers divided the battery set equally between the front and rear axles, a measure Müller felt precluded profitable battery maintenance. However this may be, Sieg concluded that for profitable pneumatic-tire maintenance the battery mass per vehicle should amount to no more than 600 kg. This in fact meant that now a *technical limit* had been set on the range of the electric cab: the field of application was now restricted by clearly defined boundaries of the technical field.[43]

While Sieg licked his wounds because of the loss of the Bedag market, AFA director Müller's wishes came true: in 1906, the Namag was founded in Bremen, with decisive support from the Norddeutsche Lloyd. The Namag was a financially strong car manufacturer that took over the Kriéger license from the Abam and wanted to target the taxicab market by setting up cab companies in several German cities. Thus, the Namag established the Hamburger Elektrische Droschken Automobil-Gesellschaft (Hedag) on 9 October 1906, with the support of tire producer Calmon, and with a capital of 500,000 marks. The new company applied for a license for a hundred electric cabs. The year after the Düsseldorfer Elektromobil-Betriebs-AG (Debag) followed, with a capital of 175,000 marks. The company announced that service would start in December with twelve electric cabs. During the two following years in Düsseldorf, where initially only electric cabs were allowed, the Debag made a modest profit.[44]

Another year later, on 9 April 1908, the Münchener Elektromobil-Betriebsgesellschaft m.b.H. (Meb) was set up. On its board were not only Namag's technical director H. S. Meyer, but also Rudolf Diesel. The Meb received the sales monopoly on all electric vehicles from Bremen. It also started a taxicab company. The electric cabs drove average daily distances of 137 km in a 24-hour service (during the *Oktoberfest,* even 140 km). Although in 1908 the Munich city council started a campaign against the gasoline taxicab, the initiator of this action, Dr. Köhles, did not succeed in excluding gasoline propulsion.[45]

In its hometown of Bremen the Namag selected a different strategy. The company applied for a concession for thirteen electric cabs, and at the end of 1908 it had two or three in operation. At the same time the company requested to be allowed to pass the concession on to a financially strong company in due course. That company was horse-cab company Bremer Droschkengesellschaft, which changed its name to Bremer Droschken-Aktien-Gesellschaft (Bdag). On 1 June 1911, the Bdag took over fifteen cars, twenty-eight AFA battery sets, and all furnishings of the Namag for 150,000 marks, after it had extended its capital from 125,000 to 375,000 marks. The contract of sale also stated the obligation to take over the tire agreement with tire supplier Peters Union, as well as the maintenance contracts with six private persons, including two women. According to contract, the Bdag could purchase new Namag cars at a discount of 25 percent, "as had been the case with the other electric-cab companies in Hamburg, Amsterdam, Düsseldorf, Munich, and Vienna."[46]

In Bremen the Namag cab fleet could flourish because the taxicab regulation, issued on 26 November 1909, contained an emphatic ban on gasoline cabs. Over the first financial year, the company made a net profit of almost 16,000 marks. The following year saw a gross profit of almost 63,000 marks, which formed a pleasant compensation for the increasing fodder and repair costs of the horse business. But problems started to develop in 1913. Due to the extensiveness of Bremen (50 percent of the kilometers driven took place without passengers) as well as the poor state of the roads (which caused damage to the cars), no dividend was paid out for the first time.[47]

In 1916, only two electric cabs appeared to be operative; there were no tires for the other cabs, due to the war situation. In a letter to the police, the Bdag wrote that within a few months it would have to take these two cabs from the streets as well. This meant that just one gasoline cab would be left. In July 1917, the company reported that both the electric and the gasoline service had ended completely.[48]

In Hamburg, the Hedag had also been in a comfortable position since its foundation, due to a total ban by the police on gasoline cabs, as had been requested by the company itself. It received a license for 100 electric cabs. In 1910, it had 59 of them in operation, all supplied by the Namag. In April 1911, 154 electric cabs were operative in Hamburg. Hedag owned 80 of them and the others belonged to 39 smaller businesses. A year later the Hedag owned 88 cabs that were active 22 hours a day in a 2-shift service. On average they covered 143 km a day and almost 55,000 km a year, a third of which they were empty. There was a backup stock of 5 percent. The life span of the positive grid plates had grown to 16,500 km in the meantime, that of the negative grid plates to 33,000 km (230 and 460 cycles, respectively—2.3 to 4.6 times the life span of the best battery at the Paris battery contest of 1899). The observer who published these figures also compared the costs of the electric cab with those of a hypothetical gasoline cab. His conclusion was similar to what had been suggested before: "as single car the gasoline car (deserves) the preference . . . , in large-scale enterprises, on the other hand, the electric car can operate cheaper

A Namag-Kriéger taxicab, built under license from Kriéger, Paris, as deployed by the Bremen firm BDAG. Although they appeared old-fashioned—landaulet body, a chauffeur exposed to the weather—they were equipped with state-of-the-art batteries, two front wheel motors, and pneumatic tires. (Focke-Museum, Bremen, Germany)

than the gasoline car, when the road-surface circumstances are favorable and the prices for electricity are low."[49]

Not until 1911 was the first gasoline cab permitted in Hamburg. This happened at a moment when, according to a member of the Hamburg senate, the city threatened to become "the only city in Germany that does not allow gasoline cars as cabs." According to the minutes of the meeting, this made an opponent exclaim: "don't let these 'stink-bombs' enter Hamburg!" The minutes secretary recorded that this led to "great mirth" among the senators.[50]

The AFA had reason to be happy too. In the beginning of 1909, it became clear that its grid-plate battery had been greatly improved and the manufacture of it much simplified, so that the company began to make a profit from the maintenance of the various electric-cab businesses in Germany. The Bedag (Bremen), the Abag (Berlin), the Debag (Düsseldorf), the Meb (Munich), and the Hedag (Hamburg) now yielded a joint turnover of battery sales and maintenance of 274,300 marks. This earned a surplus of a total of 48,900 marks. The Bedag accounted for more than half of the turnover. The Hamburg cab company was second at some distance.[51]

IN THE GERMAN TAXICAB CASE, insight into the role of local government is absolutely essential to understand the struggle between the propulsion alternatives. For the liberalization policy of the Berlin authorities had caused a flood of motorization and bankruptcies. As early as 1905, rumors circulated about a ban on gasoline cabs. Vorreiter had reacted to this by calling such a ban "a big mistake." As long as the electricity companies had not provided a network of charging stations, the gasoline cab was really essential for longer rides. But when in the middle of 1906 the Berlin city council began debating a review of the taxicab regulation of 16 February 1905 (effective as of 1 April of that year), a wave of indignation swept through the automobile trade press. According to the city council, an increase in the rates was necessary, but they wanted to exclude the gasoline cars, because they "do not enjoy the favor of the public. It would not be a shame when these gasoline cabs in their current form would disappear from city traffic." The city council therefore considered only raising the basic rate of the horse cab to 70 pfennigs (that of the electric cabs had already gone up to 80 pfennigs in 1905). Furthermore, the council would see to it "that further gasoline and spirit cabs would not be granted a concession in Berlin, and for so long, until the adverse situation that is especially and unpleasantly noticed in loud noise and bad odor, will have been cleared up by the industry." In addition to its odorless and noiseless qualities, the Berlin authorities also appreciated the electric's low speed.[52]

In a renewed attempt to end the misery, another change in the regulation followed on 22 January 1909. After five years, the liberalization of the licensing policy came to an end, at least for the gasoline taxicab. In Berlin as well as in Charlottenburg, Schöneberg, Rixdorf, and a few other suburbs, from that date on "new licenses for 'taxicabs with combustion engine' ... would no longer be issued." A storm of protest arose, although the automotive expert A. Heller remarked in the *Zeitschrift des Vereines deutscher Ingenieure* that things might not be as bad as they looked. After all, no one obeyed the methylated-spirit decree, not even the state authorities and the police.[53]

Yet even a combustion-engine proponent like Heller found that the measure "was in part justified." He pointed out that the apparent technical "superiority" of the electric propulsion stood in contrast to the often abominable state of the gasoline cars, "of which the fuel pipes are so leaky, that the smell bothers the driver inside the car, that grease leaks from the casings and bearings onto the road and makes it dirty." Others "continually emit smoke, because the engines are too richly lubricated." Smell from the exhaust gases due to incomplete combustion of the fuel, however, was in Heller's opinion unavoidable, as long as the carburetor had not been improved and could not be tuned satisfactorily enough to handle the considerably fluctuating engine loads.

Heller did not put the blame for all this on the industry, but on the management of the operating companies that neglected maintenance. In his opinion, it was a mistake "to abandon this vehicle before we have a better one." After all, the electric cab was hardly capable of providing profitable operation and, moreover, its speed and range were so low that it did not offer "complete replacement for the gasoline cab that was already often used for outings in the surroundings of Berlin." This added a new element to the discussion about the choice of propulsion. The gasoline cab, despite its greater unreliability, appeared to offer a function that its electric rival could not supply.

The absolute *Nummernsperre* for gasoline cabs was weakened on 6 April 1911. Now one gasoline cab might be deployed, if ten horse-cab licenses were handed in. This measure heralded the definitive end of the Berlin horse cab. It resulted in 2,194 motorized taxicabs in 1912, of which 277 were electric. The Neue Automobil-Gesellschaft had supplied 211 of those. The NAG also contributed 198 units to the number of gasoline-cabs in the German capital. Other brands in use were Adler (577 units), Opel (396), and Benz (153). Berlin at the time had eleven companies that owned only electric cabs (a total of 550). A further adaptation of the regulation on 30 January 1914 failed to hide the preference of the authorities. Now a new license for an electric cab could be obtained in exchange for only two horses. This made the electric-cab fleet grow to a peak of 574 units in 1914 (the year in which, three years later than in Paris and four years later than in London, the turning point of horse traction to motorized propulsion occurred). The relative flourishing of the electric cab in Berlin continued until 1924.[54]

Berlin's regulatory measures led to rumors in the international trade press. In June 1914, A. Jackson Marshall, secretary of the Electric Vehicle Association of America, asked for information about the rumor that gasoline taxicabs were no longer allowed in Berlin. He received a clear answer from the chief superintendent of the Berlin police that also explained the motivation behind the market regulation in the German capital: "This ordinance, in the first place, has not been promoted exclusively from considerations of public health. It will eventually help the introduction of electric cabs and at the same time prevent the introduction into public circulation of a number of gasoline driven cabs in excess of public necessity." Without doubt, the autonomy of local authorities in Germany boosted the diffusion of electric propulsion. For the national authorities, the outbreak of the war would soon after this turn the picture about the nature of this "public necessity" completely upside down.[55]

However this may be, American interest concerned a new phenomenon in the electric-cab business, enabled by the unique combination of local public support and a financially strong battery manufacturer. Two companies (not mentioned by name) maintained an electric-cab fleet of 600 units, together with a few private vehicles, with the exchange-charging method. One of the firms had 700 batteries and the other had 450 batteries in maintenance. Without doubt one of these companies was the AFA, which established a charging station on a site on the Chausseestrasse, rented from the Bedag. After the bankruptcy, the company took over the Bedag's big garage. Exchange charging of the taxicabs took place on the basis of time rather than kilometers driven. This led to an even distribution over the day of the charging load as well as the maintenance personnel. Another striking feature was that settlement of the costs occurred strictly on a cash basis: the driver only had a right to a second battery charge when the previous one had been paid. The maintenance man (according to the American commentator, a "high class workman") received a bonus if the plates lasted more than 10,000 km. This introduced a new phenomenon in the electric-cab business: by means of centrally located maintenance by specialists and accurate cost and income control, a division of fleet and garage ownership now became possible. It also highlighted again the crucial importance of maintenance, an insight that was no longer restricted to the electric vehicle camp, as we saw in the case of Paris. Because of this, electric cab fleets could become smaller, at least as long as the number of cab fleet operators remained sufficiently high. It made this new venture comparable to the battery service systems in the United States (to be discussed in chapter 6). Although details about this new type of management are lacking, according to American observers, private individuals and small companies were the owners of the taxicabs. They operated about twenty cabs each.[56]

THE GERMAN TAXICAB HISTORY is unique in terms of the peaceful rivalry between the two means of propulsion. It provides room for doubt about the viability of the separate spheres concept, which was to become so popular among American electric vehicle proponents of a later generation (see chapter 6). Furthermore, comparison of the two propulsion alternatives leads to the conclusion that the *size of the fleet* (the collection of all electric cabs that *independent of ownership* was serviced by one center) played an important role. A small (even tiny) fleet seemed most favorable for the gasoline cab. But for the electric cab only a large "compact" fleet (with a minimum of "empty kilometers"), combined with strict cost control, made a profitable operation possible. The question arises whether there was a critical limit to the fleet size; in other words, how close could the electric fleet size approach the ideal case of the gasoline rival? What was the minimum size of an electric-cab fleet for profitable operation, related to local circumstances, such as size of the city, population density, and state of the road system?

In the second place, the moment of foundation seemed to play a decisive role with regard to the state of vehicle technology and the competing propulsion system. In Berlin, the same thing happened what had also happened in Paris, London, and New York. The electric-car industry was still recovering from the blow inflicted by the first-generation streetcar or taxicab experiments, when the gasoline car entered a market niche, created by an increasing demand for transportation. This supports our suggestion that in a virgin market each of the two types of propulsion had a chance. Whoever was first immediately had a considerable edge over his competitor. This poses the question of how the second-generation electric cab would have behaved in a market in which it had the upper hand from the start.

In order to be able to formulate an answer to both questions, a much more detailed analysis of the costs as well as a historical situation that is opposite to that in the European cities mentioned is needed. In the following section, we will look at the well-documented electric-cab fleet in Amsterdam.

Following Critically: The Electric Taxicab in Amsterdam

Compared with companies in other European capitals, the Amsterdamsche Taxi Auto Maatschappij (Atax; Amsterdam Taxi Car Company) was a late-developer. On 28 December 1908, it was founded as a subsidiary of the Amsterdamsche Rijtuig Maatschappij (ARM, Amsterdam Carriage Company). ARM director A.C.A. Perk was in charge of day-to-day management, while electrical engineer J. F. Friderichs was appointed delegate supervisory director. In 1903, Friderichs had joined the ARM's board of directors, at the same time as H. C. Heijbroek Wzn., the legendary horse lover. In the founding year of Atax, revenues (Dfl 535,000) and gross profits (Dfl 92,700) had never been so high at the ARM. But it was clear that this result was especially due to Heijbroek's department of "Luxury and Monthly horses." The rental of coaches, however, started to yield less revenue; the public began to get used to faster means of transport, especially the electric streetcar, which was introduced in Amsterdam in 1900.[57]

Friderichs's appointment deserves some extra attention because of his decisive role in the foundation of the Atax and the choice of vehicle type. After a military education, Joan Frederik Friderichs studied electrical engineering at the University of Liège, Belgium. After his graduation he was employed by the De Laval company (steam engines) in Amsterdam. There he was in charge of the construction of a few direct-current power plants. In this position he first became acquainted with the battery, used in these plants to dampen power peaks and to compensate for brief over-consumption. This experience must have been the reason why the AFA, by then the largest battery producer in Germany, asked him to run its engineering office in Amsterdam. Friderichs started to work for the AFA on 1 October 1901.[58]

After a training period at the end of 1901 and the beginning of 1902 in Hagen (where AFA's production was concentrated), Berlin (seat of its central board of directors), and Hamburg, Friderichs became head of the Amsterdam Ingenieur-Bureau voor Nederland en Koloniën. In May 1906, the office moved to a new address, Keizersgracht 304, where a repair shop and charging station became operative.[59]

From the date of his employment at the AFA, Friderichs was present at all the so-called engineer conferences and the (usually annual) meetings of technical and commercial executives, under strict supervision of AFA founder and director Adolph Müller. In this way, he became thoroughly knowledgeable about the developments in the area of the electric-powered car, especially in Germany. At a conference in March 1908, in the middle of the foundation wave of taxicab companies outside Berlin by Namag, Müller informed his engineers that "for the moment there (is) no reason for us . . . to stimulate the establishment of new taxicab companies, or generally to encourage the sale of car batteries."[60]

Besides Friderichs, engineer Henri Maurice Enthoven was another key figure in the network behind the foundation of the Atax. Enthoven established himself in 1904 with a garage and showroom as Technisch Bureau "Holland Automobile" in The Hague, where he imported the French brands Gladiator, CGV, and the electric-powered Kriéger. He exhibited his products for the first time at the Amsterdam motor show in 1905.[61]

ON 27 FEBRUARY 1908, barely two weeks before AFA director Müller was to express his reservations about a too active electric-cab promotion, an ARM management meeting took place in Amsterdam. It was decided that a management committee would investigate "whether there could be a reason for the A.R.M. to establish an electromobile enterprise here or in The Hague." The report based on this investigation ("Report I"), signed by Friderichs, focused on a hypothetical beginning fleet of twelve cars "at such a garage lay-out that later this could easily be adapted for double the number of vehicles"—in Amsterdam, that is, for the report never mentioned The Hague again.[62]

For the variable costs, he reckoned 3 cents/km for battery maintenance, 6 cents/km for the tires, 2.5 cents/km for electricity, and a similar amount for general maintenance. In two separate appendixes, Friderichs made a cost-benefit calculation of a hypothetical Amsterdam company, based on those assumptions. He borrowed from the practice of the horse-cab business by assuming an average ride length of between 1.2 and 3.2 km. Thus, he arrived at average proceeds of between 30 and 36 cents per kilometer per cab.

Friderichs further referred to the situation in Berlin, Hamburg, and Düsseldorf, of which he knew the average proceeds (42, 43, and 50 pf/km, respectively; converted via 5 cents = 8 pfennigs: 26, 27, and 31 cents/km, respectively). He correctly assumed that the higher proceeds of Düsseldorf were due to the larger number of shorter rides. These were the most lucrative because of the relatively high basic rate. Whereas he rejected Berlin as a basis of comparison because of its deviant traffic situation, he used the difference in size of Hamburg and Düsseldorf as points of reference for an interpolation procedure. He based his figures on a few monthly overviews he apparently had received from two taxicab companies (the Hedag and the Debag). Friderichs further pointed out that in Hamburg "due to the deployment of clean and comfortable electro-mobiles an enormous increase in traffic took place, despite the extensive street-car system." This was a very interesting argument for the ARM board, as they assumed that horse traction did in fact suffer because of the competition of the streetcar. The electric car seemed to be able to turn this tendency around.

Friderichs assumed in his report that with eleven electric cabs one could replace most of the seventy horse cabs stationing in Amsterdam. He further emphasized that motorization of the taxicab was now necessary. If a competitor came up with the same idea, the ARM's garage (renting and garaging of coaches) would "suffer great damage." That is why Friderichs advised his fellow-members of the board to apply for an exclusive municipal license.

In a second report, dated 27 May 1908, Heijbroek, Perk, and Friderichs reported on their study trip to Germany, which had taken them to Düsseldorf, Hamburg, and Bremen-Hastedt (location of the Namag establishment). They had also made inquiries in Paris and paid a visit to the Trompenburg factory, manufacturer of Amsterdam's "Spyker" automobile. At the Namag they were told that the company did not agree to the original plan for renting the vehicles, but the factory did promise to buy back the cars at a discount of 20 percent "when the business does not appear to be profitable in one year's time."[63]

The investigating committee presented its conclusions in the form of theses, for the first time including the gasoline car in its considerations. This may have been the result of the situation they had encountered in Düsseldorf, where at the time ten electric cabs had to compete with eighteen gasoline cabs. The committee plainly opted for electric propulsion, however, and did so for the following reasons:

a. Pleasant for the *public,* no gasoline smell in the carriage, not necessary to start the motor in advance, no vibration by the motor, start without jolts, smooth ride, impression of safety for the user.
b. Easy to handle for the *drivers.* Knowledge of the mechanism is absolutely unnecessary. Just a few general concepts are required.
c. Pleasant for the *owner,* simpler construction and consequently less damage of the mechanism, much simpler exchange of just a few, inexpensive wearing parts, much improved safety due to the quietly and surely operating batteries and the possibility to drive with even half a battery and one motor (there are two motors that independently operate on the front wheels).
d. Possibility to make agreements with a battery factory and tire companies for the use per K.M.

The objection against electric carriages, viz. of not being able to drive more than 80 to 100 K.M. on the same battery (in Düsseldorf once a distance of 140 K.M. was reached), does not seem important for our purposes: city traffic and on occasion touring (to Zandvoort 28 K.M. and to Baarn 36 K.M.). As the speed cannot be higher than 28 K.M. per hour, not much use will be made of these longer trips, which is not deemed desirable anyway.

This report clearly was more cautious than Report I, mainly because of the doubt about the length of the daily distances that could be reached. Soon after operation had started, this doubt appeared unfounded, so that the Atax business was based on a battery-vehicle ratio of 2.26, exactly the same as that of the Bedag.[64]

Two months after the second report, during the ARM board meeting of 14 July 1908, the basic decision was taken to "start the operation of automobiles"—a remarkable decision, especially because an explicit statement on the nature of the "motive power" was now lacking. Within a week Perk had letters sent to four Dutch importers of gasoline cars (Enthoven, among others), asking for quotations. Perk's initiative should be considered against the background of the potential financiers' doubt about Friderichs's advice in favor of electric propulsion. Their doubt did not so much concern that choice itself, as the costs of pneumatic-tire maintenance.[65]

Furthermore, the increasing disquiet about the correctness of the choice for electric propulsion was also fueled by news from The Hague. There, preparations for a taxicab company based on the gasoline car were well under way. This company, the Haagsche Automobiel Taxameter Onderneming (HATO), "after thorough research" opted for the gasoline car, "as here [in The Hague] the operating costs of this business are lower than that of the electric." Enthoven was the supplier of the HATO's fourteen cars.[66]

Apparently, Friderichs managed to convince his fellow-members of the board of the superiority of electric propulsion. For, in November 1908, Perk asked Enthoven to make a Namag car available for a short trial period. At the end of that month an enthusiastic *Algemeen Handelsblad* reported on a "most pleasant introduction" to the electric cab by means of a trial ride, which showed that this taxicab complemented "a general modernization of city traffic, to which the streetcar set the example."[67]

The city council began discussing the Atax request on 20 January 1909. The request concerned the *exclusive* application of *electric* cabs, which made delegate Van Gigch wonder "why only electric propulsion is allowed, as in most big cities such as Paris, Berlin, and London gasoline has won from electricity." Councilman Delprat, chairman of the meeting, stated that requests by competitors with alternative types of propulsion would not be excluded in the future. But for now electric propulsion had been chosen "because this has proved to be the most desirable for *city traffic* (also in Berlin and Paris) as these carriages do not cause inconvenience by smelling and chugging." Delprat was completely wrong as far as Paris was concerned, whereas Berlin was not the best example of a German city. During the debate, councilor and Atax supervisory director Verschure defended the Atax vehemently. The proposal of mayor and councilmen to grant a license to the "electric monkeys" (as the Atax cabs were called by the Amsterdam population, because of the drivers' dress) was accepted with two negative votes. It was the first license for vehicles with a taximeter in Amsterdam.[68]

The maintenance contract with the AFA was signed in May. According to this contract, each battery set consisted of 44 elements (of 2 V each) of the type 5 Ky 225/4, the battery with relatively light grid plates that AFA had just put on the market. The batteries would be charged and maintained in a separate charging station under the responsibility of the AFA, while the Atax would pay for the costs of electricity. The AFA would supply the personnel for inserting and removing the battery boxes and would charge a kilometer price of 4.5 pfennigs (2.8 cents) for its services. The Atax moreover had to pay a fixed annual fee of Dfl 35 per battery set for maintenance of the battery boxes and clamps. The municipal power plant supplied a three-phase current of 3 kV that was rectified by a rotating motor generator.[69]

On Friday, 4 June 1909, at half past seven in the morning, six Ataxes appeared in the streets to occupy their stands at the Centraal Station, the Damrak, and the Leidseplein. The day before, Mayor W. F. van Leeuwen had made a trial ride. The fourth stand, on the Thorbeckeplein, was not occupied until a few weeks later, when the six remaining cars from Bremen were delivered during the month of June.[70]

The reaction of the public was overwhelming: "Da Costa taken f 2000.— in shares. Many requests for subscription. Enormous activity. Gigantic proceeds!" Perk rejoiced in his letter to fellow-director Heijbroek. But his enthusiasm was somewhat dimmed because at the same time he had to mention the first victim of this modernization: "a 86-year-old blind and deaf man could not be forced by his family to stay at home; he was hit and died. The driver was not to blame at all."[71]

THE FOUNDATION OF THE ATAX marked the beginning of a horse-replacement operation, which was actively supported by the municipality. At

The second-generation taxicab featured special pneumatic tires and a novel, reliable AFA battery. This Atax (Amsterdam) cab was supplied by Namag of Bremen, Germany; note the driver ergonomics of the extremely low steering wheel, which is reminiscent of horse carriage days. (ARM archives)

least, this may be deduced from the big difference in the stationing fee for Ataxes and for horse cabs, at the expense of the latter. The taxicab did not seem to be a threat, however, to the *voitures à grande remise* (rental vehicles) that were still operating with horse traction. The success of the electric cab increased even more when the Atax reduced its rates on 1 October 1909. This led to a considerable increase in short rides, so that the gross proceeds of the Atax fleet rose (figure 4.6). Now other companies also started to pursue a profit. The Industrieele Maatschappij Trompenburg as well as the firm M. van Gendringen (that had founded the Taxi-Auto-Maatschappij, TAM) submitted a request for a license to the city council shortly after each other. This forced the Atax board to expand its vehicle fleet sooner than it had intended.[72]

These developments brought the dilemma of "motive power" back on the agenda of the Atax directors. In less than a year the Amsterdam traffic environment had changed drastically, especially regarding the gasoline car. During a municipal meeting dealing with the new requests, it was said that "no other regulation is given than that the carriages, also regarding this issue, have to be to the satisfaction of mayor and councilmen." A year after the discussion on the Atax's request, the council did not think that the counterargument of smell was still applicable, because it "is [spread] considerably less by these well-built vehicles than earlier on, but it seems to have a certain advantage to the public that at least part of the taxi automobiles are not limited to a certain area of operation. Electromobiles certainly are. They cannot drive farther than 100 KM. and so mostly cannot be used for traffic to somewhat distant municipalities. Moreover, one should keep in mind, that every private person who purchases an automobile, is entirely free in the choice of motive power. As virtually all automobiles in private use are powered by gasoline, such carriages are already so numerous, that extension with, for example, a dozen of such cars that will obtain a stand on the public road cannot cause any appreciable increase in nui-

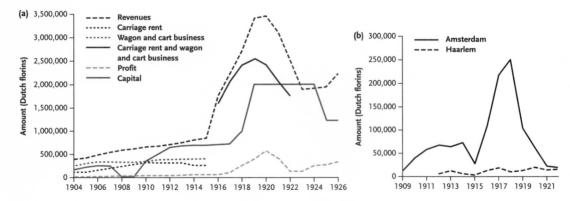

FIGURE 4.6. Turnover, capital, and profit of the ARM, 1904–26, and the Atax's share in it. (Annual reports ARM; Atax calculation based on the monthly reports in the Friderichs file, ARM archives; reprinted from Mom, "Haver- en andere motoren," 250, 259): a. ARM: from 1910, the wagon and cart business begins to exceed the rental-carriage revenues; the difference between "revenues" and "carriage rent + wagon and cart business" comprises the trade in luxury horses and the Atax.; b. Gross proceeds (revenues minus expenses) of the Atax fleet (Haarlem from 1920: gasoline taxicabs).

sance." As in so many other cases in big European cities, municipal authorities in the end had to yield to the undeniable dominance of gasoline propulsion in the general market. They now used the remarkable argument that, although the gasoline alternative was still noisy and smelly, the contribution of a few cabs to the general traffic situation would not make much of a difference. But to express its clear preference for the electric alternative, the Amsterdam municipality forbade gasoline cab owners to station on asphalt roads, to prevent damage due to "dripping fluids."[73]

In the months of March, April, and May of 1910, the second series of eleven Namags gradually arrived, so that the Atax fleet numbered twenty-four electric vehicles at the end of the spring of 1910 (figure 4.7). This meant that the electric car, with a view to the hundred or so private gasoline cars in Amsterdam at that moment (and excluding an unknown number of gasoline cabs), comprised about 19 percent of the passenger-car fleet in the capital. The bodywork of this second series was not built in Germany, but by the Amsterdam coach builder Schutter & Van Bakel.[74]

In April 1911, the milestone of a million kilometers was reached. With 30 to 40 cabs in 1912, and 57 drivers, 6 mechanics, 7 men in the battery workshop, and 32 cab stands, the Atax undoubtedly was the largest fleet manager in the capital. The biggest competitor at that moment was the TAM, with 25 to 30 cars at the beginning of 1912. The third player in the motorized taxicab business in Amsterdam was the Eerste Nederlandsche Taxi-Auto Maatschappij (ENTAM), which started a taxicab service with 16 German NAGs, with gasoline engines. In addition, there were 8 car-rental companies and a few garages with luxury cars. Some of them would later attempt to arm themselves against the "big ones" by establishing their own organization, the N.V. Automobiel Onderneming (AO). In 1914, the Atax and the TAM had similar size fleets. But the

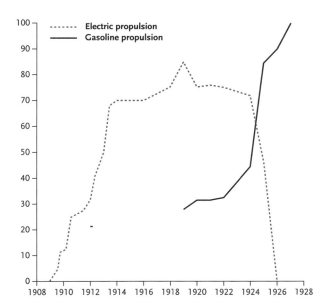

FIGURE 4.7. Development
of the Atax's stock of
electric and gasoline taxicabs, 1909–27. (Annual
reports ARM and Atax,
ARM archives; reprinted
from Mom, "Haver-
en andere motoren," 243)

relative strengths in the Amsterdam field of application can be deduced from
the results of a private traffic count by the ARM during the soccer match between Holland and Germany in the Amsterdam stadium on 5 April 1914. Of
the 541 vehicles that passed the measuring point into the direction of the stadium between a quarter past one and half past two, 109 were Atax cars, 95 TAM
cars, 26 AO cars, and 22 Entam cars. So, the gasoline cabs accounted for almost
a quarter of the rides counted and the electric cabs for 20 percent. An equal
share (116 cars) was taken up by private cars, whereas horse and carriage with
173 units still formed 32 percent of the vehicles counted.[75] This confirms our
earlier assessment about Paris and Berlin, that the first motorization wave in
the cities was for a large part driven by commercially applied vehicles.

In the summer of 1912, the second Atax charging station in the Beursstraat was finished. Until May 1913, the Atax fleet increased to sixty-two units,
some of which were deployed in the neighboring city of Haarlem. Later that
year the company purchased at least four secondhand vehicles in Munich, to
which it added one more unit as a backup before the outbreak of war. The cars
bought in 1912 and 1913 were partly provided with new bodywork. In the course
of 1913, the maximum speed of about 27 km/h was increased to approximately
36 km/h "by a change in the construction, so that the only difference to our disadvantage with the gasoline cars could be canceled out." The Atax board then
expressed a self-confident point of view to their shareholders: "The thesis, defended by us from the beginning, that the electromobile, even after a lengthy
use of already 6 to 7 years now, by exchanging cheap parts, runs like a new vehicle, in practice proves more and more to become true."[76]

The favorable judgment on electric propulsion was supported by the statistics that Friderichs managed to compile based on an accurate study of the
"monthly statistics." These statistics have all been preserved and thus allow us
to analyze the growth of the fleet in detail. They show, among other things, that

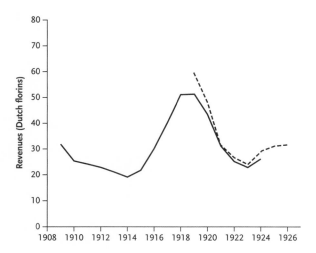

FIGURE 4.8. Revenues
per Atax taxicab per day in
Amsterdam, 1909–26;
dotted line: gasoline taxis.
(Monthly reports in the
Friderichs file, ARM
archives)

from 1910 the average daily distance per taxicab had been well over 90 km (with a peak of an *average* of 120 km in the beginning of 1917!). This proves that the fear of too short a range was unfounded.[77]

Nevertheless, the company management kept worrying about the extremely unstable, but usually quite high, tire prices. This issue was felt to be more pressing as the fleet expanded and the proceeds per car decreased because of the growing competition (figure 4.8). Shortly before the outbreak of the war the price fluctuation was eventually solved by signing a two-year contract with German tire manufacturer Continental at 7 pfennigs per vehicle kilometer. Not only the tire price, but also the higher vehicle speed led to a lower net profit per vehicle kilometer. This in fact meant that the need to compete in speed with the gasoline cab diminished the relative advantage of electric propulsion from a commercial point of view. The phenomenon was due to the faster wear and tear on the tires, but especially to the increase in the share of "empty kilometers." At the same time, probably because of increasing competition, the daily distance per car began to drop (figure 4.9). By the end of 1912, Friderichs had estimated the daily route for that year at about 89 km. At closer analysis, it appeared to be only 87.9 km, while the expectation for 1913 was that this figure would drop below 80. This fear proved correct: in 1913, the average daily route was 76.7 km and a year later it had further decreased to 69.3 km. As the number of rides per car per day also dropped (from 21.8 in 1912 to 18.3 in 1913 to 16.6 in 1914) the proceeds per vehicle in these two years decreased by 22.5 percent. Friderichs then came up with a bonus plan: drivers would receive a bonus, depending on the number of kilometers covered with passengers and the proceeds per kilometer. At an average driver's performance of about 13,000 km per year and a bonus payment that would amount to at most one-third of the surplus profit, Friderichs thought that the daily averages would rise automatically, because of a joint interest of the drivers and Atax. With 137 employees—and a telephone switchboard (from January 1914)—the Atax's daily averages rose, as the annual report for 1914 concluded with satisfaction.[78]

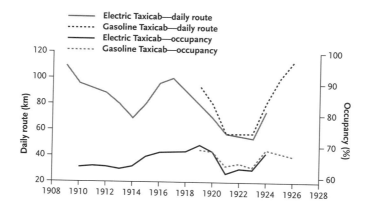

FIGURE 4.9. Average
daily distance per car and
rate of occupancy of the
Atax in Amsterdam, 1909–
26. (Monthly reports in
the Friderichs file, ARM
archives)

The outbreak of the First World War in August 1914 seemed to put an abrupt end to the flourishing of the ARM. But after a short slump, the backbone of the ARM, the horse business, quickly recuperated and a period of an unprecedented growth as well as profit dawned. Nevertheless, a number of worrisome developments began to take shape, such as the doubling of the costs of fodder. This caused the share of the expenses for "fodder and stabling" in the total revenues to increase from 24.4 percent in 1914 to 32.4 percent the following year. To make matters worse, a fungal poisoning broke out among the horses, causing forty-eight of them to die.

To compensate for this, the ARM had been pursuing further diversification even before the war. On 1 June 1912, it established, for example, the Amsterdamsche Exploitatie Maatschappij (AEM) as a continuation of the operation of gasoline cars, which Autovervoer Maatschappij Trompenburg had started barely a year earlier. With the takeover of this Spyker subsidiary—at the time intentionally set up as a competitor of the Atax—gasoline cars were included in the ARM fleet for the first time. Initially they were deployed as luxury cars to replace the rental carriages. Also, garage and repair facilities now became part of the business. Thus, as early as 1915, the ARM could claim to be the representative of the American brands Maxwell (light cars), Peerless (luxury cars), and General Motors Trucks. Through the mediation of engineer Friderichs it also became the agent for Henschel & Company, the Berlin manufacturer of electric-powered cars and special vehicles, and for Gebhardt & Harhorn, manufacturers of electric three-wheelers. So, from then on the ARM could supply electric and gasoline trucks. Even before the war six Henschels (electric-powered street-cleaning trucks) were sold to the cities of Amsterdam and The Hague. Apart from the gasoline cars, the AEM also rented out and maintained a few electric cars, among them some cars for the state railroad.[79]

At ARM subsidiary Atax, twenty-five of the 137 drivers were drafted during the mobilization. Their jobs were kept open for them, however, so that anyone who returned after the end of the mobilization was rehired. Like its parent company, the Atax also suffered from a slump at the outbreak of the World War, but the situation soon improved to "very favorable" (figure 4.10). That was especially due to two factors. First, the bonus system started to yield a profit be-

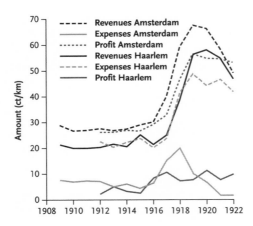

70 — ---- Revenues Amsterdam
        —— Expenses Amsterdam
60 — .... Profit Amsterdam
        —— Revenues Haarlem
50 — ----- Expenses Haarlem
        —— Profit Haarlem

Amount (ct/km)

40
30
20
10
0
   1908 1910 1912 1914 1916 1918 1920 1922

FIGURE 4.10. Revenues, expenses, and profits of the Atax in cents per kilometer driven, 1909–22. Haarlem from 1920: gasoline taxis. (Monthly reports in the Friderichs file, ARM archives)

cause the revenues per vehicle kilometer began to rise. In Amsterdam they increased from 26.5 ct/km in 1913 to over 30.2 ct/km in 1916. In two years' time, the daily route grew to almost 100 km, while the number of rides per day, the proceeds per kilometer, per vehicle, and per day, and the annual mileage of the fleet began to rise spectacularly. All this happened in spite of a drastically reduced effective fleet. Second, the Atax board for a long time managed to avoid the rising tire prices due to the "very advantageous contract with Continental." It even managed to prevent a threatening shortage of tires. By the outbreak of the war, the contract with Continental probably lost its importance, for in 1915 the company closed a kilometer contract with the Dutch Hevea factory on the same favorable basis as had been the case with Continental. This did not prevent tire maintenance from adding up to 18 percent of the revenues that year (battery maintenance amounted to 12 percent and the costs of electricity to 7.5 percent).[80]

The soaring proceeds were a sufficient basis for the Atax directors to take a rosy view of the future. For example, in 1915, they decided to buy a site on the Nassaukade "intended for the erection of the large Atax center with factory," which would make the two charging stations on the Keizersgracht and the Beursstraat superfluous. The new center would enable the garaging of more than a hundred Ataxes and allow a further extension of the electric-cab fleet. The building of this center never materialized, however, neither during nor after the war.[81]

On 11 September 1915, a strike broke out among the workshop personnel, followed the next day by the drivers. The stalemate was to last four-and-a-half months. The available documentation does not indicate a reason for the strike, but it seems as though its outbreak was not disagreeable to the Atax board. Eighty drivers were fired and, on 28 January 1916, only half of them were rehired. This number was only increased very gradually, which meant a welcome drop in personnel costs for the Atax. After the strike the spectacular rise of proceeds, average daily route, and all other parameters mentioned continued as if nothing had happened.[82]

Friderichs's life's work received international renown when the head of the municipal electricity company, Dr. W. Lulofs, described the operation of the two Atax centers in the British magazine *The Electric Vehicle*. He included a charging curve that showed a load factor of as much as 60.5 percent. This was caused by charging the batteries simultaneously only twice a day outside the peak hours of the power plant, after they had been discharged as much as possible while in use. This practice of "charging on basis of time" was probably copied from the Berlin AFA center. The life span of the tires was 6,400 to 8,000 km and that of the battery plates was 12,000 to 13,000 km for the positive plates and double that for the negative plates. A quarter of the fixed costs was spent on the batteries and 35 percent on the tires. The recuperative capacity of the lead battery manifested itself nicely when a driver drove until his battery was empty: after waiting for an hour he could still cover 12 km at a speed of about 7 km/h.[83]

In the meantime the reduced, effective Atax fleet (in 1917 still consisting of forty units) could be kept rolling only by cannibalizing chassis that were not in use. The armature (the rotating shaft with electric winding of the electric motors), the ball bearings, the carbon-brush holders, and the carbon brushes (the current collectors of the armature) were stripped from the inoperative Ataxes. The fast-wearing bodies of the first series of chassis were no longer repaired, but were completely replaced in the ARM coach factory.[84]

Meanwhile, at the level of the mother company ARM, the horse business went from bad to worse. Repeated and considerable increases in rates had not been able to stop the drop in revenues, as the higher rates also reduced the use of the carriages. Instead, they led to an increase in the demand for motorized transportation. In the Amsterdam taxicab business, this demand was guided in the direction of electric propulsion, however, when on 5 December 1917, "practically the entire motorized traffic for private persons was stopped" for gasoline cars. This driving ban was not lifted until 23 November 1918. It is obvious that this meant an enormous boost for the Atax, even though electricity was rationed. The revenues increased even more by a rise in rates that, surprisingly enough, the gasoline-car companies had applied for. Whereas the daily route and the number of rides per day—after the record of 1917 (99.8 km and 26.8 rides)—decreased the following years, the proceeds per kilometer soared again (with a record of 66 cents in 1919). The proceeds per vehicle and per day had reached a peak a year earlier (table 4.2).[85]

After the war, the ARM's turnover and profit were to grow for another two years, but from 1920 both began to fall. "The expansion that the traffic with mechanically powered vehicles will experience" was, according to the ARM management, the stimulus for the purchase of the biggest taxicab competitor in Amsterdam, the TAM, on 1 January 1919. It was taken over with its complete taxi fleet of 60 units, probably mainly Opels. The fleet was immediately reorganized: 22 cabs were sold and part of the remaining 38 were dismantled, so that at the end of that year there were 28 operative gasoline cabs and 10 chassis frames. The gasoline-cab company thus created bought 10 Renaults, an Austro-Daimler, and 2 Opels in 1920. The entire vehicle fleet, with the total adjusted for cars dismantled and sold, comprised 33 units.

The Amsterdam electric taxicab company Atax emerged from a well-known horse taxicab company with a long tradition in the Dutch capital. The body department on the second floor of the Atax factory at the Overtoom, pictured here, turned out replacement wooden bodies. (ARM Archives)

The gasoline car now penetrated the sales subdivision of the ARM, which became increasingly important—a tendency we also observed at the American EVC. Shortly after the war its list of imports had been extended with the inclusion of the Belgian brand Minerva and the Oesterreichische Daimler Motoren-Actiengesellschaft, a branch of the German Daimler. As suppliers of electric three-wheelers the Berlin company Elitewagen AG (which took over Henschel & Company in its year of foundation, 1918) had taken the place of Gebhardt & Harhorn. Trucks were also among the vehicles sold. In this way, shortly after the First World War, the first outlines of a group of companies emerged, united under the ARM umbrella. The horse business was still the most important division by far. The record profit of more than Dfl 808,000 over 1919 was generated for 58 percent by the traditional activities. And although the Atax began to feel the competition again when the driving ban on gasoline cars was lifted, its share of 17 percent in the profit generated was not negligible. The sale, maintenance, and rental of luxury cars with 13.3 percent and the gasoline cab with 11.5 percent were not trifling either. In brief, in 1919, the motorized vehicle was responsible for 42 percent of the gross profits, whereas the gasoline car accounted for almost a quarter of those profits.[86]

In 1920, the fat years seemed to be over. In the opinion of the ARM board, apart from the general "depression in trade and industry," an important factor was the new policy of the Amsterdam city council. When a firm applied for a taxicab license, the council now not only wanted to subject the rates to its judgment, but also the terms of employment of the drivers. In July 1920, the municipality set up an Auto-Centrale (AC) that from then on would distribute requests by telephone for a taxicab among the member-companies. The customer could no longer influence the choice of the company (and, remarkably enough, the type of motive power). This put an end to the prewar support by the local authorities in favor of electric propulsion in the taxicab business. Only at the cab stands was the customer free to choose his own taxi. In April 1922, the city lowered the cab rates under loud protest of the Amsterdam taxicab companies.[87]

The year 1921 was the first since 1904 that ARM did not pay out a dividend. That was a serious disappointment after the boom years of 1918 and 1919, when the dividend had reached an unsurpassed record of 12 percent. In the two years after 1921, no dividend was paid out either. In 1923, the ARM suffered a loss of more than 50,000 guilders and the taxicab business that year also closed with a deficit, due to a lengthy strike that lasted almost eight months. At the resumption of the service on 8 January 1924, new municipal regulations came into force, which enabled "an operation of the taxicab business on a more commercial basis."[88]

In the midst of this turmoil, the ARM board did not hesitate to decide the fate of the electric taxicab. In 1924, it made a beginning with a "renewal of the vehicle fleet of the Stationing Business," which was practically completed in 1925. The auditor's report for 1924 already mentioned the board's statement that "the electric vehicles and the old gasoline vehicles that are unsuitable will gradually be replaced by gasoline vehicles of a light type (Citroën)." As a consequence of this decision, the "department of electric automobiles" was liquidated. In 1925, the taxicab fleet consisted of eighty-four cars; eighty-two of them were Citroëns. A year earlier batteries had begun to be dismantled, and seventeen electric cabs were scrapped. Mid-February 1926, the last electric Atax was completing its final trips in the streets of Amsterdam.[89]

Conditions for a Successful Operation: An Analysis

Compared to other large cities, such as Paris, New York, London, and especially Berlin, Amsterdam experienced the motorization of its taxicabs quite late. Because of the ARM's secure position in the transportation sector in the Dutch capital city, the considerations for motorization could take place in peace and quiet. Only when the Namag in Bremen launched a second-generation Kriéger-Lloyd, based on a French license, did an attractive candidate seem to present itself, although its body soon became outdated. The weakest spot of the electric car, the lead battery, had at the time of the Atax's foundation developed into a reliable power source—provided it was applied in a centrally managed fleet with a relatively favorable maintenance contract, based on effects of scale. These two factors, the veteran French-German electric car with its fleet experience in German cities and a reliable battery, made the ARM a critical follower of foreign vehicle technology. In hindsight, the timing of its foundation could not have been better.

The pneumatic tire was more problematic. It was clear, even before the Atax foundation, that the costs of the tires would be higher than the battery

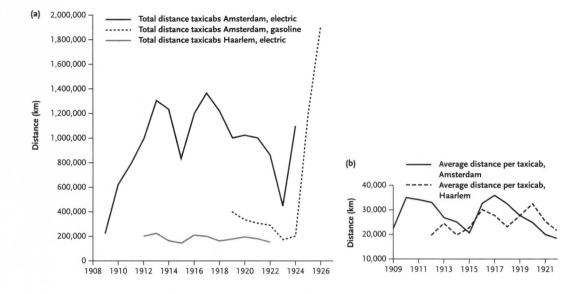

(a) Total distance taxicabs Amsterdam, electric
Total distance taxicabs Amsterdam, gasoline
Total distance taxicabs Haarlem, electric

(b) Average distance per taxicab, Amsterdam
Average distance per taxicab, Haarlem

FIGURE 4.11. Total annual distance of the effective Atax fleet and average annual distance per taxi-cab. (Monthly reports in the Friderichs file, ARM archives): a. Effective fleet, 1909–26; b. Average per taxicab in Amsterdam and Haarlem, 1909–22

costs. Consequently, signing a maintenance contract for the tires on a kilometer basis was of great importance. Without such a contract, Friderichs rightly concluded, foundation was senseless.

The decisive factor of fleet size indicates yet another element that is of importance in this analysis, which is mainly based on the meticulous records kept by Friderichs. Due to its expensive extra infrastructure, the electric car was not suitable as a luxury rental car, unless, as in the case of the Atax, a charging station was already present for another reason. For the second-generation electric cab, this leads to a conclusion opposite for that for the first generation. Because of the shaky technology of battery and tire (and the lack of a gasoline rival), the EVC tried to survive by turning to the less intensive rental business. But the second-generation electric cab only stood a chance against the gasoline competition by being used as intensely as possible. Trompenburg was one of the first to find a niche market here for the gasoline car (the luxury rental car). From this niche, the position of the gasoline car could be further reinforced after the war. This also applied to the taxicab business, as the ARM proved by the takeover of the TAM.

Looking closer at the field of application, the use per car was surprisingly high—between 20,000 and 35,000 km a year. The performance of the Haarlem taxicab was somewhat lower (except, of course, during the strike) than the Amsterdam cab (figure 4.11). During the entire period considered here, the distance per ride was approximately 4 km (Amsterdam) (figure 4.12), a value that was only established from 1912 (at a fleet size of about forty cars). Apparently, this figure forms the threshold value for the Amsterdam fleet size. Before this, the number of cab stands was so low that larger distances per taxi had to be covered to get passengers to their destinations.

Although the average number of rides per taxi generally shows the same

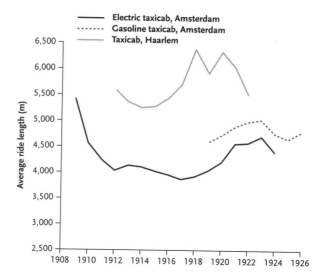

FIGURE 4.12. Average
ride length in the Atax,
1909–26; Haarlem from
1920: gasoline taxicabs.
(Monthly reports in
the Friderichs file, ARM
archives)

picture as the annual mileage per taxi, figure 4.13 shows a considerably higher
peak in the years 1916–18. There were 2,000 rides a year more in that period
than in the years before the strike, that is, five to six extra rides on average more
per day. This increase can be explained by the establishment of the bonus system that linked the drivers' wages to the "full" kilometers covered and to the
proceeds per kilometer. The maximum for 1918 was due to the transportation
monopoly of the electric cab, due to the driving ban on gasoline cars that year.

Finally, figure 4.14 and table 4.1 allow us to compare electric and gasoline
propulsion over these years. In the first place, both revenues and expenses (expressed in cents per kilometer) rose dramatically in relation to the prewar
level: the increase was more than double. Furthermore, the costs of the electric version decreased notably, possibly because parts were being taken from
the secondhand cabs from Bremen, Düsseldorf, and Munich. At the same
time the costs of the gasoline cab rose sharply, especially because of a gradual
increase in repair costs (figure 4.15; see also table 4.2), just as Friderichs tire-
lessly kept insisting ten years earlier. Eventually, the total costs of the gasoline
cab (in 1921 and 1922) were just below those of the electric cab, largely because
of the considerable drop in fuel and pneumatic-tire costs in those years. The
comparison is all the more remarkable because a fleet with about 40 percent
new gasoline cars was set opposite an old-fashioned and patched-up electric-
cab fleet, a situation which is comparable to that of the New York Transportation Company during its later years. This makes the question about the motives behind the switch to gasoline even more interesting.

The expansion of the taxicab fleet after the war can be entirely attributed
to the gasoline car. The Atax had to pay for its dependence on just one supplier—and one in a shattered Germany to boot. In France no electric-car manufacturer of any significance was active at that moment. And the United States,
with its electric Bakers and Detroits, apparently was too far away. It is true that

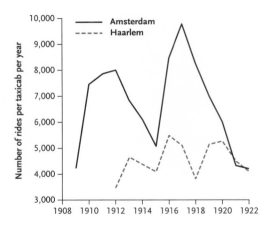

FIGURE 4.13. Average
number of rides per year
per taxicab at the Atax,
1909–22; Haarlem from
1920: gasoline taxicabs.
(Monthly reports in
the Friderichs file, ARM
archives)

in the course of the 1920s electric-car manufacturers emerged again, but by that time the gasoline car apparently was already entrenched in the ARM's taxicab fleet.

For the years after 1922, cost specifications of the gasoline cab are lacking, unfortunately. Figure 4.16, in combination with figures 4.9 and 4.11 do, however, offers some insight into the nature of this substitution process. It appears, for example, that it was not until 1926 that the gasoline-cab fleet achieved the same number of effective operating days as the electric-cab fleet did before the war, with a considerably smaller number of vehicles. This cannot be explained by the total annual mileage of the fleet, which had already been higher than that of the electric cab in 1925 (figure 4.11). Nor can it be explained by the daily route, which was longer than that of the electric cab from the very beginning (figure 4.9). So, the only possible explanation is the lower deployment of gasoline propulsion (as a result of a lower reliability).

This is truly a remarkable conclusion for state-of-the-art gasoline taxicabs in the early 1920s. It is also remarkable that their lower reliability was largely offset by higher speed. What the gasoline-car fleet, compared to the electric fleet, lost in usability, it compensated for by bringing in revenue faster. This is shown in figures 4.8 and 4.12: as the average ride length was about 6 to 12 percent higher at a virtually equal number of rides a day, daily revenues were hardly higher than those of the electric cab. Indeed, they were surprisingly comparable (see figure 4.8). This is confirmed by the average length of the paid ride, which had been higher for the gasoline cab from the start. In Amsterdam it was around 10 percent higher; in Haarlem it was even 11 to 17 percent higher than for the Amsterdam electric cab (figure 4.17). So, the larger range (that was *not* beyond the range of the electric cab) did *not* lead to higher daily proceeds, due to a remarkably comparable rate of occupancy and thus a correspondingly larger number of "empty kilometers." A share of about a third in empty kilometers seems a usual value at this stage of the second generation and in a town of this size. This was also the average in Hamburg. There is only one conclusion possible: the gasoline car as utilitarian vehicle had lost so much of its unreliability that its operation in a large fleet did not basically differ from that

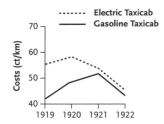

FIGURE 4.14. Total expenses for electric and gasoline taxicabs in cents per kilometer for Atax-Amsterdam, 1912–22. (Friderichs, "Vergelijkende staat Atax en Taxi Amsterdam over de jaren 1919 t/m 1922," ARM archives; reprinted from Mom, "Haver- en andere motoren," 263; see also table 4.1)

of the electric car. Whatever it was still lacking in reliability, it compensated for by its higher speed; the gasoline alternative also benefited from the well-established maintenance culture.

As discussed in the following chapter, American electric-car proponents had an aversion for the comparison between gasoline and electric propulsion on the basis of costs per kilometer. This aversion was due to the conviction that the gasoline car by virtue of its larger range would always come out favorably. That is only true, however, if the field of application allows such a range, such as an extensive city that enforces daily routes that are beyond the range of the electric car. In such a case the comparison is indeed unfair. Then the comparison based on daily costs is a suitable alternative, because in that case the difference in range is neutralized. In the case of the compact Amsterdam city area of around the First World War, the comparison on the basis of kilometers is, however, justified. It conclusively shows the greater reliability of prewar electric propulsion.

Finally, with regard to the possible technical failure factors, battery technology had advanced to the point that its maintenance and maintenance costs no longer presented any problems. But the thing that makes the Atax's foundation and flourishing so interesting is that they took place in a transitional stage of the technical and cost development of the pneumatic tire. Also with regard to the pneumatic tire, however, we must determine that this at any rate was not the cause of the failure of the electric-cab project in the Netherlands,

TABLE 4.1 Expenses for Electric and Gasoline Vehicles at the Atax, 1919–22

	Electric Taxis				Gasoline Taxis			
	1919	1920	1921	1922	1919	1920	1921	1922
Revenues	66.00	63.40	56.33	47.82	64.23	61.83	55.58	47.02
Expenses	55.56	57.65	54.07	45.67	42.40	48.22	51.77	43.22
Proceeds	10.44	5.75	2.26	2.15	21.83	13.61	3.81	3.80
Specified expenses								
Tires	11.50	8.05	5.17	4.32	11.39	8.16	5.57	4.82
Batteries/gasoline	8.11	9.83	10.28	8.46	7.44	8.87	6.66	4.32
Electricity/oil	6.82	5.95	3.44	2.50	1.09	1.33	1.21	0.56
Repair	11.90	10.95	9.65	7.19	8.50	10.74	13.05	11.68
KM costs	38.33	34.78	28.54	22.47	28.42	29.10	26.49	21.38
Drivers	8.05	10.07	12.27	12.50	6.83	8.90	12.27	12.13
Insurance	0.83	1.41	1.60	1.54	0.74	1.35	1.76	1.59
Wages	2.83	4.61	4.81	3.03	1.14	3.84	4.81	3.03
Miscellaneous	5.52	6.78	6.85	6.13	5.27	5.03	6.44	5.09
Fixed costs	17.23	22.87	25.53	23.20	13.98	19.12	25.28	21.84
Total costs	55.56	57.65	54.07	45.67	42.40	48.22	51.77	43.22

Sources: Friderichs, "Vergelijkende staat Atax en Taxi Amsterdam over de jaren 1919–22," ARM archives; reprinted from Mom, "Haver- and andere motoren," p. 263. See also figures 4.14 and 4.15.
Note: All values are given as cents per km.

FIGURE 4.15. Maintenance and labor costs in cents per kilometer for the Amsterdam gasoline taxicabs of the Atax, 1919–22. (Monthly reports in the Friderichs file, ARM archives; see also table 4.1): a. Maintenance costs; b. Drivers' wages and average wages of the other Atax personnel involved in the gasoline taxicab.

because the stimulating factors (among which especially the "emergency situation" of the war) were much stronger. Nevertheless, it is remarkable that during the war costs on a kilometer basis were much more favorable for the electric car than for the gasoline car. It leaves room for serious doubt about Meade's conclusion that the gasoline propulsion alternative was "better." Meade compared modern gas cabs from Europe with old-fashioned electrics on solid rubber tires. The Amsterdam case, however, clearly shows that, given the good Amsterdam road surface, the vehicle speed (higher for the gasoline car) and the wheel slippage (also higher, due to the jerking acceleration and gear shifting) had a larger influence on the wear of the tires than the axle load (higher for the electric car). The tire technology and the gear-changing technology of the gasoline car apparently had made so much progress in the meantime that the two types of propulsion were a match for each other as far as costs are concerned. As soon as this was the case, rational arguments were no longer dominant in the process of choice of vehicle propulsion, even in highly "rationalized" ventures such as cab fleets. One such nonrational argument was the choice in favor of gasoline propulsion after the electric cabs of the second generation had worn out.

It is, therefore, not appropriate to speak of the second-generation taxicab in terms of "failure," especially in a technical sense. This is clear not only from the Atax history, but also from that of the Hedag in Hamburg and the Bdag in Bremen. The secret of the (short-lived) electric-cab success in these cities was due to the application of electric propulsion in an intensively used, compact fleet, heavily backed by local authorities. As such, the Amsterdam second-generation electric taxicab fleet presents a powerful counterargument against the now classic example of "electric vehicle failure" of the first-generation EVC fleet.

As far as the intensity of use is concerned, with a view to the restricted life span of the battery (about five months), due to sulfating and the low mechanical shock resistance, the taxicab business was eminently suited for electric propulsion. Because of this, spectacular distances of 15,000 to 20,000 km a

TABLE 4.2 Operating Results of the Amsterdam Taxicabs of the Atax, 1909–26

Year	Average Number of Cars	Number of Operating Days	Total Distance (km)	Daily Route (km)	Occupancy (%)	Proceeds (ct/km)	Proceeds (ct/day)	Total Number of Rides	Number of Rides per Day	Length of Ride (m)
Electric cabs										
1909	10.21	2,139	229,613	107.3	—	29.4	31.60	42,385	—	5,417
1910	17.86	6,517	621,129	95.3	63.0	26.8	25.60	134,907	20.7	4,600
1911	24.00	8,748	804,358	92.0	63.5	27.2	25.00	189,410	21.7	4,247
1912	30.87	11,300	982,798	87.9	63.2	27.4	24.00	246,114	21.8	4,034
1913	46.10	16,835	1,291,044	76.7	62.2	26.5	20.30	308,648	18.3	4,183
1914	48.40	17,646	1,222,224	69.3	63.3	26.9	18.60	293,341	16.6	4,167
1915	41.70	10,558	842,505	79.8	66.1	28.3	22.60	210,413	20.0	4,004
1916	36.60	12,410	1,207,181	97.3	68.3	30.2	29.33	307,158	24.8	3,930
1917	36.90	13,458	1,343,785	99.8	68.7	40.2	40.17	351,231	26.8	3,826
1918	38.00	13,910	1,223,696	88.0	69.1	58.3	51.25	315,727	22.7	3,876
1919	34.70	12,653	981,177	77.5	70.4	66.0	51.15	241,960	19.1	4,055
1920	41.60	15,233	1,029,846	67.5	68.0	63.6	43.01	243,265	15.9	4,233
1921	48.70	17,776	964,974	54.3	61.1	56.6	30.74	209,139	11.8	4,614
1922	44.00	16,044	846,388	52.8	62.8	48.1	25.40	182,626	11.4	4,634
1923	45.00	8,391	433,115	51.6	62.1	43.5	22.45	92,308	11.1	4,692
1924	39.80	14,576	1,066,778	73.2	68.0	35.7	26.15	241,800	16.6	4,412
Gasoline cabs										
1919	11.6	4,125	384,552	93.2	68.7	64.1	59.79	84,440	20.5	4,554
1920	12.0	4,396	344,018	78.3	68.1	61.9	48.46	73,134	16.9	4,704
1921	15.2	5,550	308,704	55.6	63.2	55.8	31.05	62,977	11.3	4,902
1922	14.7	5,355	296,738	55.4	64.2	47.2	26.18	59,639	11.1	4,975
1923	14.8	2,755	151,536	55.0	62.8	43.1	23.70	30,358	11.0	4,992
1924	6.7	2,441	195,514	80.0	68.8	35.4	28.38	41,400	16.9	4,723
1925	32.6	11,897	1,204,710	101.3	67.5	30.2	30.55	256,653	21.6	4,694
1926	56.1	17,058	1,897,303	111.2	66.3	28.0	31.17	394,844	23.1	4,809

Source: Monthly reports in the Friderichs file, ARM archives.

year could be reached during a battery's life span. When used less intensively, for instance, in a private car, such a distance remained far out of reach. So, Michael Schiffer's supposition (see chapter 3) that the private electric car did not suffer so much is not correct and is in fact a wrong projection of the wear behavior of machines on the lead battery. For the second generation the exact opposite is true: the more intensive its use, the more economical its operation. As German publicist Robert Schwenke already concluded at the start of the second-generation technology: "The use of the vehicles must entail that the battery will be worn out in five months, otherwise the 2 mm thin lead grids will shortly after fall apart of their own accord."[90]

As far as the compactness of the taxicab fleet is concerned, it was clear

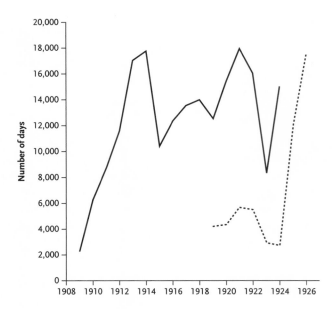

FIGURE 4.16. Total operating days of the Amster-
dam fleet, 1909–26; dotted
line: gasoline taxicabs.
(Monthly reports in the
Friderichs files, ARM
archives)

here how the electric cab made a virtue of necessity. The longer rides had to be
left to the gasoline rivals. These yielded a larger share of "empty kilometers,"
however, and thus prevented a profitable large gasoline-cab fleet before the
World War, at least in Germany. The fact that at a later date the gasoline car also
appeared capable of such profitable compactness, as the Atax history shows,
does not alter this fact. It does make clear once more, however, that one cannot look upon those phenomena as timeless properties: changes in the field of
application do not follow automatically from changes in the technical field.

Yet, in this seeming disadvantage of the gasoline cab its eventual "victory"
is also hidden, because less intensive use was an advantage. Therefore, this
type of vehicle could capture small market niches that the electric cab could
not. An example is the one-man company in Germany. With its low fixed costs
and a low purchase price of the car (and given sufficient high rates and a minimum technical reliability), such a company had a chance to survive. Another,
even more important example is the rental car. What had been an emergency
solution at the EVC and a possible reason for its demise, and what had appeared impossible at the Bedag due to a ban by the local authorities, formed
the short cut that Trompenburg chose to threaten the Atax monopoly in Amsterdam.

Why then could the electric car not extend its position (Berlin) or maintain
it (Bremen, Hamburg, Amsterdam)? Was it not the ideal "crisis car" and was
it not cheaper to maintain in a large fleet than a gasoline cab? The conclusion
can only be that the gasoline cab had an important *extra function*, which
widened its potential field of application considerably: that of the (paid) touring trip outside the city as well as the higher speed that accompanied it. That
extra function fit in with the practice of the private automobile market and
apparently provided such an important boost that electric propulsion was
doomed. This, however, was not because of an inherent technical inferiority,

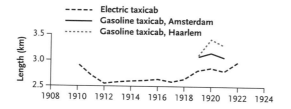

FIGURE 4.17. Average length of the paid ride of the electric taxicab in Amsterdam (1920–22) and of the gasoline cab in Amsterdam and Haarlem (1919–21). (Calculated from a graph made by Friderichs, ARM archives)

as some automotive historians argue; the later success of the Detroit taxicab fleet will show this conclusively (see chapter 7). So, when the Atax was looking for a substitute taxicab, there no longer was any dilemma regarding the choice of propulsion. The electric passenger car had virtually disappeared from the market.

In hindsight, the role of the electric taxicab can be characterized as follows: it did the pioneering work during the first phase of the substitution of horses by offering a reliable and often cheaper alternative in demanding and intensive city traffic. It fulfilled, in other words, the role of a Trojan Horse for the internal combustion engine. But at the same time, it functioned as a crucial example for the gasoline car of how to behave as a "civilized" urban vehicle.

However, electric vehicle proponents had one other trump card to play: the intrinsic technical flexibility of the internal vehicle structure, which enabled ready diversification toward other parts of the application field, all of these within the realm of commercial deployment. It is on the diversified utilitarian applications that we focus in the following chapter in our search for arguments for the "failure" of electric road propulsion.

5 The Electrified Horse:
The Commercial Vehicle in Europe

The gasoline vehicle is unclean and develops unpleasant
(terrible, to sensitive sick people) smells and gases. It moves
with noise and jerks, has jolting gear shifting, and is awkward
to turn. It is neither cheaper in use, nor in purchase, and—
last but not least—puts the sick person to be transported in
continual danger to be burned alive. After all this—and it
only concerns facts here—the question: gasoline vehicle
or electric vehicle? should have been answered for the future.
—*Physician Von Oettingen on ambulances*

THE APPLICATION FIELD of the second-generation taxicab showed that electric propulsion could be highly successful if applied in big, centrally attended
fleets under favorable local circumstances. This raises the question if the heavier vans and trucks, often deployed in much smaller fleets, stood a chance when
equipped with an electric drivetrain. In his dissertation about the history of the
battery, Richard Schallenberg expressed surprise that "the electric car and the
taxi faded, while electric vans and trucks gained popularity. The reasons for
this are not entirely clear, since the subject has not been subjected to any detailed analysis by historians of the automobile."[1] The following two chapters
are an attempt at such an analysis.

The commercial vehicle has been neglected in automotive history up until
now. It not only differs considerably from the passenger car in its technical
field, but also in its field of application. From a technical point of view, apart
from being more robust, certainly during the first decades there was no tendency to make the construction more attractive by putting the mechanism out
of sight. Such a tendency largely determines the history of the passenger car.
Also in its application, the machinelike character of this type of vehicle is
dominant, expressed in a greater emphasis on reliability and operating costs,
at a usually intensive deployment. At the same time, one finds a much greater
variety at the level of the technical field in the commercial vehicle than in the
passenger car. Enabled by a much more flexible technical structure, this enormous variety was an advantage as well as a disadvantage for electric vehicle

manufacturers. On the one hand, electric propulsion could be more easily deployed in a diversity of niches. On the other hand, however, early gasoline propulsion was not suitable for utilitarian applications and could get rid of its initial problems in one well-defined field of application, in the free space of the automobile as a sports object, not hampered by a multitude of divergent requirements. Thus, electric vehicle niches could turn into prisons instead of cradles for future dominance. Modern niche marketing strategies often seem to deny this possibility of the electric vehicle becoming an eternal "Tomorrow's Car."

During the earlier days of commercial vehicle development, however, expectations about the breakthrough to universality were already vocal. Such expectations were based on rational engineering insight into the technical superiority of electric propulsion. The commercial vehicle was a much more "logical" heir to the battery streetcar than the passenger car. The heavy commercial vehicle formed a refuge for the electric-car builders, at a time when the first-generation passenger cars threatened to end up as a failure. From this position, the second-generation commercial vehicle became an important product on a market, in many cases opened up by electric propulsion for the first time.

The market segment of the commercial vehicle was not characterized by a new transportation function. Horse traction had long dominated here and fleet management as well as a certain measure of cost-consciousness were common. As both the gasoline car and the electric car of the first generation were unsatisfactory as commercial vehicles, the second-generation electric truck did not find such a comfortable situation as the first electric cabs did. The replacement of horses by electric propulsion in most of the applications described here entailed direct competition with gasoline propulsion. This offers an excellent opportunity to study their roles in the process of social recognition of "the automobile."

It is precisely this social acceptance process that makes the commercial vehicle so important. The commercial vehicle (taxicab and bus, sprinkler cart and garbage truck, ambulance, mail van and police car, commercial vehicle and truck, fire engine) was the trailblazer for the passenger car, as it could be depicted as a social necessity, as a humane solution to the miserable horse life. The commercial vehicle also made state organizations (the mail services in the vanguard, the army as a decisively interested party soon thereafter) aware of the importance of motorized transportation. The First World War played a role in this acceptance process that is not to be underestimated.

As always in this study, it is important to make a fundamental distinction between periods, geographical differences, and the nature of the application. The last item includes the question of whether the automobile was deployed for urban or for rural tasks, and, whether—within the city—the deployment took place according to a fixed, preplanned route or whether it was unpredictable. Furthermore, the fleet size played a role in the competition between the types of propulsion. In this chapter all these aspects are dealt with, always in a different mix, depending on vehicle type and application. And always we

Heavy electric trucks of the Amstel Brewery, Amsterdam. With hub motors in the front wheels of the Lohner-Porsche system and solid rubber tires (double at the rear wheels), the reliable trucks were sold by Mercedes in Stuttgart as Mercedes Électrique. (DaimlerChrysler Archives)

will see that the balance of competition sometimes was in favor of the one type of propulsion, and at other times in favor of the other. But eventually, as in the passenger car field of application, the result was the same: gasoline traction was the "victor" in the battle between the propulsion systems.

Stop-and-Start Operation, Slow Motion, and Fixed Routes: Motorization of Mail Delivery and Municipal Fleets

After the taxi, the electric-vehicle manufacturers discovered the mail truck as the next target for guaranteed sales. In 1901, a first-generation electric mail truck became operative in Paris and there, as in several other European cities, the experiments were aborted after a while. Initially, vehicles with gasoline, alcohol, and steam engines were also tested in vain. But, on 24 October 1904, fifteen electric Mildé mail vans (of which three served as backups) were put into service, after a comparison with Peugeot gasoline cars. They would take care of all the mail transport between the headquarters and the twelve offices in the suburbs.[2]

The most important activities in this field took place in Germany, however. In Berlin the experiments with electric vehicles began as early as August 1899. None of these tests led to any purchases, mainly because of the high costs of the solid rubber tires. Interesting in this early period is the experiment with a three-wheeler that was used in Munich on a route with a length of exactly 6.4 km along 27 mail boxes, which all had to be emptied 15 times every day. Therefore, the vehicle with a weight of 430 kg could be equipped with a battery-vehicle ratio of only 30 percent. But the batteries could not cope with continually accelerating from standstill: their life span was only a quarter of what had been promised, so that economical operation did not seem possible. When, however, the three-wheeler was transferred to *Stadtverbindungsdienst* (bulk mail transport service) in 1903, where the stopping-and-starting operations were not necessary, it was perfectly satisfactory. Again it was clear that the boundaries of the technical field precluded one application, while allowing another.[3]

Just as in Paris, in Germany another series of experiments started in 1904, first in Cologne. That same year, tests were done in Berlin with a few NAG and Daimler gasoline cars. But it was not until 1908 that the decade-long involvement of the German Reichspost in electric propulsion really started. It was not a coincidence that this happened shortly after the foundation of the Namag in Bremen. In the contract between Namag and the Berlin mail authorities, a kilometer rate of 17 pfennigs was established for 31 cars. The minimum annual fee was almost 106,000 marks, based on a total yearly mileage of more than 622,000 km, or a daily distance of 55 km per car. Such a contract was possible because—in contrast to the taxicab—the daily route was largely predictable. All the expenses for the customer could be calculated in advance. For that fixed amount the Namag took care of the complete maintenance of the cars, including tires, repair of damage, and repainting. The fact that the costs were predictable and fixed—possible because of this early example of customizing—was thought more important than the question of whether the electric car was cheaper than gasoline propulsion under all circumstances. Of the latter propulsion type, one generally assumed that its costs would increase after some time anyway. In 1914, 220 electric vehicles were in use at the Reichspost (including Bavaria and Württemberg), 113 in Berlin (71 three-wheelers and 47 for parcel delivery), and 30 in Leipzig. Vienna at the time had 30 electric mail trucks.[4]

Municipalities initially were more hesitant to motorize their fleets than national mail authorities. This meant that the competition between electric and gasoline propulsion was especially fierce, because gasoline vehicle manufacturers had discovered this market as well. This competition is especially interesting because in most municipalities the fleet size was relatively small, so that the boundaries of the technical field of electric propulsion were extremely critical. The costs of infrastructure could only be divided among a limited number of vehicles, unless these vehicles were garaged in a central charging station, as was the case in Amsterdam. However this may be, anyone wanting to investigate the technical boundaries of the application of the private electric passenger car will find an important part of the answer in the archives of the municipal committees for the Sanitation Department and in the documentation for police and first aid cars. The scope of such a study would have to extend to various countries and in each country would have to include at least a few municipal archives. This means that finding the answer will be a long-term matter. For the time being, this section will present only a brief outline.

Municipal vehicle fleets were and still are very diverse and reflect in concrete fashion the public tasks that are part of municipal policy and differ considerably by geographical region and in time. The garbage collection service, for instance, has a different technical and organizational tradition than the

Part of the package delivery fleet of the Berlin post office—the largest electric vehicle fleet in Germany—in 1908. Namag in Bremen built all these front-wheel-drive vans under license from Kriéger. (Johann Heinrich von Brunn, *Ein Mann macht Auto-Geschichte* [Stuttgart, 1972], 103; reproduction Charley Werff)

street-cleaning service. The first should be considered in the wider context of garbage separation and processing. In the context of this study, it should also be seen against the background of the question of whether these tasks were carried out by municipal services or under contract by third parties. Furthermore, the way in which the city dwellers collected and stored their own waste played an important role, for example, whether they used the exchange-barrel system or a different method. Even in a purely quantitative sense there are big differences: in the period concerned American citizens generated an average of 390 kg of waste a year, in Great Britain citizens generated almost half of this amount, and in Germany even less (145 kg). In the United States, cities began to organize fleets of garbage trucks during the First World War, after most cities had abandoned the unsatisfactory contract systems with third parties. In Germany, a Verband kommunaler Fuhrparksbetriebe (Association of Municipal Vehicle Fleets, VKF) was established in September 1912. Its task was to concern itself with the deployment of vehicles. The first garbage truck became operative in Fürth in 1911, pulled by an electric tractor. Dortmund followed shortly after. Due to the war, this development gained momentum, especially in places where a system of exchange barrels already existed.[5]

Street cleaning became necessary when cities systematically started to pave roads and public hygiene became more than just a technical issue. Particularly in America, a real flood of mechanical solutions for sweeping, washing, and sprinkling problems emerged. In 1905, 85 percent of the cities with more than 25,000 residents and 55 percent of the smaller towns used some

Electric street-cleaning machine with central pivot steering, allowing an extremely small turning curve. The front driving unit, built by the German Elitewagen, could be decoupled from the cleaning machine and used for several other highly specialized municipal service functions. (ARM Archives)

form of (horse-drawn) street-sweeping machines. This was a particularly welcome new market for electric-car builders. The very low speed of these vehicles kept the gasoline competitors at bay, as controlling the combustion engine at these speeds was problematic.[6]

Initially street-cleaning motorization was especially significant in Europe. Typical for this application was the "tractor," which Kriéger also used for his first electric taxicab. This solution was particularly interesting for municipal vehicle fleets, not only because existing equipment could be readily maintained, but also because one tractor could be used for pulling different apparatuses. The tractor represented the last stage of the horse and the first stage of the car as utility vehicle. In fact, in Germany this vehicle type was called *Elektrisches Pferd* (electric horse).[7]

Other municipal commercial vehicles were the police car and the ambulance. After initiatives in 1900 and 1901 in Akron, Ohio, and Hartford, Connecticut, respectively, in 1904 an electric patrol car became operative in Los Angeles and in 1907 in Oakland, near San Francisco. They were also equipped for the transportation of sick and injured people; in 1909, the police in Baltimore mentioned that of the 21,000 requests for assistance, 2,200 were "ambulance calls."[8]

In the case of the ambulance, an important argument in support of electric traction was its smooth acceleration and the absence of smell and noise. The pneumatic tire was very important, especially because of the poor state of the spring technology of those days. In Berlin the Verband der erste Hilfe (First Aid Association) put an electric ambulance into service in 1905. It was built by KAW and provided with heavy pneumatic tires. The military medical doctors present at the occasion were very positive about the design of the transportation section, but did not mention the type of propulsion. After the war most ambulances in Germany were equipped with a gasoline engine. A typical example of this

This electric front-driving unit, built by German Elite, proved popular among municipal fleet owners because they could employ it in a host of functions, including the pulling of old horse-drawn equipment. (ARM Archives; reproduction Charley Werff)

development (and of the contrast between higher and lower authorities, emphasized throughout this study) is the city of Dresden, which purchased two ambulances in 1911 and 1912. In 1914, the army confiscated them, and in 1915, the city was assigned two old gasoline ambulances. Those were taken out of service in 1917, because the solid rubber tires were completely worn. After the war Dresden bought three electric ambulances from the Namag.[9]

At the outbreak of the World War, 53 of the 58 German municipalities with more than 80,000 residents had a joint fleet of 590 vehicles (unfortunately, not categorized according to type of propulsion). The majority of these vehicles consisted of fire engines, but 113 cars belonged to the categories *Strassenreinigung* (street cleaning) and *Strassenbesprengung* (street sprinkling). Berlin owned 42 of them, all electric. The competition with gasoline propulsion must have been particularly fierce in Berlin. For, in 1927, L. Betz (in his authoritative book about "special vehicles," which was completely geared toward gasoline propulsion) could not suppress a sarcastic remark about the so-called noiselessness of electric cars with their iron hoops around the wheels. But even Betz had to admit that electric propulsion had played a pioneering role, especially in the cleaning sector. This type of propulsion had received an extra boost, because during the war the army confiscated all municipal vehicles with gasoline engines. Betz even praised the "surprisingly good construction" of the electric vehicles, but then concluded with satisfaction that "the problem of the type of propulsion . . . now has been definitively and completely solved and . . . with no exception has led to the combustion engine." If that is true, it had only been so for a short while, for in 1919, Otto Barsch mentioned in another handbook

An electric ambulance, praised by medical doctors because of its smooth performance and the absence of gasoline odors. Mercedes-Électrique type hub motors, originally designed by Ferdinand Porsche in Vienna, drove the rear wheels. (DaimlerChrysler Archives)

in this field "that many councils of big and mid-size cities have not yet decided on the deployment of automobile propulsion in street cleaning, because adequate gasoline vehicles *are not yet* available." Here the same mechanism can be observed that cropped up earlier in this study: waiting for the gasoline car, even though another alternative was available, which was in many respects even "better." Many municipalities preferred "to stick to the irrational horse traction, however, in order not to expose themselves to the disadvantages of electric propulsion." Meanwhile, the electric version had become more expensive to use than its gasoline competitor, except in places "where the city concerned is in the possession of a garbage incinerator. Herewith steam is abstracted from the resulting heat energy, and from this electricity." This is one of the first expressions of a "chain consciousness," according to which the pros and cons of an alternative are not measured on the basis of the artifact alone, but on the whole "chain" of production and use, including the production of the "fuel."[10]

The Poor Man's (and Woman's) Automobile: The Bus

For public passenger transportation by rail-less vehicles, as for taxicabs, contests initially served as the touchstones of quality. For the first-generation city bus it was again electric traction that was dominant, as appeared from experiments in Berlin, Paris, London, and New York. In New York City the Fifth Avenue Coach Company put electric buses into service in 1900, followed in 1905 by a number of hybrids. In 1912, they were replaced by gasoline buses.[11]

The motorization of the bus started in most big European cities around 1905, and only really took off after the First World War. It was often initiated by private enterprise as a direct attack on the transportation monopoly of the municipal electric streetcar. Later those streetcar companies themselves also became involved in response to such attacks, or simply to create lines that were thought too expensive for electrification. In Berlin the history of the motorized bus started before 1898, when a few Benz and Daimler vehicles with combustion engines appeared on the road. In that year the Allgemeine Berliner Omnibus Actiengesellschaft (Aboag), a private company and owner of the largest number of horse-drawn vehicles in Berlin, organized a trial ride by electric bus. Two years later it put two of those buses into service. But electric traction was not deemed powerful enough for this application.[12]

In London motorization of the bus also started with electric experiments, such as those of the Electric Motive Power Company and the Ward Electric Car Company. The latter company, founded as early as 1888 and based on Radcliffe Ward's patents, became especially known because of the establishment of a new enterprise in May 1896, the London Electric Omnibus Company Ltd. This company announced the plan to put 125 electric buses in operation. Just like all grand schemes from this first-generation period, this initiative failed; only three or four experimental buses performed trial services in 1899. According to a contemporary, they even transported passengers free of charge, because no concession could be obtained.[13]

The competition on the London bus market differed considerably from that in Paris and Berlin, if only through its sheer size and its wide, level roads. In 1909, London had 1,100 motorized buses, versus Paris with 150 and Berlin with about 100. The London bus situation to some extent resembled that of the taxicab: from 1905 a large number of bus companies were operative, of which quite a few did not survive the recession of 1907. In one year the number of motorized buses in London dropped by 35 percent, while sixteen companies completely disappeared from the scene. A major difference from the London taxicab development was that the electric bus and the steam bus also played a role in the competition for the market. In 1909, the Metropolitan Steam Omnibus Company operated about twenty steam buses, while the London Electrobus Company had eighty electric buses in operation. The latter company had started a service between Victoria and Liverpool Street on 15 July 1907. This occasion had been preceded by a presentation to the press and prompted some journalists to write remarkably hostile commentaries. The threatening reentry of the electric bus in London led to a fierce controversy in the British trade journals and newspapers about the usefulness of these "electrobusters." The debate was pervaded by what one might call a "Bersey syndrome"; the references of the automobile magazines to the Bersey debacle of shortly before the turn of the century resulted in the withdrawal of several financiers. Nevertheless, in the fall of 1907 there were six electric buses and one hybrid in operation. British Thomson-Houston had put the hybrid on the road.[14]

Tudor (grid-plate batteries) and Gould (Planté type) supplied the batteries of the electric buses. They accounted for 37 percent of the net vehicle mass of 5,400 kg and enabled a range of 5.25 times the length of the line. But the batteries were exchanged after four rides totaling 51 km. Initially one electric motor was used with a double armature winding and two commutators. Later the company switched to a separate drive for each rear wheel, with chain transmission. Charging occurred in a block station of the Westminster Electric Supply Corporation. After a year 3 million passengers had been transported: 6.8 passengers per vehicle kilometer that yielded 0.86 French francs per vehicle kilometer, as calculated by a French contemporary. This was considerably more than the electric streetcar, which only generated 0.70 francs with 6.4 passengers per kilometer. And it was far more than the gasoline buses of the London General Omnibus Company (LGOC), with 3.6 passengers and a revenue of 0.50 francs. The 80 electric buses each drove a remarkable average of 39,000 km a year and cost 0.58 francs per vehicle kilometer: 29 percent for battery maintenance, 26 percent for the tires, 14 percent for electric energy, almost 9 percent for repairs, and 20 percent for taxes and other expenses. In March 1910, the company ceased operations. But after that electric traction kept playing a certain role due to the introduction of the well-known Tilling-Stevens buses with hybrid drive. They remained operative until the middle of the interwar period.[15]

In Paris, in 1905, Eugène Brillié won a contest for industrial vehicles, organized by the French automobile club. A Serpollet steam bus and two *pétroleo-electriques* (one de Dion and one Kriéger) also participated in the contest. The Kriéger hybrid used a generator (by polarity reversal) to start the combustion engine, an early application of the electric starter motor. Brillié then signed a contract with the CGO, the Paris monopolist in the field of public road transportation. He appeared to be the only person willing to enter a maintenance contract on a kilometer basis. He was allowed to supply 150 gasoline buses, which became operative during the two following years, shortly before he tripped over the crisis of 1907. Despite a lot of criticism on their functioning, by 1911, each of them had covered a total distance of 200,000 km.[16]

Here and there, municipalities responded at an early stage by introducing the electric trolley bus. In the winter of 1904–5, the Berlin Aboag experimented with such a vehicle, supplied by the AEG. Later trolley buses became operative in the German town of Halle a.d. Saale (1905), in Birmingham and Manchester in England (1907), in various larger provincial towns in France (for example, Lyon and Nantes in 1908), and around Vienna in 1910. The trolley bus was to disappear from the European scene shortly before the First World War, but emerged again afterward. It was especially popular in Britain, where it reached its peak shortly after the Second World War.[17]

For intercity passenger transportation the problem of the choice of propulsion did not play a role, except for the regional railroad and in some cases the trolley bus. Yet, this development is important for the electric car, because gasoline traction for heavy-duty applications was severely put to the test for the first time. Developed, as a result of this, into a universally deployable vehicle, it was about to join the attack on the electric city car that had already been instigated by the gasoline passenger car. An example of this is the *Überland-omnibus* (transit bus) that was developed in Bavaria, Germany, from the middle of the first decade. The same is true for the intercity bus in the United States

Electric buses saw service far beyond the paved streets of European cities. In this scene in a German village in about 1910 two early examples of a trolley bus—a Lohner-Porsche-Stoll system, manufactured by the Daimler Motorengesellschaft—exchange passengers, because they use the same feeder line. (DaimlerChrysler Archives)

in the 1920s, which provided a means of transport for nonmotorists, especially women.[18]

Big Men, Small Fleets: The Motorization of the German Fire Department

The extensively documented electrification of the German fire engine shows a number of decisive similarities to and differences from the vehicle types discussed earlier, particularly the taxicab. For, just as the electric cab, the fire engine initially was an urban phenomenon. Here, too, there was a centrally maintained (although considerably smaller) fleet. And the initial need for horse substitution played an important role as well. In contrast to the electric cab, however, the pneumatic tire as potential failure factor was absent: the vehicles were far too heavy for these tires. Moreover, vehicle speed—another argument often used against electric propulsion—was not a decisive factor, at least initially. The paramilitary nature of the fire department, particularly in Germany, further guaranteed a high measure of discipline among the drivers. Given adequate training, they handled the vehicles according to the regulations. Furthermore, if the fire department was going to use electric traction on a relatively large scale, the aspect of gender (the supposed female character of electric propulsion) cannot have played a significant role. The fire department was an exclusively male business and its strictly disciplined, heroic macho character is beyond dispute. Also, initially the selection criterion of reliability had a higher priority than the (purchase) costs, so that the argument of the higher purchase price of electric traction did not play a role of any importance.

And, finally, certain technical properties were in favor of electric traction in fire fighting, such as the quick start, the high torque at acceleration from standstill, and the absence of flammable fluids like gasoline, often used as an argument against the use of gasoline trucks near fires.

So, the fire engine forms an ideal subject for a case study of the opportunities and barriers of electrification of the automobile. If a rational choice, based on considerations of business management, for or against electric traction can be expected anywhere at all, it must be at the fire department. For there only one goal was important: increasing the effectiveness of firefighting in the ever-expanding cities.[19]

The electrification of the German fire engine stands out as a unique phenomenon, especially when placed against an international background, for in other countries it hardly occurred. Only in Vienna did a comparable electrification take place, while in Amsterdam the Berlin example was followed on a small scale. In France, Britain, and the United States the electrification experiments were mostly restricted to a few first-generation instances. These—just as those of the taxicab—had a paralyzing effect on later attempts.

The reason for the uniqueness of the German fire department is found in its professionalism and militarism. From the middle of the nineteenth century, urbanization, with its accompanying increase in fire hazards, led to a demand for a greater efficiency in urban fire control. This caused the emergence of professional brigades, a process that was characterized by two contrasting trends. First, the fact that the fire brigades originated from the *Turnvereine* (gymnastics clubs) led to a militarization of the organization. It had a strict hierarchy, with drills directly derived from the army, and a system of punishments, imposed by the one-man, authoritarian leadership of a *Branddirektor* (fire chief). Second, university-trained engineers joined the higher authoritarian and elitist ranks of the professional brigades shortly after the turn of the century. Initially these were mostly civil engineers (because of their knowledge of building technology), but later mechanical engineers took the lead.

The militarist and scientific approach to firefighting contributed to the character of the first professional brigades in Germany, much more so than in the United States, where the macho character of firefighting departments took on a different shape. Started on the basis of steam propulsion, soon heavy gasoline-propelled ladder wagons were racing through the big American cities, making themselves heard heroically by loud sirens and bells. The German brigades, leaning on a rigorous authoritative structure yet striving in vain for state recognition and influence on legislation until after the World War, tried to free themselves from a culture of voluntary fire-control systems. These tendencies also determined the history of the Verband deutscher Berufsfeuerwehren (Association of German Professional Fire Departments, VDB), established in 1900 after the example of the architect and engineer associations. Practically all officers who qualified for membership joined in. Yet, this vanguard of the firefighting sector in Germany formed a small minority, although it had a disproportional influence. The 26,620 nonprofessional firefighting units far outnumbered the 59 professional brigades in the 31 German towns with more than 100,000 residents. Of the 1,441,770 firemen in the entire German Empire, only 3,726 were part of a professional brigade.[20]

The German controversy over what type of propulsion the fire engine should have was closely linked to Maximilian Reichel (1856–1924). On 1 April 1900, Reichel became fire chief in Hannover, where he soon made public a *Denkschrift* (explanatory memorandum), a plea for the introduction of a complete *Automobil-Löschzug* (motorized train of fire engines). The motorization of the Hannover fire department began in 1903 and was the first to be completed. Reichel's first motorized convoy consisted of a steam-driven steam fire pump, an electric-driven carbonic-acid fire engine, and an electric *Hydrantenwagen*, carrying the fire hoses that could be connected to a fire hydrant.[21]

The electric vehicles were equipped with AFA's large-surface plate batteries (as usual with negative *Kastenplatten* [double grid plates]). The fire department took care of the maintenance. To keep the battery in good shape and to prevent its plates from becoming sulfated, it was necessary to drive a few kilometers every day. This explains the large number of "practice kilometers" of later vehicles, which on an annual basis often exceeded the number of "emergency kilometers." The batteries were charged in the main fire station, as direct current was available there, so that an expensive converter was not necessary. The vehicles had a range of 25 km at a speed of 16 km/h.

In many respects, Reichel's vehicles displayed first-generation construction principles. Their central pivot steering is an indication of the fact that he had primarily pursued the electrification of horse traction. Whereas everywhere the first-generation electric vehicles failed in other fields of application, fireman-engineer Reichel managed to finish his project successfully after twenty-one months. Because of this very remarkable feat (and unique in the early history of the electric vehicle), on 15 June 1905, Reichel was appointed chief of the Berlin fire department. Obviously, with proper engineering stamina and a strict maintenance culture, even the technical field of first-generation electric propulsion could be widened so much that a very specific application field was possible, albeit with a heavy battery concept that would not have functioned properly in much lighter passenger cars. Less than a year later, on 1 March 1906, the city of Berlin agreed to spend 50,000 marks on the building and testing of two experimental vehicles. This decision was based on Reichel's second *Denkschrift* that in its objectivity can serve as a model plea in favor of the motorization of firefighting practice. In its clarity, it also was an example of how to deal with the complexity of the choice of propulsion, against the background of the state-of-the-art in automotive technology in 1905. Reichel's motivation for starting yet another costly experiment was based on the argument that "local circumstances" did not allow indiscriminate transfer of the lessons from Hannover to the situation in Berlin. Moreover, there was still so little experience with the new automotive technologies that had become available that a profitability calculation was not yet possible.[22]

In his second *Denkschrift* Reichel showed himself an overt opponent of the gasoline engine as propulsion choice for heavy vehicles (he emphatically excluded luxury cars from his judgment). He had more problems with the choice between electric and steam traction. It is true that the range of the electric vehicles was small, he argued, but due to the size of the city and the density of the fire-station network, this did not present a problem. Thus, the advantage of the "immediate take-off" with electric propulsion seemed to overshadow steam traction. Yet he thought steam propulsion more reliable. As the fire department already employed 121 engineers and stokers to operate the twenty steam-driven firefighting pumps (which were pulled to the fire scene by horse traction), he found the operation of steam traction "extremely simple." On the other hand, the steam boiler required permanent supervision, whereas the batteries did not, he argued. Reichel's freedom of choice, however, was considerably limited by an important precondition of the Berlin city council: the horse-drawn steam fire pumps and fifteen ladders, purchased in the past couple of years, had to be reused as much as possible. Reichel interpreted this condition as an extra argument against gasoline-engine propulsion, as the mechanical transmission between engine and driven wheels was an obstacle to reusing the old equipment. This supports our earlier conclusion that gasoline car manufacturers had much more trouble motorizing *existing* vehicles, a fact that forced them toward new vehicle structure designs from an early date. Faced with this techno-political constraint, Reichel then started to work on the design of an electromobile carbonic-acid fire pump and a steam-propelled steam fire pump. A committee, including a few fire department officers and representatives of the automobile and battery industry, assisted him.[23]

If Reichel's second *Denkschrift* is an excellent example of problem definition, his report on the trial rides can be read as an equally well-documented rejection of steam as possible alternative for urban fire-engine traction. During the tests of the two prototypes (received after a delay of about seven months due to construction and delivery problems), Reichel worked with a thoroughness and a methodology that must have been enviable in the infant automotive industry. As if it concerned a modern automobile testing program based on the principle of "time compression," each car drove a distance of 10,000 km between 29 April and 21 September 1907. According to Reichel, a life span of ten years was simulated in this way, with a view to the annual mileage for horse traction of about 1,000 km, which he took as a standard. The electric vehicle made two rides every day, achieving an average of 61.5 km per ride, a longest daily distance of 154 km, and a maximum speed of 36 km/h. This surprisingly high speed was reached at the cost of range, due to the inversely proportional relation between capacity and discharge current of the batteries. During the five months of testing, ten malfunctions occurred, not of the motor or battery, but of the tires and ball bearings. The average energy consumption was 580 Wh/km, a remarkably low value and a proof of Reichel's engineering talent. At a kWh rate of 16 pfennigs, this resulted in an overall cost of 9.2 pf/km.[24]

The steam truck caused the most problems. Although Reichel boasted about the longest distance of 206 km, driven in 9.5 hours without stopping, the vehicle appeared much too heavy to zoom through the villages at full speed. The solid rubber tires simply were not able to withstand such brute force. A cost comparison with the electric version, moreover, revealed that the latter required 1,205 marks in maintenance and operating costs on an annual basis. Of this amount, 561 marks was spent on battery maintenance and electricity and a similar amount on tires and *Gleitschutz* (antiskid covers around the tires). Maintenance of the steam car cost 500 marks more. Due to Reichel's special design, this amount was considerably lower than other brigades spent on steam traction, but the conclusion was inevitable: for the city, electric propulsion was superior. For Reichel this did not mean that steam propulsion would be entirely abandoned. He emphatically reserved this type of propulsion for the rural application field, but had to admit that its definite technical field had yet to be defined. So, he used the test report to propose the purchase of a new experimental truck equipped with a steam-electric hybrid propulsion system. The electric propulsion system would ensure a quick start. After about ten minutes, the boiler (meanwhile under pressure) would take over. At a speed of 50 km/h the electric motors, after polarity reversal, would be able to recharge the batteries. Climbing a steep slope, the electric motors could serve as boosters; they would be switched on to support the steam propulsion system.[25]

To the surprise of many, Reichel, who was widely known as a steam proponent, then bought a set of vehicles *fully* powered by electricity, for the sum of 133,500 marks. The new vehicles became operative on 14 September 1908 in the new fire station no. 4 on the Schönlankerstrasse. The vehicle train was comprised of a carbonic-acid pump, a tender (for the transportation of men and equipment), a steam pump, and a ladder truck, all on an electrically driven chassis. The chassis of all four vehicles was standardized and constructed in such a way that the steam fire pumps and ladders for horse traction could be mounted without any problems. All vehicles had front-wheel drive with Lohner-Porsche wheel-hub motors. According to contract, the chassis had to be purchased from the Deutsche Mercedes-Verkaufstelle in Frankfurt. Thus, Daimler Motorengesellschaft (DMG), to which the Frankfurt branch belonged, had raked in an important order. Moreover, Reichel followed in the footsteps of the Viennese brigade, which had forty electric vehicles in operation, all with the Mercedes-Porsche (the former Lohner-Porsche) system.[26]

The new vehicles were special in yet another way. The standard structure also included the relatively light AFA battery based on the grid-plate concept (type 3 Ky 285/4, weighing 880 kg per vehicle), a clear indication of second-generation design. It is curious, however, that the test report and the substantive annual reports thereafter provide little information about the experiences with this battery—a lacuna that is all the more noticeable in the light of the experiences of other brigades. At the fire department of Charlottenburg, near Berlin, for example, the new battery type turned into a nightmare that lasted for years. The Charlottenburg firemen daily experienced what was known among AFA engineers as "rotting" of the positive plates, a phenomenon that caused the metallic lead of the grids, meant for conducting current and for mechanical strength, gradually to deform and convert into lead dioxide, which has no mechanical strength and is a bad conductor. Moreover, the capacity as well as the life span suffered from the battery's stronger self-discharge and especially from the sulfating of the lead plates. This was caused by a typical feature

A first-generation electric fire engine for the Vienna fire brigade. The vehicle combined a steam-driven pump and front wheel-hub motors designed by Lohner-Porsche. (DaimlerChrysler Archives)

of the firefighting truck: its long periods of inactivity (and as such was an indication of its restricted usability for individually owned passenger cars). In Charlottenburg, hundreds of short-circuits occurred with this new battery type, caused by the poor insulation of the wooden separators. So, in July 1908, hard-rubber plates were inserted between the separators and the lead plates. Two or three men worked full-time on the maintenance of the batteries.[27]

In Schöneberg, also in the vicinity of Berlin, the problems with the grid-plate batteries were insurmountable. Whereas the large-surface plate batteries only had to be replaced after six-and-a-half years (7,140 km), the life span of the grid-plate batteries was continually threatened by sulfating. This led to their replacement by the more robust type in May 1910. On the other hand, the electric ambulances, purchased in 1908, as well as an electric service car, were equipped with grid-plate batteries without any problems. Later it appeared that "because of the daily use of this battery sulfating did not occur." Nevertheless, after three years of operation, the whole vehicle train appeared to be 35 percent cheaper than the horse-traction fleet, despite the expensive steam car, which was part of the train. In 1911, the Schöneberg fire department decided to have the supplier convert the traction system of the steam-driven engine into an electric propulsion system.[28]

With some effort, an explanation can be found for the curious fact that the very detailed Berlin reports mention so little about any grid-plate battery problems. First, as these reports are very frank about the occurrence of defects, it is not likely that something was swept under the carpet. That leads to the assumption that the delivery date of the Berlin vehicles (14 September 1908) coincided with the moment that the major initial problems of this new technology had been overcome by AFA, its manufacturer. Second, the difference between the well-equipped and disciplined Berlin brigade on the one hand and the smaller brigades in the two neighboring municipalities on the other sup-

Fully equipped second-generation electric fire engines (*Löschzug*) as employed by the Berlin fire brigade. The Berlin fleet, by far the largest in Germany, traced its electric propulsion to the enthusiasm of fire chief/automotive engineer Maximilian Reichel, who had begun his career as a partisan of steam power. All vehicles employed front wheel-hub motors; large hoods protected the massive batteries. (DaimlerChrysler Archives)

port our earlier conclusion that the *quality of the maintenance* was decisive for the application of the grid-plate technology. This was true when these batteries were applied in taxicabs or individually owned passenger cars, but it was certainly the case at the fire departments with their long periods of inactivity and short bursts of activity. A few years later, when electric propulsion had become more widespread among the German fire departments, the battery problem became the subject of long discussions within AFA. These eventually resulted in the decision to discard the wooden separators in the fire-engine batteries and to install the far more expensive hard-rubber partitions instead. The AFA engineers preferred the large-surface plate batteries for this application, but they often had to supply grid-plate batteries, because the fire chiefs were afraid of too low a range. This is a good example of how technology is the result of both internal (technical possibilities and constraints) and external (cultural preferences, negotiations between relevant actors) factors. Eventually it was decided in 1911 to recommend large-surface plates to brigades with a range of 40 km or less, and grid plates for brigades traveling a larger distance.[29]

However this may be, the Berlin firefighting vehicles were a success in every respect. Reliability was more than satisfactory, and the costs of energy consumption were acceptable, despite the high electricity rate of 15 pf/kWh. The number of heavy, electric-powered fire engines, all with the same standard chassis, grew at a steady pace, initially with one, and later even with two complete vehicle trains a year. In 1913, there were 49 vehicles, part of a total vehicle fleet of 74 that included 25 passenger gasoline cars, but no steam cars. In Reichel's opinion, another 71 heavy vehicles were needed to round off the Berlin motorization project (figure 5.1). At that moment all electric vehicles together, during a total of 25 years of operation, had covered a distance of virtually 100,000 km. That meant an annual mileage of 4,000 km per vehicle—

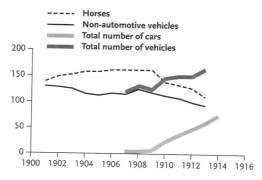

FIGURE 5.1. Horse substitution or new functions? Motorization at the Berlin fire department, 1901–14, in each case as of 1 April. Until 1908, seventy of the "nonautomotive vehicles" were "constantly horsed" for the "front service" as part of twenty firefighting trains; the rest were "backup." From 1909, eighteen bicycles were also included in the "nonautomotive vehicles." The number of electric vehicles in 1910 has been raised by four from 1911, because their delivery date was 10 March, probably after the report over the year 1909-10 had been finished. At the outbreak of the war, seventy-one extra vehicles were planned besides the forty-nine mentioned. (*Bericht Berlin over 1901–14*)

four times as much as what had been usual for horse traction. On average each vehicle cost 1,389 marks in maintenance and operation, versus 5,478 marks for a horse vehicle, so that a total amount of 400,000 marks was saved, according to Reichel. The cumulative costs for tire maintenance (including *Gleitschutz*) amounted to 25 percent, for electricity 24 percent, and for battery maintenance 43 percent of the total of 137,029 marks.[30]

DUE TO REICHEL'S PUBLICITY offensive in Hannover, the motorization of the German professional firefighting community, although starting slowly in 1902, really took off in 1906 after his propagandistic and technical successes in Berlin. The most noticeable item in the statistics of the German fire engine is the abrupt end of steam traction. After the fall of 1908, except for a single steam fire engine in Leverkusen in September 1911, no steam vehicles were purchased (figure 5.2).[31]

But Reichel's successes gradually generated a fierce and heavily debated controversy. During this controversy, two camps emerged—one of "electric cities" and one of "gasoline cities." The dividing line between the two camps cannot be explained purely by local circumstances, although most participants in the debate simply used this argument to justify their particular choice. Frankfurt, Aachen, Stettin, and Breslau (as second city of Prussia) were typical "gasoline cities." Berlin, Hamburg, Magdeburg, and Hannover belonged to the "electric camp." It seems as though the demarcation line was largely drawn on the basis of fire chiefs' egos, as massive and outspoken as Reichel's. This made the controversy a fascinating clash of opinions, unique in the history of (auto)-technical choice.

The debate entered a new dimension with the intervention of Johannes Schänker (1866–1950). Cut from the same Prussian wood as Reichel, and like

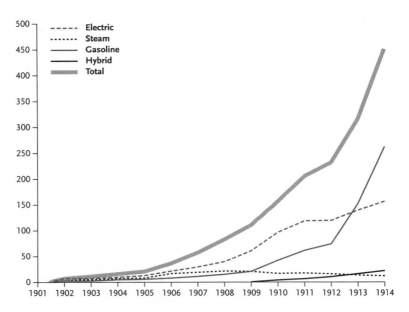

FIGURE 5.2. The motorization of the German fire departments, 1902–14. ("Zusammentstellung der seit dem Jahre 1902 bei den deutsche Feuerwehren in Betrieb gestellen bezw. Definitiv in Auftrag gegebenen Automobil-Fahrzeuge bis zum Anfang des Betriebsjahres 1911" [appendix to VDB *Verbandstag* 1912]; Hüpeden, "Statistik der Kraftfahrzeuge im Dienste deutscher Feuerwehren," *Feuerpolizei* [1914], 113–17; "Automobilgeräte und Fahrzeuge bei Deutschen Feuerwehren; Stand vom 1. April 1914" [VDB *Verbandstag* 1914, 26–28]; the latter supplemented with data from the fire brigade trade journals about Berlin, Mainz, and Offenbach)

him molded by his training in the Berlin brigade, this mechanical engineer was appointed *Brandmeister* in Hamburg in 1894. After a period in Dortmund, where he became fire chief in 1903, Frankfurt am Main recruited him in 1906 in the same position. There he remained for a quarter of a century, until his retirement.[32]

In 1907, Schänker adapted a (gasoline) Daimler truck chassis to *his* local circumstances. These were different from Berlin, indeed. By the relatively early annexation of towns in the vicinity, the application field of the Frankfurt brigade had extended to 13 km, including country roads of poor quality. As the urban fire departments were legally obliged to help in case of fire in neighboring villages, the dilemma of choice between propulsion technologies gradually began to shift. At the annual conferences of the Verein Deutscher Berufsfeuerwehren (VDB), the controversy between Reichel and Schänker (the latter fiercely supported by the Aachen fire chief Wilhelm Scholz) became increasingly more outspoken. In 1911, this led to the resignation of Reichel as chairman of the crucial Automobile Committee. Scholz became the new chairman.[33]

In the midst of the verbal abuse at the VDB conferences, which really made the propulsion choice issue into much more than a mere technical or economic problem, it became increasingly difficult for the fire chiefs in the smaller towns (and in most of the bigger cities that had not yet completed their motorization projects) to justify their particular choice of propulsion system.

For some of these doubters an attractive alternative emerged that could only be characterized as both a technical and a social compromise: hybrid traction. Elberfeld and Kassel were among the towns that implemented such a combined traction type before 1912. The degree of compromise of this solution becomes especially clear by the meaning attached to this type of propulsion: whereas the electric-car proponents described the hybrid as merely an extended electric truck, the gasoline proponents regarded it as an adapted gasoline truck.[34]

Just when the share of electric traction in the vehicle stock of the German fire departments began to drop, the short-lived and modest hybrid-traction boom began. In 1912, only one of the twenty towns that had made a decision about motorization opted for electric propulsion, seven for gasoline traction, and the remainder—the majority—for hybrid propulsion. The conflict about the right fire engine now threatened to result in fleets with an increasingly complex structure, due to the long life span of vehicles because of their robustness and low annual mileage.[35]

Meanwhile, the main point of the debate about the choice of propulsion had shifted from electric or gasoline to hybrid or gasoline. What at the time was a compromise, in historical perspective appears to have been a transitional form. It seems as though hybrid traction pops up in automotive history whenever the electric vehicle threatens gasoline propulsion dominance, not when the electric is just a possible alternative that occupies a niche. The chances for the hybrid to get such a role were especially great in a period of uncertainty among gasoline manufacturers regarding the question of whether gasoline propulsion could be made to function reliably under utilitarian circumstances. This makes the hybrid, besides a technical and a social compromise, also a strategic compromise. In the context of the German fire-engine propulsion debate, the conflicting interests proved too great for reconciliation. Aachen fire chief Wilhelm Scholz, in a presentation at the annual VDB conference of 1913, fiercely attacked the hybrid propulsion concept. A year earlier Scholz had already argued strongly against the particularism among fire chiefs, who often hid behind the local circumstances argument. Scholz then presented a three-pronged plan.

He pleaded, first, in favor of a light standard car, which, second, he wanted to be considered for subsidy by the army—just as in the case of the subsidized truck manufacture (see the next section). And to cap it all, he demanded "the most far-reaching application of the combustion engine," not only for urban but also for country application. After his attack on the hybrid in 1913, Scholz's presentation was increasingly criticized as "too scientific," a reproach he considered as a compliment. But the reproach referred to a new element in the debate, put forward by what had become known as *Automobilwissenschaft* (automotive science). For, at the end of 1912, Professor A. Riedler of the Technical University Berlin got involved in the debate with an article in an automobile magazine that was also published in the prestigious and professional firefighting magazine *Feuer und Wasser* (Fire and water). In the article he accused the electric carmakers of "clever advertising" directed at "outward appearances" and "fashion," and of promoting a propulsion type that could not stand the test of scientific criticism. He also implicitly referred to the VDB debate with the remark that one had tried there "to save the system by artificial contrast of 'science versus practice.'" Scholz (who was to get his doctorate degree on an analysis of the propulsion problems for firefighting vehicles two years later) felt that Riedler's remarks supported him against "the men of practice." In their own tests, these adversaries were often not able to compare the alternatives because the cost calculations were usually contrasted with a hypothetical horse economy for the benefit of the local authorities. Thus, for this comparison they could only use qualitative arguments.[36]

In his critique, Riedler also referred to his colleague Wehrle, "Vorstand der Münchener amtlichen Prüfungsstelle für Kraftfahrzeuge" (chief of the Munich municipal experimental station for motorized vehicles). The Munich authorities had asked Wehrle for help. The chief of the professional fire department (in contrast to many big German cities, Munich also had an extensive voluntary firefighting brigade) had published a *Denkschrift* in the fall of 1912, in which he pleaded in favor of hybrid propulsion. But in a second *Denkschrift* of January 1913, after a discussion in the city council, he suddenly proposed a ladder truck with electric propulsion, and gasoline propulsion for all other vehicles. Wehrle also suggested in his report—discussed in great detail by the local press—that due to the looming "switch to gasoline propulsion" the electric-vehicle makers "hoped to profit from the revival of 'mixed systems' to further exploit their factories, geared towards the production of electric machines." This dismissed the fire-department engineers who were in favor of electric and hybrid versions as misled amateurs.[37]

Scholz was not the only one to feel supported by the scientific interference. The automotive industry also expressed its doubt in lectures to the Automobil-und Flugtechnische Gesellschaft (Society for Automobile and Aircraft Technology) about the value of hybrid propulsion, "the more, as the gasoline-electric system has not yet participated in a military experimental contest." And so the army's preference for gasoline propulsion was effectively used against the further expansion of electric propulsion within the (militarist) firefighting community.[38]

It is remarkable in this context that meanwhile in the debate—by its mixture of technical and emotional arguments assuming an increasingly grotesque character—both parties had begun to define the electric car from the perspective of the gasoline car. This was especially obvious in the approach to the battery, where the cry for a "miracle battery" in these kinds of debates is never far away, as we have seen earlier. "We will see each other again in 10 years' time," firefighting engineer Prölls from Hannover exclaimed. "Hopefully by that time the right battery with a larger capacity will have been invented . . . If that succeeds, gentlemen, the electric propulsion system will dominate again under all circumstances." Remarkably, his opponent Scholz was entirely of the same opinion in this respect: "Only the creation of a light energy-efficient battery will keep the clean electric-motor propulsion at the height of its present application and will further stimulate it." Thus, slowly but surely, and entirely from the point of view that the range of gasoline propulsion was "superior," electric propulsion was put aside as an "inferior technology." The idea of the inferiority of electric propulsion was fed by the fact that the expectation of a "miracle battery" did not materialize. And this expectation, in turn, was totally derived from the application field of the gasoline-driven passenger car.[39]

In 1914, Reichel treated the German emperor to a grand parade with his electric-vehicle fleet. But meanwhile the front of electric carmakers had steadily crumbled. In 1911, some fire chiefs wrote: "If the victory train of electrical engineering at the fire departments has not been interrupted by the introduction of the gasoline engines, it has certainly been seriously threatened; hopefully this will be a reason for further improvements of the electric vehicle, because only then it will be able to maintain itself next to the gasoline vehicle." A third-generation electric firefighting truck (and a hybrid possibly derived from it), did not materialize, however.[40]

IN 1914, THE GERMAN fire-engine fleet had increased to about 400 in 90 cities. Except for 16 hybrids and 38 steam vehicles, 148 were electrically powered. Ninety-three of those 148 vehicles were found in 11 large fleets with 10 or more electrics; 13 other towns had an electric fleet of between 3 and 10 vehicles and another 13 towns had just one vehicle each. Meanwhile, there were also 195 gasoline cars in use, 94 of them in the 11 largest fleets. Furthermore, there were 19 towns with a fleet of between 2 and 10 vehicles and as many as 41 brigades with only one gasoline vehicle. According to the last statistics, drawn up before the war by Scholz's Automobile Committee, the total automobile fleet (including passenger cars) at the German fire departments was even bigger: 94 brigades in cities with more than 50,000 inhabitants together had 452 vehicles in service and 42 were on order. Except for 83 ladder trucks, among those vehicles were also 141 *Motorspritzen* (fire engines propelled by internal-combustion engines). And although these data concerned the professional firefighting community (and so were limited to the bigger cities), they showed a general trend, reminding us of what happened at the taxicab application field: for the smaller brigades gasoline traction was more attractive.[41]

Yet, the "conquest of the countryside" by the internal-combustion engine did not happen as dynamically as especially Scholz had hoped. Two factors played a role here. First, standardization of products was hard to achieve. The automotive industry, liberated from the fireman-engineer because of Scholz's victory within the VDB, now put its own designers to work to outsmart the competition. Adler, for example, seen by many as the pioneer of the "light standard vehicle," declared by the end of 1913 that "cheapness should not be a major factor, but only reliability and functionalism." Shortly after, Adler supplied a fire engine (with gasoline-engine propulsion) with a total mass of as much as 6,500 kg to the Flensburg brigade. Although lighter vehicles were proposed as well, the tendency in the automotive industry appeared to point to a strengthening of the multifunctionalism of the gasoline vehicle, rather than to a specialist lighter vehicle geared to use in the countryside, retaining the heavier gasoline and electric urban fleets. In 1914, Daimler, for instance, presented a sprinkling truck with gasoline-engine propulsion for the municipal sanitation department that could also serve as a fire engine by adding a pump. Thus, instead of an atomization strategy favored by (or forced upon) electric vehicle manufacturers, the burgeoning gasoline vehicle industry opted for universality. The fact that at the basis of this strategic difference lay a difference in the technical field (i.e., a difference in structural flexibility) has vanished from public memory since then.[42]

Second, the enthusiasm for motorization among the leaders of the voluntary firefighting community was not exactly overwhelming. Symptomatic of the state of affairs shortly before the World War was the press row that broke out after what happened at an instruction course for drivers of voluntary fire departments of the firefighting association of the Rhine province. It was given for the sixth time in the fall of 1913 in Cologne and that year Scholz was in charge. In the presence of the 135 amused students from the provinces, the gasoline engine of the *Motorspritze* broke down after 30 seconds during a demonstration. Neither cranking up the engine fifteen times nor Scholz's personal interference could get the machine to start again. These voluntary firemen had a good reason to be suspicious of motorization: the higher efficiency had already led to a process of reorganization in the cities. In the trade journals this prospect was also painted now for the countryside: motorization could lead to a reduction of the number of fire brigades and this reduced number could subsequently make use of one centrally accommodated fire engine.[43]

For many small municipalities an automobile fire engine was far too expensive. As the technical journal *Feuerwehrtechnische Zeitschrift* concluded shortly before the World War: "Real regional vehicles that in the first place are geared to the needs of the rural areas and maintained by a combination of municipalities or districts are not yet present in Germany."[44] Thus, a remarkable phenomenon could be observed within the German firefighting community: the victory of gasoline traction took place *without* the support of the countryside, a factor that had played such a prominent role in the field of expectation defined by Scholz. Just as for the taxicab and the privately owned passenger car, the competition for vehicle propulsion was decided *within the city*. The powerful, big-city brigades (with Schänker as most important spokesman) determined the eventual technical structure of the fire engine. It was the expectation of a wider application (supplied by Scholz) that was a decisive impulse for the chiefs of these powerful brigades. In this, they implicitly used the separate spheres concept, which would later play such a decisive role in the expectation field of American propulsion proponents.

Appetite for Standardization: The Truck and the First World War

Contests were organized at an early date not just for taxicabs and buses, but also for trucks. Just as at the Paris taxicab contest, the Concours de Poids Lourds (Contest for trucks) initially brought to light the undeniable cost advantages of electric propulsion. But, compared to horse traction, according to the jury of the contest of 1898, this advantage only held for a fully loaded vehicle. In most comparative tests, steam traction came out as the preferred propulsion system. In May 1898, for instance, the Liverpool Self-Propelled Traffic Association organized a "trial of heavy motor-vehicles," where especially the performance of the heavy steam vehicles of Thornycroft and the Lancashire Steam Motor Company were noticeable. Even seven years later, during a long-distance truck contest between Paris and Marseille, the gasoline versions did not occupy the three top positions, but Serpollet's steam cars were the winners. Of the twenty-two starters, only nine reached Marseille.[45]

Nevertheless, two impulses gave the gasoline version a decisive edge at the trilemma of vehicle propulsion: the phenomenon of the long-distance bus, mentioned earlier, and the army's interest in the new type of traction. The role of the army in the replacement of horses and especially in the choice in favor of gasoline propulsion has up to now hardly been investigated by transportation historians.

Interest in motorization among the European armed forces started much earlier than is generally recognized in automotive historiography. As early as 1898, the Austrian army experimented with one electric and four gasoline vehicles, one of the last covering a distance of 900 miles (including a mile or two through potato fields) in ten days. In an American overview from 1901 of early automobile tests by the armies in Europe, experiments in Germany, Austria, Italy, France, Russia, Belgium, Switzerland, and Norway were mentioned. The German army compared steam, gasoline, and electric propulsion, rejecting the latter alternatives because of a lack of speed, distance, and hill climbing capability. When the Ministry of War organized a competition for a tractor, one of the requirements was as follows: "The fuel supply must suffice for at least 2 day marches with full load, the cooling-water may be replenished daily." Russia boasted of being the first with its experiments with steam traction engines, used for transport in the Russo-Turkish war in 1870. And France sent a dozen vehicles to its military academies, unfortunately not specified as to propulsion type. It seems that the Boer War fueled this interest in motorization. In any case, the British army, not mentioned in the American overview, after its successful deployment of steam traction engines during this war, set up a Mechanical Transport Committee in 1901.[46]

The Automobile Club in France organized a contest in 1905 with a separate category for military trucks. The majority had gasoline traction, but Kriéger introduced a hybrid truck equipped with a big searchlight. In 1908, the German army developed its first plans for an automobile department of its own. That same year a subsidy scheme instigated an effort at enforced motorization. The production numbers of such *Subventionslastwagen,* standardized 9-ton trucks with a maximum net mass of 4.5 tons, were not spectacular. In 1909, the scheme in principle covered 2,004 vehicles, not even 9 percent of the German vehicle fleet, but until the end of March 1914 only a total of 917 trucks was subsidized. Their symbolic value (part of the field of expectation) was, however, of much greater significance, as it entailed a new recognition of the (gasoline) automobile as heavy-duty utility vehicle. France followed Germany with a subsidy scheme in 1910, Britain in 1912. The German scheme, granting a subsidy if the vehicle met certain standards and if one was prepared to hand it over to the army in case of war, assumed a life span of five years. AEG subsidiary NAG supplied the largest share of the annual number manufactured, but the Namag also concentrated on the military application of its products at an early stage: during the First World War Hansa-Lloyd grew into a large supplier of the German army by producing 200 (gasoline) trucks a month.[47]

LIKE ITS EUROPEAN COUNTERPARTS, the U.S. Army also started experiments at an early stage. The army clearly preferred electric traction; the first three vehicles purchased by the Signal Corps in 1899 were electric. But, like in Europe, they were rejected for military purposes after extensive testing in 1900 and 1901 because of the difficulty of recharging in the field. Later tests on steam and gasoline automobiles led to the conclusion of the superiority of the latter because of its lower fuel and water requirements, but for trucks steam propulsion was not rejected. Although the higher echelons of the War Department did not show much interest in motorization, some individual branches (such as the Corps of Engineers and the Ordnance Department's Board of Ordnance and Fortifications) continued experiments during 1903 and 1904 on a limited scale. And although the army ever since 1896 had been flooded with offers from the industry to construct cars and trucks for military use, all attempts in this direction before the end of the first decade failed and no subvention program comparable to those in Europe was ever developed. Nor was the plan to establish a Volunteer Motor Corps ever realized.[48]

Part of the explanation for this can be found in the policy shift in 1905 to use a standard automobile design instead of specialized commercial products, a policy that lasted until 1916. Apparently, the military wanted to steer the same course as the fire department chiefs in Germany had done at about the same time: developing their own automobile design on the basis of their own expertise in the field. By July 1907, the Quartermaster Department had bought and put in service about twenty motorized vehicles (sixteen passenger cars, two ambulances, and two trucks—both steam-propelled), but the first extensive report on this topic clearly explained the hesitance in the American army: the automobile was only applicable for transport of persons and goods "over city streets or well-kept roads," and did not lend itself to becoming a substitute for "any of the present means of Army transportation." At the end of 1908, the Quartermaster Department and the Signal Corps owned thirty of the thirty-two automobiles in the army, most of them passenger cars.[49]

Only around 1908, when interest among European armies in the automobile had visibly increased, did the Unites States Army start a long-lasting internal discussion about the possible benefits of motorization. In 1909, military attaché reports started increasingly focusing on the use of automobiles in European armies. At the same time, automobile technology had improved to such a degree that both passenger cars and trucks were deemed fit to negotiate the much harsher road conditions in the United States (compared to those in Europe), without falling apart. From this moment on, it was increasingly realized that economic considerations, which dominated the testing of automobiles until then, would not be decisive in case of war; reliability apparently had another meaning from a military point of view. From the moment that the artifact functions properly—delivers what its designers promise it to deliver and what its users expect it to deliver—costs are not a primary argument in a military environment. Then it becomes important to assess its possible functions and to investigate whether the replacement of older artifacts leads to new functions that support the army goals more effectively. Apparently, shortly after 1908 this seemed to be the case in the United States, although expectations about the application of the automobile within the army did not reach beyond the supply function. In Europe also, the transportation of high-ranking officers was practiced. It is important to stress here that, contrary to the consensus in automotive historiography about the "conservatism" of the military and their supposedly irrational belief in the future of horse traction, the assessment of the usefulness of gasoline traction for military purposes was based on solid and continuous testing since the beginning of the automobile era. The change in attitude toward the reliability of gasoline propulsion around 1908 is consistent with our findings in other application fields.[50]

From 1912 on, the Quartermaster Department seriously and systematically started to study and institute post and field motorization, for instance, by instigating a cross-country test with four trucks in that year. But as late as 1915, only fifty vehicles were to be found in the Quartermaster Depots. Norman Cary, who performed the most detailed study of the U.S. Army's motorization at the level of the artifacts, explains this by a "misreading [by Army authorities] of the European practical experience." Pointing at the still despicable roads in the American countryside (and even in many towns), an analysis from as late as 1916 concluded that, "notwithstanding these fine roads in France, the combat and field train of combat units in their entirety, as well as a large portion of their corps train (our division trains) remain animal-drawn. There seems to be no doubt that when we consider the road conditions in our possible theater of operations we will not be able to change to the motor truck until a much later date than the European Army." The war in Europe by then had already become immobilized as a static trench war; however, a potential war with Mexico would necessitate a very mobile campaign. The European experience, Cary argues, "confirmed the view of the more conservative in American military circles, most particularly that the mule and the horse would always remain the sinews of Army supply."[51]

AT THE OUTBREAK OF THE FIRST WORLD WAR, Germany (where, in 1913, 1851 trucks were built versus 12,400 passenger cars) appeared to be at a considerable material disadvantage in comparison to the Allies. During the German offensive in Belgium and France, a couple hundred trucks were deployed, but vehicle-fleet management was badly organized and 60 percent had broken down before the start of the battle at the Marne. The catalyzing effect of the "mechanized war" on the stock of army vehicles was also true for Germany, however. Whereas the German army had 9,738 trucks and 4,000 staff cars, ambulances, and motorcycles at the outbreak of the World War, by the end of the war 25,000 trucks, 12,000 staff cars, 3,200 ambulances, and 5,400 motorcycles were registered as German army vehicles.[52]

Initially, the staffs of the other armies also thought that the train and horse traction would form the backbone of the war. The British army, for example, sent 60,000 horses and only 1,200 trucks to the front in August 1914. This number was rapidly increased, at first by impounding city buses, passenger cars, and extra trucks, but soon by stepping up national production. By the end of the war there were 32,000 British trucks in France, of which 800 were equipped with steam propulsion. The French brought a total number of 54,000 trucks (Berliet, Renault, Peugeot) into the war and Italy supplied almost 46,000, most of them manufactured by Fiat. The effect of the war on the American truck industry, especially on the electric truck, is discussed in the following chapter.[53]

Yet one cannot maintain that the horse substitution in the war situation had been completed by the motorization of the nations at war. On the contrary, during the entire war Great Britain shipped more tons of animal fodder to the front than ammunition and soldiers' food. In the United States the war seemed to stimulate the production of horse wagons: the domestic production of such vehicles had risen to 700 a week in 1917 and buggy sales were booming, because these vehicles were also used for utilitarian purposes. But it cannot be sufficiently emphasized that the First World War was an important catalyst for the automotive industry, as well as for the importance that has been given to the gasoline versions among the three alternatives since then. The war also played an important role as a gigantic driving school for young men, who after the war looked for jobs as trained drivers in the growing road-haulage sector. And finally, the war, "with its large appetite for identical vehicles," stimulated the process of standardization of vehicle structure. According to one of Daimler's directors, "the war may have shown ever so many terrible things, but for automobilism it was the best propaganda one can imagine."[54]

THE DECISIVE ROLE OF THE WAR in the struggle between the propulsion systems became increasingly clear in the German firefighting fleet. There, motorization of the urban professional brigades was largely completed in the first half of the 1920s. Also, in the surrounding countries the "electric phase" seemed definitely over. In Vienna, for example—next to Berlin the model city of electric propulsion—the fire department bought 30 gasoline vehicles from the army after the war to supplement a fleet of 69 vehicles, of which 57 were equipped with electric and 3 with hybrid propulsion. From 1922, the electric fire engines were used as backups there. Between 1924 and 1927, all remaining electric trucks were converted to hybrid traction. In 1938, they became inoperative.[55]

In the German countryside, where the number of fires since the war had risen at an alarming rate, the wave of motorization (which Wilhelm Scholz had pinned his hopes on) did not materialize in the 1920s either. But, due to a shortage of horses and the surging purchase and maintenance costs of those horses, the pressure to motorize kept increasing. In the years between 1936 and 1939, Daimler-Benz for the first time managed to produce an automobile fire engine in a series of more than 100 units. This was the result of the struggle for power between the Reichsluftfahrtministerium and the Department of the Interior. The outcome of the struggle was that the German firefighting community was turned into a police service and was assigned an important task in air defense. The tiresome debate about the right type of propulsion, still raging after the First World War, ended abruptly when the federal government passed a general ordinance on 22 August 1935. It prescribed the application of "standard commercial truck chassis" with a maximum axle load of 3 tons or more and provided with a diesel engine.[56]

And so, from the beginning of the First World War, the stage seemed to be set for a complete dominance by the gasoline propulsion alternative and a secondary role in several niches for the electric version. Within these niches, however, a third generation developed, which at the technical level as well as in its appearance was inspired by the prevailing gasoline vehicle. This led to new hopes for electric vehicle proponents, especially in the United States, which was not fighting a war on its own territory.

THE THIRD GENERATION
AND BEYOND

The Serious Side of Mobility:
The Electric Truck in the United States

The more scientific motor-trucking becomes, the more electric
trucks will be used, at least for city work.
—C.W. Squires, General Vehicle Company, 1915

MORE THAN WITH THE PASSENGER CAR and the taxicab, the distinction
between a second- and a third-generation truck is difficult to make. This is
mainly due to the weak state of knowledge of the history of technical truck
structure. We need more academic research on the history of trucks, their use
and their production, in order to be able to make a fair comparison between
the propulsion alternatives. Nevertheless, and despite the black hole in our
knowledge, in this chapter we will see the emergence, in the 1920s, of an
American third-generation truck, as its structure develops at the same pace as
general automotive technology. More important, however, than the technical
field is the field of expectation, which we will witness here to be blown up to
grotesque proportions by the American proponents of electric propulsion.
This remarkable phenomenon was brought about by the fact that initially, in
the field of applications (just like the initial situation with the passenger car in
this country), the dominance of gasoline propulsion was not as clear-cut as it
was in Europe. This generated hopes for a very bright future, certainly for the
commercial parts of this field, and even led to a renewed conviction of the
eventual universal breakthrough of the electric.

On the other hand, we will witness in this chapter how the adventurous
character of the gasoline automobile culture was never far away when it concerned the choice between alternative propulsion systems. Although costs and
expected benefits were the basis of the choice trilemma (between horse, gasoline, and electric) in Europe as well as in the United States, in the latter country

this seemingly very serious side of mobility became influenced by an extremely potent field of expectation, built around the adventure machine. But unlike Europe, in the United States the unique emergence of an intermediary societal force under the guise of the Electric Vehicle Association of America (EVAA) occurred, which promised to bring about the long awaited breakthrough.

It all began very quietly, according to some, even with the "Dark Ages." And although it is true that during the first decade of automotive history (from, say, 1896) few commercial vehicles were sold in the United States, as of 1903 the situation started to change. Initially single truck users or very small fleet managers preferred the electric truck because of the unreliability of the gasoline version, and because it fit so well in horse-paced delivery practice. The fields of application nearly overlapped completely, and while the team could use a break at the place of delivery, the battery benefited from a rest now and then as well. But in bad weather or extreme heat, or when an expansion of the delivery area was pursued, electric traction for a while reigned supreme. In many of these cases it was not so much the costs or the speed that mattered, but the filling of a niche where the horse could not operate. At the same time the possession of an electric truck was a clear symbol of modernity.[1]

Then, from about 1905, this development gained momentum because of two factors. Vehicle manufacturers succeeded in interesting a few large fleet operators in replacing horses with electric propulsion, and some central stations began to develop a structural interest in electric transportation for their own needs, as they attempted to keep in pace with the growing cities where they were located. Some operators, who had put the trucks they bought in 1901 or 1902 in storage, started to use them again around 1905. That year more than half of all commercial vehicles in the United States were electric powered. Of the heavy-duty commercial vehicles the electric version was uncontested: 66 percent were electric powered. As an expression of the renewed interest in the truck, two new magazines appeared in 1906: *The Commercial Vehicle* and *Power Wagon*.[2]

The groundwork of this movement was laid by the central stations. They had a good reason for their interest in the electric vehicle. At the end of the nineteenth century, Samuel Insull, director of the Chicago Edison Company, had begun to spread the gospel of "load leveling," that is, smoothing out the production peak on behalf of electric lighting at night. Whereas their arch rival, the gas industry, could use its reservoirs for this purpose, the American electricity producers initially hardly owned enough batteries to create a comparable effect. In 1895, the total installed energy volume of the central-station batteries in the United States amounted to a meager 4,000 kWh. In 1904, this buffer capacity had meanwhile increased a hundredfold.[3]

In 1898, Samuel Insull became president of both the National Electric Light Association (NELA) and the Association of Edison Illuminating Companies (AEIC). This was the result of a long struggle between these organizations, both founded in 1885. While the AEIC emerged around a group of personal friends of Edison and his former secretary, Insull, NELA initially recruited its members from among owners of non-Edison utilities. And while the AEIC was dominated by the so-called Six Cities' firms (New York, Philadelphia, Brooklyn, Detroit, Boston, and Chicago Edison) and worked closely with General Electric, the NELA was much broader in scope, and based on smaller firms that were not dependent on GE equipment. When the group around Insull managed to get control of the AEIC as of 1893, the lingering conflicts were not yet solved. An important role in these conflicts was played by the alternatives to the large urban central station systems, such as isolated plants in individual apartment buildings and factories and the so-called neighborhood system, such as street-car stations and city-owned firms, both selling their surplus electricity to others. Most equipment sales to these alternative systems between 1895 and 1906 were by Westinghouse rather than GE. The electric truck was to play a remarkable role against the background of these still lingering differences of opinion about the degree of centralization of the electricity grid.[4]

It was Insull who directed the attention of his fellow central-station managers from production to consumption. By offering multistep rates, they stimulated the electric-car owners to charge their batteries at night. In the hours between 10 o'clock in the evening and 7 o'clock in the morning, cheap off-peak electricity could be supplied. This energy no longer needed to be stored in the batteries of the central stations themselves, but could now be transferred to other people's batteries *outside* the station. Thus, the central stations not only shifted the charge and discharge losses of about 20 to 30 percent to the customer, but also battery maintenance. Moreover, while charging its batteries, very steadily and without peaks during about seven to eight hours a day, the electric vehicle formed a typical long-hour load. And finally, the electric car gave a low-demand load: whereas eighty-six electric irons were necessary to generate the same annual income for the central station as a truck with a capacity of two tons, the truck required only 6 kW versus almost 52 kW for the irons. Electric current was sold by kilowatt-hour (an energy unit), whereas production capacity was determined by the power (in kilowatt) of the central station's machinery. Thus, an imaginary mini central station that would serve the truck could be almost nine times smaller than a station for feeding the equivalent number of irons.[5] The eagerness of the central stations to offer off-peak energy led to a narrowing in their field of expectation of the electric propulsion technology in two respects. First, it now seemed obvious to support the public garages, at the cost of charging at home. In this way one charging unit could not only supply more energy, but the central station also had an effective means with which to fight the isolated plant. "It is to the central stations' advantage to make provisions to furnish current where it can control it from the start and not allow this business to drift into the hands of a competitor in the form of an isolated plant." As late as 1916, in the United States such plants, which were not part of the electricity grid, still supplied twice as much energy as the central stations. By offering very low rates—just above cost—to public garages, the tendency to start private charging units could thus be discouraged. With a view to the supposed off-peak load this was no problem, for the energy production during this time of day cost far less than the daily average. In that respect the electric vehicle was basically the horizontal variety of the electric elevator, which Insull identified as the first target in the competitive struggle with the isolated plants.[6]

Second, the central station was more interested in the delivery van and truck than in the passenger car. On the one hand, this can be explained by the fact that a truck used 400 to 933 times more energy than a light bulb and 75 to 179 times more than an iron. A "pleasure vehicle" (as the utility companies persisted in calling the passenger car), however, consumed only 107 times more energy than a light bulb and 20 times more energy than an iron. On the other hand, the commercial vehicle was used and charged at fixed and verifiable periods.[7]

The reemergence of the electric truck also brought back on the stage those men who had earlier been involved in the Pope Manufacturing Company and the unfortunate first taxicab experiment. Robert Lloyd, for example, founded the Vehicle Equipment Company, together with Lucius T. Gibbs (not to be confused with W. W. Gibbs of the Electric Storage Battery Company). His declared purpose was to manufacture chassis, for which American coach builders had to provide the bodywork. Hayden Eames was involved in this project as well. When the plan was about to be realized, the coach builders apparently were no longer interested, so that the company decided to start its own production. To this end, the General Electric Company took over the company in 1905 and reorganized it as the General Vehicle Company (GeVeCo), located in Long Island City, New York. This made GeVeCo the American counterpart of the German NAG, founded by AEG, with the difference that GeVeCo restricted itself to the manufacture of trucks.[8]

With the GeVeCo in the lead, a movement developed toward a second-generation truck with a single-motor system, a differential, and a chain drive, because the gearing at the wheels could not deal with the higher motor torques. Shortly before the World War it also substituted the chain drive for a transmission by means of a drive shaft, thus laying the groundwork for a third-generation electric truck. As a token of their regained self-confidence after the "Dark Ages" between 1900 and 1905, electric vehicle producers established their own Association of Electric Vehicle Manufacturers in 1905. Little is known about this organization, but it likely did not have much impact during the following years.[9]

Around 1910 most American cities with a considerable electric-vehicle fleet had largely passenger cars on their grids. But in New York, with its extensive harbors, chain stores, and express companies, the electric commercial vehicle was in the majority. In the harbors electric propulsion was preferred, as the warehouses did not allow gasoline traction in their buildings because of fire hazards; also, the fire-insurance premiums were lower for electric vehicles.[10]

Besides the Lansden Company in Newark, New Jersey, GeVeCo was the main supplier of such trucks. The latter company revealed in 1910 that it had sold more than 1,600 trucks since its foundation. The company's success can largely be explained by its willingness to offer its customers a contract that limited the annual maintenance costs to a maximum amount. The contract was vitally important because maintenance costs for the more expensive electric truck were an immediate concern, compared to those of the gasoline truck. The breweries were among the biggest customers of GeVeCo. A list of its customers, in 32 types of business, included 63 brewers, 27 customers in the sightseeing business, 26 department stores, and 25 gas and electricity producers. The reason why breweries were especially interested in electric transportation was that in those days beer was perishable. On hot days it had to be transported quickly and in large quantities to the saloons in the neighborhood.[11]

Conflicts of Interest: Foolproof Technology and the Importance of "Service"
The initiative to institutionalize these scattered activities originated in Boston, where the Edison Electric Illuminating Company had obtained control of all its competitors in 1901. Three years later, with general superintendent W. H. Atkins in charge, Boston Edison established a modest fleet of two electric trucks and two electric service cars for the superintendents. Furthermore, the company published lists with charging stations in New England and directed electric-car drivers to its charging station with garage by means of enameled signposts. To expand these activities to third parties, on 11 March 1909, the Electric Vehicle and Central Station Association (EVCSA) was founded. At a meeting, preceding the foundation and organized by Atkins, representatives of the ESB and General Electric were present, as well as many local agents of the vehicle manufacturers. William P. Kennedy, an old acquaintance from the EVC period, represented Studebaker. The initiative received a boost from the publicity campaign that battery producer ESB—with the support of its EVCSA partners—launched in the course of 1909. Moreover, in the fall of 1909 the local Edison central station set up the Boston Electric Garage Company, "the largest electric garage in New England, and . . . the only all-electric garage of any size in this territory." In a period of nine months the stock of electric vehicles in Boston doubled, while the number of garages increased from two to eight. Almost simultaneously with the EVCSA in Boston, on the opposite side of the country, the Pacific Coast Electric Vehicle Association was established "quite spontaneously." Furthermore, in that same month of March 1909, meetings with a similar purpose were announced in Cleveland and Chicago.[12]

As a representative of the very large central stations, the Association of Edison Illuminating Companies appointed a Committee on Electric Vehicles in 1910, chaired by James T. Hutchings, representative of the Edison station in Rochester. From the report that the committee issued later that year, it became clear which way the ideological wind was blowing. While invoking Samuel Insull's ideas, it pointed out the considerable importance of the electric vehicle for the load curve of the central stations. It therefore recommended that the association continue the committee.[13]

However, it was not the AEIC but the much broader NELA that set up, on 1 September 1910, the Electric Vehicle Association of America (EVAA), with twenty-nine members. The motives behind this shift of emphasis toward an organization with many small stations as members have not been documented, but it does not seem too far-fetched to suppose that the NELA was considered to be a more promising battleground for an attack on isolated plants, and to use the electric truck as an efficient weapon in this attack. The foundation of the EVAA was the immediate result of a meeting on 8 June of that year, held in the office of Arthur Williams, the former NELA president and general inspector of the New York Edison Company. Thomas Edison was present as well. Hardly a month after the foundation, the ESB transferred the organization of its publicity campaign to the EVAA. Due to the entry of the Boston EVCSA, membership increased to 120. But whereas during the inaugural meeting of the EVCSA and the subsequent monthly meetings in Boston, the vehicle manufacturers and the fleet operators that were not associated with the central stations were amply represented, on the EVAA board the representatives of the central stations were dominant. This situation was to leave a decisive mark on the development of the electric-powered commercial vehicle in the following years.[14]

With the foundation of the EVAA the center of the "movement" shifted to New York, the city with the largest share of electric trucks. The headquarters of the organization were accommodated in the offices of The New York Edison Company. This company had set up an Automobile Bureau in 1908. It kept records of each electric-vehicle owner in the state, approached potential truck users personally, and actively advertised the electric vehicle in its own magazine, the *Edison Monthly*. In 1909, this central station established an Electric Vehicle Department for its own fleet—which had grown from two vehicles (in 1901) to sixty-three delivery vans and trucks. In New York City alone, the number of electric trucks doubled in the three years after 1906 (when more than a hundred were in use). The following two years the number almost increased fivefold, to nearly a thousand units (figure 6.1).[15]

In organizational terms, the short, six-year history of the EVAA can only be described as a success story: it effectively managed to maneuver itself into the role of intermediary between producers of vehicles and electricity on the one hand, and providers of service (like the garages) and the end users (mainly large fleet managers) on the other. Membership increased rapidly. The local departments, often housed in the building of the local central station, carried the burden of the organization. They organized meetings with lectures and settled domestic matters, tried to involve the representatives of the vehicle manufacturers in their activities, and arranged demonstration rides and truck parades. Moreover, all the departments presenting a report at the annual convention of 1915 mentioned the existence of an Electric Vehicle Department at the local central station. Such a department tried to interest other parties in electric driving by advertisements in the press, but especially by setting up public garages. Some departments even went further, like in New York, where a company (the New York Electric Vehicle Association) was set up with a capital of $50,000 "to purchase, sell, operate, let for hire, repair, maintain, store and care for electric automobiles." At a national level daily business was mainly supervised by permanent and temporary committees. In 1914, there were sixteen such committees, varying from a Convention Committee, to a Committee on Good Roads, to a Committee on Insurance, on Legislation, and on Educational Courses, to a Committee on Garages and Rates, on Operating Records, and on Standard-

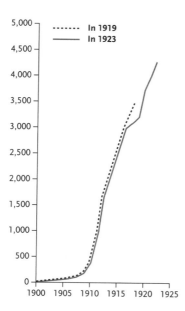

FIGURE 6.1. Reconstruction of the development of the electric-truck fleet of New York City, 1902–23. Extrapolated from 1919 and 1923 back to 1902, for which years the age structure of the fleet has been given; so, for this reconstruction, it is assumed that all trucks purchased during the period under consideration remained in service, an assumption that is not unimaginable for the years after 1908. (1919: F.F. Sampson, "The Proper Application of the Commercial Electric Vehicle," NELA 1919-V, 98–127, here: 101; 1923: NELA 1923, 512)

ization. When, mid-1916, the EVAA became part of the NELA, membership numbered 1,138.[16]

Most members expected the organization to develop into a "general publicity office," or in the more colloquial words of William Blood, the first EVAA president (and later NELA president), "a veritable boosters' club." In his inaugural speech at the first annual convention, Blood defined EVAA's aim as "(to inform) the public that the perfected electric vehicle is an accomplished fact." But despite these generalist intentions, the central stations and their personnel formed a vast majority within the EVAA, organized in seventeen local departments. More than half of the members belonged to the four largest departments of New York, Chicago, New England (Boston), and Philadelphia, all part of the Six Cities' Firms that originally founded the AEIC. Also, most of the electric-vehicle manufacturers (especially those of passenger cars) saw the EVAA as "a sort of truck association."[17]

Blood's remark about the "perfected electric vehicle" contained a hidden agenda, indeed. For it is remarkable that in the deluge of articles and speeches at the local and national organizations that emerged within the EVAA during the following years, the contributions about automotive technology constituted only a small minority. In the opinion of those responsible, second-generation technology apparently did not present any problems for a successful expansion of the electric truck. The drive-train configuration had indeed been subjected to several adaptations since its "stabilization" around 1906. An overview, presented at the EVAA convention of 1913, mentioned that the mechanical vehicle parts, such as transmission, gear wheels, and ball bearings, had become 30 to 50 percent lighter during the past ten years. At the same time their efficiency had improved by half, whereas their life span had doubled or even grown fivefold.[18]

Tires were another consideration. Their life span had doubled, whereas their price was halved. Electric vehicles were equipped with soft tires, which had the same effect as ascending a slope. This was due to "tractional wave," which occurred when the rubber (which was initially too soft) bulged out before the spot where it made contact with the road surface. As electric trucks up to ten tons had solid rubber tires (above this weight limit rubber was utterly useless), special rubber compounds with a low rolling resistance had been developed. These compounds required lower deformation energy that did not seriously affect the life span, but were not so hard that the vehicle mechanism suffered. The application of such new solid tires stretched the boundaries of the technical field (in this case, the fragile energy system) of the electric truck considerably.[19]

Also, the electrical system had meanwhile improved. The series-parallel circuits of the batteries had largely disappeared, and the electric motor had evolved toward a continuous torque design, so that accelerating became much smoother. The most important technical innovation, however, concerned the battery technology. In 1903, the Exide battery, based on the ESB's grid-plate technique, appeared on the market as a practicable energy source. After its introduction, this battery had become increasingly reliable because of the implementation of a number of small adaptations. As a result of this, and supported by the improvements in drive train technology, the range on one charge had nearly doubled during the first decade of the twentieth century (to 65–80 km). The field of expectations regarding the further expansion of the electric vehicle's range received another boost when Edison, after taking his type E alkaline battery from the market because of technical problems, launched a new alkaline type (type A) in the winter of 1908–9. Based on Edison's patent of 1905, the "Improved Edison" was now equipped with a robust positive tubular plate. With an energy density of about 28 Wh/kg, the life span was several times that of the lead battery. The battery, moreover, was more resistant to high temperatures and the active mass, protected by rubber tubes, was not as likely to fall out because of gas formation. This made it eminently suitable for "boosting," or quick charging, which extended the range considerably. For the typical American habit of "boosting during lunch," a charging current of as much as 300 A for a quarter of an hour did not present any problems. Furthermore, the "steel battery" did not suffer from sulfating. When discharged, it could be put aside and, after charging, could be used again several months later, without affecting the life span. On the other hand, the charging efficiency was considerably lower than for the ESB's Exide (50 percent versus 73 percent). It was also two to three times more expensive, which was especially due to the complex way of production. Also, the Edison battery was more sensitive to low temperatures and the voltage decrease at a high discharge current was considerable. This led to a strong decrease in vehicle speed when ascending a slope.[20]

The Edison battery presented a serious threat to the (lead) battery monopoly of the ESB. Since the Exide appeared on the market in 1903, energy density had gradually grown to over 16.5 Wh/kg, a value that increased to about 21 Wh/kg during "prolonged formation" when in use. For a larger range, moreover, a version with thinner plates had been developed (the "Hycap," launched in 1910) with an energy density of 18.5 Wh/kg (increasing to 24 Wh/kg while in operation). Finally, a "Thin-Exide" was put on the market with an energy density of 20 and 26.5 Wh/kg, respectively. This model was probably launched as a reaction to the "Philadelphia thin plate battery," produced by one of the few remaining competitors, the Philadelphia Storage Battery Company. It offered a range of 40 percent more than the conventional grid-plate battery. This was the result of the application of an oversize lead plate with a fine grid structure, the "Diamond Grid." The shorter life span of the thin plates (expressed in number of charging/discharging cycles) was compensated by the high energy density. On balance, this turned out to be more maintenance-friendly, because each cycle enabled a higher range. Apparently, the quality of grid technology had improved so much that a mutual comparison based on maintenance costs made sense. Thus, a newly defined technical field had emerged in comparison with the Paris battery contest of 1899, when the life span of the thin grid plates appeared to be an insurmountable barrier.[21]

A fierce competition between the Edison battery and the ESB developed for the electric-truck market—even involving spies in each other's factories. In the end ESB opted for the tubular-plate technique too. After all, this technology had proved to be among the better concepts at the Paris lead-battery contest. In 1906, the ESB subsequently obtained the rights on the Phénix battery with tubular plate technology. With engineer Edward W. Smith in charge and in cooperation with the American Hard Rubber Company, the company's extrusion technique for the production of fountain pens was adapted to lead-battery application. The hard-rubber tubes, placed around the lead pins of the positive plate, prevented the loss of active mass. At one stroke, the life span of the positive plate no longer was the determining factor of the life span of the lead battery. The ESB even had to make a slightly thicker version of the negative plate, so that the life span of this plate would become equal to that of the positive plate. After five years of development work, the ESB wanted to unveil its new tubular plate battery at the New York automobile show in the beginning of 1911. But the EVAA directors persuaded the company to do so at the January meeting of the EVAA in New York. By means of a clever pun, the new battery showed its indebtedness to the Edison battery even in its name: although the ESB battery did not contain the least bit of iron, it was launched as the "Ironclad-Exide" on 1 January 1911. With an energy density of the same order as the thin-plate concept and about 40 percent more expensive than the conventional Exide types, the Ironclad's attraction was mainly its low maintenance. "Cleaning is practically

The captain of American invention, Thomas A. Edison, with his second-generation nickel-iron battery with positive tubular plates—an item of front-cover interest to Scientific American in the spring of 1911. The battery had gone on the market the preceding year. (Scientific American, 14 April 1911)

eliminated," said ESB man Bruce Ford in 1912, which in his opinion made a cost reduction of about 50 to 55 percent possible. Even including the purchase costs, a cost reduction of about 40 percent was left.[22]

The race between ESB and the Edison Storage Battery Company was particularly important, because a choice in favor of the alkaline type—due to its voltage and charging characteristic—made application of the lead-sulfuric-acid type impossible, unless one opted for two different charging units. Whereas the lead battery remained dominant in the passenger car, in 1913 a third of all electric trucks were provided with an Edison battery. The General Motors Truck Company, for example, charged 30 to 40 percent more for an electric truck with an Edison battery than for a truck with a lead battery. The price difference became smaller as the truck became heavier, an extra argument why the Edison battery was especially interesting for application in heavy trucks.[23]

Nevertheless, on the electric-truck market both types of batteries had a poor reputation. This had to do with the contradictory information the competitors presented to their potential customers. The battery producers belonged to the most notorious "knockers" within the EVAA, which also led to confusion among those in charge of the central stations. Although they thus threatened to jeopardize the intermediary function the EVAA officials deemed so important for a successful organization, they could not be missed: their new batteries fit in seamlessly with the propaganda for the "maintenance-friendly" electric vehicle of the second and third generations. And that was not all: as the new batteries were better able to cope with boosting, they constituted a welcome argument against exchange charging, a form of charging that—apart from the early EVC days—had never been popular in the United States. At the EVAA convention of 1911, it was said that with the two new battery types, "two of the greatest drawbacks of electric vehicles have been overcome, namely, inadequate mileage and short battery life, leaving but one of any consequence; that is, loss of time in recharging. The need of battery exchange stations becomes less as this is reduced."[24]

The matter of boosting during lunch led to a fierce discussion within the EVAA and exposed its ideology, but also made crystal clear that the use of a battery was not as simple a matter as EVAA officials would have liked to have it. The ESB had initially opposed boosting. It was true that the Ironclad was better able to withstand "the abuse of overcharging," but—if treated in an incompetent way—too high a temperature and gas formation in the plates could still seriously shorten its life span. Initially others within the EVAA supported the ESB, but eventually the Philadelphia Storage Battery Company put the cat between the pigeons. According to the representative of this company, James M. Skinner, a boost really was conducive to the life span. As there was a tendency to long daily distances, the alternative was much worse: "it is going to hurt your battery far less to boost it at noon than to over-discharge it and try to make your last mile on very little current." Moreover, the high current during boosting was an effective means against sulfating—an argument that was especially true for the thin grid-plate battery of his own company.[25]

Ultimately the ESB had to give in to a culture of which maintenance-friendliness was one of the mainstays. This surely was related to the surprising comments—surprising also to the ESB—of the truck users about the Ironclad's life span. In the fall of 1914 some of its batteries had already lasted a distance of almost 37,000 km. Half a year later a life span of nearly 50,000 km was reported. The ESB then began to grant a guarantee of two years on its new battery and to publish tables with maximum amperage for boosting. During a boost of an hour, one could, for instance, apply three times the nominal value of the discharge current at a five-hour discharge. A battery, discharged for 50 percent, in this way could recover 30 percent of the nominal battery capacity. Thus, according to an EVAA committee, the range of 56 km of a five-ton truck could be extended by 34 km. With two boosts a day, one of an hour and one of half an hour, the original range could even be doubled.[26]

Because of the ideological power of the maintenance-friendly image of the electric vehicle, a fourth essential element of the intermediary function of the EVAA (next to the central stations, the vehicle, and the battery manufacturers) threatened to get into a tight corner: the garage sector. Small garages usually could not or would not pay for the cost of a charging unit. Many such garages were hooked to an alternate-current grid, and to serve electric vehicle drivers they either had to install an expensive rectifier, or they could opt for a small electricity-generating apparatus, an option EVAA was not inclined to promote. This led to an additional emphasis on the large public garage with a considerable number of customers.

In the area of electric-car maintenance a battle took place that would to a large extent not only determine the fate of the electric vehicle and the EVAA, but also touch upon the core of America's "own way" of electric vehicle propulsion. For the EVAA had turned service into one of the cornerstones of its propaganda, entirely in the tradition of the cultural turning point Insull had caused within the NELA. During the first EVAA convention in 1910—the year that the Edison battery was already on the market, but the Ironclad still had to appear—President Blood announced that the batteries were now "rapidly approaching a condition of being practically 'fool-proof.'" Supported by the vehicle manufacturers, the EVAA subsequently closed its eyes to the fact that the battery in 1915 was still called "the most perishable feature of the truck unit." An important reason for this was a technical phenomenon: the problem of determining the charging state of the battery and of the moment that charging should be stopped. Automotive technology at the time held that the best way was measuring the density of the electrolyte (to be done every half-hour toward the end of charging). But the ideology of maintenance-friendliness enforced a less accurate method, based on the use of the ampere-hour meter. This meter was mounted in the vehicle and indicated how much energy had been drawn from the battery during the ride. Placed on the switchboard of the charging outlet, it could also be used to measure the supplied energy. The Ah meter did, however, not indicate the *state of charge* of the battery. Aside from the fact that during boosting such a method did not provide any information on gas formation and temperature rise, it was especially inaccurate because battery capacity is dependent on battery age, temperature, and the magnitude of the charging current. Particularly the Sangamo meter, put on the market in 1910, was the subject of much discussion. It appeared that the garages often connected these meters in the wrong way. In the opinion of some officials within the EVAA, the Ah meter formed the tailpiece of a development that virtually automated battery maintenance.[27]

The garage sector, especially the influential one in Chicago, was the one to put their finger on the sore point. According to its representatives, the three other partners (car, battery, and electricity producers) in the EVAA had given the false impression that maintenance of the electric vehicle was so simple that unskilled workers could do the job. At the EVAA conventions there were frequent complaints that battery maintenance was neglected, that there was a shortage of "able attendants," and that battery maintenance was not uniform. Whereas the central stations blamed the vehicle manufacturers for this, even the maintenance work for third parties in the public garages—set up by the central stations themselves—was restricted to cleaning the vehicles and charging the batteries. In a report of the Garage Committee for the EVAA convention of 1913, none of the reporting central stations mentioned battery maintenance.[28]

Particularly the public garages that were not owned by the central stations had a hard time. They did not receive any discount on parts because the manufacturers had exclusive contracts with the local dealers. In smaller towns the garages felt robbed of their livelihood, because the central stations' public garages appropriated the maintenance work. Such central stations sometimes also had agencies, which aggravated the frustrations about the competition even more. Initially, these public garages did not get a discount on the charging current (in an interesting horse metaphor, called "the hay and grain of the battery"). This unfairness (that gradually disappeared under the influence of the EVAA from 1913) was all the more painful, because the gasoline-automobile garages did receive a discount on fuel and oil and were generally actively supported by the manufacturers. Moreover, as the customers wanted their cars back at the most unexpected times, part of the charging current had to be delivered during the expensive peak hours.[29]

The mixed garages (maintaining gasoline as well as electric vehicles) in particular encountered the distrust of the central stations. C. E. Michel, representative of a central station in St. Louis, formulated this distrust—entirely according to the dominant EVAA ideology—as follows: "The interests of the central station and the owner of a car are identical—the maximum service at the minimum cost. This is so only to a limited extent in the case of the dealer. . . . The dealer handling both gas and electric cars almost invariably gives the preference to the gas car. His mental process is such that, with few exceptions, the electric receives scant attention, and the business naturally suffers." Being vehicle users and electricity producers at the same time, the dominant large central stations apparently had a lot of trouble defining their exact role within the EVAA. The plea for an "intelligent supervision" of especially the battery eventually resulted in the proposal for establishing exclusive, large electric-vehicle garages. These garages could spread the cost of a charging unit over many customers. Moreover, they could demand a discount for the large quantity of charging energy purchased. In the course of 1916, the heated conflict in Chicago was solved when, after a few meetings, a cooperation between manufacturers, dealers, and garages came off the ground. That happened a year after the Electric Garage Owner's Association in Chicago had accepted a resolution about collective entry into the EVAA.[30]

That was the last that was heard of it and on a national scale a fruitful cooperation between manufacturers and central stations did not seem to take off either. Attempts to set up cooperative garages—and even a cooperative sales campaign—failed, due to the EVAA's emphatic refusal to decide in favor of one or more brand names. The manufacturers, not satisfied with the support of the central stations, then wrote a detailed plan, entitled "A complete plan for establishing and operating a vehicle section and how we will help," and sent it to 1,400 central stations. No reaction followed, probably because they had directly interfered in the optimistic service concept of the EVAA.[31]

These complex conflicts of interest often determined the tense nature of discussions during the annual conventions of the EVAA. The controversy made an outsider comment that within the EVAA there clearly was an "apparent hostility between the manufacturers and the central station men." This hostility is like that of the German fire engine market, where manufacturers were also confronted with a very knowledgeable and powerful customer, the local fire department and its autocratic fire chief. In Germany the automotive industry, with the help of academic "automobile science" and some related fire department chiefs, managed to wrestle the power on structural design out of the hands of the big "electric egos." In the United States, however, in the case of the electric truck, it seems that one branch of users was very well organized and confronted with a much weaker, atomized manufacturing group, in the process pushing both smaller users and garages to the side. Dominated by the central stations, the EVAA management thought it sufficient if its specialized vehicles were user- and maintenance-friendly. For the truck manufacturers, however, a further expansion of the electric truck remained blocked without a standardized series model. Thus, the EVAA seemed to become a place where two of the main players held each other hostage. The central stations were caught in a dilemma, determined by their triple role of customers of specialized trucks, suppliers of electricity from a centralized grid, and propagandists of electric propulsion in general.[32]

Propaganda and Separate Spheres: The Attack on the Urban Horse

The very confusing triple standard of the dominant actor within the EVAA tended to block a further expansion of the association's intermediary function. This became increasingly clear as, with the prospect of a saturation of the market of large fleets, further expansion of the electric truck market would mainly take place in the smaller towns and with the small-scale horse-traction operators. As these were not generally represented within its membership, the EVAA had to devise means to reach them. In 1913, this perspective was already painted at an EVAA conference: "The greatest demand of the future for commercial vehicles will come from the smaller bakers, grocerymen, laundries, and other small merchandising concerns, where their economic status is the highest, where delivery distances are short and stops frequent. It is among the smaller cities, therefore, that the greatest percentage of increase, in both pleasure and commercial cars, will be found in the future." Whereas the large fleet managers were meanwhile convinced of the necessity of motorization, now a large-scale information campaign and sales strategy were necessary to win over also the smaller motorization candidates. According to the EVAA officials, not the gasoline truck, but horse traction was the real target.[33]

In the opinion of the EVAA management, such a sales campaign had to be approached in a scientific way. And who were better equipped to do so than the central stations, which had put such an approach into practice for more than ten years when promoting electric irons, ventilators, and industrial electric motors? For the sales of such products a transportation engineer was needed, a "combination man," who—entirely in the NELA tradition—had to combine commercial and technical knowledge. The cutthroat sales tactics copied from the gasoline-vehicle business had to be abandoned.[34]

Thus, besides "service," "education" also became a catchword: an ignorant customer had to be educated about the "scientific method of transportation."

The innocent customer usually wanted something different from what he really needed, opined editor E. Foljambe of the *Commercial Car Journal*. The new salesman, therefore, must "not sell what the man wants—he must sell the man what he needs. And nine cases out of ten what the man actually asks for is not what he actually needs." And someone else added: "A number of truck manufacturers market their trucks on the same basis that they have marketed their pleasure vehicles. He [the salesman] is interested in making money and he will sell a truck to climb a tree if a man wants it to do that." The new salesman, however, was "a near engineer," whose work did not really start until *after* the truck had been sold. Whereas such a salesman was not necessary for the sale of the gasoline vehicle, because "people were standing in line to get them," the electric truck had to be sold by means of businesslike persuasion.[35]

Education was not only necessary to convince the butcher and the grocer; the lower ranks at the central stations—precisely the domain of potential users—had to be addressed, as they appeared a virtually unassailable bulwark of opposition against electric propulsion. Quite surprisingly it was the tricky theme of vehicle speed that interfered with the EVAA's concept of the electric vehicle as a "power-consuming device." These lower ranks were the men, "who, perhaps, are more interested in the joy rides they may be anticipating or the noise they may be able to make with an open exhaust in a gasoline car." That was the reason why there was hardly a central station without gasoline cars, with the argument (and according to some, the pretext) that they were essential for "emergency or territorial work." The Detroit Edison Company, at the lion's den of the gasoline-automobile camp, reported that, apart from one exception, "every one of our salesmen preferred a gasoline car to an electric." Even the first *Electric Vehicle Handbook* of the EVAA contained a tirade against the "minor executives . . . to whom . . . authority has been delegated to select and purchase machine equipment" and who were counted among the "greatest offenders." So, when a naive participant in the debate wondered, "If there is that demand, why not supply a high speed car?" he received the familiar response: "I think in this, as in a great many other things, it is best to educate the public as to what is best for them, and not always give them what they want." Most participants in the debate agreed: "Unfortunately," one of them said, "the glamor [sic] of the gasoline pleasure car still infects the business. The joy rider and the speed maniac are still with us. It is essential, therefore, to secure for the electric vehicle of either type [passenger car and truck] a utilitarian consideration."[36]

At the annual NELA convention of 1915, this campaign against "joy riding" reached its culmination. With all their authority (of president and secretary, respectively), John F. Gilchrist and A. J. Marshall attacked those managers who did not fight the "destructive opposition" of most of their subordinates. "If it be the policy of the company to permit employees to use its property after hours on week days and on Sunday for personal pleasure, those so favored should not attempt to unduly influence the purchase of any particular make or type of car, unless such favored type will first serve, *and in the best manner possible,* the interests of the company. . . . Now if 'joy-riding' is the predominating feature in the minds of the representatives of central station companies, then it is up to the management to remove that 'joy-riding' possibility." In the opinion of both speakers, 98 percent of the transportation work could be carried out with electric vehicles. Even in the rare cases when the total costs would be somewhat higher, Gilchrist and Marshall recommended the purchase of electric vehicles. Those extra costs could be written off from the advertising budget "for the good of the cause." Thus, the EVAA tried to break the link between professional use during the week and touring for pleasure in the weekend and during the holidays. This link had already existed since the turn of the century among the American physicians and formed a forceful argument in favor of gasoline propulsion. The issue was, in other words, to fill in those 2 percent of the rides that could *not* be executed by an electric car. The Boston Edison Company—as always in the vanguard whenever the issue was boosting electric propulsion—suited the action to the word: in 1912, it purchased the sporty Bailey for its employees, a true third-generation electric passenger car.[37]

This refusal to submit to the customer's wish for adventure opened up the perspective on a remarkable coalition with the gasoline-automobile trade, which, after all, since the crisis of 1907 vehemently tried to "invade" the city as a reliable means of transportation. For, according to tire specialists, the speed-loving driver was the largest cost factor that even the best tire technology could not cure: "It makes no difference how well solid tires are made, or to what extremes the manufacturer goes to fortify them against the incessant knocks of road travel, they will not withstand the abuses of the speed maniac. Speeding is an evil that can result in but one thing—decreased tire mileage and increased tire expense." The central-station manager's trade journal that published this complaint hopefully added: "The only way to clip the wings of the speed maniac is to furnish him with a truck that is geared for low or moderate speed and in which the power is limited, that is to say, furnish him with an electric truck." It will not come as a surprise that the low speed, determined by the inversely proportional relation between energy density and power density of the battery, was projected as an advantage of the electric truck. But the other camp had different ideas about it: there the manufacturers and fleet operators, confronted with the same complaints, put speed-limiting devices on the market that fixed the maximum speed of the gasoline truck a notch higher than that of the electric truck. This seemingly "joint interest" of both vehicle types confirmed the EVAA officials in their conviction that the rationality of scientific fleet management should be opposed to the cutthroat methods of the passenger-car trade. "Do not try to get a customer to put in an electric *machine* when a gasoline motor driven vehicle will do his work better," was one of the proposed rules for the electric-truck salesman. The new salesman should not direct his energy against the gasoline truck, but—preferably in cooperation with the gasoline-truck salesman—against the urban horse. "Uniting against the horse" became the motto.[38]

It is remarkable that the treatment of the electric truck as a somewhat oversized electric iron, as a giant wall plug on wheels, had to be forced upon the users. Even this seemingly rational field of application was decisively

In 1912 the Electric Vehicle Association of America (EVAA) hired advertising agencies to promote electric driving. The ensuing ad campaigns portrayed the electric vehicle as nearly maintenance free—and so simple that women and "even a child" could drive it. (The Literary Digest, 13 July 1912)

influenced by the adventurous character of mainstream (passenger) automobile use. The question now arises, on what the EVAA based its idea of peaceful coexistence with the gasoline truck.

The EVAA prepared and executed the attack on the urban horse along two trajectories. First, it adopted the initiative of the very active Edison station in Boston and turned it into a national matter. In the beginning of 1911, Boston Edison President Charles Edgar announced that he would provide $100,000 for an education campaign and for the replacement of its 114 horses and 38 gasoline cars by electric vehicles. Following this example, the EVAA set up a well-oiled propaganda campaign, based on rudimentary marketing principles and designed by McJunkin Advertising Agency, located in New York and Chicago. This agency had supervised previous campaigns for power stations. Thus, entirely in the style of the NELA tradition, the advertising agency orchestrated a monthly flood of advertisements and editorials (written by the agency itself) in the national trade press. It was supplemented by numerous contributions of the local divisions in the regional and local press. The attack on the gasoline rival was carefully avoided and the emphasis was on horse substitution, "electric thinking," "scientific" fleet management, and each feat that could possibly be ascribed to the virtues of the electric vehicle.[39]

According to the Publicity Committee, established specifically for this purpose, the emphasis of the campaign was on three properties of the electric truck: "simplicity, economy, and serviceability." Samuel Insull was a member of the committee. To support the salesmen, the *Electric Vehicle Handbook* was published. It first appeared in 1913; ten years later, the eleventh edition was published. In 1914, with support from the ESB, a dramatized instruction film was completed entitled "Selling Electric Vehicles," with EVAA members as actors.[40]

The funds for these annual campaigns were provided by three of the four partners within the EVAA. In fiscal year 1912–13, more than $42,000 was collected, of which the central stations provided 55 percent, the battery and parts manufacturers 32 percent, and the vehicle manufacturers 13 percent. Although the central stations liked to see themselves as "the logical advocate of the electric truck," a closer inspection of the contributions tells a different story. On average, the eighty-one contributing central stations appeared to have provided $285 each, whereas the thirteen vehicle manufacturers each paid $420. The eight parts suppliers (with the ESB undoubtedly in the lead) had collected almost five times as much as the central stations ($1,681 each). This shows that in this very important part of the EVAA activities (according to the central stations themselves), the reproach was unjustified that the other partners did not seriously support the matter. On the contrary, the average contribution of the stations, considering their size and their interest in the electric vehicle, was very meager indeed. In fiscal year 1913–14, the central EVAA organization further pruned the sparse campaign budget, as it had met with financial trouble. The following year it appeared impossible to organize an advertising fund, so management was stuck with a deficit. The vehicle manufacturers could not be blamed for this. They had proposed to double a new fund that the central stations and the battery producers should collect up to a maximum of $50,000. Unfortunately, they only collected $29,500 that year.[41] One can really wonder how the electric truck manufacturers could have let themselves be lured into this one-way dead-end street: forced to please one of their major customers, their desire to expand their market seemed repeatedly frustrated, and their EVAA membership indeed tended to become a prison rather than an opportunity.

The second route of the intended horse replacement concerned the scientific foundation of the propaganda campaign and involved collecting detailed operating costs and drawing up a standard cost calculation. Some large-fleet operators, such as the Ward Bread Company, had already set up such an analysis themselves. But during a discussion about the Ward system at an EVAA meeting in New York in April 1912, a fleet operator with about 1,200 vehicles (mostly horse wagons) mentioned that his head was spinning from all the contradictory advice from vehicle manufacturers and battery producers.[42]

Here, too, the Edison station in Boston had taken the initiative. In May 1911, Boston Edison invited Professor Harry F. Thomson of the Electrical Engineering Department at the Massachusetts Institute of Technology (MIT) to carry out a research project for a "liberal sum of money." The research involved a comparative study of the three propulsion alternatives (electric, gasoline, and horse traction), based on a collection of operating costs of truck companies. To this end Thomson, in cooperation with director of the Electric Engineering Research Division, Harold Pender, collected data in mainly the eastern states, and particularly in the electric-vehicle cities (New York, Chicago, Boston, Philadelphia, Washington, St. Louis, Denver, and a few smaller towns). Interim reports appeared in March and October 1912, and on a few other dates. The research project was finished in the beginning of 1914 and in the fall of that year the final report was submitted.[43]

The survey was based on a sample of 1,181 trucks and 5,787 horses in 107 companies. It provided a flood of information that largely supported the views of the EVAA. First of all, the researchers emphasized that "the human element, particularly the drivers and caretakers, may be the most important factors in the operation of a trucking or delivery service." The driver's "mental attitude," especially his start-and-stop behavior, largely determined the costs. Furthermore, the life span of the electric truck (the real length of which was "merely a guess," as the researchers admitted) was considerably longer than that of the gasoline truck. Whereas a brewer in New York still had electric trucks in operation that dated from 1901, the oldest gasoline truck the researchers had been able to find was seven years old.[44]

As far as the costs were concerned, for the electric truck in the application field of urban delivery service costs were 7 to 24 percent (depending on vehicle load) cheaper than horse traction, which had been taken as a basis of comparison. The costs of the gasoline truck varied between 14 percent cheaper and 11 percent more expensive than the horse. For parcel delivery, the electric truck—compared on the basis of the delivery distance *per parcel*—was cheaper than the gasoline truck on all counts. But horse traction left both motorized alternatives behind as far as costs were concerned on delivery distances of up to 3.2 km (2 miles) in the urban coal delivery service. Remarkable was the conclusion that up to a delivery distance of about 12 km the horse was cheaper per parcel than the gasoline truck. Equally remarkable were the results with regard to the competing battery concepts: whereas the lead battery used far less energy per vehicle mile, the total costs (in dollar cents per vehicle mile) were considerably lower for the Edison battery. The solid rubber tire was the largest expense item for the heavy trucks.

The most important conclusion, however, was the division of the delivery service into "relative fields" of application, or "separate spheres." Although the researchers arrived at a clear cost advantage for the electric truck for the delivery services investigated, they concluded in one of their interim reports: "Other services, particularly where the distances travelled per day are in excess of 45 miles, . . . in many cases show a decided advantage in favor of the gasoline truck, and for very small daily distances, 12 miles or less, the horse wagon would probably prove the cheaper." This defined the electric truck's field of application in quite a literal sense as a circular area around the charging station with an inner diameter of about 19 km and an outer diameter of 72 km.[45]

EVAA President Arthur Williams felt supported by the results of the MIT research in yet another important aspect: the cost of electricity played only a minor role in the total costs of the electric truck, with a share of 6.5 to 8 percent. That was not only lower than the fuel costs for the gasoline truck (which amounted to 8 to 10 percent), but in his opinion it especially meant that the energy costs *in general* were of only minor importance. Thus, the EVAA boasted "that the cost of current for charging is very seldom as high as 10 per cent of the total cost of operation, including proper interest and depreciation charges." This argument was deployed effectively against those vehicle manufacturers that accused the central station of greed. For instance, in Chicago the public garages paid between 2 and 6 cents per kilowatt-hour, depending on purchase, but there was a snag here. Such prices were only meant for major customers; in 1910, most "home chargers" still paid a rate of 10 dollar cents. The MIT researchers had taken a kWh price of 3 to 4 dollar cents as a starting point for their calculations, which clearly showed the limitation of their research. It was not only based on the fleet operators' operating figures, but the results of the research also only related to the large consumers.[46]

For 1912, a survey of ninety-one of the biggest central stations, located in American cities with more than 25,000 inhabitants, showed that the charging-current rates meanwhile had further decreased to an average of 3.0 dollar cents for public garages and 5.3 dollar cents for private garages. Some central stations, however, persisted in charging 12 dollar cents per kWh for battery charging. Others opposed a further decrease of the rates with the argument that especially the public garages would become so lavish in providing charging current that this would lead to overcharging and destruction of the batteries. However this may be, the consumer often did not notice the price difference between the public and private garages. Often it was mainly meant to enable

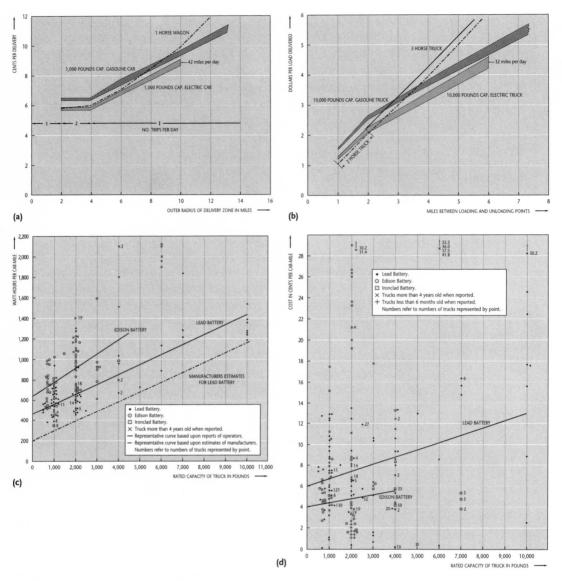

(a)

CENTS PER DELIVERY

1 HORSE WAGON

1,000 POUNDS CAP. GASOLINE CAR

42 miles per day

1,000 POUNDS CAP. ELECTRIC CAR

NO. TRIPS PER DAY

OUTER RADIUS OF DELIVERY ZONE IN MILES

(b)

DOLLARS PER LOAD DELIVERED

3 HORSE TRUCK

32 miles per day

10,000 POUNDS CAP. GASOLINE TRUCK

10,000 POUNDS CAP. ELECTRIC TRUCK

2 HORSE TRUCK

MILES BETWEEN LOADING AND UNLOADING POINTS

(c)

WATT-HOURS PER CAR-MILE

EDISON BATTERY

LEAD BATTERY

MANUFACTURERS ESTIMATES
FOR LEAD BATTERY

- Lead Battery.
⊙ Edison Battery.
□ Ironclad Battery.
× Truck more than 4 years old when reported.
— Representative curve based upon reports of operators.
— Representative curve based upon estimates of manufacturers.
  Numbers refer to numbers of trucks represented by point.

RATED CAPACITY OF TRUCK IN POUNDS

(d)

COST IN CENTS PER CAR-MILE

- Lead Battery.
⊙ Edison Battery.
□ Ironclad Battery.
× Trucks more than 4 years old when reported.
+ Trucks less than 6 months old when reported.
  Numbers refer to numbers of trucks represented by point.

LEAD BATTERY

EDISON BATTERY

RATED CAPACITY OF TRUCK IN POUNDS

the public garages to make some money: the electricity price for the consumer in that case was usually the same as the private price. So, it may be concluded that the MIT value was set at an unjustifiably low level. As far as the electricity prices were concerned, Thomson had allowed himself to be guided by the EVAA management's *wish* and *expectation*.[47]

Not everyone appreciated the MIT survey. Veteran William P. Kennedy was the most cutting in his criticism. Since his time at EVC he had worked as a consulting engineer, later calling himself "consulting transportation engineer." He had been sales manager with Studebaker and in the beginning of 1912 Baker Motor Vehicle Company in Cleveland hired him as chief of the Bureau of Service Efficiency. In Kennedy's opinion, the survey had produced unreliable figures because the managers interviewed were "necessarily amateurs"

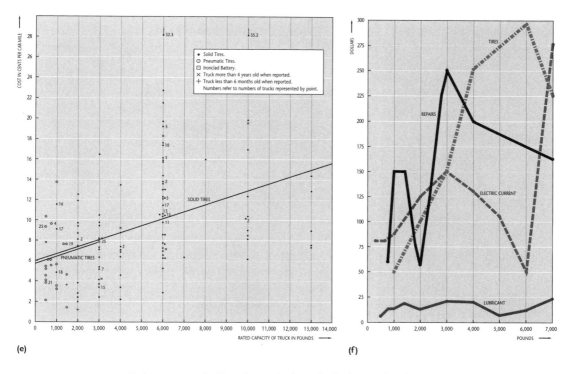

(e)

(f)

During 1911–1914, the Massachusetts Institute of Technology conducted a comparative study of horse traction, electric trucks, and gasoline trucks (the results appeared in Thomson, Relative Fields of Horse, Electric, and Gasoline Trucks, 1914). The survey examined: a. Parcel-delivery costs per delivery over a distance between two and nearly fourteen miles; b. Coal-delivery costs over a distance of one to slightly more than seven miles; c, d. The comparative benefits among lead, Edison, and Ironclad (tubular) batteries in terms of energy consumption (watt-hours per vehicle mile in trucks of ascending weight capacity). Although the Edison battery used more energy, it remained considerably cheaper because it required less maintenance. e. Solid- and pneumatic-tire maintenance costs per vehicle mile. At the lighter-delivery van end of the scale, the difference turned out to be negligible, the advantage of the air-filled tire lying almost exclusively in its greater comfort and the higher vehicle speed it permitted. f. Maintenance expenses by various cost items as load capacity increased to seven thousand pounds. Battery-charging and repair costs were heaviest among light vehicles; (solid) rubber tires costliest among the heavier trucks.

and had not been schooled in the scientific analysis of transportation costs. Therefore, the detailed cost data displayed such a large spread "as to destroy rather than to create any confidence in them."[48]

Kennedy developed a number of standard forms, which he sent to more than a thousand fleet operators. He regularly published the results and his analysis of the figures in the trade journal *Electrical World*. The final result was an idealized picture of cost factors, specified as to load category. It was based on more than 3,000 electric trucks and was adjusted after feedback from the companies concerned. Kennedy also adjusted for fleet size by omitting all quantity discounts from his cost specifications, but he did adhere to a uniform low kWh rate of 4 dollar cents.[49]

In his idealized model, the required high investment costs were particu-

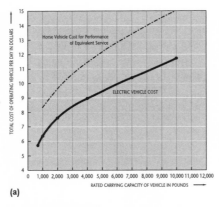
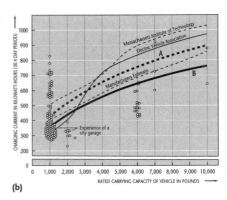

Two graphic arguments for the electric vehicle based on a survey William Kennedy conducted shortly before the
United States entered the Great War. (Committee on Operation Records, Garages and Rates; NELA 1916-IV,
pp. 160, 165): a. A cost comparison between horse traction and the electric truck as carrying capacity increased
from about a quarter ton to five tons; b. Average energy consumption in kilowatt hours over 30.4 days, in half-
ton to five-ton trucks, as determined or estimated in various surveys. Curve A represented a test by "a large
Eastern city garage" during 28 days; curve B recalibrated that data for a 30.4-day period. Kennedy aimed his calculations to form one basis for a future "transportation science."

larly noticeable, just as the high variable tire costs. But his decision to make his
model independent of annual mileage was very surprising indeed; his
overview does not make any reference to the distance covered. Kennedy was
not so much concerned with comparative cost specifications that could be immediately useful in the sale of the electric truck, but rather with a long-term,
strategic goal. By inculcating cost consciousness within the horse economy of
freight transportation and by pushing the principle of transportation engineering, he was convinced that the purchase of the electric truck would impose
itself as a natural necessity. Although he paid lip service to the concept of the
separate spheres, he hoped—as can be read between the lines of his contributions to the debate—that the *rationality* of the vehicle choice would eventually
be victorious. And he expected that the electric truck would not only be able to
claim its own sphere, but that it also would check the progress of the gasoline
competition. In this respect the separate spheres concept was not contradictory to the universality claim within the EVAA, as explained below. The electric
truck's own domain had to have the function of a "protected area," an early example of what is today called strategic niche marketing. In this perspective,
Kennedy for the first time tried systematically and scientifically to influence
the field of expectations around electric propulsion. In order to do so, he emphasized the average vehicle speed instead of the adventurous maximum
speed. This idea was strongly reminiscent of the concept of "schedule speed"
(the average, effective speed, including stops) of freight transportation by railroad. An investigation in 1915 into the average speed in New York's busy traffic
showed that it was 9.3 km/h for the electric truck (72 percent of its maximum
speed) and 9.8 km/h for the gasoline truck (58 percent of its maximum speed);

that the difference was so small was ascribed to the higher acceleration of the electric truck.[50]

The showpiece of the new transport rationality was the "largest modern electric garage" of the New York department store Gimbel Brothers. It was built in 1910 on West Twenty-fourth Street, west of Tenth Avenue, one of the most densely populated city quarters in the United States. Three years later, eighty-three electric trucks and forty-three gasoline trucks were in use, "each of these types (used) in its proper class of service." Entirely according to Kennedy's expectations, in the course of the following years Gimbel replaced many of its gasoline trucks with the electric version, especially the heavy trucks. For very short delivery service, the company even bought some horses. The electric trucks made two trips a day, at an average daily distance of almost 50 kilometers. Among the forty-five employees was a "night battery man," who inspected the charging state of the batteries every half-hour. Moreover, the model fleet was subjected to an exemplary, anticipatory maintenance schedule that was strongly reminiscent of that of the successful European taxicab fleets. A total annual overhaul per truck was part of the schedule.[51]

Another model company was the garage of the big bread factory, Ward Bread Company. By means of an accurate card-index system per vehicle, the company put into practice Kennedy's scientific fleet management. The company especially stressed the saving in drivers' costs, because the bread sellers were also the drivers of the electric trucks. Other large-fleet operators reported similar confirmations of Kennedy's ideas. "One of the largest department stores in New York City," for example, made known in 1912 that the reliability of the gasoline truck ranged between 70 and 90 percent, but for the electric truck the percentage was 98 percent. These figures were based on accurately kept daily reports. What is remarkable is not so much the high reliability of the electric truck, but the fact that the gasoline truck at that moment had largely made up for its inferior reliability. This was not the general picture, however, for five years later a brewer in Chicago mentioned that he employed only two men to maintain his fifty-eight electric trucks (some of which drove 83 km daily). His fifteen gasoline trucks "of the finest European make" were in the garage for repairs for 30 percent of the time.[52]

WITH A VIEW TO THE PRESENT STATE OF RESEARCH, it cannot be proved—although it is likely —that the growth of the American electric-truck fleet after 1911 was largely due to the horse substitution campaigns and the support of the EVAA. From all kinds of contributions at the EVAA conventions, it can be deduced that many smaller towns did not have a sales network of truck manufacturers. Despite their desire to expand to the general truck market and the small user, being a member of the EVAA "prison" was based on a very understandable reason, indeed. It is very likely, then, that the growth in those towns was especially due to the activities of the local central stations.[53]

In New York City the electric-truck fleet also kept growing after 1911 (when there were almost 1,000 electric trucks, 20 percent of all the trucks in that city).

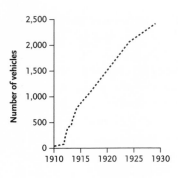

FIGURE 6.2. Development of the electric-truck fleet in Chicago, 1910–28. (Various EVAA and NELA Proceedings)

It became two-and-a-half times as big in 1915, and in 1916 the fleet numbered about 3,000 units. In greater New York, with 2,150 electric trucks in 1914, 39 percent of all trucks were electric powered, whereas this fleet accounted for a quarter of the total stock of electric trucks in the United States. In 1912, this national stock of 8,000 units formed a quarter of the entire American truck fleet. Even in Chicago, the electric passenger-car city, comparable growth had taken place, from 80 in 1910 to 1,076 in 1917 (figure 6.2). In New England the electric-truck fleet doubled in two years: in 1914, there were 341 electric trucks in private garages, whereas 520 were in use at the central stations.

There were only a few central stations among the large-fleet operators: the stations of the ten biggest cities owned a third of all electric passenger cars and commercial vehicles in use at the American central stations. But the largest fleets were in the hands of the express companies and a few other companies, such as breweries. These saw their joint electric-truck fleet grow fivefold to 1,000 units between 1911 and 1916. Another classic example was the removal sector, where the dogma of the separate spheres had been proclaimed since 1912. There, according to its proponents, the electric truck could be deployed up to a range of 32 km and the gasoline truck for the longer hauls, with 80 percent of all distances driven belonging to the first category. In general it can be concluded that most of the large American truck fleets in 1912 consisted of electric trucks.[54]

In 1914, the electric-vehicle market reached its first peak. According to an optimistic estimate of the German AFA, exactly 34,075 electric passenger cars were in use in America by the end of 1913 (3 percent of the total passenger-car fleet of 1,159,322 units). The share of the electric truck (with 17,687 units) amounted to 22 percent of the total number of 77,996 commercial vehicles, a relative decrease compared to 1912, when this share was still a quarter. Nevertheless, the market (picking up in absolute numbers) even enticed the truck division of General Motors to get involved in electric propulsion. To this end it set up an Electric Division of its own in 1911, where J. M. Lansden, who had worked earlier on the project of Edison's inexpensive electric truck (see below), was in charge. General Motors thus was in the position to offer a truck type of its own for all the distinctive fields. In the years 1912–16, General Motors produced 682 electric trucks.[55]

But the optimistic presentations at the EVAA conventions could not hide the fact that market expansion to the smaller owners seemed to have come to a halt. Already in the beginning of 1912 about forty owners possessed 40 percent of all American electric trucks. That was different for the gasoline truck: that same 40 percent only represented 10 percent of all American trucks, and 7,000 smaller companies owned the remaining 90 percent (the gasoline trucks).[56]

An important part of the expansion of the electric-truck fleet consisted of reorders. In 1912, the Baker Company, for instance, sold 70 percent of its vehicles to existing customers. And according to another source, this percentage even applied to the entire electric-truck market for the years 1912 and 1913. Two years later this situation had not really changed. Truck manufacturers apparently proclaimed such reorders as a token of the "superiority of the electric truck principle."[57]

The GeVeCo also experienced a golden era. In 1915, according to its own specification, the company accounted for 70 percent of the world production of heavy trucks and in the beginning of 1916 it even announced an extension of its sales to new customers. To be on the safe side, this company also extended its range of products to the gasoline truck. In 1915, it began to manufacture these trucks under Daimler license—a new confirmation of our earlier conclusion that GeVeCo was the American equivalent of the NAG. The electric-truck manufacturers had little to complain about: with working days of 20 hours, their average sales were considerably larger than those of the gasoline-truck manufacturers. Their EVAA prison was lined with velvet, indeed. Most electric trucks were produced by about ten brands, whereas almost 300 manufacturers operated on the gasoline-truck market.[58]

But for the smaller prospective truck user the choice of propulsion was not merely a question of direct operating costs. For him the (higher) expense for the conversion of horse traction to motorized propulsion was equally important, just as the hiring of "competent men to take care of the batteries." If his fleet was smaller than five vehicles, a cost advantage was impossible. The only thing left for him was to go to a public garage in the neighborhood. His preference for the gasoline truck even grew stronger when the First World War broke out in Europe and the need for gasoline trucks on the battlefield led to an expansion of the American truck industry. Although the electric-truck manufacturers also benefited from this sudden increase in demand, for many within the EVAA it was clear that the electric-truck manufacturers had to formulate an answer to that expansion.[59]

Following the Gasoline Trail: Cheap Electric Trucks and Exchange Charging
In order to enforce a drastic expansion of the electric-truck market, two options were possible: lowering the initial purchase costs or enlarging the range—preferably a combination of both. As far as the first option is concerned, again there were two possibilities. First, one could opt for a light and inexpensive vehicle construction. William Kennedy, for instance, suggested building in fewer batteries. The resulting lighter delivery van could be put on the market for $750. Kennedy thus followed in the footsteps of Thomas Edison and Charles Steinmetz, although his idea had been put into practice years before in Europe, in the three-wheelers of the German mail service. Edison had worked on a cheap electric truck ever since his improved battery appeared on the market at the end of 1909. Since 1904, when he finished the first version of his battery, he had subsidized the Lansden Company in New York. In 1908, he took over the entire company. After 1910 J. M. Lansden worked in vain on the Edison truck and was later recruited by General Motors. Steinmetz had also been active in this area. In a violently contested presentation at the EVAA convention of 1914, he proposed the building of a $500 electric passenger car. Almost simultaneously—at the occasion of the Electrical Show, held

in New York in October 1914—the Ward Motor Vehicle Company at Mount Vernon in New York State presented a light delivery van (with a load of about 375 kg). It was offered for sale for $875, half to a third of the price of the cheapest and lightest electric trucks on the market. This manufacturer herewith proved, in the words of the journalist James McGraw, that there was no reason "to wait for a Ford or an Edison, a Pope or a McCormick." According to Ward, it was now clear that the electric truck was "no longer a delicate mechanism, but a hardy and powerful machine, capable of fighting its own way over the roughest roads and through mud and rain."[60]

The development of the Ward Special (a clear example of a third-generation electric vehicle) had been focused "on the quantity production of a serviceable low-priced delivery wagon" and was especially geared to use as a delivery van by department stores and central stations. With its speed of about 18 km/h and a range of 60 to 70 km, it fit Kennedy's concept exactly. As its name indicated, it was a typical example of a customized vehicle, but one meant for serial manufacture. The Special, however, received the same criticism as the Kennedy idea: the small baker or grocer—if he was in for motorization at all—wanted a higher performance from his car than that of horse traction. The owner of several horse wagons was especially restricted in his flexibility if he replaced them with just one car.[61]

The second possibility open to the EVAA for lowering the initial costs was the establishment of a battery rental system. For example, the very active Hartford Electric Light Company, located in Connecticut, set up a system in 1910 that offered free garaging and charging current during the first six months. And in 1916, 150 cars were driving around in Chicago; a public garage maintained the batteries at a fixed amount per month, with an additional charge per kilometer driven. This lowered the financial threshold for the small-scale vehicle user, because the purchase price could be about 20 to almost 40 percent lower. Within the EVAA this system was thought to be derived from the European taxicab practice, but there was an important difference: the batteries were not maintained on behalf of one and the same operator, but maintenance took place in a public garage, set up as an independent enterprise. The only known application of this new type of business organization in Europe was the AFA garage in Berlin, operative shortly before the First World War.[62]

The second option for enforcing the expansion of the electric-truck market (i.e., enlarging its range) really was a further development of the battery rental system. Here the central station in Hartford again took the initiative. The Battery Service System (BSS) had been in use on a small scale since June 1912. It became known nationwide when at the end of 1914 the Hartford station started a cooperative venture with one vehicle manufacturer (GeVeCo) and one battery supplier (Edison) to realize the system on a larger scale. The exchange-charging system, which the vehicle manufacturer also promoted separately as the GeVeCo Trucking System, or "the Trouble-proof Method," actually was a further development of the specific experiences with the battery rental system in winter, when a second battery occasionally was necessary due to the high road resistances in the snow.[63]

In many respects the system was reminiscent of the EVC principle and the European electric-taxicab practice. The garage was accommodated in an old battery substation of the utility company, linked to the direct-current part of the grid. The GeVeCo trucks were provided with a standard battery box, attached by hooks to the underside of the chassis, which could be removed hydraulically in two to two-and-a-half minutes, "less than the time required to fill up the gasoline tank of a gasoline truck." The batteries were charged at night. The rates consisted of a fixed part for service and a variable part per load capacity and mile driven. The fixed part was between 15 and 60 dollars a month, the mile price between 1.5 and 7 dollar cents. It was also possible to take out a more expensive subscription for complete vehicle maintenance, but in 1916 only 15 percent of the trucks made use of this option. The public garage purchased the batteries. The mileage recorder, the only instrument in the "cabin" and the basis for determining the variable costs, also belonged to the garage. The truck owner could go to the garage for a new battery at any time. In practice, 90 percent of the trucks appeared to use exchange charging once a day, which required a battery stock of 1.4 times the number of trucks. This was remarkably close to what the EVC had practiced more than ten years earlier and considerably lower than that of the second-generation European taxicab fleets. This difference in maintenance culture indicates an important distinction between the American and the European electric-vehicle fleets. The latter (as far as they were deployed as taxicabs) were used much more intensively than the American EVC cabs and the BSS trucks. Obviously, in America the extra investment in batteries would have reduced the cost advantage compared to the gasoline rival.[64]

Nevertheless, the project was a success in various respects. From its beginning, the average daily distance per truck increased (figure 6.3). That must be the explanation for the remarkable fact that 63 percent of the trucks using the BSS scheme were purchased as replacements for gasoline trucks. In mid-1912, only 14 trucks were serviced with a total monthly distance of almost 19,000 km, but in October 1916, 88 trucks appeared to have covered a joint monthly distance of 90,000 km. Since the beginning of the project in 1912 more than 3.2 million kilometers had been sold, at an average energy consumption of 700 Wh/km and a maximum average during the winter months of 1,033 Wh/km. In the spring of 1915, an exchange-charging system was set up in Boston as well. But here, besides GeVeCo, the ESB was involved as partner and not the Edison Storage Battery Company. The excluded manufacturers, especially the General Motors Truck Company and the battery producers USL and Edison, formed a protest committee that insisted on standardization, so that the system would benefit all producers.[65]

In the fall of 1916, systems comparable to the one in Hartford appeared to exist in Spokane, Baltimore, Harrisburg, San Francisco, Los Angeles, Worcester, Fall River, and Wichita. A total of 200 vehicles were involved. Four different methods of "Electricity by the Mile" were applied. For instance, the department store R. H. Macy in New York had adopted the Hartford system, just as the Amoskeag Manufacturing Company in Manchester, New Hampshire,

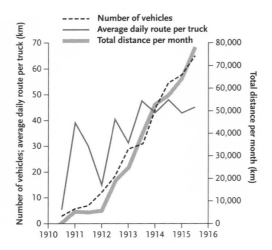

FIGURE 6.3. The Battery
Service System (BSS)
of the General Vehicle
Company (GeVeCo), 1910–
15. (Adapted from Kirsch,
"The Electric Car and
the Burden of History,"
245)

on behalf of internal transportation. Couple-Gear Trucks utilized a technique
of their own in Boston and Chicago.[66]

In Boston it appeared that small companies switched to an electric vehicle
ten times sooner than in cities without an exchange-charging system. "Merchants who deliberately operate gasoline cars at a higher cost than electrics,
rather than buy charging equipment and charge the electric on their own
premises, waive their past objections when offered service on the per-car-mile
basis." The central station in Philadelphia carried it a step further and started
an experiment, in cooperation with its own maintenance department, to rent
out batteries and wheels to replace the ones that were being repaired. Moreover, it made maintenance contracts with small users for the whole truck that
included maintenance and battery charging on the customer's premises. And
in Boston the Electric Vehicle Club Commercial Section proposed the foundation of an independent "electric motor trucking, express and parcel delivery
company" with a capital of $100,000. Unique to this proposal was the fact that
it concerned a cooperative enterprise, where truck maintenance would have to
be carried out by all affiliated producers.[67]

Now that, in the eyes of the EVAA, the last barrier toward the "perfect electric" (the longer charging time in comparison to the gasoline version) had
been removed, one should expect the expansion of the electric truck to take off
in earnest. But despite its local success, the exchange-charging system did not
result in a lasting expansion to the small-scale users and to smaller towns. This
became clear, for instance, in a survey in 1915 of electric-truck ownership of
the central stations. Only 292 of the 6,000 stations appeared to have electric
trucks in operation; they owned a joint fleet of 1,228 units. That was 10 percent
of the total American electric-truck fleet of 12,000 units, of which by far the
greater part was found in the east. In Chicago, for example, a quarter of the
910 electric trucks was in use at one company and 80 at 14 companies. Forty-
two companies possessed a fleet of 2 to 4 trucks (120 units), whereas only 62
companies had one truck in operation.[68]

Why did the electric truck remain largely restricted to large fleets at a small number of companies in the big cities? And why did the breakthrough to the small-scale user and the smaller towns not succeed? Several reasons, both economical and ideological, can be given for this state of affairs.

First, due to the higher profitability of electricity generation in densely populated areas, the electricity grid was far less evenly distributed across a vast country like America than it was in Europe. In 1917, 7 million of the 22 million houses in the United States were connected to an electricity grid. But less than one percent of those houses were "wired for *complete* electric service," so also for power and heating. Six years later only 40 percent of houses had electricity, whereas only 6 percent had electric appliances. The smaller electrical devices had not yet penetrated the small towns and villages. Moreover, the American utilities initially neglected the private market in favor of the industrial customers; compared to Europe "electrification of homes spread more slowly" in the United States. With this restricted coverage of less densely populated areas in mind, there was the possibility that by promoting electric propulsion in smaller towns, the spread of the isolated plant would be stimulated. In rural areas it was popular to generate one's own electricity, and even by the end of the 1920s only 600,000 of the 6.5 million farms had electricity, half of them with energy supplied by an isolated plant. In most cities isolated plants were the most common electricity supplier until the mid-1910s and in much of rural America isolated plants were the only form of electric service before 1930. As late as 1935, only about 11 percent of the American farms had electricity, versus more than 95 percent in the Netherlands, 85 percent in Denmark, and 90 percent in Germany and France. In this situation the choice of many central stations obviously was in favor of the electric iron and against the electric truck.[69]

Second, initially the exchange-charging stations were not profitable for the garages and central stations that had to buy the batteries and set up the infrastructure. This was because of a crucial difference from the European exchange-charging practice for the taxi. In Europe *all* revenues of the taxicab service went to the garage, because it was part of the taxicab company. With the BSS system in America, the charging infrastructure had to be written off based on miles sold, which was only one of the many expense items for the truck company. In Boston, for example, the vehicle manufacturers and battery producers earned a good income from the system, but the central station had to invest so much capital in the company that it was still in the red in 1919.[70] Given the dominant position of the electricity producers within the EVAA it was understandable that this organization did not push a system that favored vehicle and battery manufacturers.

Third, a standardized exchange-charging system, based on universal battery box sizes, did not exist, thus diminishing the cost advantages due to economies of scale. The Hartford system, for instance, was entirely geared to the GeVeCo trucks and the Edison batteries. Such initiatives must have aroused the mistrust of some central stations, maybe inspired by the fear that the exchange-charging companies (when large enough) would take care of their own energy production by means of an isolated plant. This idea was not really far-fetched, for as early as 1910, when an exchange-charging system did not yet exist, the NELA (counting among its members many small-scale electricity producers) outlined two possible applications of such a system: for a long daily route or for the truck owner who generated his own energy. In the latter case, the batteries had a function identical to that in the central station: that of load leveler and buffer for peak loads.[71]

A fourth factor was related to the (initially still European) war. By establishing exchange-charging stations, the central stations would be forced to assume the responsibility for the quality of someone else's product. They had to set up enterprises in an area that was not concerned with energy production, but with energy consumption. And they had to do so in a climate in which the central stations were exposed to criticism because of the obscurity of their business operations and their rates. The mayor of New York, who in 1918 expressed himself in favor of municipalization of the central station, received a cutting reply: "It is difficult to speak with respect of endeavors of this kind exerted at a time of national crisis, or to regard with other than contempt the ridiculous arguments and misstatements advanced by the advocates of this brand of socialism." Suggestions about annexation of central stations by municipalities had already been made when the first signs of an energy crisis became apparent in the United States. The looming crisis was caused by a coal shortage due to the explosive increase in the production and transportation sectors in relation to the provisioning of Britain and France. The central stations were confronted, for instance, with a local regulation of their rates, so that they could not keep the exchange-charging stations going by raising the rates.[72]

Preaching a Doctrine: Eschatology, Sectarianism, and the War

As of the fall of 1914, the electric-truck market seemed to collapse. But although 1915 was described as a sluggish year, at the same time a promising future was painted, in which the electric truck "will gradually occupy a broader and broader field until it has become the dominant method." And that was not all: the chastening of that year had only strengthened the belief in the future: "We have emerged from [it] with a faith stronger than ever."[73]

Such statements were common. A close and critical reading of the leading officials' contributions to the EVAA conventions yields a long list of metaphors that evoke the image of a religious movement. They closely fit in with the NELA ideology, where already at an early date "apostles of modernity" preached the "gospel of consumption." But in the EVAA they reached their culmination at the time of the crisis. "Spreading the gospel," "converting the public," "preaching . . . the sermon of the electric truck"—such imagery formed an integral part of the communication at the conventions until the end of the decade. The first EVAA president had set the tone during the first EVAA convention in 1910. According to William Blood, who described himself as "a firm believer," the EVAA was established "to sing the praise of the electric vehicle as such." At the same convention, the propaganda for the electric vehicle by the central stations was described as "preaching a doctrine." In 1913, each central station was called upon to devote itself to the cause and become a "warm-hearted missionary for the truck." Persuading the potential customer was presented as missionary work, with the purpose of making the customer a believer too.[74]

Initially such sayings could still be dismissed as innocent but apparently appealing imagery. In fact, such imagery was rooted in a long American tradition of "technological messianism" that depicted machines as "gospel workers" and that—as far as transportation is concerned—would later culminate in the "religion" of early flight. As the situation became grimmer, however, and the outside world—including a considerable part of the EVAA's "own" central stations—showed less interest than expected, evoking faith in the business and insisting on the importance of the missionary work became increasingly impassioned. In 1914, for example, an "Electric Day" was proposed, which would be "generally celebrated" with "electric vehicle parades." Originally, the idea was that it should take place on Edison's birthday. That same year the hope for the "salvation of the industry" was expressed and some vehicle manufacturers were accused of going "after false gods" by persisting in their choice for specialist constructions. The missionary work not only required "devotion," but also "real sacrifice." And when in 1916, most central stations had not yet converted to the electric truck, it was claimed: "The electric vehicle has proved itself worthy of our work, worthy of our faith, and what we have got to do is to put our soul into it . . . and count confidently on results."[75]

Those results (the sale of electric trucks and the "conversion" of the remaining central stations) were described in the same style by the "converts" themselves. Thus, a central station manager answered the question whether he could support the purchase of his electric trucks by figures as follows: "Well, we have not the figures, we simply have a conviction." And when a fleet operator with a strong preference for the gasoline truck was persuaded to change to the electric truck at an EVAA meeting in New York, the chairman called upon "the angels," who "rejoice over one sinner converted, etc." Cities that replaced the "unhygienic" horse by electric propulsion were described as "being purged of sins."[76]

The culmination of this religious movement of the electric truck was reached mid-1915, when a large number of preeminent central station managers signed "A Creed," of which the first lines read:

> "WE BELIEVE
> In the electric vehicle.
> That the electric vehicle is destined to
> supersede other forms of transportation methods
> at least for city and suburban work."[77]

And thus it continued for several stanzas. Clearly the movement for electric propulsion had reached a stage where it would no longer be susceptible to rational counterarguments. The piece was written after an impressive performance by Charles P. Steinmetz at the NELA convention of 1914. This event triggered a large publicity wave across the country. Thomas Edison's presence added to the effect. So it is not true that such repeated confessions about the future of electric propulsion originated from the lower ranks. On the contrary, on the waves of this movement the prophets themselves were the ones to paint an ever brighter future. According to Steinmetz, one could expect a market of two million electric passenger cars within ten years. And so the EVAA could now add a new prophet to its ranks: from then on Steinmetz's saying was quoted together with Thomas Edison's prediction from 1910 that within fifteen years more electricity would be sold for electric vehicles than for industrial purposes.[78]

Especially Steinmetz, with his intelligent analysis of automobile culture, had put the EVAA on a questionable track. He realized that sales of the electric passenger car might founder because of the adventurous character of the gasoline car. But instead of defining a proper application field for the electric as a city car, he linked this insight to the promise of a heavenly future of the electric as a universal vehicle. Basing himself on the development of the bicycle sport, Steinmetz emphasized the temporal nature of the "decadent sport of automobiling." When this fad disappeared, the electric car would get new opportunities. After all, those "who did not care for the sport of acting as engine driver and tender, but who wanted a car for general use, as a business and pleasure vehicle," were the real motorists of the future. In that future, "the advantage will be most decidedly with the electric, and the gasolene car will practically disappear from the field."[79]

Thus, the field of expectation within the EVAA degenerated into an ideological, quasi-religious confession of faith.

WHAT *REALLY* WAS THE MATTER HERE? What was going on in the minds of the EVAA leaders at a time when the First World War in Europe was raging and the gasoline truck market seemed to be in for golden times? The material available does not allow a definitive explanation of this strange process of sectarianism. It is not even quite clear whether those two years of 1915 and 1916 really were such bad years for the electric-truck industry. Yet, two factors influenced this defeatism that was hidden by an almost ecstatic belief in the future.[80]

First, the central stations were used to play the intermediary role of midwife to technical development. For by creating the right conditions (brought about by education and service), the "good technology"—electrical engineering—had flourished in a seemingly automatic way as a source of lighting technology, despite the competition of the gas sector. This technological determinism—surprisingly enough voiced by the historical players themselves—received a painful blow by the unexpected success of the gasoline truck due to the European war. The equation of the incandescent gaslight with the gas truck, however, obscured the view of the extra function of gasoline propulsion that electric propulsion could not offer. The insistence on the "speed maniac" and the mantralike repetition of the universality claim of the electric truck point to a blindness for the differences between lighting technology and automotive technology. In addition, electricity no longer functioned as the one and only icon of modernity. In 1916, Kennedy informed his NELA audience that "this position of superiority no longer exists in contrast with the automobile business, for the latter in a few short years, by the remarkable qualities of skill, energy and cumulative finance with which it has been endowed, has reached the volume of $1,000,000,000 per annum, which is more than twice the annual value of the entire central station business."[81]

Second, the feeling of doom became even stronger because several important market segments—clearly within the separate sphere of the electric truck, as proclaimed within the EVAA itself—threatened to fall into the hands of the arch-enemy. For instance, after the American government had set up a mail service in 1913, the EVAA started a cooperation with the NELA in order to conquer this crucial market for the electric truck. It must have been a serious disappointment for the EVAA representatives—gathered in the Committee on Parcel Post Delivery, especially set up for that purpose—when it became clear that the Post Office Department in Washington had forgotten the electric-truck manufacturers when it invited tenders. Nevertheless, here and there gasoline trucks were replaced by electric trucks and in New York the mail vans were included in an exchange-charging system, where they appeared very suitable for the "scientific routing of mail trips." But in the smaller towns with their longer routes the gasoline truck was preferred.[82]

INSISTING ON A RESTRICTED NUMBER of religious truths against one's better knowledge, the eschatology of the "deliverance" by the potential mass market, the revelation of the universality principle by the "priests" on the EVAA board and by "prophets" like Edison and Steinmetz, the continual emphasis on the importance of the "conversion zeal" of the salesmen-missionaries to educate an uninitiated, hostile outside world, the sectarian battle between the manufacturers and the central stations—in this amalgamation of ideological focus on the good, reasonable, serious technology of the electric truck, the concept of the separate spheres constituted far more than a few circles drawn on a city plan. The concept also referred to the isolation of the EVAA ideology in a figurative sense. Furthermore, it was nothing less than a "protected area," a strategic peace, based on the unwavering conviction that electric propulsion eventually would be victorious. The invasion of the gasoline truck into this protected area caused this self-assuredness to falter. And little EVAA reacted by looking for help from big brother NELA. And indeed, it was the NELA—numbering 14,000 members and thus uniting the employees of most of the American central stations—that came with an official request for a merger early in 1916. On 10 March 1916, the EVAA board complied with this request: the EVAA was continued as Electric Vehicle Section of the NELA.[83]

Commentators suggested that the EVAA had collapsed due to a lack of cooperation between central stations and vehicle manufacturers. But another factor was at least as important. To all appearances, the "crisis of the electric truck" was especially perceived in comparison to the increase in popularity of the gasoline truck. Although the figures available are somewhat ambiguous, the assumption seems justified that the years of 1916 and 1917 saw a second revival of the electric truck. So the merger with the NELA did not coincide with the collapse of the market of the electric city truck, but with the collapse of the EVAA's universality claim. Whereas in the previous chapters the technical field was often a barrier to or restriction of the field of application, here the field of expectation degenerated and threatened to block the field of application. Instead of a peaceful coexistence with the gasoline competitor, war broke out. For, in 1917 it was decided that the United States would take part in the World War. As a direct, "patriotic" consequence of this situation, all activities of the NELA committees were aborted. The discussion about the future of the electric truck seemed closed.[84]

AMERICAN PARTICIPATION IN THE WORLD WAR brought about the definitive breakthrough of the gasoline truck. As far as the field of application is concerned, two different roots of the gasoline truck can be distinguished. As urban truck, it supported the industrial decentralization of the cities, but the railroad transportation crisis also offered the gasoline truck a golden opportunity to take over the intercity traffic.[85]

It is not correct, however, to explain the fate of the troubled electric truck by an invasion from the countryside. According to this picture, the American gasoline truck—after its conquest of long-distance transportation—seized the urban market. But as long as the roads between the cities were of poor quality (a situation that did not change drastically until the 1920s), the competition between the two truck versions was mainly an urban affair. According to a study from 1912, it was true for the truck in general that its presence depended on the number of residents in a city and on the condition of the urban roads. A direct relation could be demonstrated between the number of trucks and the length of the paved roads in cities with more than 30,000 inhabitants.

But this changed because of America's participation in the war. What Henry Ford was for the gasoline passenger car, the war was for the gasoline truck. The Quartermaster Corps, responsible for the purchase of trucks, estimated its initial requirement for 70,000 trucks, 6,000 passenger cars and light delivery vans, and 250,000 horses. Not a single electric truck was included. The truck order represented a quarter of the existing gasoline-truck fleet in the United States, estimated at 280,000 units at the time. According to the same estimate, the order required a tripling of the existing production capacity. Of the 160 factories investigated, 20 were selected for the production of these trucks.

A count by the Japanese government, shortly before the United States appeared on the European battlefield, showed that the Entente powers possessed 170,000 motorized vehicles while Germany and its allies had 130,000 units. Before April 1917, the United States had an army of 80,000 men and 1,000 trucks. To these numbers it added 124,800 trucks and 21,780 motorbikes, while on the home front another 33,700 trucks and 10,170 motorbikes were deployed. Because of the shortage of trucks in Europe, purchasing agents bought trucks in the United Kingdom, France, and Italy. All in all the American Expeditionary Forces spent about $500 million for 275,000 vehicles, of which the American industry produced less than half (118,838); 51,554 were sent overseas and 67,284 were still in the United States when hostilities ceased. Maintenance of this large fleet cost the United States another $150 million. Most vehicles were disassembled and shipped in crates, for which purpose "easily the largest and best equipped shops in the world" were established in Baltimore (Maryland), Atlanta (Georgia), and San Antonio and El Paso (Texas). The factories (in reinforced concrete, the laborers-soldiers in tents) required an unsurpassed logistics approach, including "scientific" cost control and time control based on clocks with a decimal time division. Special schools turned out an army of 2,000 mechanics per annum.[86]

When the armistice was a fact, the army cancelled the order of yet another 200,000 trucks. More than 30,000 of these were delivered anyway, however. Half of them were handed over to federal institutions, mostly to the Bureau of Public Roads, which used them for road construction. So the problem was not that the war surplus flooded the market and would thus corner the electric truck, as many electric-truck proponents had feared, but the enormously increased production capacity, which had grown fourfold, to 128,000 units a year.[87]

At the same time the truck had undergone a decisive structural change because of the war. The army had imposed a standardization of the "Liberty truck," subdivided into lighter "Class A" and heavier "Class B" trucks. After the Mexican campaign in 1916, the Quartermaster Corps had put all its cards on the Standard "B" truck, an "in-house design" meant to reach complete interchangeability of parts and to be built under the explicit instruction to avoid the use of any commercial design. Full-scale production of this truck type started in April 1918 and lasted until the armistice on 11 November of that year (the lighter type "A" and "AA" trucks never reached full-scale production). But when the United States, three months after the Punitive Expedition, as the Mexican campaign became known in military jargon, entered the European war scene, the army's supply bureaus had already rejected the standardization approach of the Quartermaster Corps. Nevertheless, after the war such trucks appeared preeminently suitable for farm-to-market transportation, initially flourishing in a wide circle around the cities. This type of peri-urban transportation became the backbone of the postwar truck industry, together with the intercity traffic, of which the distances kept increasing as the national road system (started in the 1920s) was expanded. Both applications were a direct inheritance from the war. The transportation of agricultural produce was the civilian form of the most important type of transportation during the war: the transportation of troops and supplies to and from the railroad yards in Europe, usually in about a 50 km radius.[88]

Long-distance transportation had also been practiced in a war situation: during the American campaign against Mexico and during the spectacular transportation of troops and materiel during the battle of Verdun in February and March of the same year, both before America's participation in the World War. The Punitive Expedition from May to August 1916 against Pancho Villa in Mexico, was, in the words of the military historian Marc Blackburn, "a laboratory to test the capabilities of motor trucks in an operational setting." When access to the Mexican Northwestern Railroad was denied by President Venustiano Carranza, the Quartermaster Corps purchased 55 motor trucks from the White Motor Company and the Jeffery Company, both steam-propelled and both "winners" of a cross-country test in 1912. By June 1916, campaign leader General John J. Pershing (who would play a crucial, leading role in the American presence at the European war theater only a year later) had bought a total of 588 motor trucks (including 67 specialized vehicles), 75 passenger cars, and 61 motorcycles. But the inexperience of men and officers with the new vehicles and the much higher unreliability of the vehicles than expected made a special maintenance depot necessary.[89]

The transportation to Verdun was a special demonstration of transportation engineering Kennedy would have particularly admired. Le Petit Meusien, the narrow-track railroad between Bar-le-Duc and Verdun, only allowed a flow of goods of 800 tons a day, ten times less than what was needed. Faced with this infrastructural bottleneck, the French army, under the leadership of Captain Doumenc, decided to use the departmental road from Bar-le-Duc, 7 m wide and 67 km long. During 10 months, 24 hours a day, 3,500 trucks, 200 buses, 500 tractors, 800 ambulances, and 2,000 liaison vehicles were driving to and fro, one every 14 seconds, transporting 90,000 men and 50,000 tons of material per week, at an average speed of 15 to 20 km/h. Defective vehicles were put aside and repaired by groups of mechanics. In total, 2.4 million men and 2 million tons of freight were transported; 700,000 tons of limestone were spread on the Voie Sacrée (the Holy Road) and more than 10,000 workers shoveled stones under the trucks' tires. It was the conveyor belt of the mechanized war. In Verdun at the end of February 1916, gasoline traction in the shape of the truck celebrated its victory. A similar victory of the gasoline passenger car with the *taxis de la Marne* had preceded it. By the end of 1919, the American truck fleet had increased to 800,000 units, almost three times as many as in 1917.[90]

Automotive historiography has dedicated many pages to the failed standardization of the artifact itself, and has, implicitly or explicitly, derived from this that the impact of the war was only modest. It is true that, in contrast to the Europeans, the American military establishment had failed to incorporate trucks into war planning. The federal government owned several hundred trucks for use by various departments and for transportation at naval shipyards, but on the whole the American public sector was a distant observer of the motor vehicle market. It cannot be denied, however, that the American army, at least until 1908–10, had a good reason for its skepticism about the reliability of gasoline propulsion, based on thorough and regular testing.[91]

But the real inheritance of the war was the insight that the gasoline automobile, despite its much higher degree of technical unreliability, could be brought to dominance by approaching it from a systems point of view. In doing so, military gasoline truck and car fleet operators had to copy the centralized maintenance systems of prewar electric taxicab fleets. Viewed from this perspective, the garage infrastructure, which after the war was crucial for the successful diffusion of the gasoline passenger car, deserves much more study than it has hitherto received.

ALTHOUGH BARRED FROM THE BATTLEFIELD, the electric truck did get a new chance because of the war. With regard to the special applications, the deployment of the electric truck in ammunition transportation was especially noticeable. Electric propulsion was also preferred on the harbor piers and in the navy depots. This gradually developed into a type of transportation, of which the first signs had become visible before the war: that of the industrial truck. As part of the transportation system on factory sites and especially in factory buildings, in the depots, and in the harbors, the industrial truck not only provided an effective horse substitution, but also allowed the formation of the "American principle of integrated manufacturing processes."[92]

The range of such vehicles was small. A large gasoline-car manufacturer, not mentioned by name, had a fleet of forty-nine small electric tractors in operation in 1917. They each drove about 6 kilometers a day. The electric mini truck came in all kinds of versions (truck cranes, tractors, forklift trucks, platform carts for the railroad platforms). They did not replace horses, but manpower (on average ten men per truck), which the electric-truck lobby saw as patriotic during the war: it made available more "men for service." The mini trucks also helped to ease congestion in the harbors, so that the federal government bought 1,500 of them during the war.[93]

In 1919, about 5,000 electric industrial trucks were produced by 12 manufacturers, while estimates of the total fleet ranged from 6,000 to 15,000 (the manufacturer who supplied these figures reckoned 8,000), most of them built during the war. Ten percent of these trucks operated on the basis of an exchange-charging system, while a quick boost during lunch was also usual. As their battery sets were small (but in proportion to the vehicle mass bigger than that of the "street truck") and because many factories still took care of their own energy production, this market was less attractive for the central stations. In this way electric traction finally seemed to have found a separate sphere, relatively safe from the gasoline competition. As one of the many "apparatuses," it became increasingly incorporated into a "scientific" system of internal transportation, further consisting of electric cranes, elevators, and conveyor belts. Thus, the vehicle formed a substantial part of a "new industrial era," geared to strict cost consciousness. And when in the spring of 1920, Charles P. Steinmetz finally appeared on the scene with his long-awaited cheap electric vehicle, one of the two versions turned out to be an industrial truck.[94]

In 1923, 42 percent (10,192 units) of the American electric-vehicle fleet consisted of industrial trucks. Besides GeVeCo, companies such as the Automatic Transportation Company, the Crescent Truck Company, the Elwell-Parker Electric Company, the Lakewood Engineering Company, and the Yale & Towne Manufacturing Company also belonged to the manufacturers. But most renowned was the Baker R & L Company, created by a merger of the passenger-car brands Baker and Rauch & Lang.[95]

Flourishing Niches and Growing Fleets: The Interwar Years

When Chairman Mansfield opened the submeeting of his Electric Vehicle Section at the first postwar NELA convention in May 1919, he had to conclude that "the public is not interested in our proposition." With the bankruptcy of the separate spheres concept and the end of the patriotic role of the gasoline truck on the battlefield, the road was open to a more aggressive approach of the gasoline competitor. Now that the electric-truck proponents discovered that their adversaries in the urban truck market did not adhere to the demarcation lines of the separate spheres, they also threw their inhibitions overboard. The shattering of the religious hopes gave way to anger: when the gasoline-truck salesmen showed figures to their customers, as it was said at the NELA convention of 1918, "they lie like thunder."[96]

As the electric vehicle was thrown back on its own urban domain, the entire prewar propaganda campaign started again. Symbolic of this was the establishment in 1920 of the Electric Vehicle Bureau (later, Electric Truck and Car Bureau), which became part of the Commercial Section of the NELA. The electric vehicle was thus placed on one line with other electric artifacts that required extra propaganda efforts, such as the electric iron and especially, since the beginning of the 1920s, refrigeration. This put an end to the Electric Vehicle Section, a token of the disinterest on the part of the central stations' transportation departments. For, in 1920, as NELA President Foster motivated the reorganization, the electric-vehicle manufacturers sold "less than 2 per cent of their products to or through central stations." Nevertheless, Foster, in one and the same breath, also announced the revival of the former local EVAA sections. Whether he succeeded is doubtful, but it is remarkable in this connection that the Hartford Electric Light Company pops up again in the conference reports. Besides the exchange-charging system for trucks (of which the fleet size in 1924 was still about ninety units, mostly for coal transport), the central station in Hartford meanwhile had discovered another profit maker. It was also owner of 1,025 batteries that were rented out as lighting batteries (for gasoline cars without a starter motor) at 50 dollar cents a week and as starter batteries at one dollar a week.[97]

In accordance with the spirit of the times, a Transportation Engineering Committee was established in early 1919 that began to propagate the electric truck as an element "in the general transportation scheme." The committee stimulated the organization of drivers' courses and started an active, successful campaign to establish chairs in Transportation Engineering at the universities. But for a renewed cost comparison with the gasoline truck, the prewar data of Kennedy and the MIT survey had become useless, due to a considerable increase in wages and other costs, among other things. It meanwhile had become increasingly clear that an idealized cost model à la Kennedy was an illusion: the costs of electricity consumption often differed 100 percent between cities. This was due to the fact that smaller towns, with less ideal conditions of road surfaces and slopes, now also figured in the overall picture. In another study, among thirty central stations, the costs of the two largest fleets (both with more than fifty electric trucks, but one with an average daily distance per vehicle of 17 km, the other of 34 km) were almost inversely proportional with the daily distance driven. This was a second proof of the bankruptcy of Ken-

nedy's method. This understanding now also started to penetrate the handbooks of truck transportation: "the daily cost of a team [horse traction], irrespective of daily mileage, is practically constant, whereas the daily cost of a motor truck is absolutely dependent on daily mileage." This proved once more how much even Kennedy with his "modern" theories of fleet management, had still been caught up in the horse-traction era.[98]

Nevertheless, the electric-truck market seemed to pick up in 1919 and a third wave of interest in this vehicle type seemed to be on the way. In New York City electric-truck sales increased fourfold. That year was depicted as "the most prosperous" up to then. The year 1920 was also praised as "one of the most active in the development of the electric vehicle," although it was immediately added that the growth of the market had "not been spectacular." The year 1922 was called a record year again. According to Rodney K. Merrick, representative of the truck manufacturers, "our position [is] far stronger than it has ever been before." That position was based on a third-generation electric truck, the most important innovation of which was the closed propulsion system, whereas the more modern body closely resembled that of the gasoline truck. Compared to the second generation, the improvements were mostly small, such as a longer life span of the battery, a decrease in purchase price and maintenance costs, and a lighter construction.[99]

Clearly, the electric-truck manufacturers profited from the weakened grip of the electric utilities on the market. But if one looks beyond the still virulent bombast and the pep talk of the "religious movement," a turning point seemed imminent. Electric-truck manufacturers had waiting lists and new competitors emerged, such as the Auto Car Company, which had built gasoline trucks for a quarter of a century and now launched a series of electric trucks. One of the largest bakeries in New York started to replace its Ford trucks with electric trucks in 1922. That same year a new campaign started among the central stations to follow suit. It is not likely that this campaign was successful. Yet in Chicago—the long-time electric passenger-car city par excellence—2,075 trucks were driving in 1924, double the number of those before America's participation in the war. And around the same time 4,675 electric trucks were in operation in New York City, versus less than 2,000 in 1913. Whereas the national electric-vehicle fleet still comprised 10,457 (or 29 percent) electric trucks in 1922 (almost as many as the number of industrial trucks), two years later it numbered 12,500 units.[100]

Meanwhile the electric truck had to retreat for an important part in a limited number of sectors, such as the transportation of cheap products with small profit margins. Here, the transportation costs formed a considerable part of the turnover, such as at bakeries (20 to 30 percent), dairies (25 percent), ice-cream makers (25 to 39 percent), and laundries (25 to 28 percent). Also, the industrial laundries, in the United States numbering 7,500 with more than 200,000 employees in 1925, began to motorize on a large scale as of 1923. The light electric truck was very attractive to these laundries because their delivery work often took place along fixed routes with heavy traffic congestion, and as such resembled the European practice of mail delivery. According to a national survey, reported at the NELA convention of 1925, their average daily route was 22 to 40 km, with an average of 100 to 200 stops.[101]

But the largest electric-truck users were the express companies, which had approximately as many electric trucks as gasoline trucks in their fleets in 1915 when the official vehicle count for the first time made a distinction as to propulsion type. Horse traction, however, remained the dominant type of propulsion for a long time here as well (figure 6.4). By far the most important was the American Railway Express Company (AREC), the result of a merger of the four biggest express companies (Adams, American, Southern, and Wells Fargo & Company). This happened shortly after the American government took control of the railroads, including the express wagons, on behalf of the transportation of military goods. The AREC was to carry out the express transport for the federal government and after the war remained under control of the Railroad Administration. Not until 1929 was this carrier taken over by the railroad and continued as the Railway Express Agency. In 1928, the company had 1,528 electric vehicles in use (mostly of the Baker, Commercial, GeVeCo, and Walker brands) and about 400 industrial trucks.[102]

Edward E. La Schum, general superintendent of motor vehicle equipment of the AREC, grew into the role of a new William Kennedy in the course of the 1920s. When he introduced his company at the NELA convention in 1919, his motorized fleet (started in 1907) already had 1,058 electric trucks in twenty-three cities. La Schum's fleet, the largest in the world, proved to be the exemplary realization of the idea of the separate spheres and of the deployment of the electric vehicle as antiadventure machine. His truck operator was not a driver, but an "express man," he told his audience. The electric truck was the paragon of safety and "entirely free from excessive speed . . . , there being no chance for the operator to perform in a spectacular manner, therefore no sudden bursts of speed which are very likely to result in trouble." Steep slopes (as in Kansas City, St. Louis, and Denver) no longer were problematic for the third-generation electric truck. Moreover, the low insurance premiums and charging-

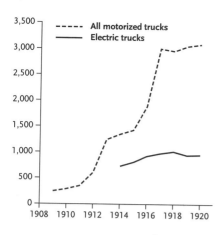

3,500
---- All motorized trucks
—— Electric trucks

3,000

2,500

2,000

1,500

1,000

500

0

1908 1910 1912 1914 1916 1918 1920

FIGURE 6.4. Motorization of the express companies, 1909–20; the number of horses in service at these companies did not decrease during the whole period under consideration, but fluctuated around 18,000. (Raburn, "Motor Freight and Urban Morphogenesis," 91)

current rates he had managed to get in all cities (not more than 2.3 dollar cents per kWh, in some cities even 1.5 dollar cents) did not form a barrier to large-scale deployment. In New York the firm utilized an exchange-charging system. La Schum demonstrated conclusively that in an urban congestion area with a diameter of about 65 km the electric truck could not be beaten. He even dreamed of a container system that would reduce loading and unloading time considerably. In 1924, the AREC, with more than 1,800 electric vehicles (including 1,225 trucks), had in fact a monopoly on parcel delivery (besides the national Post Office). In addition, the company owned another fleet of 8,200 horse wagons and 2,500 gasoline trucks.[103]

That year La Schum published his handbook on electric goods transportation, based on the AREC fleet. The book was one long praise of the electric truck that six manufacturers were still producing at the time. La Schum emphasized the importance of the "stops," which accounted for 76 percent of the daily delivery time per vehicle and thus represented a much more important factor than the maximum speed. He also drew the remarkable conclusion that the purchase price of the electric truck was lower than that of the gasoline truck, if one made an "honest" comparison: whereas the gasoline truck cost on average $3,500, the average price of the electric truck was only $3,030, after subtraction of the battery price ($970 on average). The average life span of the gasoline trucks was 6 years, whereas the oldest electric trucks in use at the AREC were 12 to 14 years old. They probably lasted about 20 years. La Schum's cost comparison, based on 1,000 trucks of both types of propulsion, produced a clearly favorable result for the electric truck (table 6.1). So his conclusions were entirely along the lines of Kennedy's: "It is becoming more and more evident that the electric truck will eventually narrow the range of the gas trucks to routes over 40 miles long. . . . As to the electric truck," he concluded his bible, "it will probably prove in the future to be the most economical unit on frequent-stop routes quite irrespective of length. Battery-charging devices and better batteries will increase its range until it can cover up to 65 miles a day. Today its normal range is 30 miles, and with that distance it has no rival in economy. It is also the speediest of trucks where stops or delivery are frequent and traffic congested.

"The electric truck is essentially a city delivery unit, and especially economical for house-to-house delivery. It is not the best unit for most of the work the gas truck has taken from railroads or created for itself. It is the most economical transportation unit sold for practically all the work still done by horses in cities and large towns. It is the most economical transportation unit for much of the work the gasoline truck has taken from the horse."[104]

But the only explanation La Schum could give for why only as little as 4 percent of urban goods transportation took place by electric truck was a reference to the ignorance of the carriers. They were not aware of the significant share of the transportation costs in their turnover and simply did not believe

TABLE 6.1 Average Costs of Operating 1,000 Electric Trucks and 1,000 Gasoline Trucks at the American Railway Express Company, 1923

	Electric Trucks	Gasoline Trucks
Gasoline/oil		1.61
Charging current	0.513	
Battery maintenance and replacement	0.617	
Subtotal	1.13	1.61
Garage (including wages)	1.15	1.41
Repairs	0.86	2.05
Total	3.14	5.07

Source: La Schum, *The Electric Motor Truck*, pp. 218–21.
Note: All values are given in dollars per day.

the spectacular cost reductions that could be realized. Remarkably, another handbook, largely geared toward the gasoline truck, confirmed this suspicion. Yet the figures did not lie: in 1923, the national truck fleet had increased to 1.4 million, a year later to 1.6 million vehicles. In contrast to what is sometimes claimed, more than half of those vehicles were deployed in the city, "in which about 85 per cent of the routes are short haul or frequent stops." So, it was not so much a matter of a closed technical field for the electric truck, but of other factors, such as unfamiliarity with the problems of transportation costs and the lower purchase costs of the gasoline truck—arguments that especially played a dominant role for the small truck user. For such a user was not in the luxurious position of being able to deploy each of the three alternatives in its own separate sphere, accompanied by an accurate analysis based on transportation engineering.[105]

A survey of the United States Bureau of Public Roads from the second half of the 1920s supported these findings. It showed that 71 percent of the trucks in California, 72 percent of those in Ohio, and even 81 percent of the trucks in Pennsylvania and Connecticut covered daily routes that were shorter than 48 km. As late as 1935, only 7 percent of all trucks were used for interstate hauling. Also the dominance, in 1926, of trucks with a load capacity of a ton or less (almost 78 percent for the joint truck fleet in the United States and Canada) is an indication of the great significance of urban delivery work as well as the even more important role of the truck in farm-to-market transportation. It was exactly the same peri-urban market niche that the gasoline taxicab had used in Europe to conquer the taxicab market. In 1926, the largest group of truck users was made up of farmers, followed by "grocers and other dealers in food products." After yet another 8 groups that especially used the truck for transportation to and from the city, the bakeries followed in the tenth position and the laundries in the fourteenth position. Of the 420,503 motor truck fleets in America in 1926, 65 percent consisted of only 1 or 2 vehicles. The opposite was true for the electric alternative. Whereas in 1923 1,454 users owned 10,018 electric trucks (6.9 per user), in 1928 these numbers had increased to 1,581 users with 14,423 trucks (9.1 per user). Most electric trucks were found in the big cities. In Chicago, for example, their number increased from 1,800 in 1924 to 2,356 in 1928. But most revealing were the figures about the fleets of the central stations: in 1928, they owned 1,229 electric trucks, hardly more than in 1917. But these vehicles were in use at 100 companies, 30 less than in 1917. Only the central stations in Chicago and New York made use of electric trucks for 100 percent on their routes up to about 65 km.[106]

The small truck user, however, was mainly found in the smaller towns, where the routes were longer and the state of road surfaces and slopes often much more problematic. That truck owner wanted a fast truck and thus opted for the gasoline version. In Idaho and Utah not a single electric truck could be found, in Montana two. Also in the other states in the Northwest, such as Oregon and Washington, the electric truck was largely absent. And so the electric truck appeared to be consigned to very large fleets in the big cities in the east and far west of the United States.[107]

All this is not to say that the automobile industry did not consume electricity. To the contrary, in 1928 the automobile industry belonged to the largest electricity consumers of the central stations—the gasoline-automobile industry.[108]

Separate versus Protected Spheres: Comparing the United States and Europe

Although comparable in a quantitative sense (in percentages of the national vehicle fleet), the second- and third-generation electric truck clearly followed a different path in Europe than in the United States. Whereas in Europe electric propulsion could thrive in a protected area at the municipal level, in the United States this was not the case. Because of this, the structural flexibility of the electric drive train could be fully exploited in Europe, as appears from the emergence of all kinds of special vehicles, each characterized by at least one advantageous aspect of this propulsion type. The background of this important difference has not been fully investigated, but it is possible to formulate an explanation, based on the reconstructions presented in this and the previous chapter.

First, a distinction between lower and higher levels of government is important—a notion that has been lacking in automotive historiography up until now. At the level of national authorities, the choice of propulsion in Europe as well as in America was soon in favor of gasoline traction. The military importance of this type of propulsion played a decisive role here, just as the fact that from the bird's-eye view of the state, gasoline propulsion had already won a generation before. The contacts at this level with the national automobile industry and its institutional representatives had meanwhile led to legislation that was geared to the gasoline passenger car. In the attempts in this phase to set up a national road system, electric propulsion did not play a role either.

For the local authorities the situation was fundamentally different, at least in Europe. In the compact European cities—already the scene of resistance against an uncritical introduction of transit technology at an earlier date—the argument of noise and smell pollution played a more important role than in the United States. Examples of a sympathetic attitude of the local European authorities regarding electric propulsion have been discussed earlier in this study, with respect to the taxicab and the fire engine. That attitude received an extra boost by the financial and often lucrative stake of the municipal councils in local electricity generation. Moreover, the paved city roads required more intense maintenance, which was preeminently (and in the beginning, exclusively) possible by means of electric propulsion, as the emergence of the sweeping and sprinkling cars in the municipal vehicle fleets showed. In that respect the municipal councils in Europe played a role comparable to that of the American central stations: there, too, a clearly economic self-interest was present in establishing a fleet of their own, fed by their own, cheaply generated electrical energy.

In the United States a sympathetic attitude of the municipal and regional authorities toward electric propulsion was not so widespread. That is shown by the failed endeavors to introduce electric propulsion in fire engines, cleaning cars, police cars, mail vans, and ambulances. The "adventurous" road conditions in the cities and their sheer size created circumstances that were comparable to those that the gasoline car met outside the cities—circumstances that presupposed a larger energy reserve of the propulsion system. Thus, a "protected area" in the United States could only be developed by private enterprise, led by the biggest interested party of all: the central stations. They combined the social forces that pursued an electric-vehicle market and created a unique intermediary force, a platform of negotiation between all societal actors on topics varying from internal vehicle structure, standardization, and maintenance to propaganda and lobbying.

On the basis of this analysis, it is understandable that the electric commercial vehicle in America did not assume the shape of a fire engine or a sweeping-sprinkling truck. Instead, a commercial vehicle for the "free market" was pushed, especially by the manufacturers, meant for urban goods transportation. Hence, the importance of public garages that also were an extra stimulus for the central stations to get involved in electric propulsion. It surely is no coincidence that the big central stations—with their large fleets, considerable self-interest, and extensive knowledge of electric propulsion—took the lead here. In their effort to downplay the importance of maintenance, against the wishes of the vehicle manufacturers and the representatives of large public garages, they erected an formidable barrier for further expansion of the market toward smaller trucks and smaller users. At the same time, as big customers, they held the vehicle manufacturers hostage. Furthermore, they restricted competition between these manufacturers, because vehicle brand names were a curse in the multitude of propaganda publications. The EVAA seemed to be more interested in selling electricity than in selling vehicles.

It was the expansion to the smaller town and the smaller user that encountered not only technical and organizational boundaries, but especially ideological ones. The opposition to electric propulsion meanwhile had assumed mental, or if one wishes, religious characteristics. The choice for or against electric propulsion was no longer only determined by considerations of "scientific fleet management," but reflected an opinion of what "a car" should look like and how it should behave. This opinion was even stronger on the "free market" of generally deployable vehicles, where the gasoline passenger car dominated. The successful electric mail van in Germany, introduced on the basis of purely business-economic considerations, is a striking counter-illustration of this. As long as the definition of what a car should be remained outside the discussion—a definition that also for the electric truck was strongly influenced by the adventurous character of gasoline propulsion—the second- and third-generation electric truck, deployed in a fleet, was completely satisfactory.

The question now arises how the tiniest fleet in the hands of the tiniest fleet operator—the privately owned passenger car—can be interpreted in the light of what has been described here. So it is time to continue our main quest for the failure factors of the electric passenger car. But before we do that, we have to follow up on all other loose ends that we left during the previous chapters, such as the third-generation electric taxicab in the United States and the mail trucks in Germany. It is a story of super utilitarianism and shattered hopes, and of a revival of the electric passenger car after the Second World War, and enables us to bridge a time span of half a century to the present time, where we will encounter the same question over and over again: will the electric automobile remain a Car of Tomorrow forever?

7 Off the Road and Back:
Utilitarian Niches or New Universalism?

It is interesting to note that our friends, the gas car
manufacturers, now advertise their self-starting cars to be "as
simple as an electric," and some of them show with pride how
far their cars will run on the electric starter motor with its
storage battery. If we carry their advertisements one step further and substitute a larger battery in the space now occupied
by the gasoline engine, we have the ideal vehicle which we
advocate—The Electric Car.
—*W. H. Blood Jr., EVAA president, 1912*

THE PRINCIPLE OF THIRD-GENERATION electric vehicle technology could
not have been formulated more clearly—and more strategically—than EVAA
President W. Blood did in 1912: the technical field of the electric car in this third
phase can rightfully be defined as a subsection of the general technical field of
the gasoline automobile. Just substitute the combustion engine and you have
your electric. This convergence of technical fields led to serious identity problems for the electric alternative and ultimately resulted in the identification of
the electric as an environmentally friendly secondary car. But universal ambi-
tions were never far away.

This chapter offers a description and analysis of the third-generation electric passenger car. As the closing chapter of our narrative we also round up our
stories about the taxicab and the European truck, showing that electric propulsion disappeared from the road, in a double sense: several application fields
simply vanished (like those of the electric cab and the electric passenger car in
private use), but other fields emerged elsewhere, especially in the factories,
railway stations, and airfield platforms. Thus, although electric road propulsion may have been "defeated," it never wholly disappeared from the automotive application field, and certainly not from the field of expectations: electric
drive stayed on as an alternative, as a permanent promise (for the proponents)
or a threat (for the opponents) called "Tomorrow's Car." No wonder, then, that
after the Second World War, when problems of environmental nuisance and
energy shortage arose at an unprecedented scale, we witness the emergence

of a fourth-generation electric vehicle, propagated as the final solution of personal mobility in a highly motorized world.

A Shifting Electric Car Culture: Electric Tourism and Third-Generation Electric Passenger Car Technology

In 1918, the annual automobile contest of the Chicago Athletic Association took place for the final time. According to co-organizer and automobile journalist Chris Sinsabaugh, this was "for the simple reason that the cars were so well built and foolproof, that it was almost impossible for anything to happen that would destroy the coveted 'perfect score' which was so hard to make in the early days." In 1920, most American automobile manufacturers disbanded their expensive racing teams. They did so, not because they had converted to the purely utilitarian deployment of the car as means of transportation from A to B, but because the gasoline car as *universal car* had adopted the positive characteristics of its competitor (even the relaxed "electric" driving). Moreover, it had retained the sporty character of the early touring car, and—clearly visible to everyone—had incorporated it under a long hood. The tamed gasoline car seemed to move even when it stood still. The influence of electric propulsion even made itself felt at the level of the automobile system: pioneered by the taxicab experiments of the early days, in the course of the 1920s a garage infrastructure emerged in every Western country, where the technical problems with the still not satisfactorily reliable gasoline propulsion system could be delegated. Dirty hands were no longer necessary for enjoying the sensation of speed.[1]

With the gasoline passenger car undoubtedly in the lead by now, the proponents of the electric alternative desperately tried to secure their scattered fields of application, but not without important changes in the technical field. In Europe the shift of the center of the electric-car culture from France to Germany became crystal clear in the results of an investigation that AFA staff member Dr. Beckmann made public shortly before the outbreak of the First World War (table 7.1). Three years before, the AFA estimated the national electric-vehicle stock at 1,200 to 1,500 units and a year later at 1,500 (2 percent of the total vehicle fleet). At the time of the investigation, the total European electric-vehicle stock appeared to amount to 3,170 units, half of them passenger cars. That means that virtually half of the total European fleet as well as half of the passenger cars were in operation in Germany. It also reveals that the German fleet had hardly grown since 1912. Moreover, the figures for the Netherlands show that the number given for the passenger cars also included taxicabs. If this was also the case for Germany (with more than 550 electric cabs), France (with about 100 hybrid Kriégers), and Britain (190 electric cabs in 1916), only about 700 privately owned passenger cars existed in Europe at the outbreak of the war. So the shift of the electric-car culture from France to Germany also implied an adjustment: apparently, a relative success of electric propulsion in the second phase in Europe seemed to be realizable, mostly in the shape of the commercial vehicle.[2]

With a view to these figures, it becomes understandable that the large German manufacturers of electric passenger cars, such as Namag/Hansa-Lloyd,

TABLE 7.1 Electric Vehicles Counted in Europe on March 20, 1914, according to an AFA Survey

	Passenger Cars	Trucks	Three-Wheelers		Total
			Private	Commercial and Mail	
Germany	862	554	3	270	1,689
The Netherlands	70	38	1	6	115
Denmark	2	21	0	5	28
Sweden	2	2	0	2	6
Austria-Hungary	132	117	1	15	265
Belgium	1	0	0	0	1
France	100	190	0	28	318
Russia	3	4	0	1	8
England	201	62	0	25	288
Switzerland	131	69	0	0	200
Rumania	0	0	0	1	1
Spain	12	0	0	1	13
Italy	60	173	0	5	238
Total	1,576	1,230	5	359	3,170

Source: P. D. Wagoner, *European Development of the Electric Vehicle Industry,* EVAA-1914.

AEG/NAG, and Siemens/Protos, also incorporated gasoline cars in their programs even before 1914. AFA was the major protagonist of electric propulsion in the years before the World War. Almost all electric vehicles mentioned in table 7.1 under Germany, the Netherlands, Denmark, and Sweden were maintained in a charging station of one of AFA's engineering departments. At the time, Germany had thirty-nine charging stations. Thirteen of those were meant exclusively for electric taxicabs, an equal number for mail vans, and only eleven for private car owners. The last number mentioned apparently reflected the demand, for already in 1908 the AFA recorded that the largest European manufacturer, Namag, had "hardly put any private vehicles on the market" via its local subsidiaries.[3]

The war reinforced a utilitarian deployment of electric propulsion, just as in the case of the gasoline car. Consequently, the electric city-car market practically vanished after 1917 and electric propulsion was almost exclusively applied in the commercial vehicle and the *Elektrokarre* (industrial truck).

WHEREAS THE EUROPEAN electric passenger car slowly but surely disappeared in the swamps of the World War, not to emerge again afterward, in America this type of vehicle had its heyday in the years before the country's involvement in the war. By the end of 1913, the electric passenger-car fleet comprised 20,000 units, two-thirds of a total of 30,000 electric vehicles. It represented 1.7 percent of the entire American vehicle fleet, which at that moment

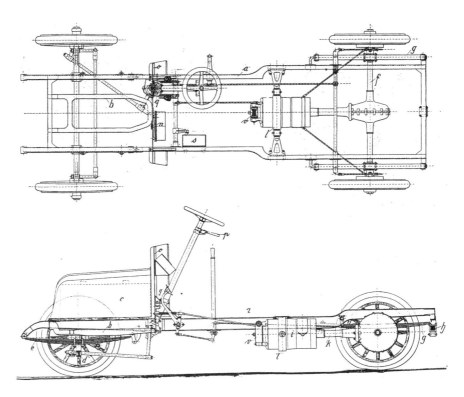

An early example of an electric automobile of the third generation. Introduced at the Berlin Auto Show of 1907, this Siemens had all the structural characteristics of its gasoline rival—a light steel chassis, propulsion by one central electric motor, and a differential at the rear axle. Such design "mimicry" owed to technical advances in tire and battery design as well as growing preferences in overall automobile design. The lighter chassis and low-friction drive trains directly benefited the electrics. (*Centralblatt*, 20 August 1908, p. 122)

consisted of 1,159,332 vehicles. A third of the electric cars were sold in 1912. To "feed" these vehicles, EVAA in 1913 counted 6,123 charging facilities in 128 cities: 5,491 private stations and 632 public stations. It was noticeable that the number of "home kept electrics" was inversely proportional to the size of the urban population (figure 7.1). This phenomenon was explained by the fact that in the big cities many car owners lived in apartments without a garage of their own, so that potential electric-car drivers would have to go to a public garage. This remarkable fact indicates that the electric-car owner meanwhile had adapted to the mainstream culture: he preferred a car in front of his own front door. Chicago had one electric car per 750 residents, but Cleveland (with a quarter of Chicago's population) numbered one electric car per 340 residents.[4]

In absolute numbers, however, Chicago had the largest fleet, with 2,200 "pleasure vehicles," followed by Denver (850), Washington, D.C. (654), New York City (498), Greater Boston (282), and Hartford, Connecticut (250). Chicago—where the Commonwealth Edison Company sold charging current for $200,000 in 1912—was a very popular electric-car city because of its level roads and its ring road around the center. Moreover, due to Insull's active policy, the

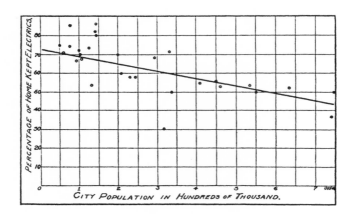

CITY POPULATION IN HUNDREDS OF THOUSAND.

charging-current rates were very low and a network of garages had been established. The building of big public garages there was particularly stimulated by Commonwealth Edison: at a purchase of 50 kW or more and a garaged fleet of 75 to 150 electric cars, Commonwealth offered its charging current for 1.75 to 2 dollar cents per kWh. Because of this measure the company supplied more than 85 percent of all the charging current in the city. "The successful operation of electric vehicles depends largely upon the daily care given them" was the slogan.[5] Whoever wishes to study the electric car culture in more detail should focus his or her attention on Chicago.

In 1915, according to the American Census, production in the United States amounted to 4,715 electric cars (not even one percent of the total vehicle production), versus 3,835 according to the 1909 census (3 percent of the total production). During the four years before 1915 the price of an electric passenger car became higher and higher, usually without any significant changes in the models (figure 7.2). As a typical luxury product, the electric car was expected to sell itself without too much advertising. Its pricing was such that it yielded a considerable profit margin. In the beginning of 1914, the average price of an electric car was $3,000, being the average of a price range between $1,600 and $5,100. This range was much narrower than for the gasoline car, where prices varied between $500 and more than $6,000. The electric car functioned as the affluent suburban family's second car, preeminently suitable for motorists who were "not mechanically inclined," or at least thought so themselves. For example, Jane Parsons Maynard described in *Suburban Life* how her husband used the gasoline touring car to drive to his office in New York City (a 20 km drive from the suburb where she lived), whereas she herself used an electric brougham. A survey by the Waverley Company among its clients revealed that most of them drove 3,000 to 4,000 km a year. This was less than half the annual mileage of the average physician in the survey of 1906 (discussed in chapter 3). Monthly these customers spent about 2.5 to 5 dollars on charging current and a similar amount on pneumatic tires. The battery lasted two to five years, with an average of 2.5 years.[6]

The high price of the electric car triggered a reaction that ended in the con-

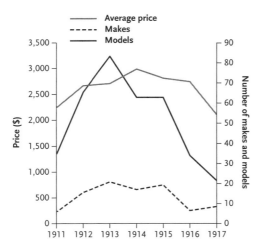

FIGURE 7.2. Development
of the number of brands,
the number of models, and
the average price of electric passenger cars, 1911–
17. (Calculated on the basis
of several volumes of EV
[U.S.]; see also figure 3.1)

clusion that there was no good reason "why an electric roadster should cost over $1,250." In September 1915, some manufacturers decided to lower their prices, a result of the falling sales due to the outbreak of the European war. Most manufacturers stuck to their opinion "that there is no buying class that wishes an electric car costing $600 to $1000, because that value would probably represent the buyer's first car and he would want speed and mileage." But others really tried to develop cheaper electrics, designed along the lines of gas car tradition. In 1914, for instance, the Ward Motor Vehicle Company of New York—known for its electric commercial vehicles—introduced a coupe, developed by Charles A. Ward and Hayden Eames, which can be considered as an intermediate form between the second and third generations. It had a hood, a closed body, and a slightly longer wheelbase than usual. The Standard Car Manufacturing Company in Jackson, Michigan, presented a similar model for $1,990 in the fall of 1914. Apparently, these events really awakened the gasoline car industry. This was dramatically illustrated by Henry Ford, who hit the headlines when he announced in the beginning of 1914 that he would put an electric car with a range of 160 km on the market for $900. His son Edsel was going to supervise its construction. This "Ford Electric," as the press called it in analogy to the Detroit Electric, appeared to be the brainchild of Ford's friend Thomas Edison. In an interview Edison even mentioned a price of $500 to $750. According to Michael Schiffer, the Ford management considered the project as a kindly gesture to Henry Ford's friend Edison, but the contemporary reactions tell a different story. These reactions were overwhelming. The Ford factories repeatedly had to issue statements that the inexpensive car was "Mr. Ford's personal project." The E-Ford project, however, remained limited to the construction of a few prototypes. In 1917, official Ford communications mentioned that the electric car was still "in an experimental stage." The E-Ford never progressed beyond this stage.[7]

It is plausible to assume that Henry Ford's remarkable maneuver was more than an insignificant gesture toward a friend. Edison's initiative was probably inspired by his disappointment about the fact that the passenger-car

makers received his alkaline battery with much less enthusiasm than the truck manufacturers. According to
Michael Schiffer, this was partly the result of the manufacturers' hesitance to adjust their cars to the bigger
battery set and to raise the price per car by about $200
to $600, without making much extra profit. But it was
also caused by arch-rival ESB, which had lured many
manufacturers into exclusive contracts. Detroit Electric, for instance, was Edison's largest customer, but the company realized that this blocked the application of the ESB battery. Therefore, the company decided to make its own lead
battery. The contract Detroit signed with Edison in 1912 was not exclusive,
however. It contained the odd clause that Detroit would receive all Edison batteries reserved for pleasure vehicles, except those claimed by "Col. Bailey and
General Healey."

In 1911, "Colonel" E.W.M. Bailey of the S. R. Bailey Company in Ames-
bury, Massachusetts, launched an aerodynamic runabout. This car can be considered as the extreme manifestation of the third-generation electric car: it had
radically broken with the exterior of the closed-body city car and could not be
distinguished from a small gasoline touring car. It was provided with Edison
batteries, which enabled a range of 130 to 190 km, according to the manufacturer. This was demonstrated by all kinds of stunts, such as a twelve-day run
of 2,000 km from Boston to Chicago in 1913 (driven at an average speed of almost 30 km/h). Another one was a race between Boston and New York in May
1914, a distance of over 400 km, driven in slightly more than twenty-three
hours. Almost half that time was spent on five boosts. "Has the day of the electric touring car arrived?" *Electric Vehicles* wondered about this contest.[8]

Bailey's stunts took place in a climate where some were eager to believe in
the future of a revived electric tourism. In 1914, for example, the New York
Electric Vehicle Association published an "Electric Touring Book." It contained
detailed route descriptions in a radius of 100 miles around New York City, including an overview of garages and charging stations along the way. In 1915, a
similar list of charging stations around Philadelphia appeared, compiled by
the Philadelphia Light and Power Company, an active EVAA member. The
Goodrich company published a map of the eastern states that year, containing
the "charging-station routes" from New York up to Moosehead Lake in Maine,
near the Canadian border. Such lists showed that the network of charging stations in the urban eastern part of the United States was sufficiently dense to
enable electric tourism. Whether that really materialized is the question. The
culmination of such initiatives was reached when the EVAA published a map
of the Lincoln Highway across the entire United States, indicating the charging possibilities. It is not very likely that large numbers of electric-car motorists drove the 5,475 km-long route. The gaps between charging-current facilities would have made this difficult: the longest distance between two such
facilities was 197 km.[9]

The trade journals wrote in great detail about this electric-tourism movement that started in 1914 and continued until the beginning of 1916. In the

summer of 1914, for example, a Detroit Electric drove the distance from Washington to Philadelphia and back. A month later it covered the distance between Boston and Philadelphia with all the high-tech means available at that moment: the Philadelphia thin-plate battery and Goodrich's new Silvertown cord tires. In 1915, the Gould Storage Battery Company established a new "world record" by covering 1,600 km in ten days (160 km each day). Only one battery charge per day was necessary. "Electric no longer town car," a self-assured headline in *Electric Vehicles* read in the beginning of 1916. But in the years to follow, the journal would prove to be right in quite a different sense than intended. For, already in the summer of 1916, the journal began to qualify the importance of touring with arguments that sounded remarkably modern. "Students of automobiling know that touring is a delusion," and it asserted, "that of the millions of cars now running in this country, only a few thousand ever take trips over a hundred miles in length." And Woods (who stuck to the solid rubber tire until 1916 and thus was the most conservative of the American electric-car makers) mentioned "that statistics prove that 98 per cent of all motor trips fall within a radius of sixty miles."[10]

Ford and Edison were not the only celebrities who ventured the design of an inexpensive electric car. Charles P. Steinmetz of General Electric was at least as renowned. In 1915, he founded the Dey Electric Vehicle Syndicate (in 1916, changed to Dey Electric Corporation), in cooperation with Harry E. Dey, an electrical engineer from New York City. The smart design made use of one single electric motor, of which both the armature and the field (each connected to a driven wheel) could rotate, so that a differential was not necessary. In a full-page article in the *New York Times,* Steinmetz declared that his concept "would dethrone the gasoline car." The Dey appeared on the market in 1917. It weighed only 636 kg and cost $985.[11]

If Henry Ford may have been alerted by this short burst of "electric tourism" enthusiasm, the threat from the electric vehicle camp did not last long. Most prospective motorists apparently did not share Steinmetz's opinion, for meanwhile a wave of mergers among the electric-car makers had started. For example, the exclusive Baker brand (presenting itself in advertisements as "The Aristocrat of Motordom" and supporting this image by hiring a French couturier for the design of its interiors in 1914) merged with the equally exclusive Rauch and Lang brand. The company name became Baker R & L Company, often referred to as Baker-Raulang. In the fall of 1915, General Electric (also dominant in electric-truck builder GeVeCo) took a stake in the new enterprise, which then doubled its capital to $5 million and took over three GE directors. After it ended its electric-car production in 1921, Baker-Raulang was to build industrial trucks until well into the 1950s.[12]

And yet, the merger mania of 1914–15 led to a last revival of the electric passenger car, which manifested itself in the reinforcement of two contradictory tendencies. On the one hand, the luxury city car had been improved by further perfecting the ease of operation. For example, the Ohio Electric Car Company in Toledo implemented the "Electric Magnetic Controller." This device contained a multifunctional turning knob for electromagnetic operation

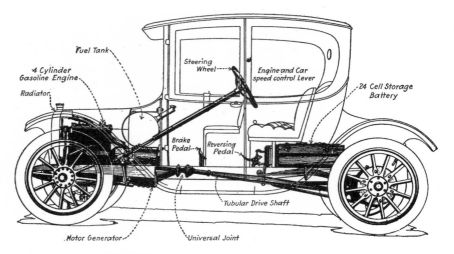

One of the best-known hybrids in the United States during the Great War: the Woods Dual Power, controlled from the back seat and, in a typical third-generation version, not to be distinguished from a gasoline car in appearance. (Pagé, *The Modern Gasoline Automobile*, p. 103)

of the controller and it had a push button for the electric bell as well. Another example was Detroit Electric with its "Duplex Drive car," which one could steer from both the front and the rear seats. A new supplier was the Milburn Wagon Company, a coach builder dating back to 1848. In 1914, this company decided to start manufacturing a light electric vehicle, of which it sold more than 2,500 during the first two years. This made Milburn the second-largest electric-car manufacturer after Detroit. Most competitors did not even reach half of these production numbers. The Milburn—used by the Secret Service under President Wilson—had been designed for exchange charging. The battery set was placed on rollers: "Simply roll out the discharged ones and roll in the freshly charged set."[13]

On the other hand, new hybrid concepts appeared on the market, of which the Woods "Dual Power" was the most renowned. The Woods concept, which promised "to [bid] central stations good bye," became "the hit of the 1917 auto shows." But, just as most other manufacturers, Woods tripped over America's participation in the World War, because it did not succeed in securing war contracts.[14]

THE CONVERGENCE OF THE TECHNICAL fields of the gasoline car and electric passenger car during the third generation was illustrated dramatically by the development of tire technology. The manufacturers, who were afraid to offer a pneumatic tire for electric cars, put solid rubber tires on the market with special resilient rubber compounds. Such expensive "cushion tires," made of very pure Para rubber, often had a special shape, so that their geometry also supported the elasticity of the tire: a deep incision was made in the middle of the tire that ran along its entire circumference. The tire walls were also given incisions. In this way, they could be depressed more deeply than a tire with the

In about 1915 a mileage contest in and around Los Angeles featured thirteen locally produced electric Beardsleys. On a single charge, the cars traveled an average distance of 101.1 miles. Ten of them were equipped with Goodyear cord tires, and five of the highest-mileage contestants (from 103.9 to 112.1 miles) rode on those tires, first developed for electric vehicles and the predecessors of modern bias-ply tires. (Goodyear Archives, University of Akron)

same rubber mass, but with a circular cross section. But the cushion tire had two major disadvantages. First, it was not as resilient as the pneumatic tire. Second, it reacted more slowly to tiny bumps in the road, so that at high speeds the wheel experienced an upward acceleration. So, as the speed of the electric car increased, this vehicle could not do without the pneumatic tire either, just as its gasoline rival.[15]

As the cotton carcass could only be woven in a cylindrical shape, the fabric folded around the rim when it was being mounted on it. This caused an irregular tensile stress in the carcass and thus an irregular resilience. Moreover, the pneumatic tires for gasoline cars, with their sturdy walls and thick tread, caused far too much energy loss to apply them to the electric car. Therefore, special thin-walled pneumatic tires were developed with a coarse-meshed carcass fabric and a thin tread. These, however, were very sensitive to a tire pressure that was either too high or too low, and thus were antithetical to a maintenance-friendly car culture, especially in the United States. The special electric passenger car tires suffered from the scissorslike effect of the crossing cotton threads, causing rapid heat development. Thus, high resiliency and high durability came into conflict, so that regular maintenance became a crucial factor, just as for the lead battery. Such tires were not very strong, but they were very expensive, because "absolutely the best cotton" and the purest natural rubber had to be used.[16] In other words, the technical field seemed absolutely closed for the application of sturdy and not too expensive pneumatics on both the electric and the gasoline alternatives. But in this case there was a way out of this dilemma, and the role of the electric vehicle appeared to be crucial in this development.

The solution consisted of a typical crossover phenomenon. The conflict between resilience and durability had been solved for the bicycle by the application of the so-called cord tire. In the cord tire's construction two layers of rubberized parallel cords were put across each other. The textile threads were not crossed, but rolled into a cord and vulcanized in the rubber side by side, in a nonperpendicular angle with the rim edge. This was done after each layer of cords had been separately rubberized first in order to further lower frictional heat buildup. The London Palmer Tyre Ltd. produced the best-known cord tire. John Fullerton Palmer patented it in America in the fall of 1892 and half a year later in Great Britain. This concept became the standard in bicycle technology.[17]

The electric car served as a guinea pig in the translation of this technology into automotive technology. Goodyear wrote in a retrospective account: "The early *cord tires* were used only on electric vehicles, where the demand was for a tire of extreme resiliency so as to get the maximum mileage per battery charge. Many designs were tried and discarded, but always Goodyear had a *cord tire* of some design to offer the electric vehicle owner. It took years to attain the desired combination of resiliency and long tire life, but always *cord fabric* was the carcass material used in the experiments."[18]

Almost all large tire manufacturers tried to put such tires on the market, but usually without enduring success. Gasoline cars were not equipped with them, although they did increase vehicle speed and improve riding qualities, as the early applications in racing proved. "But these tires will puncture more easily, are more difficult to repair, and in the end will not give as low a tire mileage cost as the gasoline type of tire. The tire mileage cost is the principal factor considered in the design of a gasoline tire, but becomes secondary in the case of tires for electric vehicles, owing to the other great savings accomplished by the electric tire." P. W. Litchfield, operations manager of the Goodyear Tire and Rubber Company in Akron, Ohio, explained it very clearly: the gasoline tire was designed for low mileage costs, but with a view to the narrow energetic margins for the "long-distance electric tires" it could not be applied to electric cars. The price the electric-car owner had to pay for it was high. The gasoline car had so much energy reserve that extra energy loss by using the conventional fabric tire did not present a problem, especially in America with its low gasoline prices. The gasoline-car driver could use an electric tire if he wished, but the electric-car owner did not have a choice. The universality of the gasoline car appeared to manifest itself even in the smallest details, up to the choice of tire type.[19]

The rolling resistance of the cord tire was a quarter lower than that of the tire with cotton carcass. So, it will not come as a surprise that in the beginning of 1915 a cord tire was launched that was also suitable for gasoline cars: Goodrich's Silvertown Cord Tire. All the racing cars that reached the finish line in Indianapolis that year were equipped with these tires that were based on the Palmer patents. By the end of that year, the first cord tires appeared on the expensive gasoline passenger cars, the beginning of the large-scale commercialization of what in later automotive technology would be called the bias-ply tire. The Silvertown was the first sturdy tire—both resilience and tire life were considerations in its design. The introduction of the cord tire during the First World War—and generally applied from the beginning of the 1920s—solved the problem of internal overheating. Its increased sturdiness also enabled

ELEKTRO-
FAHRZEUGE

Siemens-Schuckert

Siemens-Schuckert

SIEMENS-SCHUCKERT

After gasoline propulsion established its predominance, electric drive persisted in specialized vehicles for indoor use, such as these Siemens-Schuckert shop- or warehouse-floor "industrial trucks." (Rödiger, *Der elektrische Kraftwagen*, p. 155)

a considerably lower tire pressure, which improved comfort, especially as it absorbed small bumps in an otherwise smoothly asphalted road surface.[20]

In 1917, Goodyear developed a cord tire for trucks. The principle of the cord layers enabled application of pneumatic tires to such heavy vehicles for the first time. Thus, the gasoline vehicle received a significant boost in tire life and comfort, based on a technology that originally had been developed for the electric car. Just as in the case of the battery, one cannot claim that it narrowed the electric's technical field so far that application became impossible or was seriously handicapped. But the field of application was considerably restricted: just as the battery, the pneumatic tire meant an extra increase in the initial costs.[21]

IN 1916, THE ELECTRIC CARS PRODUCED in America amounted to 4,000 units and at the beginning of 1917 the electric-car fleet in most of the big cities had further increased. Chicago meanwhile had 3,000 units, Cleveland 1,750, Detroit 1,500, and Washington, D.C., 800. New York, with only 450 electric passenger cars, followed behind Kansas City (876), Denver (750), and St. Louis (590), all cities with an active central station as EVAA member. A city like Memphis, with 245 electric passenger cars and a "white population of 90,000" (as the journal *Electric Vehicles* expressed it), seems to confirm the assumption that the distribution of passenger cars did not follow that of electric trucks. The electric passenger car seemed to prefer the medium-sized town in the Midwest.[22]

America's involvement in the war, however, abruptly put an end to this trend. The manufacture of electric passenger cars was considered as one of the nonessentials. Therefore, it received far fewer war-production orders than the gasoline competition and was continually plagued by shortages of material. The Electric Vehicle Section of the NELA abandoned all propaganda for the passenger car during those years, because "no one has any business to buy a pleasure car today; the money ought to be put into Liberty Bonds or Red Cross or some other branch of patriotic service." It is yet another example of the central-station managers' ideological blindness for the emerging utilitarian character of the passenger car.[23]

In 1920, it appeared that "the electric vehicle business (was) practically confined to a very few of the largest cities," a situation Chairman Foster of the Electric Vehicle Section of the NELA ascribed to the much better garage infrastructure in those cities. More detailed local research would have to show whether his explanation was valid and not just an expression of the EVAA ideology. In electric-car city Chicago, a thorough investigation unearthed, however, that the number of private charging units in the area of the Public Service Company of Northern Illinois (also comprising the suburbs of Chicago) was double that of what one had assumed up until then. And although between

1 January 1919 and 1 April 1920, 413 electric passenger cars were sold in Chicago (and 156 trucks), not a single one was sold in New York City (but as many as 308 trucks were sold there). The limited distribution of the passenger car in New York was blamed on the large distance between the center and the suburbs, as well as on the steep hills in the surroundings, which were problematic to electric tourism.[24]

With the technical potential of the third-generation electric passenger car in mind, it becomes increasingly enigmatic why the electric alternative did not enjoy a much wider field of application. This is all the more true if we consider the taxicab application field, where third-generation alternatives clearly showed their technical superiority vis-à-vis the gasoline rival.

From Reliable to Unreliable Technology: The Third-Generation Electric Taxicab
Even within the field of taxicab application, which after 1914 just as the privately owned passenger car application field in most countries was dominated by gasoline propulsion, electric propulsion sometimes found a remarkably profitable niche, especially in the third-generation version. These few applications also shed a revealing light upon the degree of reliability of the gasoline car in heavy-duty applications.

Shortly before the outbreak of the First World War, the Detroit Taxicab & Transfer Company (one of the oldest gasoline-cab companies of the country and probably founded in 1907) found out that the manufacturers of passenger cars were not interested in the building of a robust automobile that would cost an estimated $2,000 to $2,500. At a rate of 70 cents for the first mile and 40 cents for each following mile, operating costs of between 30 and 35 cents per mile, and proceeds of about 33 cents, the profit margins for this company (with a "zero occupancy" of 40 to 50 percent) were very narrow indeed. Moreover, the gasoline cabs became noisy after a few months of intensive use, while there were also complaints about grease and gasoline smell—complaints that were remarkably reminiscent of the Berlin "taxi crisis" seven years earlier. "It, therefore, became necessary that we obtain equipment that could be operated more cheaply than gasoline cabs," I. S. Scrimger, manager of the Detroit firm, concluded. News about the large Berlin electric-cab fleet began to penetrate the United States in 1913–14. Impressed, but apparently less so by the designs of renowned electric car brands like Baker and Detroit Electric, the Detroit company decided to build its own spacious luxury electric cab. The second version of this cab can definitely be characterized as a third-generation model, because it was much indebted to both the structure and design of the gasoline-car (including a pseudo hood).[25]

On 25 June 1914, at 2 o'clock in the afternoon, the first car became operative at hotel Pontchartrain. Company engineer W. J. Behn was the designer. In January 1915, 10 or 11 cabs were in use, each covering about 2,200 km a month. In May 1915, the first cab had driven almost 20,000 km. The company considered this as proof of the reliability of its design and held out the prospect of a fleet of 70 units. If that was realized, it meant that electric cars would have replaced almost all of the company's gasoline-cab fleet: "Only a very few gasoline cars for very long runs will be retained." In the beginning of 1916, the fleet consisted of 47 electric cabs manufactured by the company itself. That same year 75 cars were in service and in the summer of 1917 almost 100. At that moment—shortly after the involvement of the United States in the World War—women were being hired as cab drivers. There was an overwhelming response of more than 500 "patriotic women." The women selected were given a few hours of instruction and they received the same wages as the men. They only worked during the day.[26]

The cabs were such a success that customers apparently were prepared to wait for half an hour or more, even though the company's gasoline cabs were waiting at the stand. Although the cars were equipped with Goodrich's expensive Silvertown pneumatic tires, operating costs remained below 20 cents, 10 to 15 cents lower than the company's gasoline cabs. This result was obtained without exchange charging. Instead, the company made use of special charging poles at the cab stands, where the range was extended by means of "curb boosting." New York, Philadelphia, Chicago, and St. Louis were interested in following the example of the "Motor City." In the first two cities it probably did not get beyond the planning stage. In New York, for instance, after a test with one of the taxis from Detroit, engineer F. G. Peck announced an electric cab fleet of 150 units in the beginning of 1916, complete with twenty-hour service and curb boosting at the cab stands. The latter system was considered more economical than the European exchange-charging system. The electric cab in New York would be adjusted down to a maximum speed of 40 km/h. By means of charging outlets at the stand, it could reach a daily route of 120 km, whereas the average daily distance of the gasoline cars in that city was around 65 km. The New York Edison Company and the New York Electric Vehicle Association supported the plan.[27]

In Chicago a comparable initiative made it beyond the drawing board. After a visit to the Detroit company, Guy Woods, owner of the American Motor Livery Company in Chicago, decided to start a fleet of twelve electric cabs. Milburn Wagon Company built these elegant five-passenger cars. Even more than the cabs in Detroit, they were decidedly third-generation with their speed of 40 to 50 km/h and their range of 160 km on one charge. Such taxicabs met the demand, characteristic for the American version of the third-generation electric cab: "It is necessary for the electric taxi to equal gas standards of speed and mileage and to operate more economically." The twelve taxicabs became operative in April 1917. They were meant as a supplement to the twenty gasoline cabs, but "practically displaced the gas cabs as patrons constantly requested them." A few months later six electric cabs were introduced in St. Louis as "the nucleus of a fleet . . . The mileage of these electrics is practically unlimited, as facilities for curb boosting are situated in various parts of the city, so that the cab may be charged while it is waiting, even though it may be only a few minutes."[28]

It is not known how much impact these impressive experiments made on the leading circles within the gasoline-car industry. In any case, they were very localized and incidental, and we have to wait until the 1920s (when the national taxi fleet had increased to 90,000 units with a joint annual mileage of about 3.9 billion km) before new electric-cab initiatives emerged again. This was the case, for example, at the Electrotaxi Company on Madison Avenue in New York. The company commissioned Rauch & Lang to build a special third-generation cab with a maximum speed of 40 km/h. Its projected daily route was about 200 km, to be realized by working in two shifts with an exchange-charging system that could exchange the battery in less than ten minutes. The race with its gasoline rival proved to be in vain, however. Whereas in 1915 the Chicago department of the renowned Yellow Cab Company—with a fleet of 31 gasoline cabs and 51 drivers—drove an average daily route of about 80 km, in 1922 this distance had increased to 235 km (with 1,624 cabs and 3,015 drivers).[29]

The European taxicab field of application after the war did not differ fundamentally from that in America, despite the severe regulations. In Germany, for instance, motorization of the national taxicab fleet increased rapidly. In Berlin the number of motorized taxicabs tripled between 1924 and 1928, after unsuccessful attempts to introduce *Kleindroschken* (four-wheel motorcyclelike cabs). The predominance of the gasoline alternative was established there, when on 23 February 1924, the police department announced that one electric cab could be exchanged for one gasoline cab. The flight from electric to gasoline propulsion was massive. In eight months' time, the number of electric cabs was halved, and in 1934 only a few were still in operation.[30]

Except for Berlin, the electric cab had disappeared from all German cities. An attempt in 1926 of AFA, AEG, and the Zschopauer Motorwerke to put a small third-generation electric cab in circulation failed, although they did obtain a license for 500 such cars. Serial manufacture of the car was started and taxicab company Avuag had 185 of them in operation. But after a year, the losses had become so high that all cars were scrapped. The range of this DEW (*Der elektrischer Wagen*, the electric vehicle) had increased to 80 km and its speed was 45 km/h, but the "unpleasant tinny sounds during the ride" made the vehicle unpopular. Ironically, the electric vehicle itself had meanwhile become unreliable, no doubt due to the marginal existence of the manufacturers in question.[31] A better proof of the third-generation electric vehicle being a derivative of general (gasoline) automotive technology can hardly be given.

Off the Road or Out of Sight: The Electric Vehicle between the Wars

From the middle of the 1920s, the electric car in America slowly but surely vanished from the streets. In 1919, its production still numbered 2,498 units (0.15 percent of the gasoline-car production of 1.65 million). But in 1922 this number had dwindled to 405, whereas in 1924 only 391 were counted (versus 3.19 million gasoline cars). That year, for the first time, not a single electric or steam car was exhibited at the National Automobile Show in New York. In 1925, no more than 22 electric cars were manufactured (and 3 steam cars). One more short revival took place in 1929, when 757 units were produced (versus 4.54 million gasoline cars). But after the stock market crash of 1933, when even the dominant automotive industry saw its production diminish to 1.56 million units, not a trace was left of any regular manufacture of electric cars. When Chris Sinsabaugh took stock of almost half a century of electric passenger cars in 1940, he counted 13,862 Detroit, about 12,000 Milburn, 9,300 Rauch & Lang, and 1,000 Woods cars. At that moment 1,450 electric cars were still in use.[32]

On a national scale, the interest of the central stations in electric vehicles had dropped to a minimum in the mid-1920s. At the NELA convention of 1925, electric propulsion was hardly mentioned. All attention was directed at refrigeration, understandable if one realizes that such systems used more electrical energy than an electric car. Moreover, they operated "independent of the whims and habits of householders." In 1927, the Transportation Committee decided to limit its propaganda to the twenty-eight biggest American cities. But even there a downward trend could be observed: the further growth in (gasoline) passenger-car ownership and household refrigeration systems caused a decrease in the delivery of goods by electric vehicles.[33]

Gasoline trucks gradually replaced the electric trucks in the fleets of the big cities. The best-documented fleet is Commonwealth Edison's exceptional fleet in Chicago. In 1925, it still had 350 to 375 electric trucks in operation, whereas the total number of urban electric trucks in that city increased by 50 percent between 1923 and 1931. Without doubt, the central station's active policy had brought this about. In 1933, Commonwealth Edison sold its own factory, the Walker Vehicle Company, to Yale & Towne. The fleet was then increased to 824, made up of all gasoline trucks. The last Walker stayed in operation until 1947. In most other cities electric traction had already disappeared "underground" at a much earlier stage. It vanished from the public roads and reappeared in the industrial truck and the versions derived from it (truck crane, forklift truck, platform cart, tractor). It became a small cog in the transportation system within and around factories, harbors, and the big transportation companies.[34]

Baker-Raulang, for instance, continued the prewar tradition with its Industrial Truck Division. But here, too, no protected market niche for electric traction seemed possible: in 1937, the company extended its large variety of products with a gasoline version. When, at the outbreak of World War II, the federal government ordered thousands of industrial trucks (in 1940, Baker-Raulang alone received an order for more units than it had produced in the preceding thirty years), the electric-truck manufacturers were not able to fill this order. Moreover, an increase in capacity was impossible due to a shortage of copper and other materials. It was then decided that "the present manufacturers of trucks should continue the electric trucks for ordnance and munitions work as well as for plants whose operations demanded high priority," but the gasoline-car manufacturers would execute the major part of the order. The electric-truck industry barely survived this second onslaught by a World War. After the war the new competitors put thousands of cheap industrial trucks with a gasoline engine on the market. Yet, in 1947, 12,300 electric industrial trucks were still being produced in various versions. In 1951, the fleet comprised 44,000 units (of which 70 percent were forklift trucks), excluding the 21,000 "driver-walk" trucks that were operated by a driver on foot. In 1954, the most important manufacturer of such trucks, Baker Industrial Trucks, merged with the Otis Elevator Company and was continued as the Material Handling

Division. The vertical and horizontal versions of electric-powered transportation from then on were produced under one roof.[35]

In Europe, on the other hand, the development of electric propulsion occurred less abruptly in the interwar period. The German Post Office started the electrification of its vehicle fleet shortly before the war, but the program stagnated when it reached 220 units. During the war the greater part of this fleet was lost and mail delivery took place again by horse traction. In 1924, the Post Office decided on a "motorization of the Berlin local delivery" by taking 360 electric two-ton trucks in operation. Nationally, the Post Office possessed 1,632 electric trucks (four times as many as the number of gasoline trucks) and an unknown number of electric three-wheelers in 1927. Mid-1930s, the Berlin mail delivery fleet consisted of 1,300 vehicles, of which 724 units were electric-powered and 395 were gasoline-powered. Other important fleets of electric mail vans were found in Breslau, Dresden, Düsseldorf, Frankfurt am Main, Hamburg, Cologne, Königsberg, Leipzig, Munich, Nuremberg, and Stuttgart. Together they comprised a fleet of 2,400 vehicles. Gasoline vehicles were deployed where high speed and substantial engine power were important. But in the big and medium-sized towns, electric vehicles were used almost exclusively for mail transportation within the city and at the stations, and for mail delivery. In the period between the wars, accurate tests confirmed the attractiveness of electric traction for mail delivery. They showed a higher acceleration and a higher average speed in heavy city traffic than the gasoline competitor, as long as the distance between stops was not more than 400 m.[36]

This made the Post Office by far the largest electric-fleet owner of Germany, for in the mid-1930s the total German electric-truck fleet numbered about 5,000 units. Of these, 1,500 were in use in municipal vehicle fleets as sprinkling, cleaning, and garbage trucks and 250 at breweries. The Berlin company Meierei C. Bolle, which started the electrification of its vehicle fleet in 1922, probably owned the largest private fleet, with about 100 trucks. At private companies the electric truck was especially popular for the transportation of food supplies (bread, meat, milk) and bars of ice for cooling. The lighter truck versions for city delivery work prevailed.[37]

The municipal vehicle fleet of Berlin was the largest of its kind, with 254 electric vehicles. The German municipal sanitation departments generally showed a remarkable interest in electric propulsion: in 1939, half of the joint municipal vehicle fleets appeared to consist of electric vehicles. Dresden also had a large stock of 700 (municipal and private) electric vehicles, distributed among the municipal market complex, the coal trade, shipping agencies, and wholesalers.[38]

Just as in the United States, however, in Germany the real continuation of electric propulsion took place outside road traffic. The industrial truck experienced explosive growth in the 1920s (figure 7.3). In 1932, the fleet size increased to 12,000 units according to some sources, but others even mentioned a size of 15,000. At the time the industrial trucks accounted for 44 percent and the electric street trucks for 23.5 percent of the total energy consumption of 100 million kWh.[39]

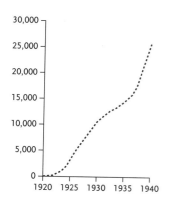

FIGURE 7.3. The development of the stock of *Elektrokarren* (electric industrial trucks) in Germany, 1920–40. (*EF*, November 1914, first page)

In the beginning of 1939, AFA estimated the total German electric-vehicle stock at 28,300 units, mostly industrial trucks, but also including 10,500 electric-powered street trucks. In addition, 1,115 *Grubenlokomotiven* (mine locomotives) were counted.[40]

In France, in the interwar period, the electric-vehicle trade did not recover from the blows inflicted by the gasoline taxicab, the economic slumps, the strikes, and the war. Following the example of Britain (see below), however, the Société pour le développement des véhicules électriques (SDVE) was established in 1922. It was located in Toulouse and four years later it started to publish a journal of its own, *Le Véhicule Électrique*. Around the middle of the 1920s, the many contests (especially geared toward direct competition with gasoline propulsion) indicated a modest revival of interest in electric propulsion, also within the established automotive industry.[41]

Due to this new interest the center of activity shifted from Paris to Lyon, where Berliet was located. Moreover, a very active municipal transportation service emerged, which on 25 December 1924, put the first of 16 electric-powered buses in operation, ordered from de Dion-Bouton. Test drives with the buses, also supplied by the Société Vétra and Renault, yielded a remarkably low energy consumption of 47 Wh/(ton.km) at a gross vehicle mass of more than 10 tons. During those tests, ranges of 145 km on one charge appeared to be possible. The Renault model was equipped with a compound motor of 15 kW, other models had 2 motors. Within a few years, 54 buses were in operation on 7 lines with a joint length of 52.4 km.

Even more striking was the side effect of these activities. For example, in Lyon the Société Lyonnaise pour l'Exploitation de Véhicules électriques (SLEVE) was founded in 1927, with a capital of 2 million francs. Two years later the company raised its capital to 5.5 million and on that occasion opened a second big garage. In 1930, 75 big and small electric trucks were parked in the 2 garages. This fleet had grown to 100 units by 1939 (in 4 garages). After the war it boasted 7 garages with 450 vehicles, while the capital was raised to 16.5 million. Seven other garages were located in a radius of 200 km around Lyon.[42]

In France the commercial vehicles were dominant, such as Jourdain et Monneret's delivery vans and those of Sovel (founded in 1925). And there, too, the electric vehicle lived up to its reputation of "crisis car": whereas in 1939 fewer than 500 electric vehicles were counted, by 1944 this number had increased to 3,500.[43]

In other countries on the European continent, the development took place on an even more modest scale. In some cases the 1920s and 1930s provided some new impulses that were often based on electric propulsion as a crisis solution. Further research would be necessary, however, to chart the various "national styles."

The situation in Switzerland was particularly noticeable. There, only one manufacturer, the Tribelhorn company near Zürich, had already supplied more than 600 vehicles in 1925. Besides three-wheeled passenger cars, these consisted mostly of heavier vehicles, such as hotel buses. In 1915, a special electric-

AMAZING SUCCESS OF
Electric Vehicles
IN GREAT BRITAIN

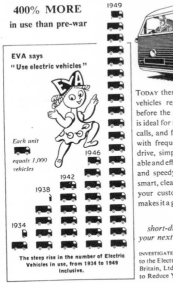

400% MORE
in use than pre-war

EVA says
"Use electric vehicles"

Each unit

equals 1,000 vehicles

The steep rise in the number of Electric Vehicles in use, from 1934 to 1949 inclusive.

TODAY there are 400% more electric vehicles registered than there were before the war. The Electric Vehicle is ideal for roundsmen's daily delivery calls, and for all short-distance work with frequent stops. It is simple to drive, simple to maintain, safe, reliable and efficient, the most economical and speedy vehicle for its job. It is smart, clean and noiseless — qualities your customers appreciate — which makes it a good advertisement for you.

If you run a
short-distance delivery service
your next vehicle should be electric

INVESTIGATE the Electric Vehicle. Send first to the Electric Vehicle Association of Great Britain, Ltd., for the free booklet on "How to Reduce Your Delivery Costs".

The Electric Vehicle Association of Great Britain, Ltd., 2 Savoy Hill, W.C.2
(E.V.1)

Until well into the 1950s, the British maintained one of the largest electric-vehicle fleets in the world. (Electric Vehicle Association of Great Britain, 1950)

vehicle journal, *Das Elektromobil,* started publication in Zürich. Shortly before the Second World War, the Swiss fleet numbered 600 electric vehicles. A hundred of these were electric trucks and sixty were in use at the Post Office, half of them in hilly Zürich.[44]

Italy also possessed a modest electric-vehicle fleet, probably mostly concentrated in Milan (where in 1930 all 37 hearses were electric) and Turin (at least 40 mail vans in 1930). In 1934, Mussolini decreed a complete tax freedom for 5 years for such vehicles. Shortly before the war, a plan "to electrify the national road network" was launched. Reminiscent of early plans from the Whitney group in the early EVC days, the Italian plan was geared toward freight transport by trolleys along a network 7,093 km in length with a constant speed of about 40 km/h. Nothing came of this plan (neither did comparable initiatives in France, Germany, and the city of Moscow). On 1 January 1941, it became obligatory to implement electric propulsion in certain types of commercial vehicles, such as mail vans and garbage trucks. Nothing is known about the effect these measures had on the electric-vehicle fleet.[45]

In Norway the electric-vehicle fleet had increased to 440 units by 1921, including 100 passenger cars, 300 trucks, and 40 industrial trucks. Oslo was the center of this modest boom. There, engineer Arthur Bjerke advocated this vehicle type in a presentation to the local Royal Automobile Club in 1915. In November 1926, Bjerke founded the Norwegian Electric Vehicle Association.

He decided to do so after the introduction of new legislation for cars that was entirely based on total vehicle mass and put an additional tax on solid tires.[46]

The most striking development, however, took place in Great Britain. There, the electric vehicle had skipped the second-generation stage altogether. And, under the influence of the American EVAA, the municipal power stations established an Electric Vehicle Committee as early as 1912. The committee published its own journal, *The Electric Vehicle,* which lasted under different names until well after the Second World War.

The boom in electric propulsion in Britain shifted the focus in European electric-vehicle history for the second time. It started at the outbreak of the war, not only because of a dire need for fuel on the domestic market, but also due to the confiscation of an important part of the British stock of horses. This forced department store Harrods, for example, to purchase twenty-eight electric trucks at once (and to order another twelve afterward). Because of the war situation, the electric trucks were liberated from their gasoline competitors in one blow. Moreover, the electric truck was the only vehicle type that did not have to fear confiscation. Most trucks were manufactured in America and imported by the London General Vehicle Company Ltd., a branch of the American GeVeCo. Others were built by Edison Accumulators Ltd., which provided them with alkaline batteries. Thus, the British market formed a short-lived and welcome outlet for the growing American electric-truck manufacture.[47]

In the interwar period, the electric truck surpassed the steam truck, which up to then was very characteristic for England. Whereas in 1927, 9,093 of such trucks were registered, this number had dropped to 5,849 five years later and to 986 in 1938. In 1941, not a single steam truck was left on the British roads. During those same years the number of electric trucks at first dropped a little, from 1,363 to 1,325. But in the years up to 1938, just as in Germany, the number soared to 4,399 (figure 7.4). The "milk float" (the milk delivery van) had a significant share in this growth, especially after World War II. Legislation contributed to this boom, as it prohibited the engine from running when the driver left the vehicle. This turned Great Britain into the world leader in electric road traffic. Germany had not been able to achieve this position because of the prevailing "vertical" way of living and the fact that milk usually was not sold in bottles there, but unpacked. Moreover, in Germany a new *Verkehrsfinanzgesetz* (traffic finance law) was imposed, based on total vehicle mass. It no longer made an exception for electric vehicles, of which up to then only net vehicle mass counted, excluding battery mass.[48]

AT THE OUTBREAK OF THE SECOND WORLD WAR electric traction seemed to be effectively pushed into some safely closed utilitarian niches like interfactory trucking and mail and dairy delivery, with an electric bus venture here and there, no doubt heavily backed by local authorities. It seemed as though the gasoline car's evolution toward multifunctionalism and universality had definitively closed the book on Tomorrow's Car.

However, developing a general multifunctionalism prevented the gasoline car from performing the same function with the same quality as the elec-

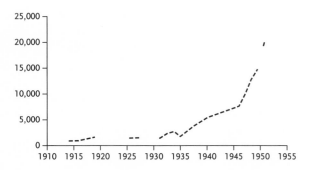

25,000
20,000
15,000
10,000
5,000
0
1910 1915 1920 1925 1930 1935 1940 1945 1950 1955

tric car. Multifunctionalism can go as far as the technical field allows: as a city car the gasoline vehicle did not stop emitting exhaust gases and it has kept on doing so ever since. And it was exactly here that the electric alternative could foster hopes of another chance.

The Fourth-Generation Electric: Hopes for a New Electric Adventure

The worldwide impulse to a postwar fourth-generation electric passenger car had different sources than had been usual up to then.[49] Meanwhile, the circumstances had changed so much that the argument of costs was no longer sufficient to force a new breakthrough. The ease of operation no longer played a role of any significance either, due to the improvements in the prevailing (gasoline) automotive technology taken over from electric vehicle practice. Instead, the massiveness of motorized traffic posed new problems. Research was done on gas-turbine propulsion, the stirling engine, whether or not in a hybrid configuration, in order to keep energy consumption and emission problems under control. Among all these alternatives, electric propulsion had become one of the possibilities, borrowing heavily, as became customary since the third generation, from mainstream gasoline car technology, especially in the technical field of materials and electronics. The necessity of its implementation, however, required new argumentation of its own, and a new field of expectations, too.

The local deterioration of the environment, especially in southern California, created such a new field of expectations, characterized by intervention of local, regional, and eventually even national authorities. It was a reaction to an increasing anxiety among the population about urban living conditions. A very concrete result of this anxiety was the establishment of pressure groups, such as the Electric Auto Association (EAA) in California. Its members could be found among environmentalists, engineers working in the military-industrial complex, and automobile aficionados, who formed their own subcultures by building electric cars and organizing annual contests and races. So, after the central stations from the EVAA era and in the tradition of the German fire brigade officers, a new actor had appeared on the American scene: the conscious, technically educated, and critical consumer, who was also capable of building his own alternative.

At the federal level this development gained momentum because of the

hearings of the American Congress on the amendments on the Clean Air Act of 1965 in the spring of 1967. The final report of these hearings concluded that the electric car was not ready for manufacture, but that "current research activities indicate that significant technical advances may be expected in the development of improved electric energy storage and conversion devices." With a view to the reconstruction given in this study, that conclusion was bizarre, to say the least, although the invocation and the hope of a "miracle battery" was nothing new.[50]

A second, new stimulating factor was necessary, however, to put the electric vehicle as "crisis car" on the political and automotive-technological agenda again. That impulse was given by the oil crisis of 1973–74. At the time the American Congress ignored President Ford's veto by passing the Electric and Hybrid Vehicle Research, Development, and Demonstration Act, which was amended once more two years later. Consequently, the Energy Research and Development Agency (ERDA) became the Department of Energy (DOE). The DOE then started sponsoring a Near-Term Electric Vehicle Program (NTEV) with the purpose of getting a commercial electric vehicle on the market within five years' time. At the same time the Department of Transportation (DOT)—opposed by the Department of Commerce—supported a development program, of which the target was clear: by developing high-performance batteries an electric car had to be created that would be "competitive with heat-engine powered cars." The federal government would purchase 7,500 electric vehicles during the following five years. President Ronald Reagan canceled the program in 1981 after a very critical report by the General Accounting Office. But before that happened, a number of demonstration programs had already started, resulting in models such as the Ford Comuta, General Electric's Centennial Electric, the Chevrolet Electrovette, and American Motors's Postal Van.

Urban air pollution and uncertainty about energy supplies elsewhere led to a global wave of such initiatives. They occurred in countries that until then had hardly played a role on the map of electric propulsion, such as Sweden, Belgium, Taiwan, the Soviet Union, and Czechoslovakia. These initiatives were characterized in most cases by the same double impulse of governmental support and local automotive-technological enthusiasm. Under the influence of these two new trends—the environmental issue and a heightened awareness of energy consumption—a different approach to electric propulsion emerged. It was now compared to its alternatives on the basis of a "life-cycle analysis." In its most extreme form, this is a determination of the total emission and the total energy consumption, calculated from the production of the artifact's raw materials and fuel up to and including the recycling at the end of its life span. Such calculations—controversial because of the large number of highly uncertain variables—yielded a varied picture of the pros and cons of this vehicle type, which was strongly dependent on the regional situation. According to an authoritative American analysis, for example, countries with cheap nuclear power or hydroelectric energy (France, Scandinavia) have a regional environmental and energy advantage. This also holds for countries that purchase their power for an important part from another state (such as the U.S. state of California). But for Germany (with its lignite power stations) or for Greece (coal) the advantages are not so obvious. Advocates, however, point to a sociocultural dimension of the new electric city car that would induce a more energy- and environment-conscious, planned use of the car. They have a remarkable belief in the sociocultural effects of technology and think that the purchase of an electric car under protected conditions (niche marketing) will force the motorist to a different behavior with regard to driving and the use of the car. And so, after the battery and vehicle manufacturers, the power stations, the garages, and the critical consumer, a sixth actor has emerged on the electric-vehicle scene: social science, especially in its policy-specific form of "constructive technology assessment."[51]

In Europe this new situation instigated a tradition of national fleet trials. The most renowned were those on the island of Rügen in Germany, in Mendrisio in Switzerland, and in La Rochelle, France. In Europe, Germany and France competed for supremacy in electric propulsion, but the center of activity shifted back to France. There the threatening overcapacity in electricity generation on the basis of nuclear power formed an extra economic incentive. With considerable support from the European Union and the very active Swiss electric-vehicle lobby, the emphasis in Europe seemed to be on the urban delivery-van fleet and the small city car. This avenue seemed closed to the American automotive industry—if it wanted to follow this tack at all—due to stricter legislation in the area of collision safety. Moreover, it is by no means certain that the small city car of the future will be provided with an electric propulsion system. Mercedes's hesitation about the Smart project, based on a Swiss design for a new vehicle concept, is revealing. Initially it was designed for electric propulsion, but it has been launched with an economical and relatively clean three-cylinder engine, ironically causing problems of "driveability" due to the raising of the center of gravity because of the elimination of the battery pack. Although the postwar development still awaits closer analysis, it is certain that the technical field has been subjected to important changes. But the fields of expectation and application show remarkable similarities to the prewar development. This can be illustrated by the most spectacular example of the past few years: the American Impact.[52]

Once more an American development made the discussion about the pros and cons of the electric car flare up. On 15 November 1990, President Bush signed the Clean Air Act Amendments. This act emphatically refers to Californian regulation, stipulating that 2 percent of the cars sold there had to be zero emission vehicles (ZEVs) as of 1998. Miracle battery or not, energy advantages or not, the California Air Resources Board (CARB), founded in 1968, reasoned that the local smog problems required drastic measures. The most important of these was the phased, partial introduction of zero emission cars. According to this plan, the proportion of such cars would have to be raised to 10 percent in 2003.[53]

In 1990, at the Automobile Show of Los Angeles on 3 January, Roger Smith, president of General Motors, unveiled the memorable Impact car. It can best be characterized as a belated version of the Bailey. This electric-powered prototype was a textbook example of what was possible with regard to decrease in weight, without violating the prevailing concept of "automobile." In that sense it was a real fourth-generation electric, in that it was based on the very frontiers of the automotive state-of-the-art. The vehicle's exterior looked high performance: an electric car had been designed in the guise of a high-tech adventure machine.

General Motors's $300-million "Manhattan Project" was meant to build "the Kleenex, the Xerox, of electric vehicles." The development of electric cars "as giant computers on wheels" appeared to be "tougher than putting a man on the moon." But while Smith announced in 1990 that he would really put "the first mass produced electric car of modern times" on the market, General Motors plunged into a deep crisis. At the same time, it did not seem likely that Californian regulations would be abolished by applying political pressure, or that the ZEV rules would be drastically mitigated. For California experienced an economic recession as a result of the end of the Cold War: 700 companies, especially in the defense and space industry, went bankrupt and 224,000 jobs were lost. Besides an ecological argument, the Californian authorities now also had an economic argument to pursue the road to an electric vehicle: it would create local jobs.[54]

What then followed was a conflict of interests, of which the details have not yet been clarified, not even by the American journalist Michael Shnayerson, who devoted an entire book to The Car that Could. On 28 March 1996, a gentleman's agreement was effected that put the first tier of the regulation (the 2 percent rule for 1998) on the back burner, but left the 10 percent rule intact for the time being. During the negotiations about the abolishment of the 1998 mandate, the automobile manufacturers declared they were prepared to build about 14,000 electric cars in 1998. As could be expected, this did not materialize. Instead, three weeks later, General Motors announced the launch of the Impact (meanwhile rechristened EV1 and much simplified) via its Saturn dealers in Los Angeles, San Diego, and Tucson, Arizona. Whereas about 100,000 electric industrial trucks and an unknown, but large number of golf carts were driving about in California, "the world's only electric vehicle that drives like a *real* car" was a fact, "sporty, aggressive and clean."[55]

Simultaneously, however, a global countermovement has emerged from the automobile industry that has revived the old idea of the "miracle battery." Besides a new wave of development efforts in hybrid propulsion, this movement is especially active in pushing the fuel cell, led by the merged Daimler-Chrysler conglomerate and the Canadian Ballard, in which Ford was also involved. In 2004, exactly a year after the 10 percent rule in California is to take effect, a super propulsion system is expected, although meanwhile some manufacturers have postponed this date to 2010. Not for the first time in the history of the electric car, this *expectation* distracts the attention from the fact that the prevailing (lead) battery technology is completely adequate for feeding (fleets of) electric *city cars*. But just as the proponents of electric tourism in France shortly after the turn of the twentieth century and in America during the First World War, these "miracle battery" makers are pursuing the *entire* substitution of the gasoline and diesel propulsion systems.[56]

Will they—at last—succeed in their efforts to become the dominant propulsion system? Will they be able to turn the seemingly eternal image of the electric as Tomorrow's Car into the Car of Today?

Epilogue

Alternative Technologies and the History of Tomorrow's Car

God doesn't want us to have full-function electric vehicles. The laws of nature don't allow this. —*John Wallace, Ford Think group, 2000*

Now THAT WE HAVE DOCUMENTED the successes and failures of the electric automobile in different applications, geographical settings, and time periods, we can deal with the issue of the failure factors of electric-vehicle propulsion in general, and those of the individually owned electric passenger car in particular. To this end, we first need to investigate three complexes of factors that are proposed as the most likely causes for the collapse of electric propulsion, especially as far as the passenger car is concerned: sociocultural (the electric car as a woman's car, in connection with the starter motor), structural (the battery and its limited energy density), and systemic (the greater infrastructural need of the electric). Translated into the analytical conceptual framework of our study, these possible failure factors are located in the field of expectations, the technical field, and the field of applications, respectively, and should each be analyzed against the background of their respective fields.

In this epilogue, we conclude that none of these complexes can fully and satisfactorily explain the "failure" of the electric vehicle, although they will reveal some interesting elements of such an explanation. In assessing the different failure factors, we will be especially interested in the technological core of these factors, because most arguments against the electric boil down to a conclusion about its technical "inferiority." In doing so, we hope to end up with a set of nontechnical factors, on the basis of which we will build our own explanation. In the second section, we round up most of the other arguments resulting from our study, after we have identified some general characteristics of automobilism during the past century. This will allow us to show that not so much the electric vehicle per se, but the interaction between this propulsion type and the dominating universal gasoline propulsion hides the key to our explanation. Hence, this explanation will also shed some more light on the quintessential question, which automotive historiography already asked itself more than a decade ago: why internal combustion?[1]

Culture, Structure, and System: Failure Factors of the Electric Car?

The first important possible failure factor is sociocultural. "Ease of operation made electric cars the favorites of female drivers," according to the section on the automotive industry in the *Encyclopedia of American Business History and Biography*. "The companies quickly picked up on the hint and began making their products resemble parlors on wheels, with overstuffed seats and luxury features. Despite their seeming advantages over gasoline cars, however, the limited range of the batteries powering electric cars ultimately doomed them to near-extinction . . . The introduction of a practical self-starter in the 1912 Cadillac meant the end of the electric as a serious contender for supremacy in the American automotive industry." Such reasoning (meanwhile belonging to the canons of automotive historiography) often does not explicitly link the electric car as a "woman's car" and the starter motor. This would require a judgment about the suitability of women for early motoring (or vice versa). Yet, the underlying reasoning usually can be read between the lines: the electric car was or became a woman's car, but failed when the electric starter motor made the gasoline car suitable for this market segment as well. A variation of this argumentation leaves out the gender aspect and concentrates on the starter motor, which would have swept the electric car from the market, because it enabled even not-"sporty" would-be buyers to handle the gasoline car without difficulty.[2]

With the emergence of an academic historiography of the automobile, especially in the United States, the femininity of the electric car has become the subject of analysis. Michael Schiffer is the most outspoken defender of the early electric car as a woman's car. "The ability of the electric car to circulate on city streets all day long made possible a new life-style in which women enjoyed unprecedented freedom of action. Male automobilists may have disparaged the electric, but women paid them no heed. They held the electric in high regard because, without a coachman, a woman could go anywhere in town anytime she desired. . . . The electric took up the liberation of the American woman where the bicycle left off, as long as she were rich."[3]

Schiffer does not present any proof for his sweeping statement about the electric car as an important support for women's liberation. But he does point to the many advertisements and articles in the popular magazines which show that "the close association of women and electric cars . . . was a faithful reflection of reality." To support this thesis he quotes—as far as is known—the only piece of evidence from automotive historiography: the clever observation of the British amateur historian G. N. Georgano, based on a photograph taken at an open day, probably in 1914, of the Detroit Athletic Club. On that occasion

Fun in the snow with a Detroit Electric, Model 43, in about 1918. Electrics—with "tiller steering" and a vis-à-vis interior layout—may have had a plausible claim on superior sociability. Adherents believed so. (Smithsonian Institution, Washington, D.C.)

the male members had invited their wives and of the thirty-five parked cars, all but three were electric, as Georgano determined.[4]

It is tempting to interpret the "elitism" and "snob appeal" (as Schiffer analyzes it) of the advertisements as proof for the supposed femininity of the electric car. But an opposite interpretation is also possible. Remarkably enough, the feminist historian Virginia Scharff proposed this for the first time. In her study of early automobilism she takes great pains to prove how indeed "driving . . . must . . . be seen as a male-dominated activity," but that the choice of vehicle was definitely influenced by women. Women, as Scharff argues, wanted to drive just as fast and far as men, and assigning feminine characteristics to the car or its function groups was a male alibi that had to cover up "woman's part in creating the car culture." In Scharff's opinion, this phenomenon was part of a general cultural trend of the "merging of female and male spheres." She adds that men, too, although secretly, welcomed the starter motor, the increased comfort, the luxury accessories, and the closed body of the gasoline car. In another context Scharff presents this thesis: "If women drivers had truly preferred electric cars, then the demand for such vehicles would have grown far more than it did over time." With a view to the small share of women in the *total* early automobilism, the question is where these women would have come from.[5]

Men's taste changed and this change was wrongfully attributed to women. Scharff's central thesis is that sexual difference is confused with the male/female opposition, or the *gender* difference. "Car makers' assumptions about gender blinded them to the potential market for enclosed vehicles." And "partly because of popular assumptions about proper masculinity and femininity, the self-starter and the American car culture took rather longer in coming than they might have done."[6]

Scharff, just as Schiffer, relies on the same grab bag of quotations in the automobile magazines of the time, as accurate figures are lacking. Scharff's reasoning is based on observing the electric car in the same uncritical way as the contemporary representatives of the prevailing technology. Whereas the introduction of the starter motor and the closed body would have been delayed by a lack of understanding of the real needs of both men and women, Scharff only wonders about the fact "not that electrics faded so early, but that they lasted so long, given their manifestly lower power, frequently higher prices, and smaller range than gas cars." Although she does not say so in so many words, her analysis assumes the "inferior technology" of the electric car—an assumption that also prevailed among contemporary car owners, who still mainly treated their vehicles as "adventure machines." But this does not mean that we can ignore the 1,500 electric vehicles in 1900, or the 20,000 to 25,000 electric passenger cars in 1914. Concerning the first figure Scharff comes up with an ingenious explanation, according to which society women did cherish the idea of the electric's femininity, although she denies this elsewhere. "There is," Sharff wrote in a discussion of Schiffer's book, "no reason to believe that women as a whole were content with, let alone preferred, such circumscribed mobility [of the electric car]. The wealthy women who swore by their silent, *underpowered* electrics held assumptions about female delicacy to a far greater extent than their middle- or working-class sisters."[7] Yet, due to this distinction between two classes of women, Scharff gets remarkably close to Schiffer's central thesis. According to this thesis, the real battle for the choice of propulsion took place as a battle between the sexes in middle-class families that could not afford two or more cars. Schiffer interprets the outcome of the battle as a victory for the "master of the house." But Scharff argues that women and men showed the same preference for the gasoline car and that the *ideology* of the gender-specific choice of vehicle delayed the development of a really comfortable gasoline car. Thus, Schiffer derives the demise of the electric car from weakly supported cultural arguments, while Scharff is satisfied with equally weakly supported technical argumentation.[8]

It is not easy to present a third interpretation in contrast to the two outlined above, because both approaches contain elements that match the conclusions of this study. But both approaches suffer from a lack of solid empirical, not to mention automotive-technological, evidence. That is why neither of the analyses distinguishes between city cars and touring cars, phases in the history of automotive technology, and geographical differences in European and American automotive history. Added to the complete lack of data, these shortcomings have led to highly speculative analyses that easily assume the character of a religious dispute. They seem to stem less from history than from a point of view in the current controversy about the social definition of technology and the gender debate.

It cannot be denied that the second- and third-generation private electric cars were clearly used as high-society vehicles. Tea parties, theaters, and a day at the tennis club were visited with such luxuriously appointed vehicles. It is also certain that the supposed femininity of the electric car formed an ambiguous category. C. Claudy, regular columnist of *Woman's Home Companion*, characterized the electric car as a "modern baby carriage" and even as a "scientific perambulator." Moreover, he described the limited speed and range as unproblematic for women, suggesting that a woman did not need to go farther away from her home than about 40 to 50 km, half the usual range. On the other hand, the Argo company in 1912 advertised with the slogan "a woman's car that any man is proud to drive." Such contrasting views expressed in a large variety of publications have led to the current confusing debate among historians. Therefore, it seems useful—against the background of the previous chapters—to first give a reconstruction of this field of expectation, before presenting an analysis.[9]

The trade journal for the gasoline-car sector, *Ignition and Accessories*, which had appeared in Chicago in 1911, changed its name to *Electric Vehicles* in May 1913. This was an important sign of a change in trends in the electric-car history and of a renewed hope for a third wave of this type of vehicle. "While the electric car," the editor wrote, "was first considered as strictly a woman's car, and without claim to speed or endurance, it is rapidly being adopted by business men who already are the owners of gas cars. The man who formerly considered the electric as effeminate is now one of its warmest advocates." At the NELA convention of 1916, publisher and editor-in-chief McGraw opined, "I think we all agree to-day that the [electric] passenger car is not just a lady's car; *it never has been such in fact*, but some manufacturers thought they were doing a great and wise thing when they promulgated that idea from coast to coast. I believe much harm has come to the electric vehicle by that sort of advertising."[10]

Although *Electric Vehicles* can be regarded as a partisan of the electric car, the historian depends on such sources for a deeper knowledge of electric-car culture, as this culture was excluded from the prevailing automobile culture. *Central Station*, the electrotechnical trade journal that was the mouthpiece of the EVAA, also shows such a shift in trends. This journal published a story that demonstrates the religious undertones of the American electric vehicle "movement." A taxicab passenger in New York, with his wife, had taken an electric cab without realizing it and was "converted." "Had I known it . . . you would never have gotten me as a passenger. I always regarded the electric vehicle as a play-toy for nervous, old women. You have not only given my wife and myself the most satisfactory ride we have ever had, but have unintentionally sold me an electric." Whom the new purchase was meant for, the husband or the wife, is not hard to guess.[11]

It is certainly true that manufacturers boosted the image of the electric car as a woman's car in their advertisements. It is very well possible that this tendency became even stronger at the beginning of the second generation. The gasoline car had experienced its first boom, the universality claim of the first-generation electric-vehicle proponents had meanwhile proved wrong, and society women seemed to prefer the electric car. A disillusioned E. P. Chalfant, member of the board of the most successful American electric-car brand Detroit Electric, suggested a different explanation in 1916: "The gasoline car dealers have branded the electric as a car for the aged and infirm and for the women, and because that is a market they did not want themselves we have accepted it, and we advertise and teach our dealers and their salesmen to talk luxurious appointments, upholstery to match gowns and liveries, coach work and finish beyond compare, a past record for building carriages and buggies, and we build up an atmosphere of ultra-refinement and picture our cars in front of palatial residences and in private parks. It is all wrong and we have deserved the false position in which we have heretofore allowed ourselves to be placed. Why create the impression one must be a millionaire to own an electric?"[12] From this perspective, the electric vehicle's historical task was to conquer a market segment that was out of reach for the dominant gasoline car as macho adventure machine.

With our knowledge-in-hindsight, it seems all too easy to accuse such manufacturers of blindness for the phenomenon of the cheaper utility car. They most certainly were aware of the threat posed by the T-Ford, but they were caught up in the technical problems that prevented the building of a lower-priced electric car. Therefore, it was understandable that most of them saw the luxury gasoline touring car as their real rival.

So, with a view to the circumstances at the time, it is not fair to accuse the electric-car manufacturers of a wrong market strategy. Nor can we put the historical task on their shoulders of breaking through the image of the adventure machine as a machine for men. This image seemed to pervade the entire automobile culture, strongly supported by the gasoline-car sector, which at that moment launched a product that seemed also usable as a city car. It is not surprising then that a representative of an advertising agency, when he was asked to design a campaign for the electric car, "seemed to think it was purely a luxury and largely for the use of women. I am afraid that is a belief which has prevailed to a great extent generally among our population." And almost two years later, F. Feiker mentioned at the NELA convention how even his "electrical friends" discouraged him from purchasing such a car: "It was called a lady's car; it was said it wouldn't run up hills; it was said it wouldn't go fast enough, and one central station man said that if I ever had sat behind a twin-six gas car I had *his* answers as to why *he* had never bought an electric."[13]

Great wealth and the electric car as a woman's car initially seemed to have gone hand in hand during a period when the automobile pioneers owned several cars. The electric version of these cars was also—and possibly with a slight preference (in the case of Chicago, even with clearly more preference)—driven by women. This preference diminished in the following phase. The emergence of the third-generation electric touring car with its slender body, copied from its gasoline rival, already indicates that the electric-car makers were trying to penetrate a different market, which they perceived primarily as masculine. The "business man," *Electric Vehicles* claimed in its first issue, "finds it [the electric car] almost as handy for town use as he does the big gas car. The 1913 electric is chiefly characterized by its 'manliness.'" In his overview of the electric-car market in 1915, former EVAA secretary A. Jackson Marshall also pointed at the "growing demand for electrics built along more masculine lines." A central station manager, asked why he did not trade in his very luxurious gasoline car for an electric, anticipated Virginia Scharff when he replied, "Well, I think it would be a fine thing, but I know my wife would be brokenhearted." Another manager gave a much less hypocritical excuse: "I would ride in an electric car if it were not for the fact that all my neighbors coming to the city pass me with their gasoline machines." And yet another hit the nail on the head when he defended his preference for the electric: "I am not seeking adventure."[14]

For the latter category of users, the electric city car was a realistic alternative, as in the case of the anonymous bank manager, who used his electric car to drive to work at half past eight in the morning. "He used it for a ride around the park and reached the bank in time for the opening of business. After that the car went to his home [how? by a chauffeur?], where his wife used it for shopping and calling, and his daughter drove it in the afternoon."[15]

This group of users, both men and women, was relatively small, which definitely was not just due to the confusion of sex and gender differences at the suppliers. Society was completely pervaded by it, especially American society, according to *Electric Vehicles*. Mid-1916 it ran several editorials entitled "The kind of car a man wants."

> The thing that is effeminate, or that has that reputation, does not find favor with the American man. Whether or not he is "red-blooded" and "virile" in the ordinary physical sense, at least his ideals are. The fact that anything, from a car to a color, is the delight of the ladies is enough to change his interest to mere amused tolerance. All this, of course, is logically absurd as it applies to the electric. It is just as much a man's car as it is a woman's. But the purchaser of a car cannot be expected to use logic in arguing himself into buying. Having imagined effeminacy into the electric, he dismisses it from his mind and buys a gas car without a struggle. . . . The lines typical of the closed body electric do not appeal to the American man. The lines of the gas car do appeal to him. Why? Is it because the gas car designers have perverted his eye so that he cannot recognize a good thing when he sees it in the shape of an electric? Some of the electric car designers appear to believe that theory; but we cannot subscribe to it. We are sure that the gas car carries out the lines it does because the men who buy want those lines—and not just because it is a gas car. And the same thing applies to the tiller steer—another thing objectionable to mere man. Man wants a steering wheel, wrong as the notion may be, in his hands when he drives. He wants the wheel not because the gas car gives it to him—the gas car maker gives it to him because he wants it. The tiller is better; but he doesn't know that. . . . It is often said that woman actuates the purchase of nearly everything. She does. But mark well the exception: Never does the woman decide the purchase of a man's car, and she never will. The designer who would build an electric to please men must forget all he ever knew about making a woman's own car.[16]

The last part of this lengthy quotation, which can be read as an eloquent rebuke of the EVAA's concept of "education" of the customer, is an especially apt description of the confusion in the electric-car camp. The gasoline car had become a multifunctional vehicle that appealed to the taste of a growing middle-class market, only because the motorist occasionally wanted to use it for touring. And in those days, apparently, a closed body was not considered to be part of this automobile culture, even in its tamed multifunctionalist version. The world seemed upside-down to those who, as Steinmetz and Edison (voicing the rational approach of EVAA and NELA), had expected that the increasing utilitarian nature of the automobile would automatically lead to the adoption of electric propulsion. Of course, these people did not have the historian's knowledge about the outcome of "the battle for the city." But the persistence with which the electric-car and battery manufacturers kept pursuing the ideal of the electric touring car is an indication that they hoped to adopt at least part of the gasoline-car functions and thus reduce the gasoline car's advantage.

The transfer mechanism, however, worked precisely the other way around. "We must conclude that the feminine influence is quite largely responsible for the more obvious changes that have been made in gas car design from year to year," *Electric Vehicles* wrote in an editorial. This was in flat contradiction with what it had claimed three months earlier, but entirely in line with Virginia Scharff's thesis. "The items of deeper and softer upholstery, easier springs, more graceful and beautiful lines, simpler control, more nearly automatic performance of the tasks of starting, tire pumping, etc., are all evidence of concession to the softer sex. . . . It is evident now . . . that the man has finally accepted the woman's viewpoint in this as in all other things. Therefore the car that a buyer selects today is a car that will please his wife or his daughter." One could also add the closed car body to this list of feminine characteristics men got accustomed to. But the lack of understanding revealed itself, as an afterthought, at the very end of the argumentation: "There is hardly a woman living who would not like an electric if she could have it. That fact alone is a pretty safe basis on which to appeal to men." Five months later, *Electric Vehicles* stopped publication.[17]

Meanwhile gasoline-car makers had fully exploited the above-mentioned transfer mechanism. "Every year the gas car becomes more electrical," *Electric Vehicles* concluded at the beginning of 1916. This electrification of the gasoline car, or (in the perception of many contemporaries) its feminization, not only included the adoption of the closed body, so that cars could be used in all seasons, but also the application of the battery and the cord tire. The cord tire could compensate for the bigger mass of chassis and closed body, just as it had earlier counterbalanced the bigger mass of the electric car. The gasoline car borrowed the electric starter motor from the hybrid version. Cadillac was the first to offer the starter motor as an option in 1912, and a year later it was a standard feature. This made the city use of this luxury car considerably easier. That same year sixty other automobile brands offered a starter motor. Thus a sporty alternative to the luxury electric car had emerged. The EVAA could hardly conceal its anger with regard to this new "exploitation" of the electric car: "The crop of self starters and coupe bodies brought out by our gas car friends in an effort to hold on to a business to which they are not entitled, bears eloquent evi-

"In years past most men believed their ideal to be a big, heavy motor car with super-speed," remarked this Anderson Electric Car Company advertisement of 1916. "But the reign of such cars is past. And now the quality most highly prized is smoothness of power flow—with, of course, sane speed. So naturally more and more men are coming to such a car as the Detroit Electric." Too late did the electric industry realize the pitfalls of focusing on women as a marketing strategy. (*Electric Vehicles*, April 1916, opposite 142; author's private collection)

dence that they realize they are being crowded off of the paved streets of the city and to the outskirts where they belong."[18]

Once more it proved to be a miscalculation: of the 611,695 gasoline cars sold during the first nine months of 1915, 261,860 (43 percent) had a starter motor. The other 350,000 cars were T-Fords. So, it is not correct to claim that the electric starter motor meant the deathblow for the electric car. But it did mean the last nail in its coffin, for the T-Ford was not provided with a starter motor until after the First World War. The electric-powered passenger car had been already driven into isolation, even before the starter motor became a standard feature in the best-sold car in America.[19]

The demise of the electric car was not brought about by a massive switch of women to the gasoline car, but by their upper-class families. The rich city-dwellers with a high standard of luxury were the last ones to join the ranks of the "sports freaks." They did so at a moment when the main feature of the automobile sport—its speed—was offered in a civilized way in the shape of the expensive gasoline car, complete with comfortable starter motor. They did not need the Model T for this switch. In other words, the starter motor was the culmination of the gasoline car's invasion into the separate sphere of the electric. With the starter motor the gasoline car had developed into a universal machine that could be deployed as adventure machine if so desired.

THE SECOND CANONIZED FAILURE FACTOR COMPLEX is the limited energy content of the battery, causing a heavy construction and a radius of action that was too small. This argument, mainly voiced within the automotive engineering community, should be differentiated by time and place, just like the other possible failure factors. After looking at the taxicab and the truck, we will focus on the privately owned passenger car, to investigate the claim of the "inferiority" of the electric propulsion system.

Technical factors were doubtless dominant in the failure of the first-generation taxicab experiments. Because of the rubber tire, and especially the lead battery, the technological margins became so narrow that the construction of a stable field of application appeared to be impossible. That was expressed in the costs, which in turn were a manifestation of life span: the electric car needed a battery of its *own*, not a miniaturized stationary battery or streetcar battery.

It is important to stress that economic indicators like costs are a *measure* of a failure factor, but in a history of technology they are not to be confused with the failure factors themselves. We must also recognize that the technical field of the first-generation electric was absolutely closed for the grid-plate and related versions of the lead battery. Maximilian Reichel's successful development of a first-generation electric fire engine fleet in Hannover on the basis of a large-surface plate battery proves this sufficiently. Thus, the identification of the life span of tire and battery as decisive first-generation failure factors does not imply that other factors, proposed by contemporaries, did not play a role. The most important arguments of such commentators were the narrow technical basis and the speculative managerial tendency in the early electric initiatives. These were characterized by megalomania, derived from the streetcar tradition and inspired by a general "electric" optimism. But such factors were not as important as the technical factors. On the contrary: fleet size and capital offered a certain protection to a technology that had not yet liberated itself from streetcar traction and industrial application. Only an application of which the mechanical unreliability belonged to its virtues (such as the gasoline car as "adventure machine") could occupy a sufficiently large field of application, a free space that appeared to be crucial with a view to the later developments.

For the second-generation taxicab, excluding the early Bedag initiative in Berlin, the grid-plate battery and the pneumatic tire were no longer dominant failure factors. The situation was very complex here, because some initiatives failed due to sociocultural and organizational factors, while others perished in the atmosphere of crisis generated by the war. But until well into the 1920s the Dutch Atax managed to flourish in a way that a gasoline-taxicab company clearly could not emulate. That success already started before the World War, so that the monopoly this war created cannot have been decisive. Second-generation electric propulsion—deployed in a large, intensively used fleet with central maintenance—was virtually unbeatable, if it was placed in a protected environment and other factors, such as accurate organization, economic business sense, and maintenance culture cooperated.

The threat was lurking at the edges of both the technical and application fields. This could not only be observed in the taxicab in extensive cities like Bremen. It was even more obvious in the German fire engine, where a virtually faultless organization failed due to the legal obligation of helping out neighbors. The fire engine is a good example of the fact that, generally speaking, the weakness of electric propulsion at the edge of the field of application was not caused by an absent infrastructure of charging stations. The issue was not the *daily distance* (important for electric tourism, for example), but the range *on one charge*. With the expanding towns, a shortcut developed, along which gasoline traction could enter. So, this "weakness" could also be traced back to a technical factor (the energy density of about 25 to 30 Wh/kg, which thus limited the maximum weight in batteries to be carried). Of course, additional technology and organization might have remedied this shortcoming (in that sense we do not point at an *absolute* failure factor here, just as everywhere else in this study). It cannot be denied, however, that contemporaries did not make any proposals in that direction. But it cannot be denied either, that this technical "shortcoming" was considerably magnified by looking at it from the perspective of gasoline-car culture.

All these factors, except range, also played a role in the short-lived boom of the electric truck in America. A solution to the problem of range had been found in the form of boosting, in combination with robust batteries. Here the wish for independence of the public garages and the fleet character of electric traction dominated, which prevented a market expansion to small-scale users and smaller towns. The economies of scale the gasoline truck profited from during the World War and the gasoline-car training that same war provided further reduced the field of application of the electric truck. Such factors put the advantages of lower operating costs, easier control, and greater reliability of this vehicle type under pressure.

So, the "failure" of the electric vehicle can be explained without referring to the long-distance properties (intercity transportation, deployment in the war) of gasoline propulsion. At most, these properties played a role indirectly, as a token of universal deployment, in the field of expectations. It also manifested itself in the application field, especially on its edges: like the occasional long trip to the outskirts of the city in the case of the cab, the farm-to-market gasoline truck found a comparable niche in the case of goods transportation. But the decisive battle was fought *in the city*, on territory where the advantages of electric propulsion were undeniable: easy speed control (sweeping and sprinkling trucks), trouble-free stop-start operations (door-to-door delivery, garbage trucks), absence of smell and noise (ambulances, transportation of food supplies). Moreover, the electric vehicle's structure was flexible, which considerably eased the conversion from horse traction and enabled a large variety of specialized versions. When there was no competition from the gasoline car (as during the First World War with regard to the British truck and the Amsterdam taxi), or when the local authorities enforced a protective area (as for the taxicab in Bremen and Hamburg or at the Berlin fire department), electric propulsion clearly flourished.

WHAT DOES THIS IMPLY for our explanation of the "failure" of the electric passenger car? And how does the battery (and the tire) fit in this explanation? Just as the German fire engine, the Dutch taxicab, and the American truck, one cannot call the electric passenger car a downright "failure." For the first generation and possibly also for the second generation in America, costs hardly played a role. For those who did not mind handling sulfuric acid and brazing lead plates on the kitchen table, the electric car was an excellent city car. And those who could afford the luxury of several cars, as a replacement of the horse stable, could also afford a retrained "stable hand." In Europe—in the third phase discussed here—the electric passenger car played only a limited role in Germany. If an individual electric-car owner lived near an AFA garage in Berlin or near the garages of the Namag taxicabs, he could go there for maintenance as well as for battery rental. Those who did not live near such a garage or did not want to make use of garaging and maintenance there for some other reason were in a much more difficult position. The reasons for this were of a structural nature: they were hidden in the technical field and were related to the battery as well as the tire.

A detailed study of the development of the AFA's grid-plate battery, published elsewhere, shows that the lead battery, as developed for the electric taxicab, was not suitable for use by private car owners. This unsuitability was due to two technical characteristics of the AFA battery.[20] First, the phenomenon of sulfating seriously threatened the life span, an effect that was magnified by a less intensive deployment. The technical solution to the problem was the use of a different composition of the lead paste on the basis of lead powder, but this made the private version more expensive than the taxicab version. Second, the phenomenon of "rotting" decreased the mechanical strength of the supporting metallic lead grid and converted it into a bad conductor. Therefore, in order to increase the life span of its battery, AFA returned to a technology that it had abandoned earlier because of the high costs. An alternative for the wooden separator was the expensive hard-rubber plates that the competitors (KAW among them) still applied. The ESB's Exide battery also incorporated such plates to prevent contact between the wood and the positive plates.

In America, the circumstances for private use of the electric car were considerably better. In the 1910s, most newly built houses came complete with a connection to the electricity grid. Especially in the Far West, the degree of electrification was high: in Los Angeles and Denver it was 90 percent and 70 percent, respectively, in 1910. Newly connected houses had alternating current, so that a rectifier was always necessary if one bought an electric car. The mercury vapor rectifier, the successor of the rotary converter, became popular as a maintenance-friendly option in those years. This, however, also increased the additional costs of keeping an electric automobile. So, instead of being a direct technical failure factor, the battery and its charging equipment had an important indirect effect on the application field: it pushed the purchase and maintenance costs of the electric passenger car above an acceptable level for middle-class gasoline vehicles.[21] Also, the introduction of the pneumatic tire had both a cost-increasing effect and a barrier-raising effect on the electric passenger car. These costs may not have played an important role for the first- and some of the second-generation car owners, but for the utilitarian user of the third generation it was an important factor indeed. Like the single butcher and grocer (who faced the same problem with the truck), he did not have the luxury of using both the electric and the gasoline car for the application fields in which they performed best.

But even if potential users surmounted the threshold of the higher purchase price, they were confronted with the dichotomy between their field of expectations and the technical field. Such users kept struggling with the non-mechanical character of the battery. This elusive character of the lead battery formed the basis for an entirely different maintenance subculture than was usual in a "machine society" and was a decisive factor in the formation of the field of application of the privately owned electric car. Its maintenance had already been a considerable problem for the professional garages. For the electric-car owner, who increasingly became a technical layman in this phase, the character of the lead battery formed a virtually insurmountable barrier. Although the implementation of automatic charging units significantly lowered the required number of charging operations, this did not solve the basic problem of the layman's inability to check the charging state of the battery. Despite the EVAA officials' protestations to the contrary, looking after the battery—as it was expressed in a typically medical metaphor—really needed "the constant attention of a physician and a trained nurse."[22]

It was Thomas Edison who, as nobody else before or after him, recognized this cultural barrier to the electric propulsion proposition. "If a man buys an electric automobile today," he told a reporter of *New York World* in 1903 in his rhetorical style, "he is furnished with a book of instructions about the battery, he must hire an expert to teach him how to run it . . . The problem, therefore, was to construct a battery, not for experts, but for general use—a battery that a man can forget and neglect, and which will plod along and do its work just the same." To define this battery one could neglect, this apparatus not intended for the technically skilled user but for the layman, Edison shrewdly applied a metaphor borrowed from the gasoline rival: "To describe this battery in a popular way is not easy, though the *machine* is very simple. It may be called an internal combustion apparatus." Others quickly picked up this metaphor to describe a battery type which indeed worked on the basis of "rusting" the iron in the negative plates, just as fuel in an engine was oxidized, and which, on top of that, was encased in shiny steel. The car magazine *Motor World* opined that Edison's battery "exteriorly looks not unlike a gallon can of gasolene." And more than a decade later even *Scientific American* commented that the new battery "looks good. A compact set of metal cans, with no chance of slopping or breaking. The battery has more the appearance of road service, and smacks less of the electrical laboratory, and this *highly mechanical look* is justified by the internal construction of the element as a working electrochemical machine." In short, Edison's battery was, as an advertisement phrased it in 1911, "built like a watch but as rugged as a battleship."[23] The launch of both the tubular lead and Edison batteries indeed considerably reduced the maintenance problem, but for the individual motorist they again increased the cost of driving electrically.

We can—with Thomas Edison—wonder why such motorists did not opt for the robust tubular-plate lead battery, based on the ESB's Ironclad and launched by the AFA as *Panzerbatterie,* or for his alkaline alternative. But even in the United States, with many more potential buyers of such a robust solution, the Ironclad and Edison's alkali type A battery were mostly used in trucks, applications with a high intensity of use that quickly offset the much higher costs. We cannot but conclude that the purchase price in this phase had become an argument against electric propulsion. Why?

Shortly before the First World War, all the measures to enhance the life span of the battery seemed to condemn the electric passenger car to the luxury-car niche—at a point in time that the general automobile market craved cheaper cars. But other factors also played a role. For example, driving about at one's leisure—for which the electric city car seemed preeminently suited—gradually became less pleasurable because of the increasing amount of traffic, especially in cities. Baroness Campbell von Laurentz complained in 1912 that the Riviera was "no place for motoring" any longer, "being quite spoilt by tramways and overcrowded with motorcars." But some electric motorists could not resist the temptation to leave the city and follow the trail of the gasoline car culture. According to a retrospective account in 1928 by W. Rödiger, an executive engineer at AFA, the electric car fell into disuse for private purposes, because it presented problems when one wanted to use it as a touring car. In this case clearly the field of expectations, deformed by the prevailing gasoline automobile culture, hurt on the boundaries of the electric vehicle's technical field.[24]

Thus, in the United States the private electric-car owner was forced to turn to the public garage. This development was at right angles to the growing phenomenon of the car as a utilitarian vehicle that had to be near at hand. So, if the second-generation battery has indeed contributed to the "failure" of the electric car, it was not because of its inadequate functioning in a restricted technical sense (although its technical properties did lead to a considerable price increase). Instead, it was due to its antimachine character, which did not fit in the dominant technical culture and certainly not in the emerging culture of the gasoline car.

THE THIRD POSSIBLE FAILURE FACTOR of electric propulsion is the much heavier infrastructure it required. This factor is closely connected to the second (the battery's low energy density), which required a much denser charging station network than the one for the gasoline car. It is self-evident that artifacts cannot be deployed as autonomous mechanisms, but can only fulfill their so-cial function when they are embedded in a larger entity. This is especially true of the car, so that the *automobile system* takes up a special place in the histori-ography of automotive technology, although it often is a neglected one.

In his pioneering study on the electrification of the West, the American historian of technology Thomas Hughes defines a system as a centrally con-trolled organization of interlinked elements that goes through various stages of development. Characteristic of one of those stages (after an inventor-entrepreneur has initiated the system and technology transfer between the various elements has started) is system growth, an irregular process that is characterized by the occurrence of so-called reverse salients. This metaphor, derived from military strategy, denotes an unbalanced area in the system, which begins to lag behind, more or less comparable (but in our case restricted to the technology at the artifact level) to our metaphor of the technical field's boundaries. A specific effort can bring it into step, so that the system better serves its purpose. Recognition of such reverse salients by system builders and their analysis yields a number of "critical problems" that can either be solved within the system, or, if that fails, form the germ of a new system that may be-come the dominant system after a period of mutual conflicts. During the sub-sequent phase, the system gets such a powerful momentum of its own that only very strong exterior forces can force it in a different direction.[25]

Others have further refined this "loosely structured model," as Hughes calls it. But critics also point out the problematic purposiveness hidden in the military analogy, as if the whole is more or less consciously directed toward a future that is more or less known. The model is also problematic because of the rather one-dimensional growth factor of the system, defined as the con-tinuous effort to maximize the "load factor." We will not pursue this discussion any further, but use Hughes's model to demonstrate that the automobile sys-tem does not meet his definition for the period described here—at least not if one considers a very strong dependence between infrastructure and medium (artifact) to belong to the very core of his definition. We can draw important conclusions from this with regard to the success of the gasoline car and the failure of the electric car.[26]

In her analysis of early motorization in Germany, Angela Zatsch calls the automobile a "gap filler" of the network of canals and railroads that had devel-oped in the course of the nineteenth century. She seems to allude to an anal-ogy she discerns (as many automotive historians tend to do) between the in-frastructural characteristics of canals and railroads on the one hand and the construction of paved roads on the other. Canal boats and trains, however, lose their function outside their "network"; they then are either docked or derailed. An automobile on the other hand, just as the bicycle, is not *unconditionally* de-pendent on its network. The boom in all-terrain bikes and off-the-road vehicles is a recent example of this. In this respect, the train fits into Hughes's electri-fication network better than the car. This is even more true of the electric street-car or trolley with overhead wires. In contrast to the steam locomotive or the steam ship, they become meaningless without a very material and continuous (and not just incidental, as in the case of the steam ship) communication with a power station.

As we have seen earlier, in the beginning of electrification there was a brief glimmer of hope of a decentralized establishment of a "network" (by the distribution of "bottles of electricity" in the shape of batteries). We also saw the struggle for dominance between NELA and AEIC involving big, centrally controlled grids versus smaller neighborhood plants and isolated plants. The strongly centralized electric grid that emerged from this struggle shows considerable similarities with the electric telegraph at the end of the nineteenth century. But the telegraph—like the streetcar—was confined to the public space and did not penetrate the private domain. The automobile on the other hand shows more parallels with the telephone—also a nineteenth-century "electric" invention. Both enabled personal communication, which was enabled, but not orchestrated by a central power station. The content of both networks (or user culture) was shaped by the users themselves.[27]

In this comparison, the early electric car is closer to the electric streetcar, especially if the electric car was deployed in a centrally controlled fleet. If Hughes's system model can be applied anywhere in the history of the automobile, it is to this special case of the electric road vehicle fleet. A system builder had to put up a central maintenance facility, organize a highly specialist overhauling schedule, and plan the logistics of the cab fleet. In this sense, J. F. Friderichs of the Amsterdam Atax cab company was the classical Hughesian system builder. The same applies to W. C. Whitney and his EVC. The private electric-car driver, however, could behave for a while as though he were driving a gasoline car, but eventually he had to return to one of the charging facilities to charge his battery. But this is not a "timeless property" of the electric car: if the dream of those few late-nineteenth-century enthusiasts about the distribution of the battery at the home had come true, a much weaker link to the power station would have been possible. Also, the practice of exchange charging in taxicab fleets, where no potential energy is charged, but mass (by exchanging the batteries, often in a much shorter time than it took to fill up a gasoline tank), in theory makes this link weaker. This would have certainly been the case if the system had also become available to private drivers through a large-scale extension of the number of battery exchange stations, certainly if decentralized electricity generation had not experienced such a formidable barrier under the guise of the megalomaniac Edison systems. As long as this had not been realized, however, the same scenario loomed for the electric car as the Parisians to their dismay were confronted with in March 1907. The City of Light for the first time was plunged into darkness due to a strike by workers of the power stations (only 1,800 employees). This also explains the army's initial reservations about centralized systems with a heavy, immobile center. If the streetcar could be stopped by such an action, wouldn't it be possible to "eliminate" the electric car by pushing a button?[28]

The steam car—at first sight more related to the gasoline car than to the electric car—was linked more closely to an infrastructure than one is inclined to think at first. Before the arrival of the "flash boiler" (a kettle consisting of small-diameter pipes, enabling the water in these pipes to reach the required temperature quickly) in the first decade of the twentieth century, it took at least twenty minutes to start the car. In freezing weather, the use of the steam car was very difficult, if not impossible. The water the American steam-car drivers needed they usually took from the water troughs for horses that could be found all over town. Finding sufficient water with a low mineral content (to prevent scaling), however, was far more difficult in some areas such as the Midwest. The dependence on large supplies of water decreased to some extent after the introduction of the condenser (which converted the steam into reusable water). But a possible breakthrough of the steam car would have required an expansion of the water infrastructure, comparable to the system that had been established along the rails for the steam train.[29]

Didn't the gasoline car depend on a network? Not in the sense of a rigid and static system that remained fixed to its site and required high investment costs. It is true that gasoline had to be produced by distilling crude oil, but its distribution took place in cans that the automobile clubs usually distributed among strategically located outlets. When the flow of traffic changed, this "feeding infrastructure" could easily follow. Once established as a standard type, gas stations appeared at strategic points. We are, however, not concerned here with the final form in which the gasoline network eventually took shape, but with the extremely flexible starting period. Such a decentralized "anarchistic" start would have been hard to realize for the electric car, even with the best organization. An exchange-charging station required considerably higher investments and was, therefore, pursued only when a large number of customers was present or at least could be expected. Charging at home was only possible if the electric grid was sufficiently spread across the country. The *minimum necessary initial condition* for the two types of automobile was different, a phenomenon that is reflected in the difference between the minimum size required for the electric and the gasoline taxicab fleet. So, the *infrastructural need* seems to be an important criterion for judging the background against which an artifact could develop, certainly an artifact on wheels. The super-fast expansion of the bicycle sport at the end of the nineteenth century is a striking illustration of this point. The centralist system of the electric car enforced a *planned* use of the vehicle, a use that was at right angles with the "anarchist," or at least individualist use of the gasoline car, even after this type of vehicle had lost its most extreme adventurous streaks.

What then was the situation with road infrastructure: isn't it just as rigid and cost intensive as that of the rail network and the canal system? A type of road infrastructure already existed before the emergence of the automobile, and even before that of the bicycle: a very finely meshed network tailored to horse pace and the human pedestrian. In most industrializing countries, during the first half of the nineteenth century roads for long-distance travel were paved to accommodate the intensifying horse carriage and wagon transport, while the latter half of that century witnessed the paving of the regional networks to function as a "feeder" to the railway network. This finely meshed network was already present when the first motorists appeared on the roads. And only when a sufficient number of automobiles, following the lead of the bicycle movement, had established the contours of a touring culture did drivers start to criticize the condition of these roads. The suction of the pneumatic tires weakened the structure of the macadam roads by dislodging the sand between the stones and spreading it as dust in the vicinity. The macadam road was an improvement in the technical structure of the paved road, but when the attention had shifted to the railroads, its maintenance had been neglected.

The enormous capital the state invested in the improvement and expansion of automobile-friendly road networks shortly before and especially after the First World War has led to the opinion that the automobile would not have been successful without this system. The total German expenses for road construction over fiscal year 1913–14 amounted to 76 percent of the expenses for the army and 56 percent of the total defense costs. The income from automobile taxes was by no means sufficient to cover those expenses. In general, the government's willingness to pay for the costs of automobile infrastructure and not burden the motorist with road maintenance indicates an important difference from the canal or railroad system. The German historian Gerard Horras is not the only one to be surprised about the "strong investment tendency" of the German state, with a view to the relatively low *economic* importance of the automobile in this period. He is especially surprised, because "here no such a strong revolutionary change could be observed as was caused by the need for capital while constructing the railroads."[30]

Despite the projection of our infrastructural bias into history, the relationship between road building and road use is not at all clear-cut. Although this relationship has yet to be thoroughly investigated, for the beginning of automobilism there is no doubt that, in most countries, road construction lagged behind the expansion of the motorized vehicle fleet. As such, this is an indication of the different type of relation the car has to its infrastructure. This *decoupled relationship* becomes even clearer if one calls to mind the American situation. There motoring in the first decade of the twentieth century had grown into a mass movement, which, keeping the condition of the roads in mind, could not have happened if we are to believe the proponents of the infrastructure hypothesis. The development of automotive technology, in fact, shows that the relations in America were the other way around. "Don't preach that motor vehicles depend on roads," *Motor Age* wrote in 1899. "They don't. They depend on good suitable construction to negotiate any kind of road surface and on perfectly reliable motors." And five years later the same journal wrote: "The common cause of delays is defective roads, not defective machines." Early American automotive technology perhaps provides the most extreme example of the individualist motor traffic, *not* tightly tied to a network at all: the robust cars with their high wheels were designed to drive on bad roads, entirely in the tradition of carriage design. In the United States, the automobile initially adjusted itself to the road (or the lack of it), not the other way around. The high-wheeled, light, but sturdy Ford Model T was the quintessential example.[31]

So, American road construction did not so much encourage the automobile sport, but it did cause the faster distribution of the car as a utilitarian means of transportation than elsewhere. A key figure in this was the American farmer. Even before roads were good, he used the T-Ford to visit his neighbors or to go into town. There is, however, no concrete evidence that the passenger car in America experienced a faster diffusion in regions with good roads. Neither can it be proved that in places where the automobile was first spread on a large scale, road construction was tackled sooner. Nowadays, the car and the road, with garages, gas stations, and signposts, need each other, but the emergence and first diffusion of the automobile was not dependent on road improvement. And when, by the end of the First World War, the relationship between motorization expansion and road network building was recognized as important, it was the truck with its destructive heavy axle loads, and not the passenger car, which provided the first impulse to consider heavier road construction (table E.1). So, the conclusion can only be that the infrastructure did not form an insurmountable obstacle for the electric car. Its distribution already came to a halt well *within* the domains that allowed its diffusion—and not at their edges.[32]

THE GASOLINE CAR WAS QUITE RELUCTANT to be tamed. The first gasoline cars were put on the market as replacements for horses. Initially the gasoline car, however, did not fulfill this promise; it had to be forced into this role by the crisis of 1907, and by its electric rival, which, by virtue of its structural flexibility, appeared to be much better equipped for the task of Trojan Horse than the gas car. Thus, only *after* the electric vehicle had broken the most ardent resistance of the horse economy could the gasoline rival invade that city. This process continued until the last rich motorists could change to the comfortable luxury of the expensive version of the gasoline car in the course of the second decade of the twentieth century. It should be stressed most emphatically, however, that this process had virtually been completed *before* the gasoline car had become so much cheaper due to economies of scale than the electric and steam-powered versions. After this, however, the result was an economic "lock-in," that is, an irreversible cost advantage in favor of the gasoline car.[33]

Automotive historiography has reached a consensus about the role of the farmer in the breakthrough of the (gasoline) car, mainly on the basis of some well-documented cases such as the diffusion of the Model T. It is, then, only one step further to conclude that the electric alternative did not have any chance because of its limited range. This, then, would be a distortion of automotive history, because the "battle of the systems" occurred in the city and to win this battle the gasoline adventure machine *first* had to be tamed into a reliable city car. It is also a rather one-sided American view on the history of automobility. Although little research has been done on this topic, it seems that the European car formed a part of an urban and peri-urban culture for a much longer time. In fact, in some densely populated European countries rural car densities (expressed in the number of cars per 1,000 inhabitants) didn't surpass those of urban conglomerations until the 1970s.[34]

But the system perspective can help us determine the balance between the

TABLE E.1 Registrations of Electric Trucks in the U.S. in 1916

Ranking in Order of Fleet Size	State	Number of Electric Trucks	Ranking by Length of Paved Roads	Total Number of Trucks in 1912
1	New York	4,060	1	7,892
2	Illinois	2,781	4	2,551
3	Ohio	2,660	5	1,171
4	Pennsylvania	2,520	2	2,664
5	Massachusetts	1,940	3	2,045
6	Iowa	1,870	22	730
7	California	1,820	6	2,198
8	Michigan	1,610	9	1,146
9	Wisconsin	1,330	10	580
10	Indiana	1,120	11	970
11	New Jersey	960	8	1,080
12	Connecticut	720	18	519
13	Minnesota	600	17	970
14	Missouri	537	7	832
15	Texas	450	12	382
16	Kansas	310	27	120
17	Rhode Island	310	15	410
18	Nebraska	300	26	220
19	Maryland	175	14	371
20	Washington	175	23	170
21	Colorado	150	28	239
22	Oklahoma	150	32	42
23	Georgia	140	25	155
24	Kentucky	125	19	146
25	Tennessee	125	13	78
26	Virginia	109	24	100
27	Maine	90	30	78

electric and the gasoline vehicle in another, unexpected way. The real Trojan Horse function of the electric is not to be found in the artifact itself (i.e., its "intrinsic" higher reliability) but in the highly organized maintenance culture that developed around it. This was due to the low life expectancy of the battery lead plates, and also got a further boost because battery maintenance as such did not fit very well in the prevailing maintenance culture. But in the end it was the gas car that benefited most from these sophisticated logistics. After all, the Verdun and Marne battles showed the first successful implementation of fleet logistics within the gasoline camp. The preparation in the United States and consecutive shipment of trucks to the European war scene was the second. And eventually, after the war, the most essential elements of this maintenance

Ranking in Order of Fleet Size	State	Number of Electric Trucks	Ranking by Length of Paved Roads	Total Number of Trucks in 1912
28	Oregon	80	20	526
29	Alabama	62	29	48
30	New Hampshire	52	36	48
31	North Dakota	45		46
32	North Carolina	40	37	96
33	West Virginia	40	34	32
34	Delaware	38	33	78
35	Vermont	35	42	34
36	South Dakota	30	41	96
37	Utah	30	39	181
38	District of Columbia	25	16	218
39	Florida	22	31	83
40	South Carolina	20	35	54
41	Louisiana	20	21	44
42	Montana	15	40	34
43	Arkansas	14	38	51
44	Idaho	8		22
	Mississippi	0		36
	New Mexico	0		29
	Wyoming	0		28
	Nevada	0		26
	Arizona	0		21
Total		27,653		29,690

Sources: EV (U.S.), June 1916, p. 187; 1912 data from CV, October 12, p. 20.
Note: According to EV, the share in the total stock of commercial vehicles was about 14 percent in 1916. The overview confirms the correlation between truck-fleet size and length of paved roads in a state. A comparable connection holds not only for the electric truck but also for the truck in general (but not for the passenger car!), as the figures for 1912 demonstrate.

culture, hitherto applied to centrally attended fleets, were introduced in the nationwide maintenance systems for the individual user, instituted as a dense "service infrastructure" where every single member of the national car "fleet" could be cared for. During the formation of this maintenance infrastructure cultural resistance was not a factor, because the technical challenge as part of the automobile sport had lost much of its appeal among the growing group of layman users, who now could delegate parts of this "sporty" function to the garage. But not all. Until well into the electronic era of automotive technology large groups of users (especially from the lower middle and labor classes) continued to enjoy the challenge of the functional side of the automobile adventure, in the process creating a new do-it-yourself culture. And although much of this culture can be explained out of economic necessity, the multitude of car enthusiast magazines, the myriad self-help books, and other paraphernalia of the car buff culture tell a different story, an analysis of which is painfully lacking in automotive historiography. But this much is clear: the taming of the adventure machine took at least another half century and was for a large part performed by the users themselves. In doing so they firmly internalized the maintenance culture as a precondition for mass motorization. The unreliability of the gasoline car propulsion system was not solved at the level of the artifact, but inspired the construction of a maintenance support system, both at the institutional level (garage system) and at the individual level (do-it-yourself system).

The success of the gas car was its *antisystemic* use and appearance. Although it was presented as the "motorization of the existing world," as the motorization of the carriage (as Benz advertised), in reality (culturally) it succeeded by motorizing the bicycle. Gradually, however, the gasoline car also became part of a Hughesian system. But this system, especially its separation of high-speed automobile flows from those of the pedestrian, the bicycle, and the rural tramway through the building of an automobile-only highway network, was *forced upon* the users and became an indispensable characteristic of the automobile culture only when this culture took on the shape of mass motorization. In this sense, not only the automobiles but also their users had to be domesticated. Quintessential in this "taming of the motorist" was the debate about road safety, which started in the 1920s in every motorizing country.[35]

Electrifying Car Technology and Culture: Perspectives for Tomorrow's Car

Anyone looking at the first fifty years of electric propulsion as presented here will find it hard to imagine how, one after the other, all initiatives failed. That was also true for the many people involved at the time, such as Fire Chief Reichel and the EVAA management. They perceived a development that did not match their rational approach to the transportation problem. Tirades against the "speed maniacs," sneers about the childish "hobbyism" and the "joy riding" of the gasoline-car proponents were the expressions of a blindness to the gasoline car's main feature: its adventurous character. The functional aspect of this adventurous character was gradually delegated to the garage, but what remained appealed more to the enthusiast's heart than to his calculating brain. So the domestication of the gasoline car did not lead to the destruction of its adventurous character, which had firmly taken shape in the "free space" of the years around the turn of the century. Since then, each new motorist has experienced the "taming" of the car during his (and her) first driving lessons. It cannot be stressed enough that it was exactly the highly *functional* flexibility (as opposed to the *technical* flexibility of the electric) which constituted the basis for the success of the gasoline car. When the motorbus and the gasoline truck appeared in the Netherlands in considerable numbers in the 1920s, it was exactly this flexible field of application that led to the collapse of one of the densest regional tramway systems in the world.[36]

If the battle between the propulsion systems was fought and settled within the city, the question becomes all the more urgent how a technology, which clearly performed better from an engineering point of view, in the end had to give way to its gasoline rival. In fact, with the recognition of the flexibility of the electric vehicle structure, we have formulated the *paradox of the electric vehicle,* as that very structural flexibility prevented the development of a standard design: for the German fire engine, the space-saving wheel-hub motor was ideal; for the European taxicab, the propulsion system with two small electric motors near the front wheels proved an unsurpassed antiskid measure; and for the American electric truck, the single-motor propulsion system with differential and rear-wheel drive meant the ultimate of low production and maintenance costs.

To wind up the above argumentation, it is tempting to refer generally to a prevailing automobile culture, a paradigm that, once embedded, had such social power that alternatives were swept away. However, such an explanatory model has a high tautological content: electric propulsion failed because the gasoline car was victorious. Despite the recent revival of the universality claim of electric vehicle builders, this study has conclusively shown that the reconstruction and analysis of the creation, development, and "failure" of the electric-powered vehicle is likely not possible within the frame imposed by the history of the gasoline car. The expectations, technology, organization, applications, and automobile culture of the two types of vehicle were and are entirely different. That makes it more than justified to give the electric vehicle a history of its own, based on its own sources and told in its own language. At a closer look, another explanation is possible that resolves the paradox of the electric car in this period. For this we should not confine ourselves to an analysis of both alternatives as separate entities, but instead focus our attention on the interrelationship between the two.

That other explanation is based on a mechanism that popped up at numerous places in our narrative. In the note on method at the back of this book, I have proposed a new name for this mechanism, the Pluto effect, indicating a technical and functional transfer between competing artifacts, a transfer that is clearly more beneficial to the mainstream technology than for the alternative, "threatening" technology. So, the Pluto effect at first sight seems to work both ways. But at closer inspection it turns out to be in favor of gasoline propulsion, not because of a paradigm effect, but due to the latter's *technical properties.* Indeed, in the course of its development, electric propulsion adopted several structural features of its gasoline rival: the single-motor system with differential; the concept of the adventure machine through "electric tourism"; and a few exterior features, such as the hood. The roadster even copied the entire shape of the bodywork, including the larger wheelbase (which was not necessary from a technical point of view because of the compact propulsion system).

The transfer flow, however, was much larger in the opposite direction. To the gasoline carmakers electric propulsion served as a real grab bag of structural features that were mostly used to "civilize" the adventurous character. They extended the technical field of the gasoline car with important elements from the technical field of the electric car, so that a larger variety of fields of application could be realized. From the point of view of an imaginary, frustrated electric-car engineer, the gasoline car even stole the entire city-car concept. Not only the closed body and the starter motor, but, in a later stage, also front-wheel drive (according to contemporary handbooks, better suited to inexperienced drivers) were used for the taming of the adventure machine. The ease of operation, the quieter engine, the controllability of the combustion engine, the suppression of noise and smell were all inspired by the electric city car. And finally, the decisive innovation of the cord tire arrived in 1915–16. Due to the Pluto effect, the advantages of electric propulsion in the city became increasingly smaller. After most of the electric's undeniable original advantages had been spread over the entire automobile culture, the problem of the choice of propulsion tended to be reduced to a matter of purchase price. The automobile had become a consumption article, and from that moment the automobile industry had to invest large amounts of money and creativity to convince potential consumers that possessing a car differed from possessing a washing machine or any other expensive commodity. It was helped in this endeavor by a consumptive leisure culture that highly appreciated escapism and adventure. The automobile fit into this culture well because it was both utility and adventure machine, a golden combination firmly embedded in our Western lifestyle for practically a century. So, only when the battle of the propulsion systems had resulted in the takeover of most of the advantages of electric propulsion, economic theory can claim that the price difference plays a decisive role. And even then there is no guarantee that this judgment remains valid. As soon as one of the undeniable technical advantages of the alternative, in a different sociotechnical setting, becomes an issue of societal concern (as was the case in California regarding its zero emission regulations) economic considerations have to be suspended again to give way to other lines of reasoning, based in this case on issues of local urban health and regional employment policies.[37]

As the purchase price became dominant in the choice of propulsion, the argument of differences in quality faded into the background. Instead, the argument of *universal deployment* began to dominate. The car owners of the first and second generations often could afford a car for different applications, as if they were the managers of a small municipal vehicle fleet. But, as more of the middle class could afford a car and had to decide about the type of propulsion, multiple car ownership became less common. For buyers of such a vehicle the higher quality of the electric city-car was no longer an argument; it simply could not be. A separate device for each task in the kitchen may give a better result, but if one considers buying kitchen equipment and has a limited budget, one would buy the universal food processor, even though it may take longer to beat the whipping cream. In other words, the issue of individual choice between technical alternatives became increasingly secondary, revealing the relativity of the rational subject from economic theory. Only when the national automobile markets became saturated did the whipping cream metaphor become invalid and automotive history seems to have come full circle: electric

vehicle proponents nowadays see a marketing niche in the fast-spreading phenomenon of the second (and third) family car.[38]

Granted, in contrast to the multifunctional gasoline car, which polluted the
city, the "specialized" electric city car performs much better. Combined with
the reemergence of the multicar family, why would the electric vehicle not de-
serve a second chance? Indeed, in a different, counterfactual social system, our
argumentation up until this point does not preclude the electric car's growing
into a universal machine. One can imagine a society with a dense charging infrastructure, in which the subculture of the civilized electric car has become the
dominant culture, and in which the car with a combustion engine is rented for
the rare trips out of town. One can even imagine an electric adventure, in which
the pleasure of driving consists of the experience of silence and the conscious
cut-back in the use of vehicle energy. One can imagine "a more benign future,"
a "dehastened" society, in which speed is a vice and composure a virtue.[39]

Falling back on the Pluto effect, within the technical field this "electrified
automobile culture" would correspond with a further evolution into an "electrified gasoline car." If the modern electric vehicle, just like its predecessors,
can be read as a materialized criticism of the dominant automotive technology,
why then can we not imagine that this process will be continued? And, indeed,
some clairvoyant managers from the industry already envision such a development. Byron McCormick, head of General Motors's fuel cell research program and one of the designers of the EV1, seemed to support the idea of the
electric's powerful role in the field of expectations: "I've been around for long
enough to see fuel cells overhyped several times because they are such wonderful devices in terms of what they can offer." And according to Bill Powers,
vice-president of research at Ford Motor Company (unlike General Motors,
known for its long lasting reluctance to test electric drive), "the most cost-
effective and efficient road to a greener world is through the gradual electrification of vehicles. That is, electrification as in adding more electrics, rather
than switching to an all-electric powertrain." Powers calls this the road to "an
evolutionary hybrid," so, rather than opting for the "high-risk technology" of
fuel cells, Ford's powertrain could develop over the next few years through evolution, and not by a paradigm shift. There are signs, indeed, that other car
manufacturers have opted for this route as well. Renault, for instance, in 2000
presented a concept car, in which the alternator (which generates electricity for
all electric functions on board) becomes so large that it also can perform a
boost function to support the combustion engine, just like EVAA officials envisioned this during the third generation. This will be possible if the battery
tension is increased from 12 V to 42 V. So, an evolutionary scenario can be
imagined that turns the dominance of the internal combustion engine gradually upside down, until hybrids appear with a very small booster on the basis
of a combustion principle, until, in the end, a fuel cell takes over all functions,
keeping intact the fuel distribution infrastructure. As in biological evolution,
the combustion principle would degenerate into a rudimentary appendix. Indeed, the fuel cell would at last enable the electric to realize a long-held fantasy

about electric tourism, because then the charging stations along the road could be taken back from the gasoline rival and a future of absolute dominance over the automotive application field could be secured.[40]

Indeed, one can imagine a utopian automobile society, because, as we have seen over and over again in this study, the field of expectations can be expanded very easily. And thus, an opposite scenario is also possible. For the dualistic nature of the hybrid as a technical and social compromise also allows the perspective that the internal combustion car, by using concepts like the "Integrated Starter Generator," integrates electric driving only during acceleration from standstill, excellently suited for stop-and-go traffic within the city.[41] Besides, the scenario depicted in the previous paragraph seems to be the same as we encountered nearly a century ago, from the very early taxicab experiments (Bixio), through electric tourism in France (Kriéger), to the sectarian eschatology of the EVAA and its prophets Edison and Steinmetz. Ever since, the miracle battery and the hybrid have been pushed to the fore as the two possible routes to dominance and now, with the fuel cell as the super-hybrid, these two routes can even be combined. After all, the fuel cell is a miracle battery *and* a hybrid. But this hope for a miracle has always diverted attention from the real-world facts that, on the basis of proven electric vehicle technology, electric taxicab fleets and truck fleets could outperform all rivals. The question, then, is, *why* we should strive for such dominance for electric multifunctionalism and universality. Or better yet: *why* is it that we prefer expectation above reality, in the meantime driving along in our gas and diesel cars? For, the potential of the existing dominant technology is still quite vast, to say the least. Alternative fuels like alcohols and hydrogen are ready for use when the fossil fuels are depleted, CO_2 reservoirs are being tested to attack the last remaining major emission problem of global warming, antinoise systems are investigated to mimic the silence of the electric, and a whole range of low- and high-tech tricks are being applied to tame the conventional combustion engine even further. Direct injection of the gasoline engine promises to lead to a convergence of gasoline and diesel engine efficiencies. Now that the CO and CH emission problems seem to be largely solved and chain calculations do not promise to offer a radically better performance for the electric alternative, the question is why we should opt for electric drive. This is especially true if one keeps in mind that the advantages of the energy and emissions characteristics of electric propulsion are not as clear-cut when viewed from the perspective of "chain calculation," including the whole life span of the artifact. The propulsion choice does not seem to be the real problem anymore: space occupancy, traffic congestion, and thousands of deaths and injuries related to road traffic accidents are the real problem, and neither the gas car nor its electric alternative seems to be able to solve this. A technological fix seems to become ever more problematic, so that sometimes the ignition key itself is pushed as the real solution: discourage automobile use through the development of a new adventure, the choice between a multitude of travel modes which the history of transportation has meanwhile to offer us. Modern traffic engineering has coined the terms *multimodality* and *intermodality* for this scenario. If this scenario comes true—why not?—the electric automobile not only appears to be a Car of Tomorrow in the past, but also in the future. As such, it contributes to the dynamics of general vehicle development, just as other nonelectric alternatives of the gasoline car.

IF ANYTHING CAN BE LEARNED from this study of electric-vehicle history, it is the notion that it makes sense to give alternatives their chances, if only because they function as a source of criticism of the prevailing technology. I am not expressing a preference here, because I do not have one. By that, I do not want to suggest that I am immune to the appeal of the "adventurous" combustion engine. But that is addictive behavior and can—continuing the fantasy—also be satisfied electronically, for example, by a digitally generated engine throb and a gasoline scent dispenser in the interior of the car, or, in a more remote future, in the virtual adventure machine installed in the attic. Instead of trying to lure the Almighty into the band of speed and adventure loving maniacs (see the beginning of this chapter), we should question our own motives and ask ourselves why we really don't want an electric car and why we keep telling each other that tomorrow, maybe . . .

Like the general history of automobility the history of Tomorrow's Car is ultimately a history of our deepest desires (although this study provides room for the assumption that these desires have a strong masculine flavor). After all, there is nothing in the law book of Nature, but there is a lot to find in our nature that prevents us from having a mobility culture based on electric propulsion. Those who sympathize with this reasoning will understand that from this perspective, the electric car has not "failed" at all. On the contrary, it was successful in its continuous threat to the gasoline car up to the present day—thanks to the Pluto effect.

For engineers working on alternative technologies and for scholars in the field of technology assessment this result at first sight seems rather pessimistic: aren't they condemned to frustrating and futile attempts to run after a forever elusive flying target? I think not. The Pluto effect tells us, on the contrary, that it makes perfect sense to develop alternatives, and that the bitterness arising from the insight that most alternatives do not become dominant is the result of a myopic focus on the alternative technology per se, instead of on the "artifactual whole," in our case "the" automobile, or, wider still, "mobility." Nevertheless, without expectations of future dominance no alternatives will be developed. Ultimately, this seems to me to be the real paradox of the electric vehicle, if not of every alternative technology. Whether electric or not, there will always be a Car of Tomorrow.

A Note on Method

LIKE THE AUTOMOBILE at the end of the nineteenth century, the analytical tools developed for and during the preceding study were not constructed in a vacuum. They emerged from an ongoing debate, both in the discipline of the history of technology and in that of economic and social history. During the course of my research the question became increasingly acute of how to approach the phenomenon of technological choice in a vast and complex field of automotive technology, spread over more than a century, developing in distinct geographical regions, and expressed in a multitude of vehicle types, national engineering traditions, and user cultures. To explain my approach, which can be characterized as contextualist and evolutionary, it is worthwhile to reconsider briefly our present knowledge of the sociotechnical dilemma of interartifactual choice.

The classical answer to the question of how technical change works when artifacts compete is a reference to the phenomenon of substitution as a special case of the diffusion of innovations. According to the bible of the diffusionist school, Everett Rogers's *Diffusion of Innovations,* this topic is—with more than 4,000 publications since the 1920s—probably the most thoroughly researched topic in the social sciences. However, in the majority of case studies in this tradition, innovation diffusion is studied as if it occurs in a virgin environment, where the spread of the innovation at the micro level is the result of a learning process that converts "nonknowers" into "knowers" of an artifact or cultural trait that fundamentally does not change during the process of diffusion. Apart from this conceptual rigidity, which is the result of a mathematical bias that only allows for a very limited number of parameters, there are several other drawbacks to this approach.[1]

First, this approach, which treats innovation diffusion as basically a process of communication and knowledge *about* technology, implicitly seems to favor a linear, one-to-one replacement concept of substitution. In most cases the entity to be replaced is simply the previous nonexistence of the innovation before adaptation, but in more sophisticated models the innovation is placed in opposition to an older technique that is treated in the same manner, that is, as an unchanged and unchangeable entity. The resulting graphs are as appealing as they are simplifying, because they tend to blur the fundamental complexity of the underlying process, which develops, as we have seen in this study, along multilinear paths and is characterized by complicated feedback mechanisms and asynchronicities. Moreover, and especially in the case of complex and variegated artifacts such as automobiles, they tend to underestimate the role of diversity due to differences in geography and application. For instance, whereas a one-to-one substitution of the horse may have occurred in big urban taxicab, fire-engine, and municipal fleets, the process of motorization generally was accompanied by a parallel expansion of the horse economy. In France, the horse population clearly decreased in Paris during the first two decades of the twentieth century, but at the same time increased in the big provincial towns due to displacement effects. In general, diffusion curves are blind to these internal displacement phenomena, unless (and this is very rarely the case) time series of data exist at the local or regional level.[2]

Second, many diffusion and substitution studies are finalist in scope, depending intrinsically on hindsight knowledge. Especially in their mathematical versions, the construction of an S-curve of diffusion is possible only after an estimate of a market saturation level. As long as diffusion does not reach saturation (as seems the case again and again in automobile diffusion) the fixation of a saturation level leads to severe misjudgments, as the history of predictions of national car fleets clearly shows. The finalist character becomes especially apparent when substitution processes are characterized by long periods of coexistence of two (or more) competing technologies, often wrongfully described as transition periods. The supposedly transitional character of these processes is based upon the juxtaposition of old and new technologies and upon an implicit assumption that new is superior to old and in the end will win. This might be true in some clear cases of obsolescence of old technologies, such as gas lighting versus electric lighting or sailing ships versus steamships (and even then there is room for doubt), but what to do with examples of seemingly continuous coexistence like the telephone versus the cordless version, or personal computers versus laptops, or even such cases as the fountain pen versus the typewriter or the PC, or coffee versus tea, Pepsi versus Coca Cola, steel versus aluminum, newspapers versus the Internet, coal furnaces versus wood fireplaces, sled dogs versus snowmobiles, bicycles versus automobiles in the city, automobiles, trains, and airplanes between cities, subsonic versus supersonic flight for long-range transport, and gasoline versus diesel vehicles? Why do some technologies, against the onslaught of an innovation, vanish while others stay?

It is true that some economic historians acknowledge that superior technologies can fail. W. Brian Arthur, for instance, referring to different case studies by himself and others, gives as examples (of inferior technologies that won) the narrow gauge of British railways, the U.S. color television system, the FORTRAN programming language, the QWERTY keyboard, the light water nuclear reactors, and, oddly enough, the gasoline car (versus the steam car). In all these cases inferior technologies won by trivial circumstances and were subsequently locked in by learning processes of early adopters.[3]

Although Arthur's corrections to the neoclassical school of economics are still as valid now as they were in a time when technical choice was considered to result, of necessity, in the selection of the superior technology, the confusion about how to define superiority and inferiority remains. For instance, if an artifact, from an engineering point of view, is clearly inferior, it can nonetheless dominate the market because this inferiority is exactly the function users appreciate (and *not* because of a historical contingency), as was the case with the gasoline-engine adventure machine in the early days of automobility.

TECHNOLOGY CHANGES THROUGH THE ACTIVITIES of people in their capacity of producer, legislator, user, or nonuser, and through their organizations, which often function as intermediary forces of negotiation on the subject of artifact design, production, and use. In most cases of technical change several solutions exist (or are thought to be possible), and people have to *choose* in order to bring about innovation. It seems, indeed, to be the rule rather than the exception that alternative technologies appear on the historical scene at about the same time; they are often a response to ongoing debates about the weak spots and reverse salients of technology at a given time.[4]

In general historiography a great deal of research has been done on the phenomenon of technical choice, although the dominance of economic historians in this domain has led to a questionable emphasis on costs and prices. The implication of this type of approach often is that user preferences follow a mainly rational and quantifiable path. Since (mainly Schumpeterian) criticism of neoclassical economics (which treated economic agents as having perfect information and, being perfectly rational, permanently seeking to maximize their satisfaction as consumers and their profits as firms) has gained momentum, several corrections of the concept of well-informed, perfectly rational choice have been put forward.[5]

A new technology can hardly be compared with the alternative(s) it is meant to replace on the basis of cost data alone, for the simple reason that these data are not available as long as this new technology is not embedded in a user culture. Instead, *enthusiasm* and *expectations* play a decisive role here, not only in terms of the economic advantages of the new technology in the long run, but also in terms of the new functions this technology promises to provide.

Thus, it is by no means clear at the moment of innovation on what criteria the choice in favor of an alternative technology should be based. In fact, while the artifact is being invented, the use of it has to be invented as well. A classic example is the American telephone, which was introduced as a better telegraph for business applications but became a medium for social communication, especially in the hands of female users.[6]

This raises the question, obviously, of why producers and users should opt in favor of a new technology, especially if one takes into account that innovation is not restricted to new technologies at all. "There is much evidence," economic historian Nathan Rosenberg asserts, "to suggest that historically, the *actual* improvement in old technologies after the introduction of the new were often substantial and played a significant role in slowing the pace of the diffusion process [of the new technology]." Yet "it is a very general practice among historians to fix their attention upon the story of the new method as soon as its technical feasibility has been established and to terminate all interest in the old. The result . . . is to sharpen the belief in abrupt and dramatic discontinuities in the historical record." It is important to stress here that for contemporary witnesses of this competition it often was by no means clear which of the alternatives was to survive as "the fittest."[7]

In order to allow for these nuances in the intricacies of technical change, I use the field metaphor, borrowed from electromagnetic theory. Applied to automotive history, the *field of application* comprises the areas of applications and prototypical, real-world experiments of a vehicle type (a passenger car, a truck, a racing car) in a certain stage of its development and at a certain geographical location. So, the contents and dimensions of such a field are historically determined, depending on its own prehistory, on technical, sociocultural, political, and economic constraints, and on *distortions* by neighboring fields. Inversely, the field of application itself also influences its surrounding fields. In the field of application the *choice of propulsion*, a central theme in this study, takes place.

Technical and social fantasies as well as images of future applications are embedded in the *field of expectation*. According to some historians of technology, it is a crucial field, because here the subjective impulses of action are formed. These constitute an important motive behind the technical change, in that they influence choice behavior. The field of expectations is often very wide: it contains prototypical near-reality ideas about artifacts and their possible applications, but also science-fiction-like constructs that are not even intended to be realized. As has been shown in this study, it also can be oriented in a direction so as to hamper seriously a further expansion of a given field of application. In technical historiography this field is often restricted to a *prevailing* field of expectations, a kind of commonly held mainstream conviction among opinion leaders in the automotive field about the desired route of development. In the German sociological approach to technology, the prevailing fantasy is indicated by the term *Leitbild,* a translation of the English *image* and a subjective form of Kuhn's *paradigm.* But the field of expectations can also be split into opposing fields, like in our case, where a gasoline culture was (and is) relatively separated from a subculture of the electric vehicle. As long as the description is restricted to these fields, the options are inexhaustible.[8]

As explained in the prologue, in between the two fields of expectations and of applications exists a cluster of fields that function as a kind of filter, a lens, a restriction of the fantasized possibilities. As we, for the purpose of our analysis, were looking for means to open up the black box of automotive technology, we focused our methodological research on the *technical field,* without, however, excluding from our analysis important parts of the other fields in this cluster. In the technical field the conversion of the schemes and expectations into a concrete artifact takes place. To study this field, the historian of technology needs a thorough technical knowledge of the artifact. The size of the technical field also determines the ease with which the conversion of a desired artifact into a realized artifact is possible (feedback also takes place here: fantasies are adapted and changed, leading to changes in artifacts).

The size of the technical field is determined by *technical boundaries* that assume a different character, depending on the distortion caused by other fields. For example, the lead battery formed the technical boundary of the first-generation electric car. The width of the technical boundaries in this case could be expressed quite accurately in the life span of the battery, which restricted the application of the electric vehicle to certain niches. For the second-generation electric car, on the other hand, this technical boundary changed significantly; the life span no longer presented a problem. Instead, the boundary was formed by problems with materials and maintenance, which made the battery too expensive for private use. Thus, the technical field can be more or less closed.[9]

The notion of fields (and the accompanying concepts of structure and system as introduced in the prologue) also implies that different types of failures can be at work in the history of technology, often not sufficiently distinguished by social-science approaches. One could, for instance, want a vehicle speed of 200 km/h in 1895 or a diesel engine in a passenger car around the turn of the century. But the failure of such an alternative is of a different nature than the failure of the American electric truck after the First World War. In terms of the field notion, in the first case the technical field was absolutely closed, in the second case it was relatively or conditionally closed. In the one case we deal with a frustrated illusion, in the other with a *realistic* alternative, as we have seen in this study. In the latter case even further distinctions are possible. For already in those days, the wish for an electric city car had a higher reality content than an electric touring car. In the terms of the field notion, the technical boundaries were too narrow for the field of application of the touring car, but for the city car they were not. The technical properties of the lead battery and the pneumatic tire were the cause.[10]

A QUESTION THAT HAS NOT YET BEEN ANSWERED by means of this model is, how can we investigate the technical field? The automobile structure is an *Idealtype,* an abstraction, as is "the automobile" (and thus "the electric car").[11]

At this point in our argumentation it is not quite clear how static concepts like systems, structures, and fields are subject to change. This raises the question of how, at the level of the application field, all these different artifacts interact. As in economic history and evolutionary theory, in the history of technology different types of interactions between alternative technologies can be distinguished. Maynard Smith distinguishes between three types: competition (where each species has an inhibiting effect on the other), commensalism (where each species has an accelerating effect on the other), and predation (where the predator inhibits the prey from developing, whereas the prey has an accelerating effect on its predator).[12]

In order to evolve by interaction a certain *variety* of artifacts has to exist. Modern evolutionary economists have recognized this and have identified different mechanisms that bring about changes in variety. These mechanisms are substitution, specialization, the introduction of completely new artifacts, and diversification. However, variety is not infinite: patterns of evolutionary development have been identified, alternately called "technological regimes" and "natural trajectories" (Nelson and Winter), "technological guideposts" (Sahal), or "technological paradigms" (Dosi). The existence of such technological paths implies that "disembodied" parts of technology constitute a "state-of-the-art" that leads to the *perception* "of a limited set of possible technological alternatives and of notional future developments."[13]

In other words, the abstract state of the art, as a "contingent result of the sociocultural evolution," is in itself a selection of concrete and fantasized artifacts and is, therefore, not identical with the abstract, "average" artifact that the historian is able to construct on the basis of the available historical evidence. It also means that variety is asymmetrical: some alternatives are favored above others, although economic evolutionary theory does not give us any clue as to how this selection process exactly functions. From this perspective, the closure concept is problematic because it implies, as Dosi's argumentation seems to suggest, that the alternatives composing a choice set are equal and that technological trajectories as "the pattern of 'normal' problem solving" favor some alternatives and exclude others. The essence of evolutionary development of technology, however, suggests on the contrary that no alternative is exactly equal to other alternatives. In reality, these (often tiny) differences in technical properties are a reflection of a different use (or of a fantasized, intended use) and as such are a representation, a materialization, of different needs and desires. The idea of a trajectory or a paradigm thus tends to place too much emphasis on the meso- and macro level, whereas a more thorough investigation into artifactual variety would suggest that internal characteristics at the local and the micro level prefigure events at higher levels: no closure without an equivalent change at the level of the artifact and at the level of the individual user and producer. This seems to me the real challenge of a history of technology: to find these small, interartifactual varieties and to explain them in their context, that is, their different fields of application. In order to do so, all possible and actual alternatives have to be included in the artifactual whole, which should be analyzed as such. Thus, alternatives should be included in the study of (mainstream) technology, not so much because they add flavor to the narrative, but because changes in the latter cannot be understood if its interaction with alternative technologies is excluded.[14]

When taking a closer look at the substitution process at the micro level, in many cases one alternative appears to have more chances of success in one field of application, whereas the other alternative has more chances in another application field. As has been said before, this makes the general idea of technological failure highly questionable. When alternative technologies compete for dominance, feedback mechanisms occur, a kind of interartifactual transfer of technology and knowledge. Mostly, if at all, this process has been described as a flow from new to old technologies, the most famous example being the interaction between sailing ships and steamships, between wooden ships and ships with an iron hull. In a hardly noticed tiny contribution to a magazine on physics, W. H. Ward coined the phrase *sailing ship effect* to describe the phenomenon "that the sailing ship developed fastest while it was being supplanted." Part of this development involved direct borrowings from the steam ship, but there were also intrinsic improvements of the sailing ship technology per se, resulting in improvements in sailing qualities, and, hence, enabling the sailing ship to compete in one of the crucial areas (transport speed), thus making the choice in favor of the new technology much less straightforward than it would seem from hindsight.[15]

In an effort to integrate this phenomenon into his evolutionary theory of technological development, Hans-Dieter Hellige proposed the term *mimicry,* which describes the effect of less successful species trying to imitate some properties of the more successful ones in order to survive the evolutionary struggle for dominance. All these, and similar, approaches seem to start with the implicit assumption of an uphill (and desperate) battle for the old technology. In the case of Ward's sailing ship effect this is all the more remarkable, because at least until mid-century the sailing technology was the dominant one. The mimicry metaphor suggests that for the inferior technology to remain viable it is necessary to deceive the selecting agents, so that they perceive this technology as being a winner. Because of this, it seems more appropriate to take a different starting point and analyze competing technologies as a part of the abstract, artifactual whole, as explained previously. This tack is even more important when it is not at all clear which of the competing technologies is new and which is old, as in the case of the automobile.[16]

The factors that bring about this change are manifold, but in this study I have stressed the importance of a mechanism that has hitherto not received the attention it deserves by historians of technology, a phenomenon that seems to govern the exchange of technical properties and user functions of competing technologies. This phenomenon is the Pluto (who got his name from a Walt Disney cartoon) effect, and can be visualized as a dog harnessed to a cart, running after the sausage held in front of his nose, reaching for it but never quite snatching it. In this metaphor the coachman represents the prevailing technology or its actor(s) and poor Pluto the alternative technology or its actor(s). Pluto literally runs after an uncatchable prey. Of course, Pluto makes the cart move. Only by means of Pluto will the coachman handling the stick with the sausage reach his goal. If the coachman does not offer Pluto an object worthy of pursuit, the mechanism goes nowhere. If the sausage is too close, he will eat it and choose his own route.

Instead of providing for only external impulses to technical change (as the closure principle and the trajectory assumptions that are based on it seem to suggest), the Pluto effect also takes the internalities of the black box into consideration. It is important to stress here that the Pluto effect covers a transfer of (technical) *properties* and (applicational) *functions,* and not necessarily or primarily a transfer of artifact parts. The introduction of the Pluto effect in the analysis of technical change has to counterbalance the closure principle in the social science approaches, which puts too much emphasis on spectacular changes that are easy to trace. It reappraises the equally important incremental changes that also continue *after* closure.[17]

The Pluto effect adds a crucial dynamic perspective to the field concept in that it keeps the evolution of the basically static concepts of system and structure going. It differs in several respects from the research tradition of artifact interaction. First, the Pluto metaphor does not approach this competition as an uphill battle by the alternative technology against the dominating, mainstream technology, as a desperate swan song of a vanishing technology. When we look at automotive history, this interpretation is possible. There are plenty of examples of mimicry. Note for instance the emergence of the hood in electric vehicle technology around 1910, when its manufacturers tried to mimic the petrol car hood, which hid the potent combustion engine and which represented the very essence of what the petrol car was all about. Note also the introduction of electric racing cars as early as 1898–99 and the marketing of slender electric touring cars during World War I in the United States, which were no longer recognizable as electric.

In fact, the Pluto effect appreciates the flow of properties and functions in just the opposite direction. It describes the "stealing" by the dominant technology of certain properties and functions, proposed by the alternative technology, which form a constant threat to the dominance of the mainstream technology. In other words, making use of the Pluto effect seems to be necessary for the dominant technology in order to remain dominant, and, as such, it is comparable to the concept of repressive tolerance in the social sciences (Marcuse's incapsulation of protest movements by the system). In fact, as we recall, Nathan Rosenberg, in his analysis of technical diffusion, already drew our attention to this phenomenon, although he narrowed his answer down to the economic consequences of a higher efficiency that would result from this process. "The imminent threat to a firm's profit margins which are presented by the rise of a new competing technology seems often in history to have served as a more effective agent in generating improvements in efficiency than the more diffuse pressures of intra-industry competition. Indeed, such asymmetries may be an important key to a better understanding of the workings of the competitive process, even though they find no place in formal economic theory."[18]

Second, the Pluto effect, because it favors continuity over discontinuity, directs our attention to incremental instead of paradigmatic changes and, as such, necessitates a thorough knowledge of the technology under investigation. However, although the analogy with evolutionary theory would suggest that technologies in interaction would behave like plants and animals (or better, their selfish genes) in their long-range struggle for survival, it must be stressed that artifacts do not copulate and that changes in them do not necessarily need to be incremental in character and, thus, do not seem to be always the result of a random process of infinitesimal mutations. And so, although biological evolutionary theory denies the possibility of interspecies breeding, cross-over phenomena are not uncommon in the history of technology. A porcupine does not develop a long neck because it sees a giraffe eat leaves from a tree, whereas an engineer can take over whole concepts from a threatening, alternative technology, such as the cord tire by gas-car engineers.

Third, the Pluto effect seems to favor technologies with a built-in potential for universality, or, better maybe, multifunctionalism. How this exactly works has yet to be established, but, as this study reveals, it seems that the broader the field of application (and, as a result of this, the lower the production costs because of economies of scale), the higher the chances for an alternative to become dominant. The same conclusion was drawn by Devendra Sahal in his study of the diffusion of the agricultural tractor: "it seems that the greater variety of tasks to which a design has been adapted, the more likely it is to serve as a guide to the general direction of technical advances." This is also true if no single artifact becomes dominant. In that case, all alternatives seem to tend toward *functional convergence,* brought about by different technical styles for different classes of artifacts. This seems not only the case with the gasoline vehicle, but also with, for instance, the digital computer.[19]

Fourth, the Pluto effect tends to relativize the concept of technological failure. As this study has argued on multiple occasions, the history of automotive technology provides many examples of outright successes for the electric vehicle which makes the failure concept nothing more than an empty ex post statement. Instead, alternative technologies, as concrete emanations of an abstract artifactual whole, can be analyzed as a materialized or frozen criticism of the mainstream technology. As such, artifacts not only can be exhibited as windows onto a culture, but as a comment (or even a criticism) on it as well. In this respect, the nearly continuous presence of the electric propulsion alternative for the dominant internal combustion engine technology is a clear case in point.

The idea of the alternative technology as a frozen criticism of mainstream technology also helps to explain why the electric vehicle retained its threatening character after the "victory" of the gasoline propulsion concept despite its marginal presence in the market. Apparently, direct economic considerations are not crucial here. Instead, the field of expectations is much more important in such cases, probably because engineers working within the tradition of the dominant technology are constantly reminded of its shortcomings by the ongoing presence of the "electric criticism" of their work within the field of expectations. How this exactly works has yet to be investigated. Part of the explanation may be found in a remarkable phenomenon that can be observed when automotive engineers present the results of their research in the many thousands of papers that annually are published by their national organizations like SAE (Society of Automotive Engineers, United States), VDI (Verein Deutscher Ingenieure, Germany), SIA (Société des Ingénieurs d'Automobile, France), IMechE (Institution of Mechanical Engineers, United Kingdom), and KIvI (Koninklijk Instituut van Ingenieurs, The Netherlands). Many of these reports start with a brief historical overview that generates the impression that the research in question started with a clean slate. At closer look, however, such often clumsy introductory paragraphs, in contrast with the style of the remainder of the paper, seem to function as an incantation, conveying a consciousness of uncertainty rather than a conviction that, here, the ultimate truth is pursued. It seems as though, in the minds of automotive engineers, the "victory" of the gasoline propulsion system is much less self-evident, as their historical incantations seem to suggest. What engineers know, and what they do not know, is a subject worthy of further research. This research, however, should not be limited to the technical field, but should be expanded far into the fields of expectations and application as well. It should certainly include the question about the knowledge and attitudes of (potential and real) users of automotive technology. A true future history of automotive technology cannot avoid a thorough analysis of the historical struggle between producers and users over the choice of the automobile's properties and functions.

Abbreviations

Titles of periodicals, conference proceedings, names of archives and other sources, have been abbreviated as follows. Because of their abundance in the text, contemporary articles from trade journals are not included in this list. They will be listed in full at first reference and cited thereafter in shortened form. The same goes for titles listed in the Bibliography.

AAZ	*Allgemeine Automobil-Zeitung*
AJ	*The automotor and horseless vehicle journal*
AK	Varta Batterie AG
AN	Archives Nationales
annual report ARM	ARM Groep
annual report AEM	ARM Groep
annual report Atax	ARM Groep
ARM archives	ARM Groep
Bericht Berlin	Conference reports and annual reports
Bericht Charlottenburg	Conference reports and annual reports
Bericht Schöneberg	Conference reports and annual reports
Cave Collection	Detroit Public Library
CGP	*Le caoutchouc & la gutta-percha*
copy book Lohner	Technisches Museum, Vienna
Crane Collection	New Jersey Historical Society
CS	*The central station*
CV	*The commercial vehicle*
EE	*The electrical engineer*
EF	*Das Elektrofahrzeug*
ENHS	Edison National Historic Site
ER	*Electrical review*
ETZ	*Elektrotechnische Zeitschrift*
EV (UK)	*The electric vehicle*
EV (US)	*Electric vehicles*
EVAA	Conference reports and annual reports
EW	*Electrical world*
FA	*La France automobile*
FuW	*Feuer und Wasser*

GAA	Gemeente-archief Amsterdam
Goodrich files	University of Akron
Goodyear files	University of Akron
HA	*The horseless age*
IE	*L'Industrie électrique*
IK	Varta Batterie AG
IRJ	*The India-Rubber & Gutta-Percha electrical trades' Journal*
IRW	*The India rubber world*
LA	*La locomotion automobile*
LK	Varta Batterie AG
Maxim Collection	Connecticut State Library
MW	*Der Motorwagen*
NAHC	Detroit Public Library
NB	*NELA Bulletin*
NELA	Conference reports and annual reports
SA	*Stahlrad und Automobil*
VA	*La vie automobile*
VDB	Verband Deutscher Berufsfeuerwehren (see Conference reports and annual reports)
VDB	Verein Deutscher Berufsfeuerwehroffiziere (see Conference reports and annual reports)
VE	*Le véhicule électrique*
Z-VDI	*Zeitschrift des Vereins Deutscher Ingenieure*

Notes

Preface

1. I would like to repeat Clay McShane's complaint about the abominable accessibility of the store of books in the Library of Congress in Washington; during two separate visits almost half of the books requested were "not on shelf" (Clay McShane, *Down the asphalt path: The automobile and the American city* [New York, 1994], vii).

Prologue

"The schoolmaster of Dearborn," *New outlook* (September 1934), 61–62, quoted in William Greenleaf, *Monopoly on wheels: Henry Ford and the Selden automobile patent* (Detroit, 1961), 138.

Werner Rammert, "Modelle der Technikgenese: Von der Macht und der Gemachtheit technischer Sachen in unserer Gesellschaft," 6 (manuscript; also published in *Jahrbuch Arbeit und Technik 1994*).

1. Gijs Mom, "De moderne elektro-auto," in Gijs Mom and Vincent van der Vinne, *De elektro-auto: een paard van Troje?* (Deventer, 1995), 11–110.

2. On imperfections, see Richard H. Schallenberg, *Bottled energy: Electrical engineering and the evolution of chemical energy storage* (Philadelphia, 1982), 56; John B. Rae, *The American automobile industry* (Boston, 1984), 13, quotation: 185–86. For a comparable opinion, see also James J. Flink, *The automobile age* (Cambridge, Mass., 1993), 10; and James M. Laux, *The European automobile industry* (New York, 1992), 13. According to common usage, "range" in this study always means the *entire* distance covered on one battery charge, and not half of it.

3. Gijs Mom, "Conceptualising technical change: Alternative interaction in the evolution of the automobile," in Helmut Trischler and Stefan Zeilinger in cooperation with Robert Bud and Bernard Finn (eds.), *Tackling transport* (London, 2003) (Artefact series: studies in the history of science and technology, eds. Robert Bud, Bernard Finn, and Helmuth Trischler, Vol. 3). Also see the note on method at the end of this book.

4. Anthonie W. M. Meijers, "The relational ontology of technical artifacts," in Peter Kroes and Anthonie W. M. Meijers (eds.), *The empirical turn in the philosophy of technology* (Amsterdam, 2000) (Resarch in Philosophy and Technology 20, ed. Carl Mitcham), 81–96.

5. Bayla Singer also concludes that the car in historiography is still too much approached as a monolithic, undifferentiated entity: "we need some additional vocabu-

lary. . . . Historians have not provided such a vocabulary, perhaps because of their customary focus on case studies." Bayla Singer, "Automobiles and femininity," *Research in philosophy and technology* 13 (1993), 37. The classification—and generally the basic knowledge about the construction of the car—in this and the following paragraphs has been taken (with a few changes) from G. Mom (ed.), *De nieuwe Steinbuch: De automobiel, handboek voor autobezitters, monteurs and technici* (Deventer, 1986 and following years).

6. On the growth industry, see D. H. Aldcroft and M. J. Freeman, *Transport in the industrial revolution* (Manchester, 1983), 97; on early steam-propelled road vehicles, see K. Kühner, *Geschichtliches zum Fahrzeugantrieb* (Friedrichshafen, 1965), passim; David Beasley, *The suppression of the automobile: Skulduggery at the crossroads* (New York, 1988); on through roads, see Philip S. Bagwell, *The transport revolution from 1770* (London, 1974), 138ff.

7. Simon P. Ville, *Transport and the development of the European economy, 1750–1918* (n.p., n.d.), 27; Klaus Beyrer, *Die Postkutschenreise* (Tübingen, 1985), 235.

8. Wolfgang Schivelbusch, *Geschichte der Eisenbahnreise: Zur Industrialisierung von Raum und Zeit im 19. Jahrhundert* (Munich, 1977), 35, 51–53, 60–61, 79ff., 121ff. Schivelbusch's thesis of an irreversible change in observation has meanwhile received severe criticism. For example, Rainer Schönhammer, *In Bewegung: Zur Psychologie der Fortbewegung* (Munich, 1991), 102–11, has convincingly shown that others, traveling by coach, had experienced the compulsion to a "sideways view" earlier. Nevertheless, the present study confirms that contemporaries experienced the automobile as a means to the restoration of the "foreground," which had been lost by the train.

9. Clay McShane, *Down the asphalt path: The automobile and the American city* (New York, 1994), 56; see also Clay McShane and Gijs Mom, "The golden age of the horse and its substitution by mechanization and motorization: A bibliographical introduction," *Achse, Rad und Wagen* (forthcoming, in a German translation).

10. Michael Evgénieff, *Droschkenbetrieb: seine Organisation und wirtschaftspolitische Probleme* (Berlin, 1934) (diss.), 21–22, 209; Erwin Knaths, *Die Entwicklung des Berliner Droschkenfuhrwesens unter besonderer Berücksichtigung seiner Motorisierung: Eine verkehrsgeschichtliche Studie* (Marburg a.d. Lahn, 1929) (diss.), 11; McShane, *Down the asphalt path*, 3–7.

11. M. J. Freeman and D. H. Aldcroft, *Transport in Victorian Britain* (Manchester, 1998), 151; Theo Barker, "Towards an historical classification of urban transport development since the late eighteenth century," *Journal of transport history*, 3rd ser., no. 1 (1980), 88.

12. E. Schatzberg, "The mechanization of urban transit in the United States: Electricity and its competitors," in W. Aspray (ed.), *Technological competitiveness: Contemporary and historical perspectives on the electrical, electronics, and computer industries* (New York, 1993), 228. The best summary of the urban "transportation revolution" in the United States and Europe is John P. McKay, "Comparative perspectives on transit in Europe and the United States, 1850–1914," in Joel A. Tarr and Gabriel Dupuy (eds.), *Technology and the rise of the networked city in Europe and America* (Philadelphia, 1988), 3–23; the final quotation is from Freeman and Aldcroft, *Transport in Victorian Britain*, 153.

13. Joel A. Tarr, *The search for the ultimate sink: Urban pollution in historical perspective* (Akron, 1996), 328; for a critique of the pollution argument, see Gijs Mom, "Competition and coexistence: Motorization of land transportation and the substitution of the horse," *Achse, Rad und Wagen* (forthcoming, in a German translation). F. M. L. Thompson, *Victorian England: The horse-drawn society* (London, 22 October 1970) (inaugural lecture); Clay McShane and Joel Tarr, "The choice of horse over steam as a mo-

tive power for nineteenth century urban transportation in the U.S.," paper presented at the XXth International Congress of the History of Science, Liege, Belgium, 24 July 1997, 23.

14. Jean Robert, *Les tramways parisiens*, 3rd ed. (n.p., 1992), 127ff.

15. A. Blondel and F. Paul-Dubois, *La traction électrique sur voies ferrées: Voie-matériel roulant-traction* (Paris, 1898), 4.

16. Eugen Weber, *France, fin de siècle* (Cambridge, Mass., 1986), 70ff; H. J. Habak-kuk, *American and British technology in the nineteenth century: The search for labour-saving inventions* (Cambridge, 1962), passim.

17. Schatzberg, "The mechanization of urban transit in the United States," 235–38; McKay, "Comparative perspectives on transit," 10.

18. D. Larroque, "Apogée, déclin et relance du tramway en France," *Culture technique* 19 (March 1989), 58. For the late electrification of the British streetcar, see Anthony Sutcliffe, "Die Bedeutung der Innovation in der Mechanisierung städtischer Verkehrssysteme in Europa zwischen 1860 und 1914," in Horst Matzerath (ed.), *Stadt und Verkehr im Industriezeitalter* (Cologne, 1996), 231–41.

19. Louis Lockert, *Les voitures électriques*, vol. 4 (Paris, 1897), 125ff.

20. This matter is extensively discussed in Schallenberg, *Bottled energy*, 222–49, on which the following paragraphs are based.

21. Ch. David, "Die Accumulatoren auf der Pariser Weltausstellung," *Centralblatt* (1 August 1900), 262–63; here: 262; on energy density increase, see *Centralblatt* (1 February 1901). In this study, the energy density is always taken on the *basis of mass*. The EPS energy densities have been calculated based on the following data: 120 Ah for a battery of 30.4 kg (1888), or 140 Ah for a battery of 10.8 kg (1901). As average discharging voltage we have taken 1.93 V, as applied at the Concours d'Accumulateurs (battery contest) in Paris in 1899. On Faure, see É. Hospitalier, "Automobiles électriques," *L'Industrie électrique* (10 May 1897), 177–80, here: 179. The expression of the energy density in Wh/kg is in fact not complete without giving the *duration* of the discharge, because the capacity (in Ah) is strongly dependent on the discharging time. The density decreases at a faster discharge, so at a higher discharging current. In this study we have decided not to give the duration, because most sources are not clear on this point. For application in the electric car, one often assumed a three- to five-hour discharge. In French electrotechnical circles, one sometimes did not give the discharging time, but the power density (in W/kg) as the measure for the rate of discharge.

22. In 1899, 204 of the electric streetcar systems installed in Europe had an overhead wire, 8 an underground wire, 8 a third rail, and 16 battery traction. "Société Internationale des Électriciens," *IE* (10 December 1896), 555–57, here: 557.

23. Gerhard Bauer (ed.), *Berliner Strassenbahnen* (Berlin [GDR], 1987), 20–22; Sigurd Hilkenbach and Wolfgang Kramer, *125 Jahre Strassenbahnen in Berlin* (Düsseldorf, 1990), 12–13, 16; P. H. Prasuhn, *Chronik der Strassenbahn* (Hannover, 1969), 19; W. Hendlmeier, *Von der Pferde-Eisenbahn zur Schnell-Strassenbahn* (Munich, 1968), 24–26; on the fierce discussion, see Johannes Zacharias, "Die Störungen im Berliner Strassenbahnbetriebe," *Centralblatt* (15 March 1900), 107–8; "Die Berliner Strassenbahnen ohne Accumulatoren," *Centralblatt* (1 October 1900), 327–28; "Berlin," *Centralblatt* (15 October 1900), 356; Johannes Zacharias, "Zur Abschaffung des Accumulatorenbetriebs in Berlin," *Centralblatt* (15 November 1900), 385–86.

24. Gijs Mom, "Das Holzbrettchen in der schwarzen Kiste: Die Entwicklung des Elektromobilakkumulators bei und aus der Sicht der Accumulatoren-Fabrik AG (AFA) von 1902–1910," *Technikgeschichte* 63, no. 2 (1996), 119–51; Schallenberg, *Bottled energy*,

90–97; Adolph Müller, *25 Jahre der Accumulatoren-Fabrik Aktiengesellschaft 1888–1913* (Berlin, 1913), 38–47.

25. E. J. Wade, *Secondary batteries: Their theory, construction and use,* 2nd ed. (London, 1908), 254. The origins of the *Revisionsorganisation* seem to predate the formation of EPS. Already in 1881, soon after Faure announced his invention, it was suggested that charged batteries could be delivered to consumers' homes each morning, just like bottles of milk. This scheme was never tried in practice. Schallenberg, *Bottled energy,* 69, 73.

26. "Zusammenstellung der elektrischen Bahnen in Deutschland nach dem Stande vom 1. Oktober 1904," *Elektrotechnische Zeitschrift* (13 July 1905), 639–40, here: 639; E. C. Zehme, "Akkumulatorenbetrieb im Vorortsverkehr auf Haupteisenbahnen," *ETZ* (8 August 1907), 791–95; Schallenberg, *Bottled energy,* 201–22.

27. Schallenberg, *Bottled energy,* 205–15.

Chapter 1: Separate Spheres

"Motive powers," *The horseless age* (February 1897), 1–2, here: 2.

1. Louis Lockert, *Les voitures électriques avec Supplément aux voitures à pétrole et Note sur les moteurs à acétylène et à alcool: Traité des véhicules automobiles sur route,* vol. 4 (Paris, 1897), 113–15, 147–51; H. de Graffigny, *La locomotion électrique* (Paris, n.d.), 40; J. A. Grégoire, *50 ans d'automobile: La voiture électrique* (Paris, 1981), 50–51; Ch. Milandre and R.-P. Bouquet, *Voitures automobiles électriques* (Paris, 1899), 185ff. Energy density calculated from Desmond G. Fitzgerald, "Electric traction on common roads in 1895," *Electrical review* (13 October 1897), 180. Fitzgerald gives 6.36 Ah/pound.

2. Gerhard Wilke (ed.), *Denkschrift Elektrospecherfahrzeuge* (Wiesbaden, 1970), 131.

3. Ernest Henry Wakefield, *History of the electric automobile: Battery-only powered cars* (Warrendale, 1994), 192; Killingworth Hedges, *American electric street railways* (London, 1894), 67–69; "Evolution of electric vehicle control systems," *HA* (11 June 1902), 693–94.

4. Grégoire, *50 ans d'automobile,* 57–58.

5. Ibid., 81; L. Baudry de Saunier, "Les voitures automobiles actuelles," *La locomotion automobile* (January 1896), 8–9.

6. Grégoire, *50 ans d'automobile,* 112–16. For the following, see also James M. Laux, *In first gear: The French automobile industry to 1914* (Liverpool, 1976), 91–93; A. T., "La voiture électrique 'Kriéger,'" *L'Industrie électrique* (16 January 1897), 113; Lockert, *Voitures électriques,* 164–66; Gaston Sencier and A. Delasalle, *Les automobiles électriques* (Paris, 1901), 253ff.

7. Graffigny, *La locomotion électrique,* 59–62; Sencier and Delasalle, *Automobiles électriques,* 191.

8. "Société des Voitures Électriques (Système Kriéger)," *Journal des chemins de fer* (26 November 1896); Compagnie Parisienne des voitures électriques (Procédés Kriéger), *Premiers Exercice du 25 Juillet 1900 au 31 Décembre 1901: Assemblée générale ordinaire du 2 juin 1902;* Société Française pour l'industrie et les Mines, *Rapport sur les opérations de la société, Année 1898, présenté à l'Assemblée Générale, le mardi 28 mars 1899;* idem, *Année 1900,* 8 (all sources in AN 65 AQ N30).

9. Thomas Parker, "Electricity on the common roads," *The electrical engineer* (London) (26 January 1900), 119–20, here: 119; Friedrich Schildberger, "Die Entstehung der Automobilindustrie in England" (manuscript, DaimlerChrysler Archives, Stuttgart), 130; I. Epstein, "Twenty-five years' progress in secondary batteries," *Electrical review* (8 December 1897), 278–79, here: 279; Walter C. Bersey, *Electrically propelled carriages* (London, 1898), 27–41.

10. A. Delasalle, "Voitures électriques Heinrich Scheele," *La locomotion automobile* (25 May 1899), 328–30; E. Sieg, "Über Automobil-Batterien," *Centralblatt* (15 May 1900), 183–86, here: 186; E. Sieg, "Über Elektromobile," *Elektrotechnische Zeitschrift* (24 December 1908), 1238–43, and (31 December 1908), 1258–62, here: 1238; *Centralblatt* (15 May 1900), 200; "Bericht über Vorträge," *Centralblatt* (15 April 1901), 128–31.

11. *75 Jahre Nutzfahrzeug-Entwicklung 1896–1971: Jubiläumsbericht der Daimler-Benz Aktiengesellschaft Stuttgart-Untertürkheim* (n.p., n.d. [Stuttgart, 1971]), 39–40.

12. Statistics courtesy of James Foreman-Peck. See also James Foreman-Peck and Masahiro Hayafuji, "Lock-in and Panglossian selection in technological choice: The power source for the motor car" (manuscript); on the "founder effect," see W. Brian Arthur, "Competing technologies, increasing returns, and lock-in by historical events," *The economic journal* 99 (March 1989), 116–31, here: 126–27.

13. Ludwig Lohner to "Georges Kellner," copy of letter no. 77 from "Heinrich, Jacob, Ludwig, Richard Lohner-Mappe," archives Technisches Museum Wien (hereafter: Copy book Lohner with number of letter); Ludwig Lohner to "Herr Trutz," Copy book Lohner, letter no. 8; Erwin Steinböck, *Lohner: Zu Land, zu Wasser und in der Luft* (Graz, 1984), 13–15; Thomas Köppen, "Ferdinand Porsche, Ludwig Lohner und Emil Jellinek—frühe Innovatoren im Elektromobilbau: Eine Falstudie über eine gescheiterte Innovation" (unpublished master's thesis, Technical University Berlin, 1987); Ludwig Lohner to "Herr von Fischer" (24 February 1898), Copy book Lohner, letter no. 93.

14. Ludwig Lohner to "Wertester Freund" (14 March 1898), Copy book Lohner, letter nos. 118 and 249. It is tempting to interpret the first quotation in the light of electric propulsion, but it is more likely that Lohner was thinking of the "petroleum engine" here (with which he meant the diesel engine) as final target for his developmental efforts.

15. John B. Rae, *The American automobile industry* (Boston, 1984), 11.

16. H. L. Barber, *Story of the automobile: Its history and development from 1760 to 1917; with an analysis of the standing and prospects of the automobile industry* (Chicago, 1917), 118.

17. Peter J. Ling, *America and the automobile: Technology, reform and social change* (Manchester, 1990), 97, 103; Wakefield, *History of the electric automobile*, 20; Rae, *The American automobile industry*, 14; Ronald A. Stringer, "The Morrison electric: America's first automobile!?!," *Antique automobile* (January/February 1984), 32–34, here: 34; Ken Ruddock, "Recharging an old idea: The hundred-year history of electric cars," *Automobile quarterly* 31, no. 1 (fall 1992), 30–47, here: 32. For the American bicycle industry as predecessor of the manufacturing technology in the automobile industry, see David A. Hounshell, *From the American system to mass production 1800–1932: The development of manufacturing technology in the United States* (Baltimore, 1984).

18. "Morris and Salom's electric wagons: The 'Electrobat,'" *HA* (November 1895); "Times-Herald contest: Rules and route of the road race; complete list of entries," *HA* (November 1895), 53–55; "The trial and postponement," *HA* (December 1895), 13–15; "The 'Electrobats,'" *HA* (December 1895), 15–17; "Thanksgiving Day route," *HA* (December 1895), 29–32; "Morris & Salom's prospectus," *HA* (January 1896), 23–25; "Times-Herald contest," *HA* (January 1866), 29–30; "Motor vehicle tests," *HA* (February 1896), 8–15; Richard P. Scharchburg, *Carriages without horses* (Warrendale, 1993), 102–3, 107.

19. "Times-Herald contest," 30; D. Farman, *Les automobiles: Voitures, tramways et petits véhicules* (Paris, 1896), 252; "Charles E. Duryea on the speed craze," *HA* (February 1896), 19.

20. Genevieve Wren, "Pedro G. Salom," in George S. May (ed.), *The automobile industry, 1896–1920* (New York, 1990), 408–9.

21. William Greenleaf, *Monopoly on wheels: Henry Ford and the Selden automobile*

patent (Detroit, 1961), 58–63; John B. Rae, "Pope Manufacturing Company," in May (ed.), *The automobile industry,* 398–99; John B. Rae, "Albert Augustus Pope," in May (ed.), *The automobile industry,* 393–97; John B. Rae, *American automobile manufacturers: The first forty years* (Philadelphia, 1959), 8–11; Henry Cave, "First pneumatic automobile tires" (1947), in Cave Collection, NAHC, Series I, Box 3, Item 24; Hermann F. Cuntz, "The Electric Vehicle Company" (1940), 1, in Cave Collection, NAHC, Series III, Box 10, Item 31.

22. Henry Cave, "Lieutenant Hayden Eames" (1941), Cave Collection, NAHC, Series II, Box 8, Item 32; on "one of the wonders," see Greenleaf, *Monopoly on wheels,* 59.

23. James Wren, "George H. Day," in May (ed.) *The automobile industry,* 120–22; H. F. Cuntz, "re Columbia Motor Cars 1895–1910" (1940), 2, in Cave Collection, NAHC, Box 9, Item 11; John B. Rae, "Hiram Percy Maxim," in May (ed.), *The automobile industry,* 327–29.

24. Hiram P. Maxim, *Horseless carriage days* (New York, 1962), 74, 129; John B. Rae, *The American automobile: A brief history* (Chicago, 1965), 11.

25. "The Columbia electric motor carriage of the Pope Manufacturing Co.," *The electrical engineer* (New York) (19 May 1897), 532–34; on the agreement with ESB, see "The Columbia motor carriage," *HA* (May 1897), 4–7; "Latest products of the Pope Company," *HA* (July 1898), 12–13; Cuntz, "The Electric Vehicle Company," 2, 6; H. F. Cuntz, "Story of the Selden case and Hartford" (August 1940), 16, in Cave Collection, NAHC, Series III, Box 10, Item 28.

26. "The Electric Vehicle Company's factory," *ER* (New York) (24 October 1900), 442–43, here: 443; "Automobiles," *ER* (New York) (4 July 1900), 6; "Electric Vehicle Company absorbs the Riker Company," *HA* (12 December 1900), 14; Rae, *American automobile manufacturers,* 70–71; Genieve Wren, "Andrew Lawrence Riker," in May (ed.), *The automobile industry,* 407–8; "The Riker electric motor cycle," *HA* (January 1896), 19–20; "Mr. A. L. Riker's new electric Victoria," *EE* (New York) (6 January 1898), 32.

27. Richard Wager, *Golden wheels: The story of the automobile made in Cleveland and Northeastern Ohio 1892–1932* (Cleveland, 1986), 221–23; Thomas Parker Hughes, *Elmer Sperry: Inventor and engineer* (Baltimore 1971); Rae, *American automobile manufacturers,* 11, 81, 85; Rae, "Albert Augustus Pope," 395–96; Darwin H. Stapleton, "Walter C. Baker," in May (ed.), *The automobile industry,* 31–33.

28. "Big electric carriage factory for Indianapolis," *HA* (June 1898), 7; "The Waverley electric vehicle," *HA* (July 1898), 4–7; "New line of Waverly [sic] electrics—steam Waverlys too, for heavy work," *HA* (February 1899), 12.

29. David E. Nye, *Electrifying America: Social meanings of a new technology, 1880–1940* (Cambridge, Mass., 1990), 92; Beverly Rae Kimes and Henry Austin Clark Jr., *Standard catalog of American cars 1805–1942,* 2nd ed. (Iola, Wis., 1989), 1515–16; C. E. Woods, *The electric automobile, its construction, care and operation* (Chicago, 1900), 21–23; "Electrical engineer in automobile manufacture," *ER* (11 October 1899), 235; "To popularize the electric vehicle," *HA* (May 1896), 22; "An electric vehicle company in Chicago," *ER* (3 May 1899), 279.

30. "A very large storage battery contract," *ER* (New York) (2 February 1898), 69; "Horseless cabs in Chicago," *ER* (New York) (30 March 1899), 199; "The Willard storage battery," *ER* (New York) (30 March 1899), 205; "Woods' electric motor vehicles," *HA* (December 1898), 9–12; "Woods' electric moto-vehicles," *The electrical engineer* (New York) (15 December 1898), 579–80; "Big electric vehicle companies in Chicago," *HA* (12 April 1899), 11; "Prospectus of the Woods Motor Vehicle Co.," *HA* (19 April 1899), 12; "Woods Motor Vehicle Co. comes strongly to the front," *HA* (4 October 1899), 7; quotation from Kimes and Clark, *Standard catalog of American cars,* 1516.

31. Stapleton, "Walter C. Baker," and Darwin H. Stapleton, "Baker Motor Vehicle Company," in May (ed.), *The automobile industry*, 33–35.

32. "The Paris exposition," *HA* (26 July 1899), 14–19, here: 14. According to an American electrical engineer, there were "probably 400 or 500" electric cars "at the most" in Europe before the turn of the century. If we take this unlikely high number and subtract the London and Paris taxi fleets (together about 170 vehicles), we are still left with 230 to 330 privately owned cars ("Automobiles in Europe," *ER* [NY] [26 July 1899], 53). In 1902, the KAW in Cologne mentioned that up until then it had sold a total of about 100 battery sets, including those sold abroad (*ETZ* [9 January 1902], 40), probably mostly to private customers. The London City and Suburban Electric Carriage Company announced that same year that it had sold about 150 electric cars up until then, most of them probably after the period described here ("The electromobile and its future," *Electric vehicles* [U.K.] [March 1921], 28–32, here: 31; this was a reprint of an article from *Electrical industries* of July 1902); on 1901, see Laux, *In first gear*, 93; on 1905, see Grégoire, *50 ans d'automobile*, 123–25.

33. Hugo Fischer von Röslerstamm, "Die Automobile," in *Berichte über die Weltausstellung in Paris 1900, Achter Band: Wasserbau, Schiffart, Ingenieurwesen, Automobile* (Wien, 1901), 112–13.

34. "Census report on automobile manufacture in 1900," *HA* (17 September 1902), 307; on production in New England, see Greenleaf, *Monopoly on wheels*, 259, note 29; L. J. Andrew Villalon and James M. Laux, "Steaming through New England with Locomobile," *Journal of transport history* 5, no. 2, new series (September 1979), 65–82, here: 73; Foreman-Peck and Hayafuji, "Lock-in and Panglossian selection in technological choice," 4, 14, 22.

35. Wakefield, *History of the electric automobile*, 122; on the traffic count, see Schildberger, "Die Entstehung der Automobilindustrie in den Vereinigten Staaten," 31; "Automobiles," *ER* (New York) (2 March 1901), 286. The 226 licenses for electrics in Chicago were not only issued to private persons. James Rood Doolittle, *The romance of the automobile industry; being the story . . .* (etc.) (New York, 1916), 364, mentions 350 licenses for Chicago in November (not specified as to traction type), of which 100 were for automobile dealers, 78 for the local electric taxicab company of the EVC, and 43 for a competing electric taxicab company of the Woods Motor Vehicle Company. "The rest [129 licenses] were issued to owners of private pleasure cars and drivers of trucks, delivery wagons and buses for Siegel, Cooper & Co., The Hub, Schlessinger & Mayer, American Express Co., Chicago American, Pabst Brewing Co., Chicago Edison Co. and Ruppert the Shoe Man." From this follows (assuming a maximum of one truck for each of the companies mentioned) that a maximum of 226 − 78 − 43 − 8 = 97 licenses were meant for privately owned electric cars in Chicago. But see below for driving without a license in electric cars by women in Chicago. "Automobile Club run," *ER* (New York) (8 November 1899), 297. On production during the first eight months of 1902, see *Sun* (21 September 1902), quoted in "The American automobile industry," *The India rubber world* (1 October 1902), 13–14, here: 13. Of the 18,135 automobiles put on the market in the first eight months of 1902 by 90 companies (including duplications for those companies that produced more than one propulsion type), 10,040 were gasoline cars, 6,260 steam cars, and 1,835 electric cars. This shows (given our generalization formulated in the next paragraph about the production numbers for the year 1900 for private persons), that the number of electric cars for *this market segment* only decreased in a relative sense during the first years of the so-called Dark Ages; in an absolute sense it increased. James J. Flink, *America adopts the automobile, 1895–1910* (Cambridge, Mass., 1970), 234. On Portland, see David A. Kirsch, "Behind the numbers: Early quantitative

data on the history of the American automobile industry," paper presented at the ICOHTEC Symposium, Prague, 24 August 2000, 5.

36. Prototypical for this reasoning are the unpublished manuscripts of the former head of the Mercedes-Benz historical archives, who mentions 39 exhibitors for the New York show in 1900: 20 gasoline cars, 9 electric cars, and 10 steam cars. One year later there were 55 gasoline cars, 58 steam cars, and 24 electric cars at the New York exhibition, while at the third show gasoline cars dominated for the first time (168 vehicles versus 51 electric cars and 34 steam cars). In 1905, there were 177 gasoline cars against 4 steam cars and 31 electric cars (Schildberger, "Die Entstehung der Automobilindustrie in den Vereinigten Staaten," 118, 129). This sequence, by the way, confirms the conclusion of the previous note about the increase (in absolute numbers) in interest in the electric car until 1904.

37. Wakefield, *History of the electric automobile,* 122. According to the Census data, 911 cars were manufactured in Connecticut in 1900 (unfortunately, not specified according to propulsion type). Greenleaf, *Monopoly on wheels,* 259 (note 29) notes that in all of New England only 197 gasoline cars were produced that year. One can assume that the share of this car type in the production of Connecticut (and so that of Pope) was small. So, up to 1901, the total production of the Columbia factory in Hartford cannot have been more than 1,500 (and probably fewer) vehicles: the 500 already mentioned in 1898 and 1899 and a maximum of 790 (half of the national production, see the text) for all of New England in 1900. According to contract this entire production was exclusively meant for the EVC. As we will see in chapter 4, a maximum of 850 vehicles was supplied to the EVC for its own operation. This leaves a minimum of 440 and a maximum of 650 for private use.

38. Michael Rauck, *Karl Freiherr Drais von Sauerbronn: Erfinder und Unternehmer (1785–1851)* (Wiesbaden, 1983); for the following, see especially C. F. Caunter, *The history and development of cycles* (London, 1955).

39. Chris Sinsabaugh, *Who, me? Forty years of automobile history* (Detroit, 1940), 19, 23, 30.

40. Wiebe Eco Bijker, *The social construction of technology* (Eijsden, 1990), 17.

41. Rüdiger Rabenstein, *Radsport und Gesellschaft: Ihre sozialgeschichtlichen Zusammenhänge in der Zeit von 1867 bis 1914* (Hildesheim, 1991), 23; Richard Harmond, "Progress and flight: An interpretation of the American cycle craze of the 1890s," *Journal of social history* 5 (Winter 1971–72), 235–57, here: 236, 242. The flight metaphor is a topos of early transport experience, dating back at least to mid-sixteenth-century coach travel. As argued also in the prologue, this again seriously questions Schivelbusch's claim about the experience of "modernity" as characteristic for the railroad age. Wolfgang Behringer, "Der Fahrplan der Welt. Anmerkungen zu den Anfängen der europäischen Verkehrsrevolution," in Hans-Liudger Dienel and Helmuth Trischler (eds.), *Geschichte der Zukunft des Verkehrs: Verkehrskonzepte von der Frühen Neuzeit bis zum 21. Jahrhundert* (Frankfurt, 1997), 40–57, here: 50; Wolfgang Schivelbusch, *Geschichte der Eisenbahnreise: zur Industrialisierung von Raum und Zeit im 19. Jahrhundert* (Munich, 1977).

42. Eugen Weber, *France, fin de siècle* (Cambridge, Mass., 1986), 4.

43. Rabenstein, *Radsport und Gesellschaft,* 26ff., 39–40; see for the next four paragraphs: 96, 69, 132, 124–25.

44. James J. Flink, *The automobile age* (Cambridge, Mass., 1993), 4.

45. Rabenstein, *Radsport und Gesellschaft,* 249.

46. Martin Scharfe, "'Ungebundene Circulation der Individuen': Aspekte des Automobilfahrens in der Frühzeit," *Zeitschrift für Volkskunde* 86 (1990), 226.

47. Claude Johnson, *The early history of motoring* (London, n.d.), 36; Scharchburg, *Carriages without horses*, 133–36.

48. Patrick Fridenson, *Histoire des usines Renault: I. Naissance de la grande entreprise 1898–1939* (Paris, 1972), 47; Henry Ford, *My life and work*, 7th ed. (London, 1925), 50.

49. Frédéric Régamey, *Vélocipédie et automobilisme* (Tours, 1898), 211; B. ten Have quoted in J. Fuchs, *Die heerlijke auto's: de eerste halve eeuw autorijden in Nederland* (Amsterdam, 1970), 38–39.

50. Siegfried Reinecke, *Mobile Zeiten: Eine Geschichte der Auto-Dichtung* (Bochum, 1986), 48; Scharfe, "'Ungebundene Circulation,'" 233 (Sombart quotation); Richard van Dülmen (ed.), *Körper-Geschichten: Studien zur historischen Kulturforschung V* (Frankfurt am Main, 1996), 12; Joachim Radkau, "Technik im Temporausch der Jahrhundertwende," in Michael Salewski and Ilona Stölken-Fitschen (eds.), *Moderne Zeiten: Technik und Zeitgeist im 19. und 20. Jahrhundert* (Stuttgart, 1994), 61–76, here: 62, 74.

51. Reinecke, *Mobile Zeiten*, 48–51; on drunkenness, see Christoph Maria Merki, "Sociétés sportives et développement de l'automobilisme (1898–1930)," in Christophe Jaccoud, Laurent Tissot, and Yves Pedrazzini (eds.), *Sports en Suisse: Traditions, transitions et transformations* (Lausanne, 2000), 45–73, here: 66 (note 50); on the novel, see Williamson, *The lightning conductor*, 13 (italics in original). The comparison between early automobile driving (especially the sensation of acceleration, and not of speed!) and drug use was also made by Italian writer Mario Morasso in his *La nuova arma (la macchina)* (1905): "Once the poet described the miraculous paradise of opium and hashish, let's try now to reveal the secret Olymp, not by using noxious poisons, but through the strengthening glorification of our own capacities of acceleration." Quoted in Attilio Brilli, *Das rasende Leben: Die Anfänge des Reisens mit dem Automobil* (Berlin, 1999), 91.

52. Peter Gay, *The cultivation of hatred* (New York, 1993) (The bourgeois experience: Victoria to Freud, vol. III) 494, 513; on softness, see Michael C. C. Adams, *The great adventure: Male desire and the coming of World War I* (Bloomington, 1990), 51; on the crisis of abundance and Féré, see Stephen Kern, *The culture of time and space 1880–1918* (London, 1983), 9, 128; on the power of the wheel, see Martin Scharfe, "Die Nervosität des Automobilisten," in van Dümler (ed.), *Körper-Geschichten*, 209.

53. Quoted by Joachim Radkau, *Das Zeitalter der Nervosität: Deutschland zwischen Bismarck und Hitler* (Munich, 1998), 65, 73, 203–7, 310–11, 321–22; last quote: Radkau, "Technik im Temporausch," 75.

54. Quoted in Angela Zatsch, *Staatsmacht und Motorisierung am Morgen des Automobilzeitalters* (Konstanz, 1993), 509, 511–12; on de Saunier, see Weber, *France, fin de siècle*, 206.

55. On fear and lust, see Rainer Schönhammer, *In Bewegung: Zur Psychologie der Fortbewegung* (Munich, 1991), 204–17; and Kurt Möser, "'Knall auf Motor'—Die Liebesaffäre von Künstlern und Dichtern mit Motorfahrzeugen 1900–1930," in *Mannheims Motorradmeister: Franz Islinger gewinnt die Deutsche Motorradmeisterschaft 1926* (Mannheim, 1996), 18–29; Hellspach: Scharfe, "Die Nervosität des Automobilisten," 216–17; on bad roads, see Theo Barker, "A German centenary in 1986, a French in 1995 or the real beginnings about 1905?" in Barker (ed.), *The economic and social effects of the spread of motor vehicles*, 26; R. J. Mercredy, "The petrol engine," in Alfred C. Harmsworth (ed.), *Motors and motor-driving*, 2nd ed. (London, 1902), 103; Swiss quoted in Merki, "Sociétés sportives et développement de l'automobilisme," 64 (note 42); on the tickling sensation, see Karl A. Kuhn, *Die Opfer des Automobils* (Berlin, 1907), 5–6 (my italics). The bicycle and the automobile do not have an exclusive claim on this fascination for mechanical unreliability. According to Uta C. Schmidt, "Vom 'Spielzeug' über den

'Hausfreund' zur 'Goebbels-Schnauze': Das Radio als häusliches Kommunikations-medium im Deutschen Reich (1923–1945)," *Technikgeschichte* 65, no. 4 (1998), 313–27, here: 313, during the very early days of the radio "it mattered less what was heard, but that one listened."

56. Lord Montagu of Beaulieu and F. Wilson McComb, *Behind the wheel: The magic and manners of early motoring* (New York, 1977), 117; on black hands, see Fuchs, *Die heer-lijke auto's*, 165; on boredom, see Christoph Maria Merki, "Das Rennen um Marktan-teile: Eine Studie über das erste Jahrzehnt des französischen Automobilismus," *Zeitschrift für Unternehmensgeschichte* 43, no. 1 (1998), 79.

57. On active traveling, see Wolfgang Ruppert, "Das Auto: 'Herrschaft über Raum und Zeit,'" in Ruppert (ed.), *Fahrrad, Auto, Fernsehschrank: Zur Kulturgeschichte der All-tagsdinge* (Frankfurt, 1993), 119–61, here: 161; Chambers quoted in Montagu of Beaulieu and McComb, *Behind the wheel*, 111–12. The only historical analysis that comes near this interpretation is Claud R. Erickson, "Electric vehicle history," *Public power* 24, no. 9 (September 1966), 29–33, here 33: "Another reason for the demise of the electric car market was that it lacked excitement which in its day was a very important factor. The electrics were actually too dependable, too quiet, and too clean. The men of that day drove a car not simply as a means of transportation but as a sport or hobby. They liked to hear an occasional backfire and they liked an occasional breakdown. This gave them a chance to demonstrate their mechanical ability." Quoted by Dietmar Abt, *Die Er-klärung der Technikgenese des Elektroautomobils* (Frankfurt am Main, 1998) (European University Studies, Series V, vol. 2295), 191.

58. Otto Julius Bierbaum, *Eine empfindsame Reise im Automobil: Von Berlin nach Sorrent und zurück an den Rhein; in Briefen an Freunde geschrieben* (München, 1979) (reprint of the first edition of 1903).

59. Sir Francis Jeune, "The charms of driving in motors," in Harmsworth (ed.), *Motors and motor-driving*, 341–43.

60. Barker, "A German centenary in 1986," 22; Montagu and McComb, *Behind the wheel*, 128.

61. J. St. Loe Strachey, "Roads: The return to the road," in Harmsworth (ed.), *Mo-tors and motor-driving*, 347–49; on Proust, see Gerard Horras, *Die Entwicklung des deutschen Automobilmarktes bis 1914* (Munich, 1982), 324.

62. Sir Francis Jeune, "The charms of driving in motors," 344; Reinecke, *Mobile Zeiten*, 36, 46, 66. In this respect, early automobile sport demonstrates remarkable parallels with the *bohème*, and especially with the bohemian type of the *flâneur*, who tried to reconcile big city culture with a nostalgic rural feeling, in his walks along the "passages" (the covered shopping malls in Paris) and his preference for the outdoor café. Klaus Kuhm recently criticized this psycho-cultural type of analysis of "automo-tive addiction," because it blinds us to power relations that lay behind the emerging car culture and because it obscures the systemic character of automobilism. Social power relations and system growth, however, are not conceivable without this fundamental and often irrational dependence of the automobile, especially during a time in which alternatives are available. Kuhm also defends Bierbaum against criticism from cultural historians, that he was simply indulging in reactionary nostalgia; instead, Bierbaum was very "modern." Bierbaum's regressive claims, however, fit seamlessly into an in-ternationally expressed exhilaration about the recapture of the foreground. Klaus Kuhm, *Das eilige Jahrhundert: Einblicke in die automobile Gesellschaft* (Hamburg, 1995), 57, 60; idem, *Moderne und Asphalt: Die Automobilisierung als Prozess technologischer In-tegration und sozialer Vernetzung* (Pfaffenweiler, 1997), 70.

63. Harmsworth (ed.), *Motors and motor-driving*, 45. Piers Brendon, in his excellent biography of the Royal Automobile Club, justly speaks not only of the "sensation of autonomous speed," but also of the "exhilaration of acceleration" in his effort to characterize early British automobilism. He also evokes the flight metaphor when he quotes a Club member who described car driving as "floating on a feather-bed between heaven and earth." Piers Brendon, *The motoring century: The story of the Royal Automobile Club* (London, 1997), 56, 58. Kern (*The culture of time and space 1880–1918*), in his analysis of fin de siècle culture, misses this point completely.

64. Martin V. Melosi, *Thomas A. Edison and the modernization of America* (n.p., 1990), 152–53; Brendon, *The motoring century*, 69.

65. Baker, "A German centenary in 1986," 37–38; Horras, *Die Entwicklung des deutschen Automobilmarktes*, 239–40.

66. März quoted in Lothar Diehl, "Das Automobil in der wilhelminischen Gesellschaft: Alltagsgeschichtliche Aspekte einer technischen Innovation" (unpubl. master's thesis, Universität Tübingen, 1990), 27; on Wilson, see David Gartman, *Auto opium: A social history of American automobile design* (London, 1994), 15 (my italics); on Dreyfusards, see Weber, *France, fin de siècle*, 122–23; on de Dion, see Gijs Mom and Vincent Van der Vinne, "Geschiedenis van de elektrisch aangedreven auto in Nederland in het licht van de internationale strijd om een geschikte tractiewijze 1880–1920," *NEHA-Jaarboek voor economische, bedrijfs- en techniekgeschiedenis*, vol. 57 (Amsterdam, 1994), 134; on Faure, see Laux, *In first gear*, 30–32.

67. Kenneth Murchison, *The dawn of motoring* (London, 1942), 23; Immo Sievers, *AutoCars: Die Beziehungen zwischen der englischen und der deutschen Automobilindustrie vor dem Ersten Weltkrieg* (Frankfurt am Main, 1995), 183.

68. C. L. Freeston, "Automobile clubs," in Harmsworth (ed.), *Motors and motor-driving*, 384–85; Diehl, "Das Automobil in der wilhelminischen Gesellschaft," 96–97. On the emergence of German automobile clubs, see Barbara Haubner, *Nervenkitzel und Freizeitvergnügen: Automobilismus in Deutschland 1886–1914* (Göttingen, 1998); and for the French Touring Club, see Cathérine Bertho-Lavenir, *La roue et le stylo: Comment nous sommes devenus touristes* (Paris, 1999).

69. Quoted in Ruppert, "Das Auto," 151.

70. Diehl, *Das Automobil in der wilhelminischen Gesellschaft*, 30, 40, 54–56; Michael L. Berger, *The devil wagon in God's country: The automobile and social change in rural America 1893–1929* (Hamden, Conn., 1979), 25.

71. Zatsch, *Staatsmacht und Motorisierung*, 513.

72. Weber, *France, fin de siècle*, 41; S. Daule, *Der Krieg gegen das Auto* (Leipzig, n.d.), 3–4.

73. *Motor cars and news of 1899 (Lloyd Clymer's historical scrapbook)* (Los Angeles, 1955), 82; on accidents with horses, see Sylvester Baxter, "How the horse runs amuck," in ibid., 41–48. On resistance by American farmers from the South against trunk-road building before the First World War, see Howard Lawrence Preston, *Dirt roads to Dixie: Accessibility and modernization in the South, 1885–1935* (Knoxville, 1991), 64–65.

74. Murchison, *The dawn of motoring*, 33.

75. "À Chanteloup," *La France automobile* (3 December 1898), 408–9; "L'électricité et la course de côte de Chanteloup," *L'Industrie Électrique* (10 December 1898), 517–18, here: 517.

76. See the articles entitled "Le record du kilomètre" in *FA* from 17 December 1898 until 9 April 1899; see also the articles entitled "La course du kilomètre" in *LA* (22 December 1898), 803, and (29 December 1898), 819, and "Le kilomètre en automobile," *LA*

(26 January 1899), 64, "Le record du kilomètre," *LA* (6 April 1899), 214. *Der Motorwagen* ([1899, Heft V], 47) notes: "specially rolled in advance and closed to regular traffic."

77. "106 kilomètres à l'heure!," *LA* (4 May 1899), 276; "Records des voitures électriques sur route," *IE* (10 May 1899), 183; "Aluminium and partinium," *HA* (20 June 1900), 27.

78. *HA* quoted in Scharchburg, *Carriages without horses*, 126; "The Providence race," *HA* (September 1896), 6–7; "From the chairman of the judges," *HA* (September 1896), 7; Rae, *The American automobile*, 30; on 5,000 spectators, see Michael Brian Schiffer (with Tamara C. Butts and Kimberly K. Grimm), *Taking charge: The electric automobile in America* (Washington, D.C., 1994), 46.

79. "The Baker electric racing automobile," *Scientific American* (14 June 1902), 419; "The Staten Island tragedy," *HA* (11 June 1902); Wager, *Golden wheels*, 207; Stapleton, "Walter C. Baker," 32.

80. "Les courses et la vitesse," *FA* (24 October 1896), 312–15; Paul Meyan, "Impressions," *FA* (16 July 1898), 244–45, here: 244.

81. L. Béguin, "Faut-il des courses?" *LA* (31 January 1901), 65–66, here: 65; Gérard Lavergne, "Encore le marasme," *LA* (29 November 1900), 765–66; De Dion: "Quelques opinions," *LA* (4 June 1903), 353–55, here: 353; L. Baudry de Saunier, "Notre cancer," *La locomotion* (6 September 1902), 561–62.

82. "La carte électrique," *LA* (21 July 1898), 461; "Echos et nouvelles," *FA* (18 June 1898), 201; Prade: "La recharge des accumulateurs des voitures électriques en France," *IE* (10 June 1898), 222–23, here 223; on *électro-tourisme*, see "La carte électrique," *LA* (26 October 1899), 691.

83. "Carte des stations électriques," *LA* (19 April 1900), 247; "La carte électrique du Touring-Club," *LA* (5 July 1900), 429; "Échos," *LA* (27 September 1900), 623; Emile Dieudonné, "Le rechargement des accumobiles," *LA* (20 September 1900), 606. In some countries this idea apparently found followers. The Dutch brothers J. W. and A. W. Scholte of the Nederlandsche Metaalwarenfabriek (Dutch Construction Factory), together with a former state railway employee and mechanical engineer P. J. Neijt, in January 1899 applied for a concession to put "electricity points" along the state roads on an average distance of 3 miles. Nothing came of this. Letter of 5 January 1899, reply: 15 February 1899, in Nationaal Archief (Dutch National Archives), 2e afdeling, Waterstaat A, 1-IIIB (1878–1905); Dept. v. Wat., Handel en Nijv., Wegen in het Algemeen; arch. 2.16.01, inv. nr. 1673.

84. "Électricité contre pétrole," *LA* (12 January 1899), 31; "Chronique de Paris," *FA* (23 April 1899), 199; "Paris-Rouen en voiture électrique," *LA* (6 July 1899), 435. As yet, no historian of the automobile has come up with an estimate of the maximum distance an average gasoline car could reach on one fill-up during this early period.

85. "Tourisme électrique," *LA* (14 September 1899), 591; "Le Tour de France," *FA* (16 July 1899), 342–43; "Le Tour de France électrique," *LA* (7 February 1901), 82; "Le Tour de France électrique," *LA* (2 May 1901), 282.

86. "Le critérium des électriques," *LA* (14 June 1900), 381; "La coupe des électriques," *LA* (21 June 1900), 386; "307 kilomètres en électrique sans recharge," *LA* (24 October 1901), 674; on 41 Wh/kg, see Adrien Gatoux, "Le record électrique de Kriéger," *La locomotion* (26 October 1901), 55–56; on Hospitalier, see "Le nouveau record des voitures électriques," *IE* (25 October 1901), 465–66 (italics in original); on the reply of Kriéger and Brault, see "Le record des voitures électriques," *IE* (10 November 1901), 492.

87. "De Paris à Rouen en accumobile," *IE* (10 July 1899), 285; Abel Ballif, "La carte électrique du T. C. F.," *La locomotion* (31 August 1901), 1–2.

88. "London notes," *HA* (27 December 1899). 11; "Le concours d'électriques en Angleterre," *LA* (22 November 1900), 761–62.

89. David A. Kirsch, "The electric car and the burden of history: Studies in automotive systems rivalry in America, 1890–1996" (diss., Stanford University, 1996), 52–53; H. P. Maxim, "Electric vehicles and their relation to central stations," *HA* (March 1899), 17–19, here: 19; "Long distance lunacy," *HA* (11 October 1899), 4.

90. By the end of 1899, lead batteries with an energy density of more than 22 to 24 Wh/kg were no longer available in the entire United States. "The storage battery for vehicles," *HA* (29 November 1899), 17. On Maxim, see "160 kilomètres en automobile sans recharge des accumulateurs," *Éclairage électrique* (13 January 1900), Supplément, XXI–XXII; on bicycles, see "Electromobile speed test," *ER* (22 November 1899), 333.

91. Of course, this is also true for the gasoline engine (but not for fuel cell technology!): the faster the engine is made to run, the more fuel it consumes—also on a relative basis (for instance, per kWh). But the contradiction between torque and power, between energy generation and the speed or the intensity of this energy generation, is much more pronounced in a chemical storage battery. During one and a half centuries of storage battery development, this contradiction has been lessened, but not resolved.

92. É. H. [Hospitalier], "L'Utilisation des accumulateurs," *IE* (25 October 1897), 451–52; "Les voitures à accumulateurs," *FA* (11 December 1897), 399.

93. Sir David Salomons, "Motor traffic," *The automotor and horseless vehicle journal* (15 May 1897), 295–305, here: 305.

94. "Les voitures électriques à l'Automobile-Club de France," *IE* (10 July 1898), 271.

95. "The status of the electric vehicle," *HA* (1 January 1902), 3.

96. For the following, see J. Blondin, "Accumulateurs pour automobiles électriques: Le concours international de l'Automobil-Club," *l'Éclairage électrique* (13 January 1900), 57–64, and the continuation under the same title by A. Bainville, (27 January 1900) 130–37, (3 February 1900) 171–17, (24 February 1900) 289–92, and (3 March 1900) 336–42.

97. Sencier and Delasalle, *Les automobiles électriques*, 81–93.

98. This is suggested at any rate by *HA* (8 November 1899), 6, which has to be quoted with caution due to the fact that it was more than a party in the conflict surrounding the "Lead Cab Trust" (see chapter 2); the one-year delay may also have been caused by the commotion around the World Fair.

99. For the following paragraph, see also M. Kallmann, *Wettbewerb und Prüfungsfahrten für elektrisch betriebene Fahrzeuge in Berlin, Frühjahr 1900: Bericht über die Ergebnisse* (Berlin, 1900); a summary in *Centralblatt* (15 October 1900), 352–54; quotation: 354.

100. Quoted in Johnson, *The early history of motoring*, 29 (my italics).

101. *HA* (6 November 1901), 654–71.

102. *HA* (7 January 1903) 3–85 and (8 April 1903) 449; W. M. Hutchinson, "The storage battery from the standpoint of the user," *HA* (31 January 1900), 17–19, here: 17; Newport: "Electromobiles in New England," *ER* (New York) (12 July 1899), 23.

103. Henry Garrett, "An electrician prefers the electric vehicle," *HA* (3 December 1902), 622; Régamey, *Vélocipédie et automobilisme*, 204 (my italics).

104. On Thrupp, see "An English road carriage," *HA* (June 1896), 20; on Jeantaud, see "Echos et nouvelles," *FA* (24 December 1898), 432; on Alexandra, see Philip Sumner, "The evolution of the electric car," *Veteran and vintage magazine* 13, no. 1 (September 1968), 18; Leopold: Friedrich Schildberger, "Die Entstehung der Automobilindustrie in Frankreich" (manuscript, Mercedes Archives, Stuttgart), 244.

105. Grégoire, *50 ans d'automobile*, 123; *Motor cars and news of 1899*, 4–6; "The great electric automobile parade in New York," *ER* (New York) (31 May 1899), 399.

106. "Gasoline carriages and feminine drivers," *HA* (12 September 1900), 9; on the "city engineer," see "Automobiles," *ER* (New York) (28 November 1900), 582.

107. Paul Meyan, "La voiture électrique," *FA* (1896, no. 26) (no date specified), 202; L. Béguin, "La voiture électrique," *LA* (7 April 1898), 210 (italics in original).

108. "Concours pour un coffret," *LA* (17 November 1898), 728; "Avis aux électriciens," *FA* (26 November 1898); "Concours pour un coffret avec prise de courant universelle pour les automobiles électriques," *IE* (10 November 1898), 470; "Le concours de coffret avec prise de courant pour automobiles électriques," *IE* (25 February 1899), 73–74; P. M. Heldt, "Electric vehicles at the Madison Square Garden Show," *HA* (7 November 1900), 42–45, here: 44.

109. Lockert, *Les voitures électriques*, 184–97; quotation: 197.

110. Bersey, *Electrically propelled carriages*, 33.

111. Clay McShane, *Down the asphalt path: The automobile and the American city* (New York, 1994), 25, 40. For the following, see ibid., 30–40; and Kirsch, "The electric car and the burden of history," 69–70.

112. "Barred out of Boston's sacred parks," *HA* (March 1899), 6; on Colorado Springs, see "Autos barred from this park," *HA* (29 January 1902), 157; "Electromobiles in Central Park, New York City," *ER* (New York) (4 April 1900), 338; Woods: "Automobiles," *ER* (New York) (3 January 1900), 21; electric buses: "Automobiles," *ER* (New York) (8 August 1900), 132.

113. On Baltimore, see "Automobiles," *ER* (New York) (13 June 1900), 627; on San Francisco, see Kirsch, "The electric vehicle and the burden of history," 70; on Philadelphia, see "Automobiles," *ER* (New York) (17 October 1900), 415.

Chapter 2: Failed Experiments

"Automobiles électriques," *L'Industrie électrique* (10 October 1896), 433.

1. "Certificate of Registration of Design," no. 231523 of April 1894 (archives of the Science Museum, London).

2. "Great Horseless Carriage Company," *The automotor and horseless vehicle journal* (January 1898), 125–26. The stock prospectus is printed in full in *AJ* (November 1896), 82–83; quotations: 82.

3. "An autocar battery," *Electrical world* (5 December 1896), 704–5.

4. For the following, see also W. Worby Beaumont, *Motor vehicles and motors: Their design construction and working by steam oil and electricity*, 2nd ed. (Westminster, 1902), 397–414; quotations: 402, 414.

5. "Electric motor-cab service in New York City," *EW* (14 August 1897), 183–86, here: 184. The tractive energy of the first electric taxicab in New York was measured at 0.92 kWh per ton at a speed of 10 km/h on a virtually level road; according to the source, this was 2.33 times higher than on rails.

6. "A propos des premiers tramways à accumulateurs," *IE* (10 March 1895), 472; Craig R. Semsel, "More than an ocean apart: The street railway of Cleveland and Birmingham, 1880–1911," *Journal of transport history* (third series) 22, no. 1 (March 2001), 47–61, here: 59 (note 14); for the following paragraph, see also Beaumont, *Motor vehicles and motors* (1902), 395; "An electrical cab service for London," *AJ* (15 September 1897), 483–87, here: 483.

7. According to the American *Electrical world* (28 August 1897), 241, the number of taxicabs in use in the first fleet was even smaller: "Only two of them are plying for hire, the others having been pre-engaged for 25 shillings a day." On "first in the world," see "Great Horseless Carriage Company," 125; on "first in Europe," see Walter C. Bersey, *Electrically propelled carriages* (London, 1898), 62; on "rental carriages," see "The press

on the electric car," *AJ* (September 1897), 507. Bersey was wrong: the New York taxicabs were earlier (see the third section of this chapter).

8. "New electrical cabs for London," *HA* (May 1898), 14; "No public electromobiles in London?" *ER* (New York) (26 July 1899), 53; "London Electrical Cab Company (Limited)," *AJ* (December 1898), 130–36; on "taken over," see *HA* (16 August 1899), 15; Malcolm Bobbitt, *Taxi! The story of the 'London' taxicab* (Dorchester, 1998), 8.

9. For this and the following two paragraphs, see "London Electrical Cab Company (Limited)," *AJ* (December 1898), 130, 131; "The London Electrical Cab Company (Limited)," *AJ* (June 1900), 487.

10. Henry Charles Moore, *Omnibuses and cabs: Their origin and history* (London, 1902), 277; *The electrician* (25 July 1899), 504; quotation: Bobbitt, *Taxi!* 9; Beaumont, *Motor vehicles and motors* (1902), 395, 414; *AJ* (April 1900), 364. An American review mentions in retrospect that the loss amounted to $30,000 "through breakage of jars [probably glass-battery boxes] and careless handling of untrained employees and damage suits" ("A short history of the electric taxi-cab," *Electric vehicles* [U.S.] [July 1916], 2).

11. *Electrician* (9 December 1898), 214; Beaumont, *Motor vehicles and motors* (1902), 414 (the first edition dates from 1900).

12. *Electrician* quoted in "Electrical cabs in London," *HA* (July 1897), 4–6, here: 4; on the "last coup," see Immo Sievers, *AutoCars: Die Beziehungen zwischen der englischen und der deutschen Automobilindustrie vor dem Ersten Weltkrieg* (Frankfurt am Main, 1995), 158–69; last quotation: Piers Brendon, *The motoring century: The story of the Royal Automobile Club* (London, 1997), 30.

13. G. N. Georgano, *A history of the London taxicab* (Newton Abbott, 1972), 40ff.

14. "Electrical Power Storage Company," *AJ* (August 1898), 439–40, here: 440; Louis Lockert, *Les voitures électriques*, vol. 4 (Paris, 1897), 199; "The Paris electricabs," *AJ* (June 1899), 457–59, here: 458; Anne Boudou, "Les taxis parisiens de la fondation des Usines Renault aux 'Taxis de la Marne' 1898–1914" (unpublished master's thesis, Université de Paris X Nanterre, 1982), 182; on the "horse taxi crisis," 316; on glanders, see L. Harmant, "L'assemblée annuelle des actionnaires de la Compagnie Générale des Voitures," *Le Messager de Paris* (27 August 1897), in AN 65 AQ Q370³.

15. "Les voitures à Paris et l'automobilisme," *Moniteur des Tirages* (27 May 1897), AN 65 AQ Q370³; for the complete text of the report, see "L'automobilsme à la Compagnie Générale des Voitures à Paris," *IE* (10 June 1897), 241–42; quotation: 241.

16. On 50 million francs, see L. Harmant, "L'assemblée annuelle"; interview quoted in "Electric cabs in Paris," *AJ* (October 1897), 17; on the requirement of 1,000 kg, see Lockert, *Les voitures électriques*, 174, 198–99; on horse taxi distance, see "The Paris electricabs," *AJ* (June 1899), 457–59, here: 457.

17. On the "humming bird," see Brendon, *The motoring century*, 30; on 100 km, see "Les petites-voitures," *Le Paris-Bourse* (31 May 1901), AN 65 AQ Q370³; on 60 km, see Lockert, *Les voitures électriques*, 174; on the EVS, see Compagnie française de voitures électromobiles, *Statuts* (Paris, 1897) (AN 65 AQ Q370³); on interest CGV, see "Compagnie Générale des Voitures à Paris," *Le journal des intérêts financiers* (11 August 1900); on the order of motors, see Nicholas Papayanis, "The development of the Paris cab trade, 1855–1914," *Journal of transport history (New Series)* (March 1987), 52–65, here: 62.

18. On the announcement of Bixio, see "L'automobilisme à la Compagnie Générale des Voitures à Paris," *IE* (10 June 1897), 242; on contest rules, see "Automobil-Club de France; Programme du Concours des voitures de place automobiles d'avril 1898," *IE* (25 March 1897), 119–20.

19. "Automobil-Club de France: Le premier concours de voitures de place automo-

biles," *IE* (25 May 1898), 201; "Le concours de fiacres de l'Automobil-Club de France," *IE* (10 June 1898), 222; "French tests of electric cabs for city service," *EE* (New York) (4 August 1898), 102–4, here: 102; on the renewed interest in electrics, see *MW* (1898, Heft IX), 89.

20. *MW* (1898, Heft VI), 51.

21. É. Hospitalier, "Concours de voitures de place automobiles," *IE* (10 July 1898), 272–86; idem, "Concours de voitures de places automobiles de l'Automobile-Club de France," *IE* (9 February 1899), 82–86; on the title of Hospitalier, see "Concours de voitures de place automobiles," 272. For an extensive summary of Forestier's report, see "Le Concours des Fiacres," *LA* (6 October 1898), 626–29.

22. On optimism, see Gaston Sencier and A. Delasalle, *Les automobiles électriques* (Paris, 1901), 10; Henri Coupin, "Voitures de place automobiles: Concours de 'L'Automobile-Club de France,'" *La Nature* (9 July 1898), 87–91, here: 87.

23. Sencier and Delasalle, *Les automobiles électriques*, 11.

24. Boudou, "Les taxis parisiens," 226; "Operation of electric cabs in Paris discontinued," *HA* (23 January 1901), 31; "The public electromobile station in Paris," *ER* (19 July 1899), 37; "Première sortie des fiacres automobiles," *Le Petit journal* (2 April 1899) (box D/b 505, Archives de la Préfecture de Police, Paris); "Les fiacres électriques," *LA* (10 November 1898), 712–14; Sencier and Delasalle, *Les automobiles électriques*, 105–10; on derivation from the streetcar, see J. Laffargue, "La traction électrique par accumulateurs," *IE* (25 May 1895), 209–11; on the test course, see "Station centrale de fiacres électrique de la Compagnie Générale des Voitures à Paris," *Le Génie Civil* (15 April 1899), 373–76.

25. The EPS was sentenced to pay a compensation of 112,000 French francs ("Les petites-voitures"). The EPS's name is not mentioned anywhere in the French sources, but has been derived from "Automobilism in favor in Paris," *EE* (New York) (9 February 1899), 160, which mentions "a little delay" in the delivery of "accumulators from England."

26. "Première sortie des fiacres automobiles"; É. H. [Hospitalier], "Les fiacres électriques de la Compagnie Générale des Voitures, à Paris," *IE* (25 April 1899), 166–69, here: 166; "Les fiacres électriques parisiens," *LA* (6 April 1899), 221.

27. É. Hospitalier, "Les voitures électriques à l'Exposition internationale d'automobile de l'Automobil-Club de France, 15 juin-9 juillet 1899," *IE* (25 June 1899), 265–66, here: 265; "The Paris electrocabs," *AJ* (June 1899), 457–59; Sencier and Delasalle, *Les automobiles électriques*, 262–69; Lockert, *Les voitures électriques*, 199, 252; *La Patrie* (3 April 1899) (box D/b 505, Archives Préfecture de Police); "Compagnie Générale des Voitures à Paris," *l'Économie Européen* (19 May 1899), AN 65 AQ Q370³.

28. "Les petites-voitures"; on the "manufacturer's hobby," see "A discredited expert," *HA* (18 October 1899), 5–6, here: 5; on "a few months," see Sencier and Delasalle, *Les automobiles électriques*, 110; number 16060: "Promotion of electric vehicles in Paris," *HA* (4 October 1899), 8.

29. A. Delasalle, "Les automobiles électriques," *LA* (13 December 1900), 797–98, here: 798; on 149 cabs, see "Les petites-voitures."

30. According to the American review in retrospect quoted earlier, the CGV taxicabs were "under-batteried": this source mentions a range of only 45 km. It also mentions an exact closing date: 10 November 1900 ("A short history of the electric taxi-cab," *EV* [U.S.] [July 1916], 2).

31. Boudou, "Les taxis parisiens," 279.

32. Sencier and Delasalle, *Les automobiles électriques*, 81–93.

33. Compagnie française de voitures électromobiles, *Rapport du conseil d'Admin-*

istration: *Assemblée Générale Ordinaire du 19 Juin 1900* (Paris, 1900), 5 and 9 (AN 65 AQ Q131).

34. Sencier and Delasalle, *Les automobiles électriques,* 14.

35. "An electric carriage company," *ER* (New York) (22 January 1896), 53; on "after ten years," see "Electric motor-cab service in New York City," *EW* (14 August 1887), 183–86, here: 183; "The Electric Carriage and Wagon Co.," *EE* (New York) (10 June 1896), 612; Lloyd: "The November meeting of the Electric Vehicle Association of America," *The central station* (December 1914), 177–85, here: 180.

36. "Prospectus of the Electric Carriage & Wagon Company," *HA* (June 1896), 26; "The Morris & Salom electric carriages and wagons," *EE* (New York) (30 September 1896), 327–28; "New vehicles of the Electric Carriage and Wagon Company," *HA* (September 1896), 18; last two quotations: "Meeting of the American Institute of Electrical Engineers," *EW* (23 January 1897), 117–18, here: 117.

37. "Electric hansoms now for hire," *HA* (January 1897), 3; "Motor cabs in New York," *HA* (March 1897), 5; "Electric carriages in New York City," *ER* (New York) (24 February 1897), 85–86, here: 86; on 27 March, see "Work of electric hansom in New York," *HA* (July 1897), 12; "Electric motor-cab service in New York City," *EW* (14 August 1897), 183–86; London *ER* quoted in "Horseless cabs in New York," *AJ* (May 1897), 327.

38. "Electric motor-cab service in New York City," *EW* (14 August 1897), 183–86, and part II in *EW* (21 August 1897), 213–16; "New vehicles of the Electric Carriage and Wagon Company," *HA* (September 1896), 18; "An electric hansom," *AJ* (14 April 1897), 247–48; on streetcar technology, see "The gearing of electric vehicles," *HA* (15 May 1901), 142.

39. "An American electric carriage," *AJ* (September 1897), 497–99, here: 497.

40. "Electrical cabs in New York City," *ER* (New York) (3 November 1897), 211.

41. On international comments, see "Horseless cabs in New York," *AJ* (May 1897), 327; Beaumont, *Motor vehicles and motors* (1902), 433. The rate was $1 for the first two miles or part thereof, and 50 dollar cents for each following mile, with a maximum of two passengers. For the rental car the rate was $1 per hour. "Electric motor-cab service in New York City—II," *EW* (21 August 1897), 213–16, here: 216; on results, see "Work of electric hansom in New York," *HA* (July 1897), 12 (my italics).

42. "Electric motor-cab service in New York City—II," 216; on exceeding expectations, see "The cabs and central station of the Electric Vehicle Company of New York," *EE* (New York), 204–11, here: 204; on the EVC, see "Big electric cab company formed," *HA* (September 1897), 2; 24 September, "certificates of incorporation," "Electric Vehicle Company," and "agreements," in Cave Collection, NAHC, Series II, Box 5, Items 1 and 3.

43. On 150 cars, see "The age of automobiles," *EE* (New York) (28 July 1898), 87; on fifty cars, see Kirsch, "The electric car and the burden of history," 111–12.

44. "Central station of the Electric Vehicle Company," *HA* (September 1898), 9–17, here: 9; on twenty cars, see Kirsch, "The electric car and the burden of history," 115; Vieweg: "Snow and mud pictures," *HA* (December 1898), 14–15, here: 14.

45. "The cabs and central station of the Electric Vehicle Company of New York," 204; "The new station of the Electric Vehicle Company," *EW* (3 September 1898), 227–32; "Central station of the Electric Vehicle Company," *HA* (September 1898), 9–17; "Big electric cab company formed," 2; "Lead cab activity in Boston," *HA* (24 January 1900), 13; quotations: "Our electric cab station," *HA* (September 1898), 7, 15.

46. "The cabs and central station of the Electric Vehicle Company of New York," 211; "New York Electrical Society," *ER* (New York) (22 February 1899), 119.

47. Beaumont, *Motor vehicles and motors* (1902), 424, even gives 56 percent for the battery/car weight proportion; *ER* (New York) (23 February 1901), 253; "Horseless car-

riage motors for the Electric Vehicle Co., New York," *EE* (New York) (6 October 1898), 333–34; on Diamond tires, see Kirsch, "The electric car and the burden of history," 115.

48. "Progress of the Electric Vehicle Co.," *HA* (January 1899), 13; Kirsch, "The electric car and the burden of history," 119–21; quotation: 121; my thanks to David Kirsch for making his notes with these daily records available.

49. Condict, "The motor vehicle in commercial operation," *EE* (New York) (2 March 1899), 252–53, here: 253; on eight manufacturers, see Kirsch, "The electric car and the burden of history," 115; interview with *India Rubber World* quoted in "Automobile tires," *EE* (New York) (16 February 1899), 206; "First pneumatic automobile tires," Cave Collection, NAHC, Series I, Box 3, Item 24.

50. "Penna. Electric Vehicle Co.," *EE* (New York) (2 March 1899), 265; "More electric cabs for New York," *HA* (February 1899), 13.

51. "New York Electric Vehicle Transportation Co.," *EE* (New York) (2 March 1899), 265; on the "giant trust," see Kirsch, "The electric car and the burden of history," 125; certificate of incorporation in Cave Collection, NAHC, Series II, Box 5, Item 3.

52. "A new and powerful electric storage battery company," *ER* (New York) (12 April 1899), 231; "An electric vehicle company in Chicago," *ER* (New York) (3 May 1899), 279; "Automobiles," *ER* (New York) (23 May 1900), 550.

53. "Electric cab service for the New York Central Railroad," *ER* (New York) (31 May 1899), 347; "New public electromobiles in New York City," *ER* (New York) (11 April 1900), 362; "General Carriage Company's electromobile rates in New York City," *ER* (New York) (2 May 1900), 464; "The last of the car horse," *ER* (New York) (6 June 1900), 582; "The General Carriage Company," *HA* (17 May 1899), 8. The bankruptcy in 1902 of the General Carriage Company apparently included forty-five electrics ("Sheriff's sale of forty-five electric vehicles," *HA* [5 March 1902], 305; *HA* [12 March 1902], 341); James Rood Doolittle (ed.), *The romance of the automobile industry; being the story . . .* (New York, 1916), 357–58; on Kriéger scheme, see "Plans of the New York Autotruck Co.," *EE* (New York) (5 January 1899), 24; "A new automobile trust," *ER* (New York) (8 November 1899), 297; "Krieger cab rights bought by the Autotruck Co.," *HA* (10 May 1899), 9; "Storage battery war," *HA* (14 June 1899), 6.

54. "Lead cab financiering," *HA* (21 January 1900), 11; H. E. Cuntz, "Harold Hayden Eames" (1940), Cave Collection, NAHC, Series I, Box 3, Item 1; idem, "The Electric Vehicle Company," 3 (Cave Collection, Series III, Box 10, Item 31); idem, "Story of the Selden case and Hartford," 3–8 (Cave Collection, Series III, Box 10, Item 28); William Greenleaf, *Monopoly on wheels: Henry Ford and the Selden automobile patent* (Detroit, 1961), 57ff.

55. Brian J. Cudahy, *Cash, tokens and transfers: A history of urban mass transit in North America* (New York, 1990), 45–46; Charles W. Cheape, *Moving the masses: Urban public transit in New York, Boston, and Philadelphia, 1880–1912* (Cambridge, Mass., 1980), 45ff.; Andrew C. Irvine, "The promotion and first twenty-two years' history of a corporation in the electrical manufacturing industry" (unpublished master's thesis, Temple University, 1954), 60ff.

56. Richard H. Schallenberg, *Bottled energy: Electrical engineering and the evolution of chemical energy storage* (Philadelphia, 1982), 260; Greenleaf, *Monopoly on wheels*, 57ff.

57. On the worldwide network, see Schallenberg, *Bottled energy*, 260; for the following, see also "Annual statement of the Electric Vehicle Company," *HA* (27 September 1899), 7–8; Greenleaf, *Monopoly on wheels*, 65ff.

58. "Census reports on automobile manufacture in 1900," *HA* (17 September 1902), 307; "Electric Vehicle Companies for every state and territory," *HA* (24 May 1899), 11; "A $200,000,000 enterprise," *HA* (10 May 1899), 6.

59. "Four thousand two hundred electromobiles ordered," *ER* (New York) (19 July 1899), 34; 12,000: "The Electric Vehicle Co. spreads out," *HA* (12 April 1899), 11; 8,000 a year: "Annual statement of the Electric Vehicle Company," 8; Riker: "The Electric Vehicle Company's factory," *ER* (New York) (24 October 1900), 442–43, here: 443; "Automobiles," *ER* (New York) (4 July 1900), 6; "Electric Vehicle companies consolidated," *ER* (New York) (12 December 1900), 651; "Electric Vehicle Company absorbs the Riker company," *HA* (12 December 1900), 14.

60. "The new repair department and charging station of the New York Electric Vehicle Transportation Company," *ER* (New York) (28 March 1900), 319; "The New York Electric Vehicle Transportation Company's new station," *ER* (New York) (4 May 1901), 545–48, quotation: 548.

61. "A $200,000,000 enterprise," 6; "Public electromobile cabs in Chicago," *ER* (New York) (6 September 1899), 156; on the charging station in 1898, see Harold L. Platt, *The Electric City: Energy and the growth of the Chicago area, 1880–1930* (Chicago, 1991), 159.

62. "An electromobile railway cab service," *ER* (New York) (28 March 1900), 323; "The Washington Electric Vehicle Transportation Company," *ER* (New York) (27 June 1900), 665.

63. "Electromobiles in New England," *ER* (New York) (12 July 1899), 23; "Lead cab activity in Boston," *HA* (24 January 1900), 13.

64. "Automobiles," *ER* (New York) (11 April 1900), 371; "Proposed dissolution of the New England Electric Vehicle Company," *ER* (New York) (6 April 1901), 422; "Doings in lead cabdom," *HA* (7 March 1900), 19; "Automobiles," *ER* (New York) (2 May 1900), 450; "Automobile news," *ER* (New York) (16 August 1899), 103; "Automobiles," *ER* (New York) (10 October 1900), 376.

65. R. A. Fliess, "The electric automobile for business purposes," *HA* (6 February 1901), 45–51, here: 50; the number for Boston is confirmed by the 245 cars that were present in the bankrupt's estate ("Lead cab funerals," *HA* [3 April 1901], 5); 109 in Chicago: "Illinois Electric Vehicle Transportation Company," *HA* (16 March 1901), 354.

66. Letter from "WFK" to EVC, 21 November 1899, Crane Collection, Box 2, Folder 8; letter from Condict to W. H. Johnson, 27 November 1899, Crane Collection, Box 1, Folder 8. My thanks to David Kirsch for making his notes available, of which the given data have been extracted. On construction changes, see G. Pellissier, "Les fiacres électriques de New-York," *Éclairage électrique* (13 January 1900), 42–50, here: 47; on wooden wheels, see Beaumont, *Motor vehicles and motors* (1902), 424; on 40 percent batteries, see "The new repair department and charging station of the New York Electric Vehicle Transportation Company," *ER* (New York) (28 March 1900), 319; on tires, see "Automobiles," *ER* (New York) (10 October 1900), 376.

67. Kirsch, "The electric car and the burden of history," 142–45.

68. Greenleaf, *Monopoly on wheels*, 71ff.; "The General Carriage Company," *HA* (17 May 1899), 8; "A discredited expert," *HA* (18 October 1899), 5–7, here: 5.

69. "A British view of accumulators and electric vehicles," *ER* (New York) (20 December 1899), 393; "Lead cab finale," *HA* (20 December 1899), 8; on a dividend of 8 percent, see John B. Rae, "The Electric Vehicle Company: A monopoly that missed," *Business history review* 29, no. 4 (December 1955), 303.

70. "Lead cab financiering," *HA* (24 January 1900), 11; "A short history of the electric taxi-cab," *EV* (U.S.) (July 1916), 2.

71. "We draw the line," *HA* (7 February 1900), 9–13; "Get together!" *HA* (27 June 1900), 9; Greenleaf, *Monopoly on wheels*, 73; "Automobile notes," *ER* (New York) (6 July 1901), 28.

72. On *Motor Age*, see Kirsch, "The electric car and the burden of history," 135.

73. "Illinois Electric Vehicle Transportation Company," *ER* (New York) (16 March 1901), 354; "Lead cab funerals," *HA* (3 April 1901), 5.

74. Irvine, "The promotion," 64, 82–87; on 95 percent, see *Centralblatt* (1 May 1903), 114; *Centralblatt* (1 May 1902), 123; Isaac L. Rice, "To the stockholders of the Electric Storage Battery Company," Cave Collection, NAHC, Series II, Box 5, Item 12.

75. "New York Electric Vehicle Company," *ER* (New York) (21 February 1900), 188; "The Electric Vehicle Company's electromobiles," *ER* (New York) (28 March 1900), 313; on electric stages, see "Lead bluff in Boston," *HA* (27 December 1899), 10; on an exclusive license, see "An electromobile omnibus," *ER* (New York) (28 March 1900), 326. The only conclusive evidence of a real taxicab field of application in the contemporary press is given by Boston, where, in the heavy snow storms of January 1900, the cab drivers (in coupes and not in the usual hansoms with their open fronts) were instructed "to accept no calls for long-distance service, and to return to the re-charging station after every three or four city calls" ("Lead cab activity in Boston," *HA* [24 January 1900], 13). David Kirsch also found an indication for the application of taxicabs in archives that had not been used before. This was in Atlantic City, the subsidiary of the EVC company in Boston that suffered heavy losses, where only *two* hansoms were operative in the summer of 1900 (Kirsch, "The electric car and the burden of history," 141–42).

76. Schallenberg, *Bottled energy*, 269; *Centralblatt* 81 (1902), 243.

77. Cuntz, "The Electric Vehicle Company," 12; on Condict as director, see "Automobile notes," *ER* (New York) (18 May 1901), 637; "Lead cab company reduces stock," *HA* (15 January 1902), 67. The Pennsylvania Electric Vehicle Company also continued to exist: it was still mentioned in the ESB's annual report of 1902; at the time the ESB owned a third of the shares (*Centralblatt* 9 [1903], 114).

78. Rae, "The Electric Vehicle Company: A monopoly that missed," 309; on the date of bankruptcy, see Cave Collection, NAHC, Series II, Box 5, Item 10; on one of the first, see Greenleaf, *Monopoly on wheels*, 185.

79. Greenleaf, *Monopoly on wheels*, 63.

80. Schallenberg, *Bottled energy*, 264.

81. John B. Rae, "The Electric Vehicle Company: A monopoly that missed," 293–311, here: 301–3. According to Rae, the EVC was an "illuminating example of business failure" based upon "errors of judgment, most of them . . . avoidable." In short, the EVC, by opting for the electric propulsion, bet "on the wrong horse." For a criticism of Rae's conclusions, see David A. Kirsch and Gijs P. A. Mom, "Visions of transportation: The EVC and the transition from service- to product-based mobility," *Business history review* 76 (Spring 2002), 75–110.

Chapter 3: Horse Power

"The importance of the storage battery," *Electrical review* (New York) (21 June 1902), 815–16, here: 815.

1. Michael Brian Schiffer (with Tamara C. Butts and Kimberly K. Grim), *Taking charge: The electric automobile in America* (Washington, D.C., 1994), 73.

2. Albert L. Clough, "The retarding effect of promised perfection," *The horseless age* (13 August 1902), 159.

3. Karl A. Kuhn, *Die Opfer des Automobils* (Berlin, 1907), 10–11.

4. Ch. De Sarcy, "Les stations de charge," *La France automobile* (12 April 1900), 225–26; Hart O. Berg, "Outlook for American automobiles abroad," *ER* (New York) (19 January 1901), 110; "La voiture électrique," in *Grand album illustré de l'industrie automobile pour l'année 1902* (Paris, n.d. [1902]), 32–33.

5. Profits that had amounted to about 209,000 francs in 1902 gradually increased

during the following years to 525,000 francs in 1905. But the year after, profits suddenly plunged to 197,000. Société des Garages Kriéger et Brasier S. A., *Statuts* (Paris, 1905), AN 65 AQ N61; *Supplément à la Cote du Marché des Banquiers en Valeurs au Comptant du 2 Avril 1906*, AN 65 AQ N61; Compagnie Parisienne des voitures Électriques (Procédés Kriéger), *Assemblée Générale Extraordinaire* (4 March 1905), AN 65 AQ N30; James M. Laux, *In first gear: The French automobile industry to 1914* (Liverpool, 1976), 92; "Garages Kriéger et Brasier," *Revue financière* (12 July 1906), and "Société des Garages Kriéger et Brasier," *Informations* (22 July 1906), both in AN 65 AQ N61.

6. H. E. Dick to Edison, 22 September, 1903 (Edison National Historic Site); "The future of the electric automobile," *ER* (New York) (15 April 1905), 620; first quotation: W. Worby Beaumont, *Motor vehicles and motors: Their design construction and working by steam oil and electricity*, 2nd ed. (Westminster, 1902), 557.

7. For the next two paragraphs, see Harry W. Perry, "Public passenger service in New York City," *The commercial vehicle* (June 1906), 124–29, here: 128; Harry W. Perry, "Garage facilities for New York passenger service," *CV* (July 1906), 151–60.

8. Hiram Percy Maxim, "The serious side of the automobile," 4 (undated lecture, Maxim Collection, Connecticut State Library).

9. *The Car* quoted by: Lord Montagu of Beaulieu and F. Wilson McComb, *Behind the wheel: The magic and manners of early motoring* (New York, 1977), 125; Clay McShane, *Down the asphalt path: The automobile and the American city* (New York, 1994), 162–63.

10. Scharff's figures actually concern a later period, but indirectly allow a conclusion about the first phase because the lists were apparently kept in historical order. Virginia Scharff, *Taking the wheel: Women and the coming of the motor age* (New York, 1991), 44. Regarding Washington: "Many young women of Washington, D.C., in diplomatic circles, are expert chauffeuses, driving their own automobiles with considerable skill, handling them like veterans" ("Automobiles," *ER* [New York] [5 December 1900], 602).

11. Mercedes archives, Folder 290, "Elektro-Daimler." The Mercedes Electrique list must have been drawn up after 1906 because of the delivery to the Bedag, mentioned in the list; it numbers 153 cars sold to private customers, 15 of whom were women. On Lohner, see Thomas Köppen, "Ferdinand Porsche, Ludwig Lohner und Emil Jellinek-frühe Innovatoren im Elektromobilbau: eine Falstudie über eine gescheiterte Innovation" (unpublished master's thesis, TU Berlin, 1987), Appendix, x–xx. Dorothy Levitt, *The woman and the car: A chatty little handbook for all women who motor or who want to motor* (London, 1909), 85, states in 1909 that "there is no country in the world in which woman may be seen at the helm of a motor-car so frequently as in England." She suggests a few reasons for this ("a greater sense of security from annoyance on public roads or simply . . . superiority of pluck") and also concludes that the same goes for horsemanship. For both Europe and the United States much more effort should be dedicated to unearthing further evidence of very early automobile use on the state, county, and even local levels. For the United States, David Kirsch made a first start with such an investigation in David A. Kirsch, "Behind the numbers: Early quantitative data on the history of the American automobile industry," paper presented at the ICOHTEC Symposium, Prague, 24 August 2000.

The thesis in the text about the few numbers of women in early automobile use and culture does not imply that women's indirect influence was negligible, as passive users (passengers), as part of the family decision process regarding mobility choices, and even as nonusers, but to assess this, much more sophisticated and subtle methods should be used, such as picture analysis and close reading of diaries and other micro documents than has been hitherto the case in automotive historiography.

12. *ER* (New York) (15 August 1900), 146.

13. "Automobiles for physician's use: Are they practical? are they desirable? are they economical? are they better than horses?" *Journal of the American Medical Association* 46, part 2, no. 14 (7 April 1906), 1172–1207 (also for the next two paragraphs).

14. All in all the poll confirmed the conclusion of automotive historian James Flink that the medical doctors in the small towns and villages were pioneers in bringing the gasoline car to the countryside (they were usually the first in their community to purchase a car). But it does not in any way affect our conclusion that the early car primarily was a phenomenon of the big city. Even as late as 1916, the more than 6 million American farmers' families only bought 300,000 cars, not even 20 percent of the total sales of more than 1.5 million that year. James J. Flink, *The automobile age* (Cambridge, Mass., 1993), 28; on 6 million farmers' families, see H. L. Barber, *Story of the automobile* (Chicago, 1917), 124; on total sales in 1916, see *Automobiles of America* (Detroit, 1961), 104.

15. Charles P. Kindleberger, *Manias, panics, and crashes: A history of financial crises* (London, 1978), 133–34.

16. James M. Laux, *The European automobile industry* (New York, 1992), 53, 18, 48; Gustave Rives, "La date du salon et la crise de l'automobile," *La vie automobile* (9 November 1907), 703–7; C. Faroux, "Les tendances nouvelles," *VA* (23 November 1907), 737–38; John B. Rae, *The American automobile: A brief history* (Chicago, 1965), 42.

17. Gerhard Horras, *Die Entwicklung des deutschen Automobilmarktes bis 1914* (Munich, 1982), 151; Immo Sievers, *AutoCars: Die Beziehungen zwischen der englischen und der deutschen Automobilindustrie vor dem Ersten Weltkrieg* (Frankfurt am Main, 1995), 89–90; "Die Automobil-Krise in Frankreich," *Stahlrad und Automobil* (21 February 1909), 1–3; Laux, *The European automobile industry*, 19–20.

18. Donald Finlay Davis, *Conspicuous production: Automobiles and elites in Detroit, 1899–1933* (Philadelphia, 1988), 43; Michael L. Berger, *The devil wagon in God's country: The automobile and social change in rural America, 1893–1929* (Hamden, Conn., 1979), 33.

19. Angela Zatsch, *Staatsmacht und Motorisierung am Morgen des Automobilzeitalters* (Konstanz, 1993), 227–35; "Schädigende Ereignisse im Kraftfahrzeugbetrieb 1909/1910," *SA* (16 April 1911) (no page numbers indicated). The long working days were partly responsible for the high "accident proneness" of urban coachmen. A German survey mentioned, for example, that 83 percent of all horse-bus coachmen worked more than sixteen hours a day. Martin Scharfe, "'Ungebundene Circulation der Individuen': Aspekte des Automobilfahrens in der Frühzeit," *Zeitschrift für Volkskunde* 86 (1990), 240.

20. McShane, *Down the asphalt path*, 176.

21. "Tissues pour pneumatiques," *Le caoutchouc & la gutta-percha* (15 March 1907), 902–6, here: 902; Montagu of Beaulieu and McComb, *Behind the wheel*, 70, 76; "Tires at the New York automobile show," *IRW* (1 February 1904), 159–60.

22. C. O. Weber, "Technical notes on motor tyres," *IRW* (29 February 1904), 79–80; Eric Tompkins, *The history of the pneumatic tyre* (n.p. [Birmingham], 1981), 21; Tom French, *Tyre technology* (Bristol, 1988), 14–15; Michelin: H. Petit, "Comment on doit se conduire avec les pneus," *VA* (13 April 1912), 235–37, here: 235.

23. L. Baudry de Saunier, "La mort par le pneumatique! . . . ," *VA* (2 July 1904), 417, 418, and its follow-up articles: (9 July 1904), 441–43, (16 July 1904), 453–55, (30 July 1904), 488–90 ("killer tire": 489), (6 August 1904), 505.

24. L. Baudry de Saunier, "La roue élastique," *VA* (8 April 1905), 216–19; Léon Overnoy, "La roue élastique De Cadignan," *VA* (15 October 1904), 668–69. For an overview of 140 designs between 1906 and 1909, see J. Rutishauser, *Roues élastiques* (Paris, 1941); on the Achilles' heel, see "Die Pneumatikfrage bei künftigen Konkurren-

zen," *Allgemeine Automobil-Zeitung* (Berlin) (17 November 1905), 28–32, here: 28; on the serious competitor, see "La roue élastique de Cadignan," *CGP* (15 February 1905), 52; on spring-absorber combinations, see "Les roues élastiques rivales des pneus," *CGP* (15 December 1905), 380; on absorption of the obstacle, see Mortimer-Mégret, "La vie en auto," *CGP* (15 January 1910), 3603–4.

25. C. Faroux, "Et le cours du caoutchouc montait toujours: le pneu tuerait-il l'automobile?" *VA* (23 April 1910), 257–58; on the crash of raw rubber, see "Le Krach des Caoutchoucs," *CGP* (15 October 1910), 4512; on tire prices, see Horras, *Die Entwicklung des deutschen Automobilmarktes*, 201–5; Zatsch, *Staatsmacht und Motorisierung*, 61.

26. Horras, *Die Entwicklung des deutschen Automobilmarktes*, 280–84, 302–3.

27. Horras, *Die Entwicklung des deutschen Automobilmarktes*, 280; on the "wooden run," see David Gartman, *Auto opium: A social history of American automobile design* (London, 1994), 21.

28. R. McAllister Lloyd, "The influence of the pioneer spirit on electric vehicle progress," *CS* (December 1914), 179–81, here: 180.

29. New York City at the time ranked second in terms of number of automobiles (Paris had the most). In 1905, 45 percent of the American cars were found in the states of New York and New Jersey, and most of them were in New York City. As late as 1910, automobile density per resident in this city was higher than the national average. Mc-Shane, *Down the asphalt path*, 174.

30. "Automobilverkehr in Paris," *AAZ* (Berlin) (21 April 1905), 51; 600: *EZ* (24 November 1904), 997.

31. David A. Kirsch, "The electric car and the burden of history: Studies in automotive systems rivalry in America, 1890–1996" (unpublished diss., Stanford University, 1996), 61, note 32; William Greenleaf, *Monopoly on wheels: Henry Ford and the Selden automobile patent* (Detroit, 1961), 178.

32. "Paris-Trouville en électrique," *LA* (24 August 1905), 88–89; Paul Bary, "Voiture électrique Védrine," *IE* (25 January 1906), 37–41; "Traction électrique: Gallia et Galliette," *IE* (26 October 1905), 217–20; Laux, *In first gear*, 94; Kriéger: "Les véhicules industrielles," *La journée industrielle* (2 December 1924). My thanks to V. Christian Manz, Madrid, for drawing my attention to this source.

33. "Kriéger Automobil-A.-G.," *ETZ* (5 October 1905), 938; Compagnie Parisienne des voitures Electriques (Procédés Kriéger), *Assemblée Générale Extraordinaire* (4 March 1905), AN 65 AQ N30; "Norddeutsche Automobil-& Motoren-Aktien-Gesellschaft," *Centralblatt* (5 July 1906), 180; Wolfgang Gebhardt, *Deutsche Omnibusse seit 1895* (Stuttgart, n.d. [1996]), 23; on Kriéger production, see "Les véhicules industriels."

34. Horras, *Die Entwicklung des deutschen Automobilmarktes*, 171.

35. *MW* (10 February 1906), 133, (10 March 1906), 210, (30 May 1906), 420, (30 June 1906), 500, (10 September 1906), 672, (31 December 1906), 1048, (20 January 1907), 46, 47, (20 February 1907), 137; *AAZ* (Berlin) 41 (1907), 71; *SA* 7 (1907), 28, 7 (1908) 22; *Centralblatt* (5 December 1908), 123. See also Norddeutsche Automobil-und Motoren-gesellschaft, "Norddeutsche Automobil- und Motoren-Aktiengesellschaft Bremen-Hastedt," in Alexander Engel (ed.), *Historisch-biographische Blätter der Staat Bremen* (Berlin, 1900–11), Bd. 3, 847–60 and 879–82; Klaus Brandhuber, *Die Insolvenz eines Familienkonzernes: Der wirtschaftliche Niedergang der BORGWARD-GRUPPE* (Köln, 1988) (diss.), 10–12.

36. "Das Kriéger-Automobil," *AAZ* (Berlin) (24 November 1905), 59–61.

37. Hans-Otto Neubauer, *Autos aus Berlin: Protos und NAG* (Stuttgart, 1983), 30–31, 83–97, 112; Horras, *Die Entwicklung des deutschen Automobilmarktes*, 167–70; W. A.Th. Müller, "Die Elektromobile auf der Internationalen Automobil-Ausstellung

Berlin 1905," *Centralblatt* (1 March 1905), 47–50, here: 48–49; on 1,500 cars, see *150 Jahre Gottfried Hagen* (Köln-Kalk, n.d. [1977]) (brochure, Varta archives) (no page numbers).

38. Köppen, "Ferdinand Porsche, Ludwig Lohner und Emil Jellinek," 69–76; Hans Seper, "Daimler und Benz in Österreich" (manuscript), Mercedes archives, Ordner "Beteiligungen Österreich," part 1/3, 70–74.

39. Köppen, "Ferdinand Porsche, Ludwig Lohner und Emil Jellinek," 76, and Appendix, x–xx; "Mercedes-Elektrique (Lohner-Porsche)" (printed list of the Daimler-Motoren-Gesellschaft, DaimlerChrysler archives, Folder 290, "Elektro-Daimler").

40. Schiffer, *Taking charge*, 115–16.

41. *Minutes of the meeting of the Association of Electric Vehicle Manufacturers, held at the Hotel Belmont, 42nd Street & Park Avenue, Borough of Manhattan, New York City, December 6, 1906, at 10 o'clock a.m.*, 30.

42. Kirsch, "The electric car and the burden of history," 118.

43. "The Studebaker electric vehicle," *HA* (30 July 1902), 129–30; "Electric automobiles," *ER* (19 November 1904), 714–16; Donald T. Critchlow, "Studebaker Corporation," in George S. May (ed.), *The automobile industry, 1896–1920* (New York, 1990), 433–37; John B. Rae, *American automobile manufacturers: The first forty years* (Philadelphia, 1959), 16, 93; Chris Sinsabaugh, *Who me? Forty years of automobile history* (Detroit, 1940), 154. On 1,000 vehicles, see Donald T. Critchlow, *Studebaker, the life and death of an American corporation* (Bloomington, 1996), 46.

44. Rae, *American automobile manufacturers*, 38; Richard Wager, *Golden wheels: The story of the automobiles made in Cleveland and Northeastern Ohio 1892–1932* (Cleveland, 1986), 208; quotations, 213–15.

45. George S. May, "Detroit Electric Car Company," in May (ed.), *The automotive industry, 1896–1920*, 123–24; Tom LaMarre, "Detroit Electric: Society's town car," *Automobile quarterly* 27, no. 2 (second quarter 1989), 161; Sinsabaugh, *Who, me?* 155.

46. "The automobile," *ER* (16 November 1901), 598; "A new departure in storage batteries," *HA* (17 April 1901), 52–53; "Another departure in storage batteries," *HA* (12 February 1902), 194. According to Schallenberg (*Bottled energy: Electrical engineering and the evolution of chemical energy storage* [Philadelphia, 1982], 351), Thomas Edison's sudden preference for the electric car was also due to his hope that in case of a large-scale implementation the switch to alternating current would be slowed down. On Jeantaud, see "L'accumulateur E. I.t.," *IE* (10 February 1905), 49–50; E. Hospitalier, "L'Accumulateur E. I.t.," *VA* (11 February 1905), 81–83; Georges Rosset, "L'accumulateur E. I.t. au plomb allotropique," *VA* (25 February 1905), 122–24.

47. *IE* (10 February 1905), 49.

48. For the following, see Gijs Mom, "Inventing the miracle battery: Thomas Edison and the electric vehicle," in *History of technology* 20 (1998), 17–45; and Gijs Mom, "Das Holzbrettchen in der schwarzen Kiste: Die Entwicklung des Elektromobilakkumulators bei und aus der Sicht der Accumulatoren-Fabrik AG (AFA) von 1902–1910," *Technikgeschichte* 63, no. 2 (1996), 119–51.

49. "Storage battery traction," *ER* (8 June 1901), 715; "Annual Meeting of the American Institute of Electrical Engineers," *ER* (25 May 1901), 665–69; on "Old Man," see Schiffer, *Taking charge*, 78.

50. "Automobile exhibit in New York," *ER* (9 November 1901), 582; "Successful automobile trip with an Edison storage battery," *ER* (31 October 1903); "A new auto record," *ER* (14 November 1903), 699.

51. "Edison contra Jungner," *HA* (28 January 1903), 189–90; "Die electrische Sensation," *AAZ* 1 (Berlin) (1904), 25–26; Schiffer, *Taking charge*, 77–78.

52. "The situation regarding the Edison storage battery," *ER* (8 August 1903), 198–99, here: 198; *Bulletin of the New York Edison Company* (August 1906), 114–15. Horse power is the unity of power, not of torque.

53. "Fourth annual automobile show at Madison Square Garden," *ER* (23 January 1904), 147.

54. E. Sieg, "Die letzten Neuerungen auf dem Gebiete transportabler Akkumulatoren, insbesondere alkalische Sammler (Jungner-Edison) mit Demonstration" (lecture at the Elektrotechnische Gesellschaft in Cologne), *ETZ* (30 March 1905), 311–13.

55. Frank Lewis Dyer and Thomas Commerford Martin, *Edison: His life and inventions* (New York, 1910), 754; *MW* (15 February 1902), v (advertisement); *ETZ* (16 January 1908), 56–57; K. Perlewitz, "Der Edison-Akkumulator," *ETZ* (29 October 1908), 1061–63, and (17 December 1908), 1232–33; *ETZ* (3 June 1909), 532–33. Quotation: Schallenberg, *Bottled energy*, 369; Garvin: "Eine Edison-Akkumulatoren-Gesellschaft in Berlin," *AAZ* (Berlin) (10 March 1905), 55; LK (30 April 1906), 41–42 (Varta archives).

56. Schallenberg, *Bottled energy*, 371.

57. "A new type of storage battery," *ER* (New York) (15 June 1901), 756; "Elektrische automobile Postkutschen in New-York," *Centralblatt* 8 (1902), 104; Schallenberg, *Bottled energy*, 269 (life span), 371; "The expiration of the Brush storage battery patent-effect on the storage battery industry," *ER* (28 February 1903), 293.

58. "A review of the storage battery situation for the year 1904," *ER* (14 January 1905), 67.

59. V. Heinz and V. Klement, *Z dejin automobilu* (Praha, 1931), 194–96 (the page numbers refer to the German translation, in manuscript, in the DaimlerChrysler archives, Stuttgart). My thanks to Stanislav Peschel of the DaimlerChrysler Archives for drawing my attention to this translation. Édouard Hospitalier, "L'Auto-Mixte," *VA* (22 December 1906), 827–30; A. Delasalle, "Voiture pétro-électrique des Etablissements Pieper, de Liège," *LA* (21 September 1899), 604–6; LK (5 March 1906), 37; LK (30 April 1906), 37; AK (17 July 1906), 54–55 (all Varta archives).

60. "Krieger's combination vehicle," *HA* (6 August 1902), 146; H. de Graffigny, *La locomotion électrique* (Paris, n.d.), 47.

61. The history of the hybrid car has yet to be written. For a summation of most early designs, see Ernest H. Wakefield, *History of the electric automobile: Hybrid electric vehicles* (Warrendale, 1998).

62. J. L. Krieger, "Un siècle de véhicules électriques en France (suite)," *La Berline* (newsletter of the "Amis du Musée de Compiègne") (February 1995), 5; patent no. 362976 of 3 February 1906 "for a vehicle with gas turbine and electric transmission." My thanks to Jean-Luc Krieger, Louis's grandson, for making available the sources mentioned here. Friedr. Warschauer, "Die Elektromobilen im Pariser Salon," *AAZ* 1 (Berlin) (1906), 22–45, here: 43–45; *La revue financière* (15 September 1906), 691 (AN 65 AQ N30); Compagnie Parisienne des voitures électriques (Procédés Kriéger), *Sixième Exercise, Assemblée Générale ordinaire et extraordinaire*, 8, 16 (AN 65 AQ N30).

63. Erwin Steinböck, *Lohner: zu Land, zu Wasser und in der Luft* (Graz, 1984), 29.

64. On Jenatzy, see "Transmission progressive pour automobiles," *LA* (3 September 1903), 571–72; on Kriéger, see Maurice Braun, "En faveur des accumulateurs," *VA* (29 December 1906), 844–45.

65. *ETZ* (2 March 1905), 213–14; R. de Valbreuze, "Quelques notions générales sur les voitures électriques (suite)," *VA* (9 June 1906), 363–67, here: 363.

66. P. M. Heldt, "Electric vehicles at the Madison Square Garden Show," *HA* (7 November 1900), 42–45, here: 43; "Brush storage battery patent expires," *HA* (4 March

1903), 312; W. H. Maxwell, "Electromobile evolution," *ER* (New York) (12 January 1901), 51–75, here: 52–53.

67. "Neuerungen und Fortschritte auf der Berliner Automobil-Ausstellung 1905," *AAZ* (Berlin) (24 February 1905), 28–32, here: 32; E. Sieg, "Über elektrische Automobilmotoren," *ETZ* (9 January 1908), 39–40, here: 40; on the "overloaded gasoline engine," see E. Sieg, "Über Elektromobile," *ETZ* (24 December 1908), 1238–43, here: 1241.

68. W. Poynter Adams, *Electric and petrol-electric vehicles* (London, 1908), 47; E. Sieg, "Über Elektromobile," *ETZ* (24 December 1908), 1238–43, here: 1240; "Elektromobile der Siemens-Schuckert Werke G.M.B.H.," *Centralblatt* (20 August 1908), 121–23.

69. Maurice Sainturat, "Sur la décadence des voitures électriques," *VA* (17 April 1909), 247–48, and *VA* (22 May 1909); on the poll, see "Notre référendum," *VA* (14 August 1909), 513–14.

70. "Les véhicules industrielles," *La journée industrielle* (2 December 1924); P. Gasnier, "l'Électricité dans les voitures automobiles: voitures électriques; voitures pétroléo-électriques," *IE* (10 January 1908), 6–10; *SA* (23 February 1908), 20; Krieger, "Un siècle de véhicules électrique en France (suite)," 5; Friedrich Schildberger, "Die Entstehung der Automobilindustrie in Frankreich" (manuscript DaimlerChrysler archives, Stuttgart), 214.

71. Laux, *In first gear*, 93; *Globe* (12 December 1907); *SA* (23 February 1908), 20; on the date of liquidation, see *Journal des actionnaires* (?, illegible title on clipping, AN 65 AQ N30). The Société des Garages Kriéger et Brasier was liquidated in 1914 (AN 65 AQ N61); on the merger, see *International dictionary on world automobile technicians* (brochure International Historical Commission of the FIA, Paris, October 1992), 32; on 400/2,000, see J. A. Grégoire, *50 ans d'automobile; 2: la voiture électrique* (Paris, 1981), 116.

72. *AAZ* 9 (Berlin) (1911), 55; *SA* (19 June 1910), 14–15; Hansa-Automobil-Werke Aktiengesellschaft Varel-Oldenburg, *Bericht über das Geschäftjahr 1913 für die erste ordentliche Generalversammlung am 22. Mai 1914*; idem, *über das Geschäftsjahr 1914 für die ordentliche Generalversammlung am 16. Juli 1915*. Both in archive 4, 75/5 HRB 178, Staatsarchiv Bremen.

73. "Electric Vehicle settlement," *ER* (New York) (26 June 1909), 1200; "Electric Vehicle reorganization," *ER* (New York) (10 July 1909), 78; quotation: "Electric Vehicle Company's affairs," *ER* (New York) (19 June 1909), 1158; date of insolvency proceedings, Cave Collection, Box 5, Item 11.

74. Critchlow, "Studebaker Corporation," 436; Schallenberg, *Bottled energy*, 288; "Wisconsin Electrical Association," *ER* (New York) (29 January 1910), 238ff., here: 240.

Chapter 4: The Trojan Horse

Ansbert Vorreiter, "Motordroschken und deren Betriebskosten," *MW* (31 December 1904), 573–75, here: 573.

1. Alexander Lang, "Die finanzielle Verwertung des Motordroschkenverkehrs," *SA* 3 (1906), 7–8, here: 7.

2. "L'automobile industriel: Les fiacres automobiles à Berlin," *La Locomotion automobile* (7 September 1906), 149–50, here: 149; "Compagnie française des Automobiles de Place," *Circulaire financière* (1 April 1906), AN 65 AQ N26; Anne Boudou, "Les taxis parisiens de la fondation des Usines Renault aux 'Taxis de la Marne' 1898–1914" (unpublished master's thesis, Université Paris X Nanterre, 1982), 67–70, 292–302; Patrick Fridenson, "*Histoire des usines Renault: I. Naissance de la grande entreprise 1898–1939*" (Paris, 1972), 55–63.

3. Pol Ravigneaux, "Les nouveaux fiacres automobiles," *VA* (19 December 1905),

801–3; Claude Rouxel, *La grande histoire des taxis français 1898–1988* (Pontoise, 1989), 38, 112–13; René Bellu, *Toutes les Renault* (Paris, 1979), 25, 112–13.

4. Herbert Bauer, "Automobile im öffentlichen Verkehr von Paris und Berlin," *AAZ* (Vienna) (16 December 1906), 37–45, here: 40.

5. "Dans deux ans les taxi-autos remplaceront les chevaux de fiacre," *La Presse* (25 June 1907), Archives de la Préfecture de Police, Paris, box D/b 505, folder "Taxis"; Boudou, "Les taxis parisiens," 293–97; Fridenson, *Histoire des usines Renault*, 57.

6. F. Girardault, *Les automobiles industrielles* (Paris, 1910), 361–82.

7. "Voitures à Paris," *Le mouvement financier* (22 August 1908), AN 65 AQ 370³; Boudou, "Les taxis parisiens," 280.

8. "Compagnie Parisienne des Taxiautos-Electriques," *Le pour et le contre* (3 February 1907); "Compagnie Parisienne des Taxiautos Électriques," *L'Économiste Français* (2 February 1907); "Les Taxautos Électriques," *Journal des Actionnaires* (3 February 1907); "Compagnie Parisienne des Taxautos Électriques (Extrait du journal *L'information* du 22 Janvier 1907)," all in AN 65 AQ Q170.

9. Boudou, "Les taxis parisiens," 267–68, 296; on tire costs, see M. P. [Marcel Plessix], "Le fiacre automobile," *LA* (10 August 1906), 81–83, here: 82.

10. Calculated on the basis of the *Annuaire statistique de la ville de Paris* over the years concerned; Fridenson, *Histoire des usines Renault*, 56–58. According to Philippe Laneyrie and Jacques Roux ("Transport traditionnel et innovation technique: L'exemple du taxi en France," *Culture technique* 19 [March 1989], 263), Paris had 6,000 cab drivers in 1911, whereas the total private vehicle fleet in the entire Paris region was owned by hardly 20,000 proprietors. The high share of the taxicab in early automobile diffusion was not typical of just Paris. For instance, at the beginning of 1907 Berlin had 1,449 passenger cars, nearly half of which (700) were in use as taxicabs. Karl A. Kuhn, *Die Opfer des Automobils* (Berlin, 1907), 18. In Amsterdam, shortly before the First World War, the situation was similar (see the third section of this chapter).

11. Froissart, "Les fiacres automobiles de Londres," *VA* (27 June 1907), 473–75; Froissard [sic], "Les fiacres automobiles de Londres (suite et fin)," *VA* (3 August 1907), 484–85; G. N. Georgano, *A history of the London taxicab* (Newton Abbot, 1972), 59–60; Nick Georgano, *The London taxi* (n.p. [London], 1985), 7–9.

12. "London electric taxicabs," *NELA Bulletin* (March 1917), 215. The two other companies were J. Stockton and Sons with fifteen and the Landaulette Company Ltd. with twenty electric cabs.

13. "Automobildroschken in England, Amerika und Frankreich," *AAZ* (Vienna) (29 December 1907), 40; Gorman Gilbert and Robert E. Samuels, *The taxicab: An urban transportation survivor* (Chapel Hill, 1982), 34; "Taximeter cabs for New York," *CV* (May 1907), 146; Harry W. Perry, "Taximeter cabs actually in use in New York," *CV* (July 1907), 174–75; "Taximeter cab situation in New York," *CV* (October 1907), 254–56. As far as is known, Boston was the only other American city where two electric cabs were operative in this period (Howard S. Knowlton, "Side light on the electric vehicle situation," *CS* [May 1910], 244–48, here: 247).

14. Richard W. Meade, "Influence of standardization on taxicab operation," *HA* (27 July 1910), 119–20; memo from Richard W. Meade, n.d., Richard Worsam Meade Papers, Columbia University, New York, Box 15, "New York Transportation Company" Folder. Both sources quoted in David A. Kirsch and Gijs P. A. Mom, "Visions of transportation: The EVC and the transition from service- to product-based mobility," *Business history review* 76 (Spring 2002), 75–110. For the Atax company, see the third section of this chapter.

15. "More cabs for New York," *CV* (May 1908), 102.

16. Harry W. Perry, "Taxicab operators and builders in America," *CV* (February 1908), 36–38; "Taximeter made in America," *CV* (March 1908), 75–76; "Electric taxicabs for principal cities," *EV* (U.S.) (May 1915), 175–76, here: 175.

17. Russel A. Sommerville, "Taximètres américains," *VA* (29 April 1911), 268; Fridenson, *Histoire des usines Renault*, 58.

18. "New York taxicab company sold," *CV* (April 1912), 35; E. H. Ritzwoller, "Taxicab service at standstill in Chicago," *CV* (May 1910), 177–78; "Telephone system for taxicab company," *ER* (New York) (10 February 1912), 294; 65 percent: A. Jackson Marshall, "Electric taxicabs," *CS* (April 1915), 319–21, here: 320.

19. Josef Hüls, "Das deutsche Kraftdroschkengewerbe" (typescript, diss. Munich, 16 June 1931),15; Michael Evgénieff, *Droschkenbetrieb: Seine Organisation und wirtschafts-politische Probleme* (Berlin, 1934), 35–36, 71; Erwin Knaths, *Die Entwicklung des Berliner Droschkenfuhrwesens unter besonderer Berücksichtigung seiner Motorisierung: Eine verkehrs-geschichtliche Studie* (Marburg a.d. Lahn, 1929), 28.

20. Evgénieff, *Droschkenbetrieb*, 35–38; Gerhard Horras, *Die Entwicklung des deutschen Automobilmarktes bis 1914* (Munich, 1982), 167; Immo Sievers, *AutoCars: Die Beziehungen zwischen der englischen und der deutschen Automobilindustrie vor dem Ersten Weltkrieg* (Frankfurt am Main, 1995), 60, 68–69, 117; "10 Jahre NAG," *SA* (7 January 1912), 12–17; *MW* (30 April 1906), 341.

21. "Bericht über Vorträge," *Centralblatt* (15 April 1901), 128–31, here: 130; "Neue Automobil-Droschken im Berliner Verkehr," *AAZ* 49 (Berlin) (1903), 12–13; Hans Fründt, *Das Automobil und die Automobilindustrie in Deutschland* (Neustrelitz, 1911), 33.

22. "Berlin," *Centralblatt* (1 August 1904), 179; Vorreiter, "Motordroschken und deren Betriebskosten," *MW* (1905), 502–12 and 537–38.

23. "Allgemeine Betriebs-Aktiengesellschaft für Motorfahrzeuge in Köln," *MW* (1905), 326.

24. Vorreiter, "Motordroschken und deren Betriebskosten," *MW* (31 December 1904), 573–75, here: 575; Berlin: Vorreiter, "Motordroschken und deren Betriebskosten," *MW* (1905), 537–38, here: 538.

25. Th. Wolff, "Automobil-Droschken" (part II), *SA* (22 December 1907), 1–4, here: 2–3; Knaths, *Die Entwicklung des Berliner Droschkenfuhrwesens*, 29, 59–60.

26. Fritz Warschauer, "Die Elektromobilen auf der Internationalen Automobil-Ausstellung Berlin 1906, 1–13. November," *Centralblatt* (20 November 1906), 286–89; "10 Jahre NAG," *SA* (7 January 1912), 12–17, here: 14; Bruno Schweder (ed.), *Forschen und Schaffen: Beiträge der AEG zur Entwicklung der Elektrotechnik bis zum Wiederaufbau nach dem zweiten Weltkrieg*, Band 2 (Berlin, 1965), 349–56.

27. E. Sieg, "Über Elektromobile," *ETZ* (24 December 1908), 1238–43, and *ETZ* (31 December 1908), 1258–62, here: 1261–62; quotations: 1261.

28. E. Sieg, "Über elektrische Automobilmotoren," *ETZ* (9 January 1908), 39–40.

29. Sieg, "Über Elektromobile," 1262.

30. A. Heller, "Motorwagen für gewerbliche Zwecke und die Kosten ihres Be-triebes (Schluss)," *MW* (20 March 1907), 201–7; Knaths, *Die Entwicklung des Berliner Droschkenfuhrwesens*, 53–58.

31. "Geschäftliche Nachrichten," *ETZ* (10 August 1905), 765; Evgénieff, *Droschken-betrieb*, 38–41; for Abam as example, see Vorreiter, "Motordroschken und deren Be-triebskosten," 511; *SA* (1906 no. 27), 34; *IK* (Techn. Teil) (24–25 June 1907), 32 and *AK* (5 March 1907), 16 (both Varta archives); *MW* (30 May 1906), 419; on Müller, see Gijs Mom, "Das Holzbrettchen in der schwarzen Kiste: Die Entwicklung des Elektromobil-

akkumulators bei und aus der Sicht der Accumulatoren-Fabrik AG (AFA) von 1902–1910," *Technikgeschichte* 63 (1996), No. 2, 127–38; on battery development at AFA, see "Protokoll über die heutige Besprechung betreffend Automobil-Batterien (22 February 1905)," 1, and "Protokoll über die Besprechung betreffend Automobil-Batterien (11 April 1905)," 1 (Varta archives).

32. "Geschäftliche Nachrichten," *ETZ* (10 August 1905), 765; "Berlin," *MW* (31 March 1906), 263; quotation from Adolph Müller, *25 Jahre der Accumulatoren-Fabrik Aktiengesellschaft 1888–1913* (Berlin, 1913), 200.

33. "Berliner Elektromobil-Droschken Akt.-Ges.," *Centralblatt* (1 March 1906), 65; Robert Schwenke, "Die Elektromobilen auf der Berliner Automobil-Ausstellung 1906, 3–8. February," *Centralblatt* (1 March 1906), 54–57, here: 56; Herbert Bauer, "Automobile im öffentlichen Verkehr von Paris und Berlin," *AAZ* (Vienna) (16 December 1906), 37–45, here: 38; on 186 taxicabs, see Fründt, *Das Automobil und die Automobilindustrie in Deutschland,* 44; on the mixture of European cars, see AK (5 March 1907), 16.

34. W. A.Th. Müller, "Die Elektromobil-Industrie," *Centralblatt* (5 January 1908), 1–3, here: 1; *AAZ* (Berlin) (1907), No. 51, 61, suggests that the Bedag opted against the Kriéger system because of the complicated control of the two compound motors; Erwin Steinböck, *Lohner: zu Land, zu Wasser und in der Luft* (Graz, 1984), 29; E. Rumpler (ed.), *Automobiltechnischer Kalender und Handbuch der Automobil-Industrie für 1907,* 4th ed. (Berlin, n.d.), 552.

35. Evgénieff, *Droschkenbetrieb,* 40; Fründt, *Das Automobil und die Automobilindustrie in Deutschland,* 44; on 250 km, see E. Sieg, "Elektrische Kraftwagen," *ETZ* (1 November 1906), 1017–21, here: 1018.

36. Evgénieff, *Droschkenbetrieb,* 39–40; Fründt, *Das Automobil und die Automobilindustrie in Deutschland,* 44; *Centralblatt* (1 March 1906), 65, (20 June 1907), 98 (5 July 1908), 103; *MW* (20 February 1906), 163 and (30 May 1906), 419. The sources show discrepancies in the size of the losses. On the strike, see AK (5 March 1907) 1, and AK (13 May 1909) 13 (both Varta archives).

37. *Tägliche Rundschau,* 4 December 1913, quoted by: [Ulrich Kubisch], "Die Geschichte der Elektrodroschken: Strom contra Sprit," in Ulrich Kubisch et al. (eds.), *Taxi: Das mobilste Gewerbe der Welt* (Berlin, n.d. [1993]), 47–56, here: 54; AK (17 July 1906), 19, 45, 47, 51; last quotation: AK (17 July 1906), 46 (all AK sources: Varta archives). Müller's remark is the only one in the available sources alluding to in-house manufacture of the Bedag cabs. However, he may have meant the change to rear-wheel drive (see later in the text).

38. Müller, *25 Jahre,* 200; AK (4 September 1912), 8–9 (Varta archives).

39. AK (5 March 1907), 17 (Varta archives).

40. AK (13 October 1906), 23; AK (4 March 1906), 2 and 5; AK (20 January 1912), 7; AK (4 September 1912), 4.

41. "Elektrotechnische Gesellschaft zu Köln," *ETZ* (9 January 1908), 39–40, here: 39; AK (17 July 1906), 6; AK (4 March 1906), 6, 8 (the two last sources: Varta archives); on tow trucks, see Evgénieff, *Droschkenbetrieb,* 39; for Müller on car manufacturers, see AK (17 July 1906), 19 (Varta archives); *Centralblatt* (5 January 1908), 7.

42. Evgénieff, *Droschkenbetrieb,* 39–40; *ETZ* (15 April 1909), 361; Müller, *25 Jahre,* 200.

43. E. Sieg, "Über Elektromobile," *ETZ* (24 December 1908), 1238–43, and *ETZ* (31 December 1908), 1258–62, here: 1260; idem, "Elektrische Kraftwagen," *ETZ* (1 November 1906), 1017–21, here: 1018; idem, "Über elektrische Automobilmotoren," *ETZ* (9 January 1908), 39–40, here: 39.

44. Dr. Albrecht to "die Polizeibehörde," 5 April 1910 (Staatsarchiv Hamburg, Amstgericht Hamburg-Handels- und Genossenschaftsregister-B 1970–47: Deeds concerning the Firm "Hedag"); Debag: *AAZ* 46 (Berlin) (1907), 74.

45. *SA* (3 May 1908), 35; *AAZ* 16 (Berlin) (1908), 73; Köhles: [Kubisch], "Die Geschichte der Elektrodroschken," 54.

46. Namag to "die Polizei-Direktion der freien Hansestadt Bremen," 26 August 1908 (archive 4, 14/1-VI. B.8.g, document no. 41–45); Carl Theilen to "Senator Dr. Lürman," 6 December 1910 (archive 4, 14/1-VI. B.8.p); Namag to "die Polizei-Commission des Senats," 8 May 1911 (archive 4, 75/5 B 94 II, document no. 72); archive 4, 75/5 B 94 II, document no. 76 (all: Staatsarchiv Bremen).

47. Attendance list for "Geschehen Bremen, den dreizehnten Juni Neunzehnhundertundzwölf" (report of the shareholders meeting of 13 June 1912) (archive 4, 75/5 B 94 II, document no. 79; Staatsarchiv Bremen); "Geschehen Bremen, den dreissigsten Mai Neunzehnhundertunddreizehn" (report of the shareholders meeting of 30 May 1913) (idem, document no. 81); "Geschehen Bremen, den fünfundzwanzigsten Juni Neunzehnhundertundvierzehn" (report of the shareholders meeting of 25 June 1914) (idem, document no. 83); idem, document no. 83, Appendix 2.

48. F. Alfes to "Sehr geehrte Herren!," June 1916 and idem, July 1917 (archive 4, 75/5 B 94 II; Staatsarchiv Bremen).

49. Dr. Albrecht to "the Police," 5 April 1910 (Staatsarchiv Hamburg, Amtsgericht Hamburg-Handels- und Genossenschaftsregister-B 1970–47: Deeds concerning the Firm "Hedag"); *Jaarverslag Atax over 1910;* on the observer, see Wendt, "Elektrische Kraftwagen und ihre Betriebskosten," *Zeitschrift des Vereins Deutscher Ingenieure* (17 February 1912), 270–72. The daily average and the total annual mileage of the fleet contradict each other. Even at an impossibly high number of effective business days of 365 a year, the fleet kilometers do not add up to more than 143 × 365 = 52,195.

50. "Gesuch der Firma Ernst Dello & Co. um Zulassung von Benzin Droschken" (archive Cl. 1 lit.N^b. No. 3 Vol 1^z Fasc. 16; Staatsarchiv Hamburg).

51. AK (30 January 1909), 7 (Varta archives), 7; turnover and surplus calculated.

52. Ansbert Vorreiter, "Motordroschken und deren Betriebskosten," *MW* (1905), 537–38.

53. Knaths, *Die Entwicklung des Berliner Droschkenfuhrwesens,* 29; also for the following, see A. Heller, "Eine Massregel gegen Motordroschken mit Verbrennungsmaschinen," *Z-VDI* (20 February 1909), 315–16.

54. Knaths, *Die Entwicklung des Berliner Droschkenfuhrwesens,* 31, 42; H. C. Graf von Seherr-Thoss, *Die deutsche Automobilindustrie: Eine Dokumentation von 1886 bis 1979,* 2nd ed. (Stuttgart, 1979), 77.

55. "Electric taxicabs," *EV* (U.S.) (September 1914), 110.

56. J. F. Gilchrist et al., *Report of Committee on Garage and Rates: Fifth annual convention of the Electric Vehicle Association of America, Philadelphia, Penn., October 19, 20, 21, 1914,* 13; A. Jackson Marshall, "Electric taxicabs," *CS* (April 1915,) 319–21, here: 320.

57. *Algemeen Handelsblad,* 20 April 1909 (Municipal Archives of Amsterdam [further: GAA]-Coll. Hartkamp, sheet 390A).

58. *De Ingenieur* (17 November 1928), 45–46; *AFA Rundschau* (1928, nos. 3/4), 170.

59. Müller, *25 Jahre,* 182; Friderichs to Dopler, 10 May 1912 (ARM archives); IK (Techn. Teil) (23–24 May 1905), Punkt 53, 4 (Varta archives).

60. AK (10 March 1908) (Varta archives).

61. Gijs Mom and Vincent van der Vinne, "Geschiedenis van de elektrisch aangedreven auto: de eerste en de tweede generatie (1881–1914)," in idem, *De elektro-auto: een paard van Troje?* (Deventer, 1995), 111–93, here: 177.

62. The report concerned, "Aan den Raad van Beheer der Amsterdamsche Rijtuig Maatschappij," has the handwritten indication "Rapport I" (Report I) and is dated 26 March 1908 (ARM archives).

63. "Den Heer voorzitter & Heeren Leden van Beheer der Amsterdamsche Rijtuig Maatschappij" ("Rapport II," Report II) (ARM archives), in handwriting (also for the following two paragraphs; my italics).

64. "Balans 'Ataxbedrijf' per 31 december 1918," appendix to *Jaarverslag Atax* over 1918.

65. Copy book without heading of the ARM, 1899–1911 (henceforth: Letter book Perk), letter nos. 284, 287, 288, 289, 290, and 305 (ARM archives).

66. *De Fiets,* 10 March 1909, 20–21; Mom and Van der Vinne, "Geschiedenis van de elektrisch aangedreven auto," 178; fourteen: Vincent van der Vinne, *Spyker 1898–1926* (Amsterdam, 1998), 117–18.

67. Perk to Enthoven, 10 November 1908 (Letter book Perk, letter no. 312); *Algemeen Handelsblad,* 30 November 1908 (GAA-Coll. Hartkamp, sheet 395).

68. *Gemeenteblad,* Afdeeling 1, No. 81: "Uitgifte van standplaatsen voor stationneerende automobielen," 21 January 1910, 83–87 (italics in original). In the Atax's license, it was emphasized that one "was not allowed to use any other motive power but electricity" (*Gemeenteblad,* Afdeeling 2, 1910, 285–91 and 323–65, here: 332–33).

69. "Contract" (ARM archives); this unsigned copy was dated for Amsterdam on 13 May 1909, for Berlin on 21 May of that year ; 3 kV: W. Lulofs, "The electric taxicabs of Amsterdam—II," *EV* (U.K.) (September 1915), 23–24, here: 24.

70. *Algemeen Handelsblad,* 2 June 1909: *Nieuws van den Dag,* 3 June 1909 (GAA-Coll. Hartkamp, sheet 395A).

71. Perk to "H. Heijbroek Esq.," 8 June 1909 (Letter book Perk, letter no. 344/345).

72. *Algemeen Handelsblad,* 23 February 1910 (GAA-Coll. Hartkamp, sheet 374A); on the lower rates, see *Algemeen Handelsblad,* 30 September 1909 (GAA-Coll. Hartkamp, sheet 396); the initial distance (at a daytime rate of 40 cents) was doubled to 1,200 m.

73. *Nieuws van den Dag,* 24 January 1910 (GAA-Coll. Hartkamp, sheet 374/374A).

74. *Jaarverslag Atax* over 1910; unidentified newspaper clipping with the heading "Gemeenteraad, zitting van 9 maart 1910" (city council, session of 9 March 1910) (GAA-Coll. Hartkamp, sheet 374/375); *Nieuws van den Dag,* 15 January 1910 (GAA-Coll. Hartkamp, sheet 374).

75. *Jaarverslag Atax* over 1911; Minute book Atax, "Vergadering van 25 April 1911" (meeting of 25 April 1911); on ENTAM, see *Het Leven,* 12 November 1912, 327; *Nieuws van den Dag,* 29 February 1912 (GAA-Coll. Hartkamp, sheet 402); TAM: "Bewijs tot stationneeren met auto's 'Taxi'" (12 December 1913) (ARM archives); AO: *Algemeen Handelsblad,* 23 April 1912 (GAA-Coll. Hartkamp); on the traffic count, see separate sheet with the heading "Alleen richting Amstelveenscheweg" in Minute book Atax; on comparable fleet size, see Van der Vinne, *Spyker,* 118, 166.

76. Minute book Atax, meeting of 16 May 1913; idem, meeting of 19 December 1913; *Jaarverslag Atax* over 1913. According to the report of the board meeting of 19 December 1913, "the batteries were equipped for the higher speed, which required an expenditure of about 300 M. per unit" (Minute book Atax, meeting of 19 December 1913). The change in the construction probably consisted of the addition of extra lead cells in the battery boxes; connected in series to the remaining cells they took care of a higher voltage of the battery set, so that the maximum motor rotation frequency could rise. Last quotation: *Jaarverslag Atax* over 1912. The indication "6 or 7 years" cannot refer to the Atax and probably refers to the year 1905, when the second wave of the electric passenger car started.

77. Friderichs, "Stand Ataxbedrijf Amsterdam, November 1912" (ARM archives).

78. On the tire contract, see Minute book Atax, meeting of 6 April 1914; the starting date of the contract is illegible; Minute book Atax, meeting of 8 April 1915; *Jaarverslag Atax* over 1914; J. F. Friderichs, "Premiën ter verhooging der gemiddelde opbrengst per KM" (ARM archives).

79. *Jaarverslag ARM* over 1915; on the AEM, see Supplement to the *Nederlandsche Staatscourant*, 29 August 1911, no. 202 and 26 June 1912, no. 147; *Jaarverslag ARM* over 1912; on dealerships, see *Jaarverslag AEM* over 1913, and *Jaarverslag ARM* over 1916; State Railroad: Board of directors to "H. H. Commissarissen der Automobiel-Exploitatie-Maatschappij," 3 April 1915 (ARM archives).

80. *Jaarverslag Atax* over 1914; *Jaarverslag ARM* over 1915; on the bonus, see J. F. Friderichs, "Rapport betreffende rentabiliteit Atax in verband met een voorstel voor eene gewijzigde afschrijvings-politiek," 6; on tires, see Minute book Atax, meeting of 8 April 1915; "Balans 'Ataxbedrijf' per 31 december 1918," appendix to *Jaarverslag ARM* over 1918; Perk to "H. H. Commissarissen der Maatschappij 'Atax'," 29 March 1916, appendix to *Jaarverslag Atax* over 1915 (all ARM archives).

81. Perk to "H. H. Commissarissen der Amsterdamsche Rijtuig-Maatschappij," 15 March 1915 (ARM archives); *Jaarverslag Atax* over 1915; *Jaarverslag ARM* over 1915 and 1918.

82. *Jaarverslag Atax* over 1915; moreover, the deposits paid by the drivers on strike also fell to the Atax as a kind of "bonus" (Minute book Atax, meeting of 30 March 1916).

83. W. Lulofs, "The electric taxicabs of Amsterdam," *EV* (U.K.) (June 1915), 24–25 and *EV* (U.K.) (September 1915), 23–24.

84. A. Perk and H. Heijbroek, "Staat van wijzigingen; balans, 31 december 1916" (ARM archives) 1; for forty vehicles in 1917, see "Toelichting balans per 31 december 1917" (appendix to "Amsterdamsche Rijtuigmaatschappij; balans Ataxbedrijf per 31 december 1917" [ARM archives]); for the coach factory, see Minute book Atax, meeting of 8 April 1915.

85. J. F. Friderichs, "Rapport betreffende rentabiliteit Atax," Appendix; *Jaarverslag Atax* over 1916; *Jaarverslag ARM* over 1917 and 1918.

86. *Jaarverslag ARM* over 1918; "Voorstel winsten en afschrijvingen 1919," appendix to *Jaarverslag ARM* over 1919.

87. *Jaarverslag ARM* over 1919, 1920, 1921, 1922.

88. "Veertig jaar tax's in Amsterdam," *Nieuws van den* Dag (?, title difficult to read), 30 June 1949 (GAA, file M 992.026); Jaarverslag *ARM* over 1923.

89. Auditor's report to the *Jaarverslag ARM* over 1924, 11; Auditor's report to the *Jaarverslag ARM* over 1925, 7; Gijs Mom, "'Äffchen' und Blockbandtaxen," in Kubisch et al. (eds.), *Taxi*, 301.

90. Robert Schwenke, "Die Elektromobilen auf der Berliner Automobil-Ausstellung 1906, 3.–18. Februar," *Centralblatt* (1 March 1906), 54–57, here: 55.

Chapter 5: The Electrified Horse

Quoted by Manfred Gihl, *Rettungsfahrzeuge: Von der Krankenkutsche zum Notartztwagen* (Stuttgart, 1986), 20–21 (no date of the quotation indicated).

1. Richard H. Schallenberg, *Bottled energy: Electrical engineering and the evolution of chemical energy storage* (Philadelphia, 1982), 275.

2. Lucien Périssé, "Voitures postales à alcool," *LA* (30 January 1902), 70–71; "Les automobiles postales système Mildé," *IE* (25 October 1904), 500.

3. "Das Automobil-System Kühlstein-Vollmer," *MW* (1899 Heft 1), 5; *MW* (1899

Heft 12), 218; Erwin Maderholz, "Elektrofahrzeuge im Postdienst," *Archiv für deutsche Postgeschichte* (Heft 2/1981), 5–33, here: 19–20.

4. "Elektrischer Post-Omnibus (Briefträger-Wagen)," *AAZ* (Berlin) (21 July 1905), 47–48; on Namag, see Maderholz, "Elektrofahrzeuge im Postdienst," 13, 17; *Lloyd-Wagen* (n.p., n.d. [1908?]) (brochure of the Norddeutsche Automobil- & Motoren AG, archives Deutsches Museum, Munich); on the higher costs for gasoline vehicles, see "Postelektromobile," *AAZ* 50 (Berlin) (1909), 35–37; for statistics for 1914, see H. Beckmann, "Das Elektromobil als Nutzwagen," *AAZ* 22 (Berlin) (1914), 23ff.

5. Martin V. Melosi, *Garbage in the cities: Refuse, reform, and the environment, 1880–1980* (College Station, Tex., 1981), 152, 155, 159–60; Gottfried Hösel, *Unser Abfall aller Zeiten: eine Kulturgeschichte der Städtereinigung*, 2nd ed. (Munich, 1990), 167–68, 178–79.

6. Melosi, *Garbage in the cities*, 141–42. It is important to emphasize here that from counterfactual argumentation based on engineering knowledge, this application field would not have been inaccessible for gasoline propulsion. It would have meant that gasoline truck manufacturers would have applied very high end gearing in their trucks, just like Karl Benz before the turn of the century had installed a reverse gear in his car, which resulted in a kind of "creeping speed" when driving backward. But for reasons that have not yet been well researched, the gasoline car producers did not follow this path of extreme specialization during the early years. The evidence presented in this study suggests that they were too busy developing a reliable universal propulsion system, which would also function satisfactorily within the city.

7. Hösel, *Unser Abfall aller Zeiten*, 178.

8. Monty McCord, *Police cars: A photographic history* (n.p., n.d.), 12–13.

9. "Das Elektromobil im Dienste des Krankentransports," *AAZ* (Berlin) (22 December 1905), 47–48; "Das Krankentransportwesen der Stadt Dresden," *Feuerwehrtechnische Zeitschrift* (5 and 20 December 1918), 162–65.

10. "Die Verwendung von Automobilen in grösseren Stadtverwaltungen," *Feuer und Wasser* (14 March 1914), 83–85; "Vorteile und Nachteile des elektrischen Lastwagen-Betriebes," *Feuerwehrtechnische Zeitschrift* (1918), 157–60, here: 157. The seventy-four units of the Berlin fire-engine fleet have not been included in the count here, as this fleet fell directly under Prussian state control. L. Betz, *Spezial-Lastautomobile, Band I: Kommunal-Automobile, Automobil-Kipper, Tank-Automobile, Sonderfahrzeuge* (Berlin, 1927), xiv–xv; Otto Barsch, *Moderne Automobil-Strassenreinigungsmaschinen* (Berlin, 1919) (Autotechnische Bibliothek, Band 63), 149, 145 (my italics).

11. Susan Meikle Mandell, Stephen Peter Andrew, and Bernard Ross, *A historical survey of transit buses in the United States* (SAE Special Publications, SP-842), 7; for a new initiative of the Electric Omnibus Corporation in New York City in 1912, however, see "Field electric omnibus for depot service," *ER* (New York) (6 April 1912), 678.

12. D. Gammrath and H. Jung, *Berliner Omnibusse* (Berlin, n.d.), 11, 16; Wolfgang Huss and Wolf Schenk, *Omnibusgeschichte, Teil 1* (Munich, 1982), 160, 200, 212, 226.

13. Henry Charles Moore, *Omnibuses and cabs: Their origin and history* (London, 1902), 120; L. Lockert, "Un omnibus automobile électrique," *FA* (14 November 1896), 341; "Electric bus of the Electric Motive Power Company," *HA* (September 1896), 20–21; for the following, see also E. L. Cornwell, *Commercial road vehicles* (London, 1960), 83–84; "American 'electro-gasoline' buses for London streets," *HA* (16 July 1902), 76.

14. Yves Guédon, "Les taxi-autos et les autobus à Paris et à Londres," *VA* (27 February 1909), 142–44, here: 143; *VA* (13 March 1909), 167–70, (3 April 1909), 217–18, (10 April 1909), 231–32; Yves Guédon, "Transports en commun: Paris et Londres," *VA* (11 May 1907), 299; Yves Guédon, "La lutte des autobus à Londres," *VA* (21 December 1907),

821–22, here: 822; *ER* (New York) (14 December 1907), 953; on hostile commentaries, see "The electrobuster?" *The India-Rubber & Gutta-Percha electrical trades' journal* (23 April 1906), 73; on the controversy, see *Centralblatt* (5 June 1906), 154, and *Centralblatt* (20 June 1906), 168.

15. "Les 'électrobus' de Londres," *IE* (25 November 1907), 562–63; "Omnibus électriques de Londres," *IE* (1909), 146; on hybrids, see J.-A. Montpellier, "Les voitures automotrices pétroléo-électriques," *VA* (28 September 1912), 625–27; J. Graeme Bruce and Colin H. Curtis, *The London motor bus: Its origins and development* (London, 1973), 26–29.

16. Guédon, "Les taxi-autos et les autobus à Paris et à Londres," 169; "Les omnibus automobiles à Paris," *VA* (3 February 1906), 69.

17. Theo Barker, "Towards an historical classification of urban transport development since the later eighteenth century," *Journal of transport history* 1, 3rd ser. (1980), 86.

18. Angela Zatsch, *Staatsmacht und Motorisierung am Morgen des Automobilzeitalters* (Konstanz, 1993), 26–40, 42, 48–78; Margaret Walsh, *Making connections: The long-distance bus industry in the USA* (Aldershot, 2000), 11.

19. This section is a summary of a much more extensive treatment of this topic in Gijs Mom, "Wie Feuer und Wasser: Der Kampf um den Fahrzeugantrieb bei den deutschen Feuerwehr (1900–1940)," in Harry Niemann and Armin Hermann (eds.), *100 Jahre LKW: Geschichte und Zukunft des Nutzfahrzeuges* (Stuttgart, 1997), 263–320.

20. Branddirektor Ruhstrat, "Zur 10jährigen Tätigkeit des Preussischen Feuerwehr-Beirates (P. F. B.)," *Feuerwehrtechnische Zeitschrift* (1919), 87–92, here: 92; D. W. Reutlinger, *Bericht über den vom 16. bis 19. Juni 1904 zu München abgehaltenen Vierten Verbandstag der Deutschen Berufsfeuerwehren* (Frankfurt a.M., 1904) (Sonderabdruck aus *Feuer und Wasser*) (Feuerwehrmuseum Fulda, document 1.5.3.2.-4/1A), 3; "Das Feuerlöschwesen der deutschen Städte im Jahre 1900 oder 1900/1901," *FuW* (22 November 1903), 399–402; "Uebersicht der deutschen und österreichischen Feuerwehrverbände 1. Januar 1902" (table), *FuW* (Festausgabe für den XVI. *Deutschen Feuerwehrtag*) (1904), 11.

21. Max R. Zechlin, "Feuerlöschautomobile," in *Handbuch der Automobilindustrie 1909*, 11. The *Denkschrift* was not printed until two years later: M. Reichel, *Der Automobil-Löschzug der Berufsfeuerwehr Hannover* (Berlin, 1903).

22. Reichel, "Mitteilungen über die Belegung der Feuerwache II zu Hannover mit einem automobilen Löschzuge," *FuW* (10 January 1904), 9–13, here: 9; "Die Uebername der Berliner Feuerwehr durch Branddirektor Reichel," *FuW* (2 July 1905), 231–32; *Die Berliner Feuerwehr 1901–1911: Als Nachtrag zur Geschichte des Korps aus Anlass des 60 jährigen Bestehens im amtlichen Auftrage bearbeitet* (Berlin, 1911), "Anhang"; *Denkschrift* in *Bericht Berlin 1905*, 6–7, 9.

23. *Bericht Berlin 1905*, 8–10.

24. Branddirektor Reichel, *Bericht über das Ergebnis der mit Kraftfahrzeugen bei der Berliner Feuerwehr vorgenommenen Versuche* (26 december 1907, appendix with its own page numbers to *Bericht Berlin 1907*) 5, 22, 26, 28, 34, 41.

25. Reichel, *Bericht über Versuche*, 15, 19, 34, 36–37, 41–44.

26. *Bericht Berlin 1907*, 5, 7–9, 13; *Bericht Berlin 1908*, 73; "Feuerwehr-Automobile," *FuW* (29 November 1908), 383.

27. *Bericht Berlin 1907*, 10; *Bericht Berlin 1908*, 9; Charlottenburg: [Wilhelm] Scholz, *Anlage zum Verhandlungsbericht über den 13. Verbandstag vom 4. bis 6. Juni 1913 in Stettin; statistische Zusammenstellung der Automobilkommission* (Aachen, n.d. [1913]), 30; *Bericht Charlottenburg 1911*, 5, 9–11; *Bericht Charlottenburg 1908*, 45; on self-discharge, see *Bericht Berlin 1908*, 7.

28. "Schöneberger Brief," *Feuerpolizei* (1912 no. 5), 74–76, here: 74; *Bericht*

Schöneberg 1903–1908, 659, 662, 664–65; "Die erste Automobil-Dampfspritze in Berlin-Schöneberg," *AAZ* (Berlin) (1905), No. 45, 38–40; *Bericht Schöneberg 1908–1912*, 28, 34–37; quotation: 37. Cost calculation based on "Vergleichende Aufstellung der laufenden Kosten für einen Automobil-Löschzug und einen bespannten Löschzug bei der Schöneberger Feuerwehr," *FuW* (19 February 1910), 57–59, here: 57.

29. Automobil-Konferenz (17 May 1911), 6, 15 (Varta archives); 1908: Gijs Mom, "Das Holzbrettchen in der schwarzen Kiste: Die Entwicklung des Elektromobilakku-mulators bei und aus der Sicht der Accumulatoren-Fabrik AG (AFA) von 1902–1910," *Technikgeschichte* 63 (1996), 134–37.

30. *Bericht Berlin 1913*, 9–12; *Jubiläumschrift 125 Jahre Berliner Feuerwehr* (Berlin, n.d.), 64.

31. Scholz, *Anlage zum Verhandlungsbericht . . . 1913*.

32. *FuW* (2 November 1902), 362; "Berufsfeuerwehr—Liste 1912," *FuW* (20 January 1912), 19–21, here: 21.

33. "Kommissionen," *Zwölfter Verbandstag*, Anlage 19.

34. Scholz, "Die Wertung der verschiedenen Feuerwehrautomobil-Typen nach dem heutigen Stande der Geräte" (VDB, *Zwölfter Verbandstag*, Appendix 16 [no page numbers]).

35. Hüpeden, "Die Weiterentwicklung des Feuerwehr-Automobil-Wesens im Jahre 1912," *FuW* (4 January 1913), 1–3, here: 2; Kaiser, "Die Betriebssicherheit des benzin-elektrischen Systems," *Feuerpolizei* 21 (1913), 329–32, here: 332.

36. Verband Deutscher Berufsfeuerwehren, *Dreizehnter Verbandstag am 4. bis 6. Juni 1913 in Stettin* (Hamburg, n.d.), 18; Riedler: "Wissenschaft und Praxis," *FuW* (16 November 1912), 376–78; Wilhelm Scholz, *Die Übergang zum Automobilbetriebe bei den Feuerwehren, seine Begründung, Durchführung und Ziele* (Aachen, 1914) (diss.).

37. "Die Automobilisierung der Münchener Feuerwehr," *FuW* (10 August 1912), 261–63; Hüpeden, "Die Weiterentwicklung des Feuerwehr-Automobil-Wesens im Jahre 1912," 2.

38. "Vorträge über Feuerwehrautomobile in der Automobil- und Flugtechnischen Gesellschaft (Schluss)," *FuW* (10 August 1912), 263–65, here: 264.

39. VDB, *Dreizehnter Verbandstag*, 16–17; Scholz, "Die Wertung der verschiedenen Feuerwehrautomobil-Typen."

40. "Die grosse Uebung der Berliner Feuerwehr vor Seiner Majestät der Kaiser," *FuW* (21 February 1914), 57–59; quotation: Von Moltke and Auhagen, "Die Elektrotechnik im Dienste der Feuerwehr," *FuW* (16 September 1911), 301–3, here: 303.

41. Hüpeden, "Statistik der Kraftfahrzeuge im Dienste deutscher Feuerwehrn," *Feuerpolizei* 8 (1914): 113–17; "14. Verbandstag Deutscher Berufsfeuerwehren am 9. bis 11. Juni," *Feuerwehrtechnische Zeitschrift* (1914), 235–36, here: 235.

42. "Die Brand- und Rauchproben im Wiesbadener Hof-Theater am 21. August 1913, II," *FuW* (20 September 1913,) 298–301, here: 300; "Die Flensburger Motorspritze," *FuW* (28 March 1914), 98–99; "Daimler-Motorsprengwagen der Daimler-Motoren-Gesellschaft, Zweigniederlassung: Berlin-Marienfelde," *FuW* (11 July 1914), 223–24.

43. "Die Frankfurter Motorspritze," *FuW* (20 December 1913), 401–6, especially 403–5, and *FuW* (10 January 1914), 11–13. Only after the course had finished did it appear that the cause was a porous carburetor float, which had interrupted the fuel supply. "Münchener Brief," *Feuerpolizei* 7 (1912), 114–15, here: 115.

44. Hüpeden, "Automobile für Landfeuerwehren," *Feuerwehrtechnische Zeitschrift* (1914), 110.

45. *Deuxième concours des poids lourds; Versailles, 1898; Rapport de la commission* (Paris, 1899) (collection J.-L. Krieger, Paris) (I am indebted to Mr J.-L. Krieger, Paris, for

making available a copy of this report); on Liverpool, see Claude Johnson, *The early history of motoring* (London, n.d.), 47; Paris-Marseille: *MW* (10 January 1907), 10.

46. Hans-Otto Neubauer, *Autos aus Berlin: Protos und NAG* (Stuttgart, 1983), 13; on arguments against electric propulsion, see "Military automobiles," *HA* (10 July 1901), 335; on the British army, see Norman Miller Cary Jr., "The use of the motor vehicle in the United States Army, 1899–1939" (unpubl. diss., University of Georgia, 1980), 6; see also E. S. Shrapnell-Smith, "Five decades of commercial road transport with inferences about its future," *Journal of the Institute of Transport* (February/March 1946), 214–29, here: 214.

47. "Concours de véhicules automobiles industriels," *LA* (27 July 1905), 28–30; "Le projecteur Kriéger," *LA* (28 September 1905), 165; Olaf von Fersen (ed.), *Ein Jahrhundert Automobil-technik: Nutzfahrzeuge* (Düsseldorf, 1987), 18; "10 Jahre NAG," *SA* (7 January 1912), 12–17; Immo Sievers, *AutoCars: Die Beziehungen zwischen der englischen und der deutschen Automobilindustrie vor dem Ersten Weltkrieg* (Frankfurt am Main, 1995), 214–16.

48. Cary, "The use of the motor vehicle in the United States Army," 11–21, 94.

49. Ibid., 29–30. In 1908, the U.S. Army bought two additional electric buses (32).

50. Ibid., 39–43, 33, 49.

51. Marc K. Blackburn, "A new form of transportation: The Quartermaster Corps and standardization of the United States Army's motor trucks 1907–1939" (diss., Temple University, 1992), 21–22, 16; see also Blackburn's book publication on the basis of this dissertation: *The United States Army and the motor truck: A case study in standardization* (Westport, Conn., 1996); Cary, "The use of the motor vehicle in the United States Army," 52; quotations: 79, 81.

52. James M. Laux, *The European automobile industry* (New York, 1992), 33, 45–49; idem, "Trucks in the west during the First World War," *Journal of transport history* 6, no. 2 (September 1985), 64–70; Helmut Otto, "Die Herausbildung des Kraftfahrwesens im deutschen Heer bis 1914," *Militärgeschichte*, 3/1989, 227–36; Kurt Möser, "World War I and the creation of desire for cars in Germany," in Susan Strasser, Charles McGovern, and Matthias Judt (eds.), *Getting and spending: European and American consumer societies in the twentieth century* (Washington, D.C., 1998), 195–222.

53. T. C. Barker, "The spread of motor vehicles before 1914," in Charles P. Kindleberger and Guido di Tella (eds.), *Economics in the long view: Essays in honor of W. W. Rostow; vol. 2: Applications and cases, part I* (London, 1982), 156.

54. On buggy sales, see Donald D. Critchlow, *Studebaker: The life and death of an American corporation* (Bloomington, 1996), 82; on large appetite, see Laux, *The European automobile industry*, 60; on propaganda, see K. Helfferich, *Der Weltkrieg*, II (Berlin, 1919), 223, quoted by Joachim Radkau, *Technik in Deutschland: Vom 18. Jahrhundert bis zur Gegenwart* (Frankfurt am Main, 1989), 240.

55. "Aus dem Verwaltungsbericht über die Tätigkeit und die Verwaltung der Feuerwehr der Stadt Wien für die Zeit vom 1. Juli 1916 bis 30. Juni 1917," *FuW* 47/48 (1918), 191; document "MA 68-B1/8/65" (Technisches Museum, Vienna); "Bericht der Feuerwehr der Stadt Wien, 1. Juli 1918 bis 30. Juni 1919," *Feuerwehrtechnische Zeitschrift* (1920), 208.

56. Leopold Merz, "Die Verwendung von Automobilspritzen in kleinen Städten," *Feuerwehrtechnische Zeitschrift* (20 December 1920), 201–4, here: 201; Udo Paulitz, *Historische Feuerwehren: Fahrzeuge und Einsätze von 1900 bis 1970* (Stuttgart, 1994), 45–73.

C. W. Squires Jr., "The future of the electric truck," *CS* (1915–16), 157–58, here: 158.

1. Gijs P. A. Mom and David A. Kirsch, "Technologies in tension: Horses, electric trucks, and the motorization of American cities, 1900–1925," *Technology and culture* 42, no. 3 (July 2001), 489–518.

2. Dietmar Abt, *Die Erklärung der Technikgenese des Elektroautomobils* (Frankfurt am Main, 1998), 169; on magazines, see James Rood Doolittle, *The romance of the automobile industry; being the story . . .* (New York, 1916), 372.

3. Thomas P. Hughes, *Networks of power: Electrification in Western society, 1880–1930* (Baltimore, 1983), 85, 218–24; Harold L. Platt, *The Electric City: Energy and the growth of the Chicago area, 1880–1930* (Chicago, 1991), 84–91, 95–108, 112, 143, 153–57, 161; on buffer capacity, see L. B. Stillwell, "Electrical power-generating stations and transmission," *ER* (29 October 1904), 705–6, here: 706.

4. Mark Granovetter and Patrick McGuire, "The making of an industry: Electricity in the United States," in Michel Callon (ed.), *The laws of the markets* (Oxford, 1998), 147–73, here: 153–61.

5. Richard H. Schallenberg, *Bottled energy: Electrical engineering and the evolution of chemical energy storage* (Philadelphia, 1982), 280; on long-hour load, see A. E. B. Ridley, "The electric automobile as an income producer for central stations," *CS* (1903–4), 85; on low-demand load, see H. W. Hillman, "Relative importance of the electric truck as compared with other classes of central station business," *CS* (1912–13), 248–52.

6. Quotation: "The November meeting," *CS* (December 1909), 116–19, here: 117; on 1916, see Fred Darlington, "Central station electric systems and railroad power," *NELA 1916-III*, 552–63, here: 552; on the elevator, see Platt, *The Electric City*, 104. Granovetter and McGuire ("The making of an industry," 166) even argue that the Insull group "after 1902 . . . essentially ignored electric cars . . . The making of policy through defensive reaction was so pervasive that several of Insull's circle even ignored their personal investments in electric car companies while pursuing trolley loads, hurting their industry, their firms, and themselves." This, however, might only be true, if the term *car* is taken literally as "passenger car." For the truck, their efforts to promote it were quite elaborate, as we will see.

7. "Report of the Committee on Electric Vehicles," *CS* (1910–11), 102–4, here: 103. This comparison was based on an annual energy consumption for the passenger car of 1,600 kWh, for the truck of 6,000 to 14,000 kWh, whereas the incandescent lamp and the electric iron used 15 and 78 kWh, respectively.

8. R. McAllister Lloyd, "The influence of the pioneer spirit on electric vehicle progress," *CS* (1914–15), 179–81; Hayden Eames, "Impediments to the general introduction of power wagons," *EVAA 1911*, 25–28; "Commercial vehicle plant in New York City," *CV* (September 1906), 236–37; Doolittle, *The romance of the automobile industry*, 367.

9. *Minutes of the meeting of the Association of Electric Vehicle Manufacturers, held at the Hotel Belmont, 42nd Street & Park Avenue, Borough of Manhattan, New York City, December 6, 1906, at 10 o'clock a.m.*

10. "The November meeting," 117; Harry W. Perry, "Motor trucking on the New York docks," *CV* (October 1906), 251–53, here: 252.

11. "The Lansden truck," *CV* (January 1910), 173; J. H. Vail, "The electric commercial vehicle at the close of 1909," *CV* (January 1910), 222–23, here: 223; "The butcher, the baker, the candlestick maker," *CV* (December 1910), 176; Day Baker, "The advance of the electric vehicle in New England," *CV* (January 1910), 146–47.

12. "The electric automobile in central station practice," *CS* (1903–4), 244; L. D. Gibbs, "Booming the electric vehicle in New England," *ER* (6 March 1909), 426; W. H.

Blood Jr., *Presidential address; EVAA 1910*, 2; Hayden Eames, "The electric vehicle opportunity," *NELA 1910-II*, 88–119, here: 114–15; "Electric Vehicle Association of America," *CS* (September 1910), 79–80, here: 80; "The November meeting," 116; quotation: Howard S. Knowlton, "Side lights on the electric vehicle situation," *CS* (1909–10), 244–48, here: 247; "The relation of central stations to the automobile business," *ER* (27 March 1909), 559; Fred T. Kitt, "Relation of central stations to the automobile business," *ER* (27 March 1909), 576–77.

13. "Committee on Electric Vehicles," *ER* (5 March 1910), 468; "Report of the Committee on Electric Vehicles," *CS* (October 1910), 102–4.

14. W. H. Blood Jr., *President's address; EVAA 1911*, 2. The EVAA was incorporated in the state of New York on 29 August 1910 ("Electric Vehicle Association of America," *ER* [20 September 1910], 616); entry EVCSA: "Boosting electric vehicles," *ER* (1 October 1910), 699; for 8 June, see "The Electric Vehicle Association of America," *CS* (June 1910), 293–94.

15. Arthur Williams, "How and where central stations can use electric vehicles," *NELA 1917-V*, 84–102, here: 100; Arthur Williams, "The central station and electric trucks," *ER* (10 February 1912), 288–89; H. Robinson, "The New York Edison Company and the electric vehicle," *CS* (1909–10), 196–98, here: 197; "Incandescent lamp delivery system of the New York Edison Company," *The Edison monthly* (February 1909), 213–15; Harvey Robinson, "Report of the secretary," *EVAA 1913*, 3–4, here: 4.

16. *Report of Committee on Central Station Cooperation; EVAA 1914*, 8–18; "Report of the New York Electric Vehicle Association," *EVAA 1913*, 32–33, here: 32; A. Jackson Marshall, *Annual report of the Secretary; EVAA 1915*, 8–9.

17. First quotation: George H. Kelly and E. J. Bartlett, "Electric-vehicle salesmanship," *EVAA 1913*, 62–64, here: 64; quotations Blood resp.: Blood, *Presidential address; EVAA 1910*, 2; "W. H. Blood Jr." *ER* (11 March 1911), 469; on the "truck association," see "Report of the Chicago section," *EVAA 1910*, 29; on local departments, see "Report of the Secretary," *NELA 1916-IV*, 5–7, 18; Jackson Marshall, "Electric Vehicle Association activities," *CS* (December 1915), 153–57, here: 157.

18. *EVAA 1913*, 69. In general, the electric-vehicle industry could apply lower-grade materials than the gasoline-vehicle industry, because its vehicles were less exposed to vibrations and shocks.

19. F. E. Whitney, "The tire question as applied to electric vehicles," *EVAA 1913*, 82–84, here: 82; Dan C. Swander, "Tires—their use and abuse," *CS* (April 1911), 280.

20. On continuous torque, see F. E. Whitney, "Electric truck troubles and means taken to eliminate them," *NELA 1916-IV*, 151–58, here: 152, 155; on Exide, see "Review of storage battery industry for 1903," *ER* (9 January 1904), 70; on the Edison battery, see Schallenberg, *Bottled energy*, 362–72; Harold H. Smith, *The Edison storage battery in service; EVAA 1912*, passim; Walter E. Holland, "The Edison storage battery," *CS* (January 1911), 196–98; Robert E. Russell, *Electric vehicle battery charging apparatus; EVAA 1912*, 13; *EVAA 1911*, 18; *NB* (December 1912), 241–42; *NB* (February 1913) 367, 376–77. For an extensive analysis of the development of Edison's battery, see Gijs Mom, "Inventing the miracle battery: Thomas Edison and the electric vehicle," *History of technology* 20 (1998), 18–45.

21. Bruce Ford, "The electric vehicle battery," *EVAA 1910*, 144–45, here: 144; "Thin plate lead battery," *CS* (December 1910), 174; James M. Skinner, "The Philadelphia thin plate battery," *CS* (March 1913) 289–98, here: 289; James M. Skinner, "Thin plate battery truck in service," *CS* (March 1916), 238–41; "Vehicle batteries," *CV* (February 1910), 76.

22. Bruce Ford, *Some recent developments in the lead battery for electric vehicles; EVAA 1912*, 3; Schallenberg, *Bottled energy*, 270–71, 284–86; "The December meeting of the Electric Vehicle Association of America," *CS* (January 1911), 191–92, here: 191; Bruce

Ford, "The 'Ironclad-Exide' battery," *CS* (February 1911), 226; "The 'Ironclad-Exide,'" *CS* (January 1911), 200. When the German AFA shortly afterward launched a virtually identical battery, ESB's pun in the name giving was lost because of the literal translation to "Panzerbatterie" (Schallenberg, *Bottled energy*, 271).

23. *The electric truck: A sketch indicating the development and present status of this modern successor of the horse* (Pontiac, Mich., 1912). This brochure mentioned an energy density for the Edison battery of 27 Wh/kg; for the thin-plate battery and the Ironclad lead battery this was 20 and 17 Wh/kg, respectively (31). On a third, see H. C. Cushing Jr. and Frank W. Smith, *The electric vehicle hand-book* (New York, 1913), 13 (advertisement).

24. "November meeting of the Electric Vehicle Association of America," *CS* (December 1912), 183–200, here: 192; S. C. Harris, "Vehicle battery practice in central station companies," *EVAA 1911*, 16–18, here: 18.

25. *NB* (October 1911), 140.

26. "Veterans," *CS* (September 1914), 75; "'Ironclad-Exide' storage batteries," *CS* (April 1915), 322; Bruce Ford, "Recent development in the lead battery for electric vehicles," *EVAA 1913*, 69–71, here: 70; "Tuesday afternoon session," *EVAA 1913*, 75; table and guarantee: "'Exide' activity," *CS* (October 1914), 116–17, here: 117.

27. Foolproof: Blood, *Presidential address*; *EVAA 1910*, 6; most perishable: "Report of the Garage and Rates Committee, 1915," in *Reports of committees*; *EVAA 1915*, 17–22, here: 21; Ah meter: Harris, "Vehicle battery practice in central station companies," 17; "The Sangamo ampere-hour meter," *CS* (February 1911), 228–29; R. C. Lanphier, "Summary of advantages of ampere-hour meters in connection with electric vehicles," *CS* (May 1914), 409–11.

28. "Report of the Garage and Rates Committee, 1915," 17; Harris, "Vehicle battery practice in central station companies," 18; "Report of the Garage Committee," *EVAA 1913*, 47–52, here: 52; on able attendants, see R. Macrae, "Ideal electric garage service," *CS* (May 1913), 370–72, here: 370.

29. "Report of the Garage Committee," 47, 52, 53; on hay and grain, see C. L. Morgan, "The proper garaging of electric vehicles," *CS* (December 1910), 167–70, here: 168; "Electric Vehicle Association of America," *CS* (June 1915), 381–96, here: 382.

30. Quotation Michel: *CS* (June 1911), 348; on intelligent supervision, see "Methods of design and operation which assure the efficiency of the electric vehicle," *CS* (January 1916), 187–89, here: 187; on the proposal for an exclusive garage, see R. Macrae, *The function of the electric garage; EVAA 1915;* on resolution in Chicago, see "The Electric Vehicle Association of America," *CS* (March 1915), 270–76, here: 271.

31. George H. Kelly and E. J. Bartlett. "Electric-vehicle salesmanship," *EVAA 1913*, 62–64, here: 63; R. L. Lloyd and John Meyer, "Electric vehicle commercial problems," *EVAA 1911*, 6–7, here: 7.

32. *CS* (1912–13), 261.

33. Maxwell Berry, "The charging of storage batteries in unattended garages," *EVAA 1913*, 54–62, here: 54.

34. *CS* (July 1911), 2.

35. Day Baker, "The advance of the electric vehicle in New England," *CS* (December 1909), 146–47, here: 146; Foljambe: *EVAA 1913*, 65; for climb a tree, see *CS* (July 1911), 21–23.

36. "The fifth annual convention of the Electric Vehicle Association of America," *CS* (November 1914), 138–58, here: 153; on the "power-consuming device," see Kelly and Bartlett, "Electric-vehicle salesmanship," 64; last quotation: James H. McGraw, *Stimulating electric vehicle progress; EVAA 1914*, 4; on the high-speed car, see *CS* (March 1912), 261; on minor executives, see Cushing and Smith, *The electric vehicle hand-book*, 11.

37. J. F. Gilchrist and A. J. Marshall, "The electric vehicle and the central station," *NELA 1915-I*, 323–400, here: 326–30 (italics in original); Bailey: "December meeting of the Electric Vehicle Association of America," *CS* (January 1913), 218–28, here: 226.

38. On solid tires, see "Speed maniac new menace to trucks," *CS* (June 1912), 367–68; on speed limiting devices, see James T. Hutchings, "The electric vehicle in heavy trucking service," *EVAA 1910*, 143–44; on "reversely proportional," see E. S. Foljambe, *Progress of commercial cars in America, with special reference to the electric truck, EVAA 1912*, 7. The regulated speed recommended was 24 km/h for the one-ton gasoline truck and gradually decreased to 8 km/h for the ten-ton truck; for the trucks with a load between one and six tons these values were about 8 km/h below the maximum speeds that could be reached. On the electric machine, see Newton Harrison, "Storage batteries for vehicles," *CS* (September 1910), 57ff., here: 58 (my italics); on uniting against the horse, see F. Nelson Carle, "The merchant, the central-station, and the electric truck," *EVAA 1913*, 33–35, here: 34.

39. E. S. Mansfield, "Central station back of the electric vehicle," *EVAA 1911*, 28–31; "Progress of Boston vehicle campaign," *ER* (15 April 1911), 732; on agency, see "E. V. A. A. advertising," *CS* (April 1912), 296; *EVAA 1913*, 80. It is interesting to establish here that the historian, studying trade journals on electric vehicles already at such an early date is confronted with sources that have been doctored so rigorously from the viewpoint of propaganda for the electric vehicle cause.

40. Frank W. Smith, *President's address; EVAA 1914*, 9; Cushing and Smith, *The electric vehicle hand-book*, and H. C. Cushing Jr., *The electric vehicle hand-book*, 11th ed. (New York, 1923); quotation: Charles L. Edgar et al., *Report of Publicity Committee on the national co-operative advertising campaign of the Electric Vehicle Association of America; EVAA 1912*, 7.

41. "Report of the Publicity and Advertising Committee," *EVAA 1913*, 76; on the logical advocate, see Carle, "The merchant, the central-station, and the electric truck," 33; 1913/14 and 1915: John F. Gilchrist, *President's address; EVAA 1915*, 3–5.

42. Charles A. Ward, "The electric vehicle installation of the Ward Bread Co.," *CS* (May 1912), 321.

43. "The electric vehicle campaign of the Edison Illuminating Company of Boston," *CS* (May 1911), 319–21 (quotation: 320); Mansfield, "Central station back of the electric vehicle," 29; Harold Pender and H. F. Thomson, "Observations on horse and motor trucking," *CS* (April 1913), 325–31, here: 325; Harold Pender and H. F. Thomson, *Notes on the costs of motor trucking; EVAA 1912*, 3.

44. Harry F. Thomson, *Relative fields of horse, electric, and gasoline trucks* (n.p. [Boston], August 1914) (Vehicle Research Bulletin No. 4), 21; Harold Pender and H. F. Thomson, "Observations on horse and motor trucking," 325; Harold Pender and H. F. Thomson, *Notes on the costs of motor trucking* (reprint of the presentation at the EVAA convention, October 1912) (Vehicle Research Bulletin 2), 9, 11.

45. Thomson, *Relative fields of horse, electric, and gasoline trucks*, 6, 44–45; quotation: Pender and Thomson, "Observations on horse and motor trucking," 331.

46. "November meeting of the Electric Vehicle Association of America," *CS* (December 1912), 183–200, here: 199; quotation: E. L. Callahan, "Co-operation between electric-vehicle manufacturers and central stations," *EVAA 1913*, 37–40, here: 38; "Chicago reduces charging rates," *CS* (August 1912), 55.

47. "Electric trucks give good service," *CS* (January 1913), 229–30, here: 229; on 1912, see "Report of the Committee on Rates and Charging Stations of the Electric Vehicle Association of America," *EVAA 1913*, 14–22, here: 15; on making some money, see *CS* (April 1913), 335.

48. William P. Kennedy, *A practical project to secure authentic cost of operating commercial electric vehicles; EVAA 1914*, 4–5; for biographical data on Kennedy, see "The March meeting," *CS* (April 1910), 217–18, here: 218; "Studebaker activity," *CS* (December 1910), 176; *CS* (February 1912), 242; *CS* (February 1914), 340.

49. "Operating Costs for Commercial Electric Vehicles," *Electrical World* 64, no. 14 (3 October 1914), 664–65; *Electrical World* 64, no. 15 (10 October 1914), 719–20; *Electrical World* 64, no. 16 (17 October 1914), 765; *Electrical World* 65, no. 20 (7 November 1914), 1282–88. See also *NELA 1915-IV*, 81.

50. William P. Kennedy, "Operating costs of commercial electric vehicles," *Reports of the committees of the Electric Vehicle Association of America; EVAA 1915*, 57–75. The EVAA leaders' fear of a cost comparison based on distances driven was understandable, considering the misleading results such comparisons had produced in the past. For example, George Craven, "Motor versus Horses for Haulage," *HA* 23, no. 23 (1909), 781–82, concluded that *per mile* the electric truck was ten times as expensive as the gasoline truck. He based this on his cost analysis of seventy-four gasoline, sixteen electric, and six steam trucks. Although this is a classic comparison of apples and oranges (city trucks and long-distance trucks), even Craven cannot but conclude that per *ton-mile* the electric truck was 2 percent cheaper, due to its higher load. Such misleading comparisons have made some historians doubtful about the fact of the lower operating costs of the electric truck in this early period (for instance, Abt, *Die Erklärung der Technikgenese des Elektroautomobils*, 179), whereas they actually supported the conclusions of Kennedy and the MIT research: although the differences were small, the lower operating costs of the electric truck justified its deployment as city truck in a fleet. New York: William J. Miller and Stephen G. Thompson, *The comparative performance of gasoline and electric vehicles in similar service; EVAA 1915*.

51. David F. Tobias and C. A. Duerr, "The electric vehicle in department store service," *EVAA 1913*, 66–69; "What becomes of the horses and wagons?" *CV* (February 1912), 36–39, here: 36; Charles L. Eidlitz, "The largest modern electric garage," *EVAA 1910*, 150.

52. W. J. McDowell, "The electric truck in brewery service," *NB* (March 1917), 213–15, here: 215; "The electric vehicle installation of the Ward Bread Co.," *NB* (May 1912), 313–18; "Electric vehicle efficiency in department store work," *NB* (1916), 922.

53. Witherby, "Central station's greatest opportunity—the electric vehicle," *NELA 1913-III*, 350.

54. "The trend of the times in truck work, Part II: A consideration of the electric vehicle," *CV* (March 1912), 24–27 and 58–59, here: 27; *NB* (October 1913), 110–11; Walter C. Reid, "Electricity as a substitute for horses in local removals," *CS* (August 1912), 49–53; "Brewery trucks reduce delivery costs," *NB* (1916), 923–24; "Of interest to central station executives," *NB* (1916), 664.

55. H. Beckmann, "Das Elektromobil als Nutzwagen," *AAZ* (Berlin) (1914), No. 22, 23–26, here: 25–26; 682 trucks: Taylor Vinson, "Auto history at NAIAS," *SAH Journal: The newsletter of the Society of Automotive Historians* 167 (March/April 1997), 7.

56. 90 %: Stephen G. Thompson, "Is central station activity in the electric vehicle field justified?" *CS* (January 1912), 198; 7,000 companies: "The electric commercial vehicle," *CS* (July 1912), 30.

57. "Baker electric trucks," *CS* (August 1912), 57; on "superiority," see E. E. Witherby, "The significance of electric truck reorders," *CS* (September 1913), 86–88, here: 87.

58. On GeVeCo, see "General Vehicle sales increasing rapidly," *CS* (April 1916), 266; "New electric truck boom," *G. V. central station bulletin* (August 1915), 9; "General Vehicle Company acquires Mercedes gasoline truck," *ER* (3 August 1912), 232; 20 hours:

E. R. Davenport, "Constructive criticism of the electric vehicle," *CS* (August 1913), 48–53, here: 48; ten brands: Stephen G. Thompson, *Conditions in the East and the possibilities of the electric vehicle; EVAA 1912*, 5–6.

59. *NELA 1914-III*, 239; C. L. Morgan, "The proper garaging of electric vehicles," *CS* (December 1910), 167–70, here: 167.

60. On Kennedy, see *CS* (October 1913), 121; on Lansden, see Michael Brian Schiffer (with Tamara C. Butts and Kimberly K. Grimm), *Taking charge: The electric automobile in America* (Washington, D.C., 1994), 131–33. Steinmetz's presentation caused so much aversion among especially the vehicle manufacturers that the convention leaders decided not to publish Steinmetz's speech, nor J. C. Bartlett's response with the sarcastic title: "An answer to Doctor Steinmetz." "The fifth annual convention of the Electric Vehicle Association of America," *CS* (November 1914), 138–58, here: 149–50. Steinmetz's speech, however, was included in the NELA proceedings *(NELA 1914-II*, 162–73) as well as Charles Steinmetz, "The relation of the automobile to the central station," printed in *CS* (September 1914), 70–74. Ward: Marshall, *Annual report of the Secretary; EVAA 1915*, 17, 19–20; "A notable electric vehicle campaign," *NB* (1915), 706–7. The prices of the electric trucks with a load capacity of less than half a ton were between $1,400 and $2,350 in 1914 (Schiffer, *Taking charge*, 163); on McCormick, see James H. McGraw, *Stimulating electric vehicle progress; EVAA 1914*, 4; Ward quotation: "733 miles in a 'Ward Special,'" *CS* (November 1915), 131–32.

61. "The success of the 'Ward Special,'" *CS* (July 1915), 23–24; Howland, "Practical ideals in electric vehicle promotion," *CS* (October 1914), 98–104, here: 100; on higher performance, see *CS* (October 1913), 122.

62. *NELA 1916-IV*, 53; W. C. Anderson, "How a central station can develop its electric vehicle load," *CS* (July 1913), 24–25; Willis M. Thayer, "The Hartford Electric Light Company's experience with the Battery Exchange System for commercial vehicles," *CS* (December 1915), 141–44.

63. Thayer, "The Hartford Electric Light Company's experience"; P. D. Wagoner, "Battery service," *CS* (August 1916), 40–42; "The electric vehicle: Battery exchange service," *CS* 17 (July 1917–June 1918), 43–45.

64. "Report of the Committee on Rates," *EVAA 1913*, 15; quotation: "The electric vehicle: Battery exchange service," 44; "New electric truck boom," *G. V. central station bulletin* (August 1915), 9; "New battery service system," *CS* (April 1915), 321.

65. David A. Kirsch, "The electric car and the burden of history: Studies in automotive systems rivalry in America, 1890–1996" (diss., Stanford University, 1996), 243–44.

66. Wagoner, "Battery service," 42; Kirsch, "The electric car and the burden of history," 246; "Report of the Committee on Rates," 15; "Report of the Garage and Rates Committee, 1915," in *Reports of committees; EVAA 1915*, 17–22, here: 22; Day Baker, "A standardized universal battery and compartment essential for making the electric truck popular," *CS* (February 1917), 179–80, here: 180; "Motor truck service in textile mills," *CV* (February 1910), 43–46, here: 46.

67. Whitney, "Electric truck troubles and means taken to eliminate them," 157–58; "Philadelphia a prospective center for electrics," *EV* (U.S.) (April 1914), 143–45; quotation: P. D. Wagoner, "Battery service—a unit in a comprehensive plan for the successful exploitation of the electric vehicle," *NELA 1916-IV*, 170–85, here: 181; Boston: "Electric Vehicle Club of Boston," *CS* (July 1912), 26–27, here: 27.

68. Marshall, "The electric vehicle and the central station," *NELA 1915-I*, 329; Chicago: "Report of the Garage and Rates Committee," *Reports of the committees; EVAA 1915*, 17–22, here: 18.

69. Less than one percent: "Looking backward and abroad," *CS* (December 1917), 196–98, here: 197 (italics in original); on 1923, see W. S. Vivian, "The power company as appliance retailer," *NELA Bulletin* (1923), 728–30; on "more slowly," see Claude S. Fischer, *America calling: A social history of the telephone to 1940* (Berkeley, 1992), 56; on isolated plants, see George J. Kirchgasser, "The electric-vehicle situation in Milwaukee," *ER* (27 May 1911), 1033–35; Granovetter and McGuire, "The making of an industry: Electricity in the United States," 152; David E. Nye, *Electrifying America: Social meanings of a new technology, 1880–1940* (Cambridge, Mass., 1990), 294–99.

70. J. B. N. Cardoza, "The electric vehicle from the salesman's standpoint," *NELA 1919-V*, 69–74, here: 73.

71. G. P. Hutchins, "The wisdom and value of duplicate batteries," *CS* (April 1910), 222; Harris, "Vehicle battery practice in central station companies," *EVAA 1911*, 16–18, here: 16.

72. "Government ownership again," *NB* (March 1918), 136; on the energy crisis, see Platt, *The Electric City*, 201; Hughes, *Networks of power*, 285–323.

73. On the collapse, see William P. Kennedy, *A practical project to secure authentic cost of operating commercial electric vehicles; EVAA 1914*, 11; quotations: Gilchrist, *President's address; EVAA 1915*, 7.

74. On NELA, see Platt, *The Electric City*, 95, 240; spreading, converting, and preaching: A. Jackson Marshall, "Report of the Electric Vehicle Section," *NELA 1917-V*, 7–14, here: 9; E. S. Mansfield, "Address of Chairman Mansfield," *NELA 1919-V*, 2–8, here: 6; *NELA 1919-V*, 162; Blood, *Presidential address; EVAA 1910*, 3; "The March meeting," *CS* (April 1910), 217–18, here: 217; *EVAA 1910*, 143; missionary: "Report of Committee on federal and municipal transportation," *NELA 1916-IV*, 114–37, here: 115; F. Nelson Carle, "The merchant, the central-station, and the electric truck," *EVAA 1913*, 33–35, here: 34. In the discussion following this presentation (35) it was observed about the prescribed write-off procedure: "We have written them [the electric trucks] off religiously ten per cent."

75. On early flight, see Joseph J. Corn, *The winged gospel: America's romance with aviation, 1900–1950* (New York, 1983), 46; on a more general level, see David F. Noble, *The religion of technology: The divinity of man and the spirit of invention* (New York, 1997); Electric Day: Marshall, *Annual report*, 21–22; salvation and gods: McGraw, *Stimulating electric vehicle progress; EVAA 1914*, 9; F. Nelson Carle, *Special applications of the electric trucks; EVAA 1914*, 3; devotion and sacrifice: *CS* (December 1914), 181; last quotation: *NELA 1916-IV*, 65.

76. Conviction: *NELA 1919-V*, 46; angels: *CS* (1912–13), 259; sins: *CS* (September 1910), 59.

77. "Electric Vehicle Association of America," *CS* (June 1915), 381–96, here: 395.

78. "Central station opportunity in electric vehicles," *NB* (July 1914), 440–42, here: 441; "2,000,000 electrics in 10 years," *CV* (15 June 1914), 12–13; "Steinmetz predicts electric as eventual vehicle," *CV* (15 June 1914), 15–17.

79. "Address of Dr. Charles P. Steinmetz," *NELA 1914-II*, 162–73, here: 165–68.

80. Even at the last EVAA convention of 1915 it was stated that, despite "the industrial disturbances which have occurred since 1913 . . . the electric vehicle load has developed as rapidly as has the power load in the same period." H. H. Holding and Stephen G. Thompson, *Comparative development of the commercial-power and electric vehicle loads; EVAA 1915*, 3; "New electric truck boom," *G. V. central station bulletin* (August 1915), 9.

81. According to NELA president Walter Johnson "electricity was . . . *destined* to supersede all other forms of energy for light, heat and power, and . . . such a condition

can be more readily and easily secured through the scientific co-ordination of all related efforts." Walter H. Johnson, "Electric Vehicle Section, N. E. L. A.," *NB* (1916), 199–200, here: 199 (my italics). William P. Kennedy, "Central station promotion of electric vehicle use," *NELA 1916-IV*, 206–13, here: 213.

82. "Report of the Committee on Parcel Post Delivery," *Reports of the committees of the Electric Vehicle Association of America; EVAA 1915*, 76–81, here: 79–80; "Electrics supplant gas cars in parcel post delivery," *CS* (January 1914), 302.

83. "E. V. A. A. absorbed by the N. E. L. A.," *CS* (March 1916), 233.

84. On the second revival, see Arthur Williams, "How and where central stations can use electric vehicles," *NELA 1917-V*, 84–102, here: 101.

85. For the following, see Robert Alan Raburn, "Motor freight and urban morphogenesis with reference to California and the West" (unpubl. diss., University of California, Berkeley, 1988), 51–59, 80, 115, 118, 145–47, 167, 177–80.

86. "The horse and the food crisis," *NB* (June 1917), 492–93; "The war and mobile transportation," *NB* (June 1917), 500; "Big automobile fleets," *NB* (March 1918), 136–37; on the Japanese, see "Enormous use of motor vehicles in war," *NB* (October 1917), 764; on shops, see H. B. Lohmeyer, "Motor transportation in the United States army," *NELA 1919-IV*, 81–108; quotation: 91; Marc K. Blackburn, "A new form of transportation: The Quartermaster Corps and standardization of the United States Army's motor trucks 1907–1939 (diss., Temple University, 1992), 38–39.

87. James J. Flink, *The automobile age* (Cambridge, Mass., 1993), 78.

88. In 1920, the average length of a farm-to-market trip appeared to be more than 70 km (Raburn, "Motor freight and urban morphogenesis," 188). On the "liberty truck," see S. V. Norton, *The motor truck as an aid to business profits* (Chicago, 1918), 473; and Blackburn, "A new form of transportation," 44–45.

89. Blackburn, "A new form of transportation," 37–43 (quotation: 37); for the punitive expedition, see also Cary, "The use of the motor vehicle in the United States Army," 96ff.

90. "Truck versus horse," *NB* (March 1920), 155; "La Voie Sacrée," *La lettre de la fondation de l'automobile Marius Berliet*, no. 94 (July/August 2001), 9. Most of the Parisian taxis were already requisitioned by the military, but the remaining 3,000 (most of these Renault Type AG 1 and most owned by the "G7" company) were given the task of transporting the soldiers (up to eight per cab) to the Marne front, all of them paying cash. Claude Rouxel, "Frankreich auf der Suche nach dem idealen Taxi: Mit dem Taxi zur Front," in Ulrich Kubisch et al. (eds.), *Taxi: Das mobilste Gewerbe der Welt* (Berlin, n.d. [1993]), 263–84, here: 268.

91. "Opportunities for demonstration," *CV* (July 1906), 176; "No trucks for army work," *CV* (August 1906), 213; "Motor preparedness," *CV* (1 February 1916), 1313; "101 makers out of a possible 200-odd bid on five classes of army trucks," *CV* (15 June 1917), 14–15.

92. "Electric transportation for explosives," *NB* (April 1917), 308; "Government use of electric trucks," *NB* (June 1918), 390.

93. "Costly congestion in freight deliveries," *NB* (1916), 287–88; "Economic electric industrial trucks release men for service," *NB* (November 1917), 820.

94. *NELA 1919-V*, 143; W. Van C. Brandt, "The storage battery industrial truck and tractor as a new source of revenue for the central station," *CS* (July 1919), 12–16; on the large proportion of battery mass, see *CS* (December 1913), 263; on internal transportation, see Zenas W. Carter, "Electric machinery for all handling—an opportunity," *NB* (July 1920), 458–72; "The new Steinmetz electric," *CS* (March 1920), 302–3.

95. "Electric industrial truck manufacturers join together in cooperative program,"

NB (1924), 395; 1923: William Van C. Brandt, "The national increase in the use of electric trucks," *NELA; Proceedings, 46th convention* (New York, 1923), 523–26, here: 524.

96. "Address of Chairman Mansfield," *NELA 1919-V*, 2–8, here: 5; *NELA 1918*, 293.

97. Foster: "Electric Vehicle Section session," *NELA 1920*, 451–56, here: 455; "Report of the Committee on Garages and Rates," *NB* (September 1920), 652–55, here: 655; "Electric truck notes," *NB* (1924), 751.

98. E. S. Mansfield, "Address of Chairman Mansfield," *NELA 1919-V*, 2–8, here: 7; "Report of the Transportation Engineering Committee," *NELA 1919-V*, 50–56; "Transportation Engineering Committee meets," *NB* (March 1919), 154–55; first quotation: "Report of the Transportation Engineering Committee," *NB* (September 1920), 655–64, here: 655; for 100 percent difference, see "Report of the Manufacturers' and Central Station Cooperation Committee," *NELA 1925*, 19, 45–46; last quotation: F. Van Z. Lane, *Motor truck transportation: The principles governing its success* (New York, 1921), 41.

99. On 1919, see "Electric commercial and industrial vehicles widely adopted last year: Extensive developments contemplated in 1919," *NB* (February 1919), 87; Charles R. Skinner, "The turn of the tide," *NB* (October 1920), 751–52, here: 751; for "most prosperous," see "Electric Vehicle Section session," 451; on 1920, see "Report of Transportation Engineering Committee," *NELA 1921-I*, 437; 1922: "Report of the Electric Vehicle Bureau," *NELA 1922-III*, 340–43, here: 341.

100. On the Auto Car Company, see "Electric Truck and Car Bureau," *NELA 1923-I*, 509–17, here: 511, 513; on Ford trucks, see *NELA 1923-I*, 520; on the new campaign, see *NELA 1924*, 620; for statistics, see William Van C. Brandt, "The national increase in the use of electric trucks," *NELA 1923*, 523–26, here: 524; *NELA 1925*, 638.

101. Edward E. La Schum, *The electric motor truck: Selection of motor vehicle equipment, its operation and maintenance* (New York, 1924), 296–97; "The electric truck in laundry service—its use and possibilities," *NELA 1925*, 650–53, here: 650; "Report of the Manufacturers' and Central Station Cooperation Committee," 20; Cushing, *The electric vehicle hand-book*, 331.

102. Raburn, "Motor freight and urban morphogenesis," 89–101.

103. La Schum, "The electric truck in modern transportation," *NELA 1919-V*, 150–62, here: 153; La Schum, *The electric motor truck*, 188, 210.

104. La Schum, *The electric motor truck*, 223, 299.

105. Ibid., 207, 295–97; Norton, *The motor truck as an aid to business profits*, 32–37, 89–90.

106. John W. Lieb, *Electric vehicles: Report prepared for the Union Internationale des Producteurs et Distributeurs d'Energie Electrique, Paris Meeting, July 5th to 10th, 1928* (n.p., n.d.) 700; 7 percent: William R. Childs, *Trucking and the public interest: The emergence of Federal Regulation 1914–1940* (Knoxville, 1985), 51.

107. "Third session, Commercial National Section; Transportation Committee," *NELA 1925*, 641; Arthur Williams, "How and where central stations can use electric vehicles," *NELA 1917-V*, 84–102, here: 96–97.

108. *NELA 1928*, 477.

Chapter 7: Off the Road and Back

W. H. Blood Jr., *President's address; EVAA 1912*, 4.

1. Chris Sinsabaugh, *Who me? Forty years of automobile history* (Detroit, 1940), 125, 134.

2. H. Beckmann, *Gegenwärtiger Stand der Technik stationärer und transportabeler Accumulatoren* (AFA brochure, Varta archives), 23; P. D. Wagoner, *European development of the electric vehicle industry; EVAA 1914*, 6–7.

3. "Zusammenstellung der in den einzelnen Abteilungen laufenden Elektromobile," *AFA Rundschau* 4 (1914), 100. The AFA apparently had a blind spot with regard to the Swiss market, for there 6 percent of all motor vehicles were electric powered in 1913 (versus 1.5 percent in Germany). It concerned 200 to 300 motorists here (Albert Kloss, *Elektrofahrzeuge: vom Windwagen zum Elektromobil* [Berlin, 1996], 108–9). "Verzeichnis der in Deutschland und Holland vorhandenen Privat-und Droschken-Ladestationen," *AFA Rundschau* 4 (1914), 98–99; *AFA Rundschau* 2 (1914), 57.

4. *AFA Rundschau* 5 (1914), 145–46; "Reviewing 1913 progress," *EV* (U.S.) (February 1914), 65; Maxwell Berry, "Charging of batteries in unattended garages," *EV* (U.S.) (February 1914), 67; J. F. Gilchrist and A. J. Marshall, "The electric vehicle and the central station," *NELA 1915-I*, 325–400, here: 377. According to another source, the stock of electric vehicles numbered 37,000 units, of which 25,000 were passenger cars. "Census shows increase in electrics," *EV* (U.S.) (January 1914), 22.

5. Arthur Williams, "Co-operation in electric vehicle progress," *EV* (U.S.) (May 1913), 3–6, here: 3, 5; W. H. Snow and David W. Beaman, "Value of electric vehicles to central stations," *EV* (U.S.) (October 1913), 201–6, here: 202; "Reviewing 1913 progress," *EV* (U.S.) (February 1914), 65; slogan: "Luncheon by the Commonwealth Edison Co.," *EVAA 1913*, 72–74, here: 72.

6. "Manufacture of automobiles," *EV* (U.S.) (March 1916), 85; for prices, see "Present prices," *EV* (U.S.) (March 1914), 97; *Suburban Life:* "Finds variety of uses for electric," *EV* (U.S.) (July 1913), 97; on Waverley, see Charles C. Havill, "Cost of maintaining a passenger electric," *EV* (U.S.) (September 1913), 171–72.

7. "Present prices," EV (U.S.) (March 1914), 97; second quotation: E. R. Davenport, "Constructive criticism of electric vehicle conditions," *EV* (U.S.) (July 1913), 93–97, here: 94; "Low price passenger cars," *CS* (April 1914), 395–96; "The low price 'Standard,'" *CS* (August 1914), 51–52; "Preparations for Ford Electric," *EV* (U.S.) (June 1914), 212; "Edison outlines Ford Electric," *EV* (U.S.) (July 1914), 34; "Seeing ahead for the electric vehicle: An interview with Thomas A. Edison," *EV* (U.S.) (February 1917), 41; Michael Brian Schiffer (with Tamara C. Butts and Kimberly K. Grimm), *Taking charge: The electric automobile in America* (Washington, D.C., 1994), 130–31, 149–50, 153–54, 166–67.

8. J. S. Codman, "Touring by electric automobile," *EV* (U.S.) (July 1914), 23–24; also published in *CS* (June 1914), 463–66; "New high-speed electric roadster," *ER* (New York) (4 November 1911), 946; "The long distance high-speed electric passenger vehicle is here," *CS* (March 1912), 268. In 1910, Babcock in Buffalo, New York, also proposed a "racy roadster" with an aluminum chassis and a steel nose, which could go more than 110 km/h, according to manager F. A. Babcock. "Babcock standards," *CS* (October 1910), 109–10.

9. "Electric touring book," *EV* (U.S.) (July 1914), 39; "Electric vehicle charging stations: Charging and garaging within a one hundred mile radius of New York City," *EV* (U.S.) (February 1915), 44–45; "Charging stations in Philadelphia and environs; Eastern charging facilities well distributed, offer excellent opportunity for touring," *EV* (U.S.) (May 1915), 169–70; on Goodrich, see *EV* (U.S.) (June 1915), 204.

10. "Washington and return in a Detroit Electric," *CS* (July 1914), 22; "Boston-Philadelphia run with Philadelphia storage battery," *CS* (August 1914), 48–49; "A remarkable performance," *CS* (December 1915), 160; "New mileage record," *CS* (December 1915), 161; "Electric no longer town car," *EV* (U.S.) (March 1916), 90; "The unimportance of touring," *EV* (U.S.) (August 1916), 51–52, here: 51; "Two percent of trips over sixty miles," *EV* (U.S.) (April 1915), 158.

11. "Address of Dr. Charles P. Steinmetz," *NELA-II*, 162–73, here: 165–68; "New

type Dey Electric announced," *EV* (U.S.) (April 1915), 137–38; "Has the cheap electric arrived?" *EV* (U.S.) (December 1916), 197–98; quotation *New York Times:* Ernest Henry Wakefield, *History of the electric automobile: Battery-only powered cars* (Warrendale, 1994), 220.

12. Richard Wager, *Golden wheels: The story of the automobiles made in Cleveland and Northeastern Ohio 1892–1932* (Cleveland, 1986), 212, 217–18; R. Thomas Wilson, *Baker Raulang: The first hundred years, eighteen fifty three–nineteen fifty three* (Cleveland, 1953), 31–33.

13. Ken Ruddock, "Recharging an old idea: The hundred-year history of electric cars," *Automobile quarterly* 31 no. 1 (Fall 1992), 37–38; "Detroit electric cars for 1914," *CS* (February 1914), 338; "Development of the Ohio Electric," *EV* (U.S.) (January 1916), 11ff.

14. "Woods electric now a self-charger; adds gasoline auxiliary power and bids central stations good bye," *EV* (U.S.) (February 1916), 37–38.

15. H. Petit, "Les pneus à cordes," *VA* (4 May 1912), 282–83; "Les succès des pneumatiques Palmer," *CGP* (15 March 1913), 7132; S. V. Norton, "Relation of tires to the efficiency of electric vehicles," *NELA 1916-IV*, 91–113, here: 93–104; W. H. L. Watson and R. J. Mitchell, "Electric automobiles in England," *EV* (U.S.) (July 1913), 103–7, here: 107.

16. "Tires for electrics," *EV* (U.S.) (July 1913), 90; Harvey Graves, "Special features of tires for electrics," *EV* (U.S.) (August 1913), 129–30.

17. Eric Tompkins, *The history of the pneumatic tyre* (n.p. [Birmingham], 1981), 22.

18. "The first ten years" (manuscript), Box 2–5 in Goodyear archives (italics in original).

19. P. W. Lichfield, "History of the pneumatic tire design," *EV* (U.S.) (June 1914), 234–37, here: 236.

20. T. D. Brewster, "Cord tires for passenger electrics," *EV* (U.S.) (March 1915), 111–12; John F. Palmer, "Construction of cord tires," *EV* (U.S.) (August 1915), 67–69; Ralph C. Epstein, *The automobile industry: Its economic and commercial development* (Chicago, 1928), 107.

21. Norton, "Relation of tires to the efficiency of electric vehicles," 102–4; "Pneumatics for trucks," *EV* (U.S.) (December 1917), 187; *Goodyear tire news* (November 1915), 6; see also Henry C. Pearson, *Pneumatic tires* (New York, 1922), 48.

22. E. S. Mansfield, "Report of the Electric Vehicle Section," *EV* (U.S.) (August 1917), 47; "Impressive electric vehicle figures," *EV* (U.S.) (March 1917), 101; "Memphis' 600,000 Kw.-Hr. vehicle load," *EV* (U.S.) (May 1917), 173–74, here: 173.

23. *NELA 1918*, 293.

24. *NELA 1921-I*, 385; *NELA 1920* (New York, n.d. [1920]), 451, 455; Arthur Williams, "The electric vehicle," *NELA 1921-I*, 97.

25. S. Scrimger, *The electric taxicab; EVAA 1915*; reprinted in *EV* (U.S.) (November 1915), 169–70; probably founded in 1907: *EV* (U.S.) (November 1915), 170; on "one of the oldest," see Marshall, "Electric taxicabs," *CS* (April 1915), 319–21, here: 319.

26. "Electric taxicab successful in Detroit," *EV* (U.S.) (February 1915), 54; "Electric taxicabs for New York," *EV* (U.S.) (January 1916), 16; "Women to drive taxicabs in Detroit," *EV* (U.S.) (December 1917); A. Jackson Marshall, "Women to drive taxicabs," *CS* (July 1917), 93–94, here: 94; "Electric taxicabs for Chicago," *NB* (1916), 475.

27. "Electric taxicabs for principal cities," *EV* (U.S.) (May 1915), 175–76, here: 175; "Harvey Robinson, electric vehicle expert sees future for electric taxi," *CS* (January 1916), 189; A. Jackson Marshall, "The electric vehicle situation," *CS* (December 1917), 173–77, here: 176.

28. A. Jackson Marshall, "Electric taxicabs," CS (April 1915), 319–21, here: 319;

"Electric taxicabs for New York," 16; "Electric taxicabs for Chicago," 475; "A short history of the electric taxi-cab: Development leading to the latest models in Chicago," *EV* (U.S.) (July 1916), 1–3.

29. Percival White, *Motor transportation of merchandise and passengers* (New York, 1923), 364–65; New York: "The new electric taxicab," *NB* (November 1922), 685; "An electric taxicab," *EV* (U.K.) (August 1922), 16–18; "United States electric vehicle market," *EV* (U.K.) (January 1923), 31.

30. R. Winckler, "'Strom statt Benzin': Akkumulator-Fahrzeuge und ihre Bedeutung als Stromverbraucher" (typescript, Varta archives, file II D 2: "1909–1935"), 5.

31. Michael Evgénieff, *Droschkenbetrieb: Seine Organisation und wirtschaftspolitische Probleme* (Berlin, 1934), 41–42, 48, 64–67, 166–68; Erwin Knaths, *Die Entwicklung des Berliner Droschkenfuhrwesens unter besonderer Berücksichtigung seiner Motorisierung: Eine verkehrsgeschichtliche Studie* (Marburg a.d. Lahn, 1929), 32–35, 76–77.

32. *Electric vehicles and other alternatives to the internal combustion engine* (Washington, D.C., 1967), 222; on the Automobile Show 1924, see Alan P. Loeb, "Birth of the Kettering doctrine: Fordism, Sloanism and the discovery of tetraethyl lead" (unpubl. lecture, 41st Business History Conference, Fort Lauderdale, 10 March 1995), 13 (I am indebted to Alan Loeb for making available the text of his presentation); on the production of 1922 and 1925, see Grace R. Brigham, "Those elusive vehicles: A history of the Society of Automotive Historians," *Automotive history review* (Fall 1995), 8; Sinsabaugh, *Who, me? Forty years of automobile history*, 156.

33. *NELA 1927*, 454.

34. Gijs P. A. Mom and David A. Kirsch, "Technologies in tension," *Technology and culture* 42, no. 3 (July 2001,) 489–518, here: 489, 518.

35. Wager, *Golden wheels*, 218; Wilson, *Baker Raulang*, 31–39 (quotation: 35); on the production of 1947 and the fleet of 1954, see "Electric trucks in America," *EV* (U.K.) (December 1951), 105.

36. F. Hubrig, "Das Kraftfahrzeug im Dienst der Deutschen Reichspost," in Allmers et al. (eds.), *Das deutsche Automobilwesen der Gegenwart* (Berlin, 1928), 2; Ketelhohn, *Das Elektrofahrzeug in der deutschen Kraftverkehrswirtschaft* (Heft 1 of: W. Schuster [ed.], *Wirtschaft und Technik* (Berlin, 1940), 28–29.

37. Winckler, "Strom statt Benzin," 6–7.

38. Wegner, "Die elektrischen Strassenreinigungsfahrzeuge der Stadt Berlin," *Das Elektrofahrzeug* (January 1936), 2; Max Pöhler, "Das Elektrofahrzeug unter besonderer Berücksichtigung seiner Verwendungsmöglichkeit in städtischen Fuhrparksbetrieben (Varta archives, file II D 2: "1938–1954"), no page numbers; "Municipal electrics in Berlin." *EV* (U.K.) (June 1944), 269; G. Klinner and R. Welzel, "Sonderfahrzeuge für Städtereinigung," in Allmers et al. (eds.), *Das deutsche Automobilwesen der Gegenwart*, 124–26; F. Krause, "Dresden hat 700 Elektrofahrzeuge," *EF* (February 1938), 5–7, here: 6.

39. G. Lucas, "Elektrokarren als Stromverbraucher," *Der Werbeleiter* (1933 Heft 10/11), 1 (Sonderdruck, Varta archives, file II D 2: "1909–1935"); H. Müller, "Das Elektrofahrzeug als Grossabnehmer elektrischer Energie," *Elektrizitätswirtschaft* (June 1932), first page (no page numbers) (Sonderdruck, Varta archives, file II D 2: "1909–1935").

40. "Bestand an Elektrofahrzeugen in Deutschland am 1. Januar 1939," *TA-Nachrichten* (Folge 1, January 1940), 6, 10; Hanns Köhlmann, "Produktionssteigerung durch Einsatz von Elektrofahrzeugen," *EF* (February 1939), 1–2, here: 2, gives 20,000 industrial trucks for the beginning of 1939.

41. "The French view of battery traction," *EV* (U.K.) (March 1922), 11; Witte, "Eine

französische Denkschrift über akkumulator-elektrische Fahrzeuge," *EF* (August 1941), 2.

42. M. C. Chalumeau, "Die elektrischen Akkumulatoren-Fahrzeuge der Stadt Lyon" (undated typescript, Varta archives, file II D 0: "1909–1935"); "Battery vehicles in France," *EV* (U.K.) (September 1934), 240; "La Société Lyonnaise pour l'Exploitation de Véhicules électriques," *VE* (October 1930), 28–32; "Electrics in France," *EV* (U.K.) (August 1947), 34–36.

43. L. Kriéger, M. Buchon, E. Lacroix, et al., *Le véhicule électrique utilitaire à accumulateurs* (Paris, 1947), 20–24; on fleet size, see "Electric vehicles in France," *EV* (U.K.) (June 1944), 286; Sovel and Jourdain et Monneret: G. Mestayer, *Les véhicules électriques* (Paris, 1941), 86–93; on the year of foundation of Sovel, see Henri Petit, *La voiture électrique à accumulateurs* (Paris, 1943), xi (advertisement).

44. C. W. Marshall, *Electric vehicles* (London, 1925), 90; Kloss, *Elektrofahrzeuge: vom Windwagen zum Elektromobil*, 109–10; "Electric trucks in Switzerland," *EV* (U.K.) (June 1942), 268–70, 275.

45. "Les applications de la traction par accumulateurs aux poids lourds," *VE* (April 1930), 1–12, here: 9 and 12; "Das Elektrofahrzeug im Ausland," *EF* (November 1936), 5; "Ein Gesetz über Elektrofahrzeuge in Italien," *EF* (November 1940), 3; "Förderung der Einführung von Elektromobilen in Italien," *TA-Nachrichten* (Folge 9, September 1940), 15; on the freight trolley, see "Een plan tot electrificatie van wegen in Italië," *Wegen* 15 (1939), 253–54.

46. "Electric vehicles in Norway," *EV* (U.K.) (October 1921), 16–18; "The new Electric Vehicle Association in Norway," *EV* (U.K.) (June 1927), 180.

47. "The electric commercial vehicle," *EV* (U.K.) (December 1914), 5–6, here: 6; Stanley M. Hills, *Battery-electric vehicles: Dealing with the construction and operation of all types of battery-operated electric vehicles and accessory equipment* (London, 1943), 7.

48. James M. Laux, *The European automobile industry* (New York, 1992), 245; "Census of mechanically-propelled road vehicles in Great Britain (up to September 30th, 1946)," *EV* (U.K.) (November 1947), 112; "Electric Vehicle Association; review for the year 1951–1952," *EV* (U.K.) (July 1952), 14; A. W. Reid, "The effective use of milk delivery transport," *Dairy Industries* (May 1952), 422–28, here: 427 (Varta archives, file II D 2: "1938–1954"); world leader: H. W. Heyman, "The electric vehicle (Address . . . at the 8th Annual Meeting of the Scottish Co-operative Transport Association . . . Glasgow . . . 11th March, 1952)," 2 (Varta archives, file II D 2: "1938–1954"); Germany: Max Pöhler, "Das Elektroauto in Vergangenheit und Gegenwart," 19 (undated typescript [1967?], Varta archives).

49. Although the fourth-generation electric passenger car is only briefly touched upon in this study, its main characteristics can be determined as follows. Like the third generation it is solidly based upon mainstream gasoline (and in the meantime also diesel) automotive technology. It is sometimes enriched with high-tech characteristics to provide it some break-away potential from the mainstream, for instance, by using space-age technology, or by trying to use ultralight materials, or even by designing completely new car types like small city cars painted in bright colors, fitting in a new urban lifestyle. Further research should decide whether this urge to be at the forefront of automotive technology provides enough ground to differentiate between a postwar fourth-generation electric ("conversion cars" such as the "Eco-Golf" by Volkswagen of the 1980s and early 1990s, the Fiat Cinquecento Elettra [1992] or the Daihatsu Cuore EV [1993]) and a fifth-generation electric (like the German "Hotzenblitz" [1994] or the General Motors EV-1). It is remarkable that a technology transfer between the alternatives is frequent. For instance, European tire manufacturers developed special, ultra-low

rolling resistance tires for these new city cars, but they were immediately taken over by mainstream manufacturers and their main characteristics are now part of "normal" tire design for gasoline and diesel cars. Also, the city-car concept itself has meanwhile been integrated into mainstream manufacturers' product range.

50. See, also for the following: David A. Kirsch, "The electric car and the burden of history: Studies in automotive systems rivalry in America, 1890–1996" (diss., Stanford University, 1996), 255–80, quotations: 259. The best analysis of this postwar stage in the American electric-vehicle history can be found in Dietmar Abt, *Die Erklärung der Technikgenese des Elektroautomobils* (Frankfurt am Main, 1998), 203–41. Abt analyzes this stage as a struggle between an artifact-centered and a system-centered approach. In the first case the passenger car gets all the attention, in the second case the (electric truck) fleet.

51. Max Pöhler, "Das Elektroauto in Vergangenheit und Gegenwart," 26–31 (undated typescript [1967?], Varta archives); Gijs Mom, "De moderne elektro-auto," in Gijs Mom and Vincent van der Vinne, *De elektro-auto: een paard van Troje?* (Deventer, 1995), 11–110, here: 30–40; for the technology of the European and Japanese electric car from this stage, see Ernest Henry Wakefield, *History of the electric automobile: Battery-only powered cars* (Warrendale, 1994), 331–36; for the technology of the German electric cars of the 1950s and 1960s, see Gerhard Wilke (ed.), *Denkschrift Elektrospeicherfahrzeuge* (Wiesbaden, 1970); and for an overview of the German fleet trials between 1976 and 1985 in Berlin, see Horst Bomke and Wilfried Porsinger, "Strom macht mobil: Elektro-Strassenfahrzeuge und Elektrowasserfahrzeuge in Berlin," 2nd ed. (Berlin, 1986) (Bewag brochure), 9–31; for an example of the social science approach, see Tarja Cronberg and Knut H. Sørensen (eds.), *Similar concerns, different styles? Technology studies in Western Europe* (Proceedings of the COST A4 workshop in Ruvaslahti, Finland, 13 and 14 January 1994) (Brussels, 1995).

52. Roland Wolf, *Le véhicule électrique gagne le coeur de la ville* (n.p., n.d. [Paris, 1995]), 47, 83, 88; Mom, "De moderne elektro-auto," 22–23; Michael Shnayerson, *The car that could: The inside story of GM's revolutionary electric vehicle* (New York, 1996), 133; "Auch Kleinstwagen muessen sicher sein," *Automobil-Revue* (31 August 1995); M. Rehsche, "Smart und A-Klasse," *Automobil-Revue* (14 March 1996); "Smart-Macher unter Zeitdruck," *Automobil-Revue* (16 May 1996); on the three-cylinder engine, see *Automobil-Revue* (3 August 1995).

53. For the following, see Shnayerson, *The car that could*, passim; Kirsch, "The electric car and the burden of history," 276–79; Mom, "De moderne elektro-auto," 11–25; Daniel Sperling (with contributions from Mark A. Delucchi, Patricia M. Davis, and A. F. Burke), *Future drive: Electric vehicles and sustainable transportation* (Covelo, Calif., 1995), passim.

54. On Kleenex, see Shnayerson, *The car that could*, 82, 104, 139; all other quotations: Scott A. Cronk, *Building the E-motive industry: Essays and conversations about strategies for creating an electric vehicle industry* (Warrendale, 1995), 53–54, 58, 101, 113.

55. 14,000 cars: *Electrifying times* 4, no. 1 (Spring–Summer 1996 edition), 37; quotations: Shnayerson, *The car that could*, 183, 274 (note 18) (my italics).

56. Gijs Mom, "Main stream and its alternatives: The electric vehicle as a critical comment on the combustion engined car," in *A future for the city: Electric Vehicle Symposium 15*, Brussels, 29 September to 3 October 1998 (on CD-ROM).

Epilogue

Quoted by Lisa Callaghan, "Publicity war breaks out over zero emission mandate," *Automotive environment analysis* 66 (July 2000), 20–21, here: 20.

1. Rudi Volti, "Why internal combustion?" *Invention and technology* (Fall 1990), 42–47.

2. Robert J. Kothe, "Electric automobiles," in George S. May (ed.), *The automobile industry, 1896–1920* (New York, 1990), 172–73.

3. Michael Brian Schiffer (with Tamara C. Butts and Kimberley K. Grimm), *Taking charge: The electric automobile in America* (Washington, D.C., 1994), 120.

4. Ibid., 139; G. N. Georgano, *Auto's uit de jaren 1886–1930* (Alphen aan den Rijn, 1986), 38 (Schiffer quotes from the original British publication *Cars, 1886–1930*).

5. Virginia Scharff, *Taking the wheel: Women and the coming of the motor age* (New York, 1991), 51, 82, 117, 127–28 (especially note 69), 167; Scharff in this connection speaks about "driving in the 1920s," but, in view of her analysis, her remark also concerns the early period. Last quotation: Virginia Scharff, "Gender, electricity, and automobility," in Martin Wachs and Margaret Crawford (eds.), *The car and the city: The automobile, the built environment, and daily urban life* (Ann Arbor, 1992), 80.

6. Scharff, *Taking the wheel*, 124, 65.

7. Ibid., 49; last quotation: *Isis* 86, no. 2 (1995), 351–52, here: 352 (my italics). Calling the electric car "underpowered" is based on a persistent misunderstanding of the technical properties of the combustion engine and the electric motor and is a typical token of an approach to the electric car from the point of view of the "prevailing technology." Whereas the combustion engine needs a large power reserve to prevent stalling while climbing a hill, the electric motor can be briefly overloaded to a few multiples of the nominal power rating. So, the *nominal* power ratings of both cannot simply be compared.

8. Schiffer, *Taking charge*, 169.

9. Ibid., 35–43; quotation Argo: 38.

10. "Starting 'Electric Vehicles,'" *EV* (U.S.) (May 1913), 17–18, here: 17; NELA, 39th convention, 22–26 May [1916], Chicago (n.p., n.d. [New York, 1916]), *Accounting Section sessions; papers, reports and discussions*, 43 (my italics).

11. "Converted by a taxi," *CS* (October 1914), 106.

12. "Electric passenger vehicles," *CS* (July 1916), 6–8, here: 6.

13. *NELA 1915-I*, 395; F. M. Feiker, "The customer's view-point on the electric vehicle," *NELA 1917-V*, 71–77, here: 72 (italics in original).

14. "The electric is everybody's car," *EV* (U.S.) (May 1913), 32; A. Jackson Marshall, "Review of electric vehicle industry and forecasts for the coming year," *CS* (January 1916), 170–73, here: 172; *CS* (January 1913), 226; *CS* (February 1913), 258.

15. "May meeting of the Electric Vehicle Association of America . . . etc.," *CS* (June 1913), 404–13, here: 407.

16. "The kind of car a man wants," *EV* (U.S.) (June 1916), 173–74, here: 174.

17. "The woman's influence," *EV* (U.S.) (September 1916), 98.

18. "The electrified gas car," *EV* (U.S.) (January 1916), 60; T. A. Boyd, "The self-starter," *Technology and culture* 9 (1968), 585–91; "Starters and starting cranks for 1912," *CV* (March 1912), 16–18, here: 16; last quotation: C. E. Michel, *Where we stand today; EVAA 1912*, 4–5.

19. That does not mean, by the way, that Ford did not try to introduce a starter motor of its own. The ENHS archives contain correspondence that shows plans for an "Edison-Hartford system," based on an Edison battery. According to a handwritten draft for a letter by Edison, "we have an order from the Ford Auto [sic] Co. for about 4 million dollars worth of battery per year for their new self starter" (undated [probably 1914] with the legend "Harry"). Two typed reports confirm work on this system ("Test on electric cranker for 'Ford' touring car" [26 February 1912] and "Report of experiment on

Ford and Edison Electric Lighting system from May 20, 1914 to May 27, 1914" [27 May 1914], both ENHS).

20. For the following, see Gijs Mom, "Das Holzbrettchen in der schwarzen Kiste: Die Entwicklung des Elektromobilakkumulators bei und aus der Sicht der Accumulatoren-Fabrik AG (AFA) von 1902–1910," *Technikgeschichte* 2, Bd. 63 (1996), 119–51, here: 135–41.

21. Schiffer, *Taking charge*, 106; Robert E. Russell, "Mercury arc rectifiers," *CS* (January 1911), 192–96, here: 192.

22. E. E. la Schum, "The electric truck in modern transportation," *NELA 1919-V*, 150–62, here: 160.

23. "Edison's new battery," *New York World* (15 February 1903) (my italics); "Baiting Edison's battery," *Motor world* (29 January 1903), 627; Joseph B. Baker, "Thomas A. Edison's latest invention: A storage battery designed and constructed from the automobile user's point of view," *Scientific American* (14 January 1911), 30–47, here: 30 (my italics); advertisement: *Scientific American* (9 December 1911), 535. See also Gijs Mom, "Inventing the miracle battery: Thomas Edison and the electric vehicle," *History of technology* 20 (1998), 18–45.

24. Lord Montagu of Beaulieu and F. Wilson McComb, *Behind the wheel: The magic and manners of early motoring* (New York, 1977), 192; W. Rödiger, "Elektromobile und Elektrokarren: Verwendungszwecke und Betriebswirtschaftlichkeit," in Allmers et al. (eds.), *Das deutsche Automobilwesen der Gegenwart* (Berlin, 1928), 131.

25. Thomas P. Hughes, *Networks of power: Electrification in Western society, 1880–1930* (Baltimore, 1983), 7–17.

26. That Hughes's model is not immediately suitable for application to early automotive history is also related to his preference for *production* systems, whereas our analysis is focused on *use*. Although Hughes, for example in his *American genesis: A century of invention and technological enthusiasm 1870–1890* (New York, 1989), 220–26 pays explicit attention to the "automobile production and use system" (220), he restricts himself to fuel *production* in the form of oil refining there. A comparable emphasis on the coordinating endeavors in system building can also be found in Hughes's analysis of the urban car-tunnel project in Boston in the 1980s and 1990s (Thomas P. Hughes, *Rescuing Prometheus* [New York, 1998], 197ff.). However, in his analysis of the development of Arpanet, the predecessor of the World Wide Web, Hughes recognizes the complete absence of a central regulator (297).

27. Angela Zatsch, *Staatsmacht und Motorisierung am Morgen des Automobilzeitalters* (Konstanz, 1993) (diss.), 515 (note 1); telegraph: Alain Beltran and Patrice A. Carré, *La fée et la servante: la société française face à l'électricité, XIXe–XXe siècle* (Paris, 1991), 45. For a comparable analysis of the air traffic system as "largely a mental rather than a physical construct," see Todd R. La Porte, "The United States air traffic system: Increasing reliability in the midst of rapid growth," in Renate Mayntz and Thomas P. Hughes (eds.), *The development of large technical systems* (Frankfurt am Main, 1988), 215–44, here: 223.

28. Beltran and Carré, *La fée et la servante*, 164–67; Eugen Weber, *France, fin de siècle* (Cambridge, Mass., 1986), 128–29. Up to now, the historiography of electricity production has restricted itself mainly to the centralist, large systems. But the decentralist developments, which were at least as important for prewar history, have been grossly neglected. For a recent, very welcome alternative approach, see Erik van der Vleuten, *Electrifying Denmark: A symmetrical history of central and decentral electricity supply until 1970* (diss., University of Aarhus, 1998), and idem, "Autoproduction of electricity: Cases from Danish industry until 1960," *Polhem: Tidskrift för teknikhistoria* 14 (1996), 118–54.

For a further analysis of the antisystemic character of early automobility, see Gijs Mom, "Networks, systems and the European automobile: A plea for a mobility history programme" (review essay for the first A^MES Workshop, Scenario 1: European Infrasystem, Torino, 2–4 November 2001).

29. Alan Loeb, "Lead in gasoline: The harmonious order" (part I of a manuscript for a book about the introduction of leaded gasoline in America; December 1994), 42–44 (my thanks to Alan Loeb for making his manuscript available). The gasoline car, by the way, initially also used large quantities of water for engine cooling.

30. M. G. Lay, *Ways of the world: A history of the world's roads and of the vehicles that used them* (New Brunswick, N. J., 1992), passim; Dominique Barjot, "Advances in road construction technology in France," in Theo Barker (ed.), *The economic and social effects of the spread of motor vehicles: An international centenary tribute* (Houndsmill, 1987), 291–312; Gerhard Horras, *Die Entwicklung des deutschen Automobilmarktes bis 1914* (Munich, 1914), 314.

31. *Motor Age* quoted by James J. Flink, *America adopts the automobile, 1895–1910* (Cambridge, Mass., 1970), 211.

32. Flink, *America adopts the automobile,* 212. For a convincing argument that it was the truck, rather than the passenger car, that stimulated early road construction in the United States, see Louis Rodriquez, "The development of the truck: A constructivist history" (unpubl. diss., Lehigh University, 1997).

33. In an advertising brochure from 1888, Karl Benz recommends his first automobile as "Replacement for horse and carriage!" quoted by Siegfried Reinecke, *Mobile Zeiten: Eine Geschichte der Auto-Dichtung* (Bochum, 1986), 28; James Foreman-Peck and Masahiro Hayafuji, "Lock-in and Panglossian selection in technical choice: The power source for the motor car" (manuscript), 16, 23.

34. George Kirkham Jarvis, "The diffusion of the automobile in the United States: 1895–1969" (unpubl. diss., University of Michigan, 1972); Gijs Mom and Peter Staal, "Autodiffusie in een klein vol land: Historiografie en verkenning van de massamotorisering in Nederland in international perspectief," in Yves Segers, Reginald Loyen, Guy Dejongh, and Erik Buyst (eds.), *Op weg naar een consumptiemaatschappij: Over het verbruik van voeding, kleding en luxegoederen in België en Nederland (19de–20ste eeuw)* (Amsterdam, 2002), 139–80.

35. Kurt Möser, "Benz, Daimler, Maybach und das System Strassenverkehr: Utopien und Realität der automobilen Gesellschaft," *LTA-Forschung,* issue 27/1998 (Mannheim, December 1998), 23. Möser uses the phrase "mechanization of the existing world," ignoring, like many historians of the automobile, that the technical history of mobility has a longer tradition than that of the automobile. For an analysis of the "domestication of the car user" by the building of roads and other traffic control measures, see Gijs Mom, "Networks, systems and the European automobile: A plea for a mobility history programme" (review essay for the first A^MES Workshop, Scenario 1: European Infrasystem; Torino, 2–4 November 2001).

36. J. W. Schot e.a., "Concurrentie en afstemming: water, rails, weg en lucht," in Johan Schot et al., *Techniek in Nederland in de twintigste eeuw,* vol. 4 (Zutphen, 2002), 19–43.

37. G. Mom and H. Scheffers, *De complexe aandrijflijn* (part 3B of G. Mom [ed.], *De nieuwe Steinbuch; De automobiel*) (Deventer, 1993), 119.

38. Michael Schiffer, "Social theory and history in behavioral archaeology," in J. M. Skibo, W. H. Walker, and A. E. Nielsen (eds.), *Expanding archaeology* (Salt Lake City, 1995), 22–35.

39. Daniel Sperling, *Future drive: Electric vehicles and sustainable transportation*

(Covelo, Calif., 1995), 2; Peter Peters, "Vergroeid met het gaspedaal: Nederland barst van de auto's," *Humanist* (February 1996), 23–28; idem, "De cultus van snelheid: Nederland barst van de auto's, deel 2," *Humanist* (March 1996), 27–32.

40. Graham Johnson, "Topping the bill," *Electric & hybrid '99: Vehicle technology international* (New Malden, Surrey, 1999), 6–9; Bernhard Schwab, "Elektrifierung des Automobils," *Automobil Revue* (29 June 2000), 12; Jim McCraw, "Leader of the Stack," *Engine technology international* (2002), No. 1, 24–27, here: 24.

41. The "Integrated Starter Generator" was introduced by Volvo in a concept car based on the S80 model in 2001. Roger Gloor, "Ein weiterer Schritt zum Startergenerator," *Automobil Revue* (13 December 2001), 22.

A Note on Method

1. Everett M. Rogers, *Diffusion of innovations*, 4th ed. (New York, 1995); on the most researched, see Vijay Mahajan and Robert A. Peterson, *Models for innovation diffusion* (Beverly Hills, 1985), 7. For an extensive critique of the quantitative approaches of the diffusion phenomenon, see Gijs Mom and Peter Staal, "Autodiffusie in een klein vol land: Historiografie en verkenning van de massamotorisering in Nederland in international perspectief," in Yves Segers, Reginald Loyen, Guy Dejongh, and Erik Buyst (eds.), *Op weg naar een consumptiemaatschappij: Over het verbruik van voeding, kleding en luxegoederen in België en Nederland (19de–20ste eeuw)* (Amsterdam, 2002), 139–80. This note on method is a shortened version of Gijs Mom, "Conceptualising technical change."

2. Arnulf Grübler, *The rise and fall of infrastructures: Dynamics of evolution and technological change in transport* (Heidelberg, 1990); more recently also Arnulf Grübler, *Technology and global change* (Cambridge, 1998); Gijs Mom, "Competition and coexistence: Motorization of land transportation and the substitution of the horse," *Achse, Rad und Wagen* (forthcoming, in a German translation).

3. Arthur concludes that a "founder effect" akin to that in genetics can be used to explain the importance of random historical events in the competition between technologies. W. Brian Arthur, "Competing technologies, increasing returns, and lock-in by historical events," *The economic journal* 99 (March 1989), 116–31, here: 126–27.

4. For the concept of the "reverse salient," see Thomas P. Hughes, *Networks of power: Electrification in Western society, 1880–1930* (Baltimore, 1983), 7–17.

5. These corrections stem from two different traditions. First, a revisionist approach within economic history introduced internal as well as external constraints on the choice process of "economic man." See Herbert A. Simon, "A behavioral model of rational choice," *Quarterly journal of economics* 69 (1955), 99–118; see also Armen A. Alchian, "Uncertainty, evolution, and economic theory," *Journal of political economy* 58 (June 1950), 211–21. On the other hand, behavioral psychology studied "the motives, attitudes, and expectations of consumers and businessmen," and introduced the concept of "habitual behavior" or "routine" as an analyzing tool, thus rescuing economic theory from the inconveniences of acknowledging the existence of "irrational choice." See George Katona, *Psychological analysis of economic behavior* (New York, 1951), 4, 49; George Katona, *The powerful consumer: Psychological studies of the American economy* (New York, 1960), 138–39.

6. Claude S. Fischer, *America calling: A social history of the telephone to 1940* (Berkeley, 1992), 10–11.

7. Nathan Rosenberg, "On technological expectations," *The economic journal* 86 (September 1976), 523–35, here: 531 (emphasis in original). However, Rosenberg also uses the (at the time: unproven) "total superiority of the internal combustion engine"

(532) to explain the "failure" of the electric vehicle. Nathan Rosenberg, "Factors affecting the diffusion of technology," *Explorations in economic history* (1972), 3–33, here: 23.

8. Steffen Koolmann, *Leitbilder der Technikentwicklung: Das Beispiel des Automobils* (Frankfurt, 1992).

9. The introduction of a measure of closeness (not to be confused with the closure concept from economic and social theories) of the technical field meets the justifiable wish of Walter Vincenti, "The technical shaping of technology: Real-world constraints and technical logic in Edison's electrical lighting system," *Social studies of science* 25 (1995), 553–74, here: 565, to devise some "kind of hierarchy of real-world constraints, depending on degree of directness and restriction." As examples of absolute boundaries, Vincenti mentions the *perpetuum mobile* and gravitation in the design of airplanes. Sometimes even the field of expectation can be a constraining factor. This was the case, for instance, when the American electric vehicle "booster" EVAA narrowed its attention to commercial vehicles in big fleets, thus hampering a further expansion into other fields of application.

10. For a taxonomy of failure factors, see Hans-Joachim Braun, "Introduction" ("Symposium on 'failed innovations'"), *Social studies of science* 22 (1992), 217–26.

11. Max Weber called such an abstraction an Ideal Type, which "belong(s) to the world of ideas" and is "inductively abstracted from reality" (quoted by Mikael Hård, *Machines are frozen spirit: The scientification of refrigeration and brewing in the 19th century— A Weberian interpretation* [Frankfurt am Main, 1994], 35). Hård distinguishes between Ideal Type and Archetype, which, unlike the former, actually exists in historical or contemporary reality, and functions as an exemplar (*Vorbild*) in a Kuhnian sense (36–37, 50–51). In this study, especially for the tentative construction of the three generations of the electric vehicle, a third category seems to do the job better: the "Average Type," which describes the common denominator of all concrete artifacts at a certain point in time. The latter is constructed from existing structural components of these artifacts, but, like the Ideal Type, it mostly has no equivalent in artifactual reality. Whereas the Ideal Type is often to be found in contemporary handbooks, the Average Type mostly has to be constructed by the historian, based on the evolutionary population concept of artifactual variety and its distribution. Thus, "generations" are successions of these Average Types. For instance: while the Ford Model T may have functioned as an Archetype (as a sturdy, cheap, high-wheeled family car) during the period between 1910 and 1925, and may have been treated by contemporaries as the Ideal Type, it certainly did not coincide with the Average Type of the car during that period; this becomes clear if the historian of automotive technology constructs the latter type for this period, and discovers that the Model T's planetary gear box, its flywheel-system with electricity generator, and its lack of an electric starter motor until the end of the second decade of the twentieth century, are no part of this average.

12. P. Paolo Saviotti and J. Stanley Metcalfe (eds.), *Evolutionary theories of economic and technological change: Present status and future prospects* (Chur, 1991), 16.

13. As quoted in ibid., 25; Giovanni Dosi, "Technological paradigms and technological trajectories," *Research policy* 11 (1982), 147–62, here: 152.

14. First quotation: Reiner Grundmann, "Gibt es eine Evolution von Technik? Überlegungen zum Automobil und zur Evolutionstheorie," *Technik und Gesellschaft* 7 (1994), 13–39, here 29.

15. W. H. Ward, "The sailing ship effect," *Bulletin of the Institute of Physics and Physical Society* 18 (1967), 169. For the history of this competition, see Gerald S. Graham, "The ascendancy of the sailing ship 1850–85," *Economic history review* 9 (August 1956),

74–88; and C. K. Harley, "On the persistence of old techniques: The case of North American wooden shipbuilding," *Journal of economic history* 33, no. 2 (1973), 372–89.

16. Hans Dieter Hellige, "Von der programmatischen zur empirischen Technikgeneseforschung: Ein technikhistorisches Analyseinstrumentarium für die prospektive Technikbewertung," *Technikgeschichte* 60, no. 3 (1993), 186–223, here: 207.

17. Hans-Joachim Braun has repeatedly pointed at the existence of such a mechanism; see, for example, Hans-Joachim Braun, "Gas oder Elektrizität? Zur Konkurrenz zweier Beleuchtungssysteme, 1880–1914," *Technikgeschichte* 47, no. 1 (1980), 1–19. In a way, the Pluto effect has also been formulated before by Wolfgang König: "On the one hand, the 'new' will be designed after the 'old,' which it aims to supplant, and in doing so will integrate elements of the 'old,' on the other hand it adapts itself into the continuing environment of the 'old,' likewise taking over elements of the 'old.'" Wolfgang König, "Technik, Macht und Markt: Eine Kritik der sozialwissenschaftlichen Technikgeneseforschung," *Technikgeschichte* 60, no. 3 (1993), 243–66, here: 248.

18. Rosenberg, "Factors affecting the diffusion of technology," 26. Indeed, the Pluto effect is not restricted to the material world. Nor is the Pluto effect, when applied to the history of technology, primarily material in nature. It very often describes a *functional* transfer, solved in different technological styles. Nor does the Pluto effect necessarily describe conscious processes of imitation, as the mimicry metaphor would suggest. Most changes described by the Pluto effect cannot be explained by direct copying, but rather by a general attitude of engineers and users to keep constantly in touch with the state-of-the-art, which functions as a basis of inspiration for similar solutions. For instance, as we have seen in this study, the desire to decrease plate distance in battery construction was solved by lead battery manufacturers by inserting wooden separators between the plates. Edison, however, achieved the same goal by transforming the plates themselves into a tubular shape. However, when lead battery manufacturers took over the tubular plate concept, they did so not to decrease plate distance, but to counteract shortcircuiting and, thus, enhancing longevity. Gijs Mom, "Inventing the miracle battery: Thomas Edison and the electric vehicle," *History of technology* 20 (1998), 18–45. In a similar vein, the Welsbach mantle in gas lighting was introduced to improve the brightness function (Rosenberg, "On technological expectations," 532).

19. Devendra Sahal, "Alternative conceptions of technology," *Research policy* 10 (1981) 2–24, here: 20; Thomas P. Hughes, *Rescuing Prometheus* (New York, 1998), 31. A comparable fitness for universality seems to be present in steel construction; here, the "increasing competition from aluminum seems to have led to the setting up of product-research and engineering laboratories in the steel industry." Rosenberg, "Factors affecting the diffusion of technology," 28. This, clearly, was not the case with wooden ship technology, as it hurt, during the struggle with the iron hull alternative, on "the natural limitations of wood construction," especially the problems of stress connected with the screw and its shaft. Graham, "The ascendancy of the sailing ship 1850–85," 78.

Bibliography and Resources

Archives, Museums, Libraries, and Private Collections

NATIONAL, STATE, AND MUNICIPAL ARCHIVES

Archives Départementales de la Seine, Paris:

Grand album illustré de l'industrie automobile pour l'année 1902 (Paris, 1902)

Archives Nationales, Paris [AN with box number]:

12 F 7713: Commission des transports postaux, Ire sous commission

19 AQ 38: Entreprise générale des omnibus

65 AQ L2231: Société Française pour l'Industrie et les Mines (l'Indusmine)

65 AQ N16: Richard (1904)

65 AQ N26: Compagnie Française des Automobiles de place (1905)

65 AQ N30: Compagnie Parisienne de Voitures Electriques, Procédés Kriéger (1897–1909)

65 AQ N61: Garages Kriéger et Brasier (1905–14)

65 AQ N193: Société française de construction d'automobiles et d'exploitation de voitures de place pétroléo-électrique (Etablissements Prod'homme et de Poorter réunis) (1908)

65 AQ N194: Société générale d'automobiles électromécaniques (1907)

65 AQ N264: L'Electrique (1898)

65 AQ Q131: Compagnie française de voitures électromobiles (1897–1911)

65 AQ Q169: Compagnie parisienne de fiacres automobiles (1907)

65 AQ Q170: Compagnie parisienne de taxautos électriques (1907–9)

65 AQ Q370³: Compagnie Générale des Voitures à Paris (1866): Coupures de presse 1894–1909

65 AQ Q368³: Compagnie Générale des Omnibus (CGO)

Archives de la Préfecture de Police, Paris

Box B/a 1383: Grèves des mécaniciens de 1892 à 1900

Box D/b 500: Circulation & transports dans Paris et le Dép^{t.} de la Seine: Voitures de place & de remise; taxis; voitures des chemins de fer; voitures de courses; pièces diverses

Box D/b 505: Circulation & transports . . . etc: file "Taxis"; file "Cochères; chauffeuses"

Connecticut State Library, Hartford, Connecticut

Hiram Percy Maxim Collection [Maxim Collection]

Gemeente-archief (municipal archives) Amsterdam [GAA]:
    Archief 635/409: Brandweer
    Archief 733: Collectie C. Poel Jr.
    Archief 809: Collectie Spijker
    Archief 5274: Stadsreiniging
        Akte 5274/170: Stukken betreffende auto's, veegmachines, sproeiwagens,
            etc. 1906–20; dossier 2A
        Akte 5274/171: idem, dossier 2B
    Collectie Hartkamp 1840–1915, port. 44, map 1 [Coll. Hartkamp, with sheet
        number]
    Cat. No. 314/320: Collectie T. A. Haijen, doos 171: Het paard in Amsterdam
    Akte M 763: Amsterdamsche Rijtuigmaatschappij
    Akte M 992.026: Verkeer; taxibedrijf algemeen
    Akte M 992.027: Verkeer; automobielverkeer, taxibedrijf, taxi's
    Akte M 992.030: Verkeer; rijtuigverkeer, ARM
    Akte M 992.031: Verkeer; rijtuigverkeer A–L
    Akte M 992.032: Verkeer; rijtuigverkeer M–Z
    Akte M 1133: N. V. Amsterdamse Rijtuigmaatschappij (ARM)
Gemeente-archief Nijmegen
Gemeente-archief Rotterdam
Gemeente-archief 's-Gravenhage:
    Haagsch Effectenblad 1909–10: articles concerning the Haagsche Automobiel-
        Taxameter-Onderneming (HATO)
Handelskammer Bremen
Landesarchiv Berlin
    Archives nos. 22, 23, 24, 49/1, 49/2, 68 and 72: documentation AFA
    Rep. 14: Deputation für das Verkehrswesen
    Rep. 15–02: Stadtbetriebsamt
    Rep. 256: Bewag
National Museum of American History, Smithsonian Institution Archives, Division
    of Transportation, Washington
Staatsarchiv Bremen:
    Archive 4, 14/1 VI. B.8.b: General-Akten der Polizei-Direktion der freien Hanse-
        stadt Bremen
    Archive 4, 14/1 VI. B.8.c: General-Akten der Polizei-Direktion der freien Hanse-
        stadt Bremen: Taxen
    Archive 4, 14/1 VI. B.8.f: General-Akten der Polizei-Direktion der freien Hanse-
        stadt Bremen: Revisionen der Droschken
    Archive 4, 14/1 VI. B.8.g
    Archive 4, 14/1 VI. B.8.k: General-Akten der Polizei-Direktion der freien Hanse-
        stadt Bremen: Droschkenfuhrwesen in anderen Städten
    Archive 4, 14/1 VI. B.8.p, 2 bis 11: General-Akten der Polizei-Direktion der freien
        Hansestadt Bremen: Droschkenordnung vom 16.11.1909
    Archive 4, 14/1 VI. B.8.q, 1 bis 10: Haupt-Akten der Polizeidirektion Bremen;
        Kraftdroschken
    Archive 4, 75/5 B 94 II
    Archive 4, 75/5 HRB 178: Handelsregisterakten Hansa-Lloyd Werke AG
    Archive 4, 75/5 HRB Fol. 175.406.462 B 94[II:] Bremer Droschken-
        Aktiengesellschaft

Staatsarchiv Hamburg:
>III–I Senat Cl. I lit.N^{b.} Nr. 3 Vol I^z Fasc. 16: Eingabe der Firma Ernst Dello & Co in Hamburg um Zulassung von Benzindroschken, insbesondere . . . (1910–11)
>231–7: Amtsgericht Hamburg-Handels-und Genossenschaftsregister-B 1970–47: Akten betreffend die Firma Hedag (1906–40)
>Zeitungsausschnitt-Sammlung A921: Droschken

ARCHIVES IN MUSEUMS, PUBLIC LIBRARIES; COMPANY ARCHIVES
The Netherlands and Belgium
ARM Groep, Amsterdam: historical archives [ARM archives]:
>The annual reports of the ARM, AEM, and Atax (published with various titles) are given as [Annual Report *ARM* (*AEM, Atax,* resp.) over (year under review)]

Historisch Museum, The Hague: library
Nationaal Rijtuigmuseum, Leek: archives
Openluchtmuseum, Arnhem: library
Tank Museum Brussel (Koninklijk Legermuseum): archives
Technische Universiteit Delft: central library
Technische Universiteit Eindhoven, central library: Collection Van Groningen

Germany, Austria, and Czech Republic
AEG, Firmenarchiv und Museum, Frankfurt am Main
Bremer Landesmuseum zur Kunst-und Kulturgeschichte (Focke-Museum), Bremen: Photo collection Namag
Berliner Feuerwehr, Feuerwehrmuseum, Berlin-Tegel
Deutsches Museum, Munich: library and archives: brochures electric-vehicle manufacturers
Deutsches Feuerwehrmuseum Fulda: archives
Hamburger Elektrizitätswerke
Landesmuseum für Technik und Arbeit, Mannheim: library
Mercedes-Benz-Archiv, Stuttgart (meanwhile rebaptized DaimlerChrysler Archiv) [DaimlerChrysler archives]:
>File "Taxi I: Geschichte; Presseveröffentlichungen"
>Files "Feuerwehr"
>Files "Fremdfirmen": nos. 13, 27, 28
>Folder 290, "Elektro-Daimler"; File "Beteiligungen Österreich"
>See also: Schildberger, Friedrich
Museum Achse, Rad und Wagen, Wiehl, Germany: archives
Museum für Post und Kommunikation, Frankfurt am Main: archives
Museum für Verkehr und Technik, Berlin: library and archives
Národní Technické Muzeum (National Museum of Technology), Prague: library and archives
Technické muzeum Brno
Technische Informationsbibliothek, Hannover
Technisches Museum Vienna:
>MA 68-BI/8/65
>>"Heinrich, Jacob, Ludwig, Richard Lohner-Mappe" [copy book Lohner with letter number]

Varta Batterie AG, Hagen, Germany: historical archives [Varta archives] (meanwhile moved to: Westfälisches Wirtschaftsarchiv, Dortmund, see: Stremmel, "Der Bestand 'VARTA Batterie AG'"). Most sources are accounted for in the notes. Stenographic reports of internal conferences 1897–1928, under various titles (often divided in "Technischer Teil" and "Kaufmännischer Teil"):
Ingenieur-Konferenze [IK with date]
Automobil-Konferenze [AK with date]
Laboratoriums-Konferenze [LK with date]
Varta Batterie AG, Forschungs- und Entwicklungszentrum, Kelkheim, Germany: library [Varta-kelkheim]
Verkehrmuseum Dresden, archives: brochures electric-vehicle manufacturers

France, Italy
Bibliothèque et Documentation de la Société des Ingénieurs de l'Automobiles (SIA), Paris: documentation electric vehicles and Kriéger
Bibliothèque National, Paris
Centre national de la recherche scientifique La Villette, Paris:
     Centre de Recherche en Histoire des Sciences et des Techniques (CRHST)
     La Médiathèque d'Histoire des Sciences
Centro di Documentazione Museo dell' automobile, Turin
     Cartella n. 82/1, busta 1: documentation STAE-Kriéger
Conservatoire National des Arts et Métiers (CNAM), Paris: library
Musée Automobile de la Sarthe, Le Mans: Kriéger-automobile
Musée National de la Voiture et du Tourisme, Compiègne: file Kriéger
Musée des Transports, Paris: archives
Renault, Paris:
     historical archives: 91 AQ 7, 9, 10, 42, 49, 51, 52

Great Britain
British Library, London
Science Museum, London:
     archives: file Bersey
     Science Museum Library
SP Tyres UK Ltd, Fort Dunlop, Birmingham

United States
American Automobile Manufacturers Association (AAMA), Detroit, Mich.: Patent Library (meanwhile transferred to the Kettering Archives, Kettering University, Flint, Mich.)
Detroit Public Library: The National Automotive History Collection [NAHC]: library archives: "Henry Cave Collection" [Cave Collection, NAHC, with series, box, and item number]
Edison National Historic Site (ENHS), U.S. Department of the Interior, National Park Service, West Orange, N.J.: The Edison Archives [ENHS]
Engineering Societies Library, New York:
     EVAA-Proceedings (see Conference reports and annual reports) (collection meanwhile moved to Linda Hall Library, Kansas City)
The Free Library of Philadelphia:
     Automobile Reference Collection
Henry Ford Museum & Greenfield Village, Dearborn, Mich.

Library of Congress, Washington, D.C.
New Jersey Historical Society, Boston:
    William F. D. Crane Collection [Crane Collection with box and folder numbers]
New York Public Library, Science Division:
    EVAA-Proceedings (see Conference reports and annual reports)
Van Pelt Library, University of Pennsylvania, Philadelphia
University of Akron, Ohio, central library:
    Goodyear Files:
        D-1–8, Box 7–8, 7–9 (*India Rubber Journal*)
        D-1–8, Box 7–10 (*Goodyear Tire News*)
        Box 7–12 (Annual Reports)
        Box 2–5: Folder 16 (History), "The first ten years" (manuscript)
    Goodrich Files: Series A, B, H, J, K, N

PRIVATE COLLECTIONS, ADVICE, AND CORRESPONDENCE

Ariejan Bos, Arnhem, Netherlands
Lucien Chanuc, Soulac, France
H. Duparc, Delft, Netherlands
Erik Eckermann, Seeshaupt, Germany
Paul Erker, Freie Universität Berlin, Institut für Wirtschaftsgeschichte, Berlin,
    Germany
Hans Fogelberg, Göteborg, Sweden
Patrick Fridenson, Paris, France
Helmut Friese, Kerpen, Germany
J. M. Fuchs, Netherlands
R. J. Grogan, Birmingham, England
J. van Groningen, Leiden, Netherlands
Helmuth Herth, Frankfurt am Main, Germany
F. J. M. Hennekam, Aarle Heide, Belgium
B. Hinskens, Nijmegen, Netherlands
Jean-Michel Horvat, Paris, France
Mme. Anne Hurard-Boudou, Epouville, France
David A. Kirsch, Washington, D.C.
Ulrich Knaack, Braunschweig, Germany
Andreas Knie, Berlin, Germany
Thomas Köppen, Wiehl, Germany
J.-L. Krieger, Paris, France
Jacques Kupélian, Brussels, Belgium
Peter Kurze, Bremen, Germany
Alan Loeb, Washington, D.C.
V. Christian Manz, Madrid, Spain
Franco Maggiolini, Milan, Italy
Tapani Mauranen, Helsinki, Finland
Clay McShane, Boston
Ladislav Mergl, Prague, Czech Republic
Christoph Merki, Bern, Switzerland
Gijs Mom, Eindhoven, Netherlands
Kurt Möser, Mannheim, Germany
Hans-Otto Neubauer, Hamburg, Germany
Nicholas Papayanis, New York

Claude Rouxel, Bordeaux, France
Thomas Saal, Cleveland, Ohio
Michael Brian Schiffer, Tucson, Arizona
Hans Seper, Vienna, Austria
W. J. Simons, Amsterdam, Netherlands
Paul Simsa, St. Goarshausen, Germany
Mrs. M. Tompkins, Birmingham, England
H. M. Vellekoop, The Hague, Netherlands
Vincent van der Vinne, Nijmegen, Netherlands

Journals and Periodicals

If no year is given, the periodical concerned has been used only sporadically; in such cases its location is mentioned in the notes.

AFA-Ring (8. Jahrgang, Heft 5; September 1941) (Varta archives)
AFA Rundschau (3.-12. Jahrgang; 1913–30) (in-house periodical, Varta archives)
Annuaire statistique de la ville de Paris (1890–1914)
Archiv für Feuerschutz, Rettungs- und Feuerlöschwesen
De Auto (Netherlands)
L'auto d'Italia (special issue on the occasion of the automobile show in Turin, 16 February–3 March 1907)
Automobil-Revue (Switzerland)
L'automobile (Italy) (1906)
The automotor and horseless vehicle journal (1896–1900) [*AJ*]
Algemeen Handelsblad (Netherlands)
Allgemeine Automobil-Zeitung (Berlin edition) (1905–13) [*AAZ* (Berlin)]
Allgemeine Automobil-Zeitung und Officielle Mitteilungen des Oesterreichischen Automobil-Club (Vienna) (1900–1912) [*AAZ* (Vienna)]
Bescheiden betreffende de geldmiddelen (Statistiek van het Koninkrijk der Nederlanden) (1st–29th document, 2nd part; 1861–1903) (The Hague, 1861–1904) [*Bescheiden betreffende de geldmiddelen* with date]
The Brooklyn Edison
Bulletin of the New York Edison Company
Le caoutchouc & la gutta-percha; organe mensuelle du caoutchouc, de la gutta-percha & des industries qui s'y rattachent (fils et câbles, amiantes, fibres vulcanisées, durci, etc., etc.) (Ire–I^e Année; 1904–14) [*CGP*]
The central station (vols. 3–20; 1903–21) [*CS*]
Centralblatt für Accumulatoren-und Elementenkunde; Organ für Wissenschaft und Technik, mit besonderer Berücksichtigung des Accumobilismus (Berlin) (Erster-Neunter Jahrgang, 1900–1908) (as of 1902: *Centralblatt für Accumulatoren-, Elementen- und Accumobilenkunde*; as of 1904: *Centralblatt für Accumulatoren-Technik und verwandte Gebiete; Internationales Organ für die Industrie der Primär- und Sekundärelemente und ihre Hilfs- und Anwendungstechniken*; as of 1908: *Centralblatt für Accumulatoren und Galvanotechnik*) [*Centralblatt*]
The commercial vehicle (vols. 1–12; 1906–15) [*CV*]
L'Éclairage électrique (Tomes XXII–XXV; 1900)
The Edison monthly (1902–25)
Electrifying times
The electric vehicle: The official organ of the Electric Vehicle Committee (London) (vols. I–XXXVI, 1914–52; New Series (quarterly): 1954–55) (as of vol. XIV: *Electric vehicles*

and batteries; as of vol. XXV: *Electric vehicles: A journal devoted to battery electric transport*; as of March 1955: *Electric vehicles and industrial trucks*) [*EV* (UK)]

Electric vehicles (vols. 3–12, 1913–18), continuation of *Ignition and accessories* [*EV* (U.S.)]

The electrical engineer (London) (vols. XXV–XXVII (New Series); 1900–1901) [*EE* (London)]

The electrical engineer: A weekly review of theoretical and applied electricity (New York) (vols. 21–27; 1896–99) (in 1899 taken over by *Electrical World*) [*EE* (New York)]

Electrical review: A journal of scientific and electrical progress (New York) (vols. 25–53, 1894–1908) [*ER* (New York)]; continued as: *Electrical review and western electrician* (vols. 53–61, 1908–14) [*ER* (New York)]

Electrical world [*EW*]

The electrician (London) (vols. XLII–XLV; 1898–1900)

Das Elektrofahrzeug [*EF*] (1936–41)

Elektrotechnische Zeitschrift (XXI.–XXXII. Jahrgang, 1900–1911) [*ETZ*]

La France automobile: Organe de l'automobilisme et des industries qui s'y rattachent (1896–1900) [*FA*]

Feuer und Wasser; Zeitschrift für moderne Brandschutz (7.-34. Jahrgang; 1900–1927) [*FuW*]

Feuerpolizei: Zeitschrift für Feuerschutzwesen (Bd. XII–XVI; 1910–14)

Feuerschutz: Zeitschrift des Reichsverbandes deutscher Feuerwehr-Ingenieure (1.-13. Jahrgang; 1921–33)

Feuerwehrtechnische Zeitschrift (I.–IX. Jahrgang; 1913–21)

De Fiets (Netherlands)

General electric review (vols. XII–XVI; 1909–13)

Le Génie Civil (1898)

Goodyear tire news (vol. 4, no. 3–vol. 5 no. 1, March 1915–January 1916)

G. V. central station bulletin (1914–16)

Harper's weekly (1896–98)

The horseless age (vols. I–XI, 1895–1915) [*HA*]

Ignition and accessories (vols. 1–3; 1911–13), continued as *Electric vehicles*

l'Illustration (supplément) (4 October 1941)

The India-Rubber & Gutta-Percha electrical trades' journal (London) (vols. XXV–XXXII, 1903–6) [*IRJ*]

The India rubber world (vol. XXIV, no. 3–XXXII, no. 1; June 1901–April 1905) [*IRW*]

L'Industrie électrique (Tomes III–VIII, X–XX; 1894–99, 1901–11) [*IE*]

De Ingenieur (Netherlands)

The light car and cycle car (1914–15)

La locomotion: Automobilisme, cyclisme, tramways, aérostation, yachting, etc. (Première-Troisième Année, 1901–3) (continued as *La vie automobile*) [La locomotion]

La locomotion automobile: Revue universelle illustré des voitures, vélocipèdes, bateaux, aérostats et tous véhicules mécaniques; publiée sous le haut patronage du Touring-Club de France (Première-Treizième Année, 1894–1906) [*LA*]

Motor (Heft-Ausgabe von Braunbeck's Sportlexikon) (June 1913) (DaimlerChrysler archives, file "Feuerwehr; Prospekte; 1911–1912–1913. deutsch")

Motorkampioen (Netherlands)

Der Motorwagen (1898–99, 1902, 1906–7) [*MW*]

NELA Bulletin (vols. 2–6; New Series, vols. 1–13; 1908–26) [*NB*]

La Nature (1898)

Nederlandsche Staatscourant, Supplement

Le Petit Journal (1899)

Revue des transports parisiens (1898)

Scientific American

Stahlrad und Automobil (1906–13) [*SA*]

TA-Nachrichten (1940, Folge 1–12 and 1944, Folge 1; January–December 1940 and February 1944) (in-house, typewritten AFA-periodical, Varta archives)

Varta-Ring (April 1962–1977/2; a few issues)

Le véhicule électrique (4me Année; April and October 1930) [*VE*]

La vie au grand air (1898)

La vie automobile (Troisième-Quatorzième Année, 1903–14) (continuation of *La locomotion*) [*VA*]

Wegen (Netherlands)

Western Electrician (Chicago) (vols. XXXVI–XLII; 1905–8)

Zeitschrift für Elektrotechnik (1898)

Zeitschrift des Mitteleuropäischen Motorwagen-Vereins (1909–10)

Zeitschrift des Vereins Deutscher Ingenieure (Bd. 51–56; 1907–12) [*Z-VDI*]

Conference Reports and Annual Reports

Bericht über die Tätigkeit und Verwaltung der Charlottenburger Feuerwehr für das Jahr 1908 [*Bericht Charlottenburg 1908*] (Feuerwehrmuseum Berlin)

idem, 1909, 1910, 1911 [*Bericht Charlottenburg 1909, 1910, 1911*] (Feuerwehrmuseum Berlin)

Bericht über die Verwaltung der Feuerwehr und des Telegraphen von Berlin für das Etatsjahr 1901 (1. April 1901–31. März 1902) (Berlin, 1902) [*Bericht Berlin 1901*]

idem, 1902, 1903, . . . 1913 [*Bericht Berlin 1902, 1903, . . . 1913*] (Feuerwehrmuseum Berlin)

Dritter Verwaltungsbericht der Schöneberger Berufsfeuerwehr; 1. April 1903 bis 31. März 1908 (Sonderabdruck aus dem dritten Verwaltungsbericht des Magistrats der Stadt Schöneberg) (Schöneberg, 1910) [*Bericht Schöneberg 1903–1908*] (Feuerwehrmuseum Berlin)

4. Verwaltungsbericht (1908–1912) der Feuerwehr Schöneberg (typescript, Feuerwehrmuseum Berlin, document 146i) [*Bericht Schöneberg 1908–12*]

Electric Vehicle Association of America:

"The first convention of the Electric Vehicle Association of America," *The central station* (vol. 10, no. 5, November 1910), 137–50 [*EVAA 1910*]

W. H. Blood Jr., *Presidential address* (Electric Vehicle Association of America, First annual convention, Madison Square Garden, New York City, October 18, 1910) (New York City, n.d. [1910]) (Free Library of Philadelphia, Automobile Reference Collection) [Blood, *Presidential address; EVAA 1910*]

Electric Vehicle Association of America; Second annual convention: Papers, reports and discussions, New York City, October 10th, 1911 (New York City, n.d. [1911]) (reprinted from the November issue of *The central station*) (Engineering Societies Library, New York) [*EVAA 1911*]

W. H. Blood Jr., *President's address* (n.p., n.d. [New York City, 1911]) (Free Library of Philadelphia, Automobile Reference Collection) [Blood, *President's address; EVAA 1911*]

Electric Vehicle Association of America; Third annual convention, Boston, Mass., Oct. 8–9 1912 (contributions to convention published as separate brochures and cited as such) (Free Library of Philadelphia, Automobile Reference Collection) [*EVAA 1912*]

Electric Vehicle Association of America; Fourth annual convention; papers, reports and discussions, Chicago, October 27th and 28th, 1913 (New York City, n.d. [1913]) (reprinted from *The central station*) (Engineering Societies Library, New York) [*EVAA 1913*]

Fifth annual convention of the Electric Vehicle Association of America, Philadelphia, Penn., October 19, 20, 21, 1914 (contributions to convention published as separate brochures and cited as such) (Free Library of Philadelphia, Automobile Reference Collection) [*EVAA 1914*]

Sixth annual convention of the Electric Vehicle Association of America, October 18 and 19, 1915, Cleveland, Ohio (contributions to convention published as separate brochures and cited as such) (Free Library of Philadelphia, Automobile Reference Collection) [*EVAA 1915*]

National Electric Light Association:

Twentieth convention (Niagara Falls, N.Y., June, 8th, 9th and 10th, 1897) (New York, 1897) [*NELA 1897*]

21st convention, 7–9 June 1898, Chicago (New York, 1898) [*NELA 1898*]

22nd convention, 23–25 May 1899, New York (New York, 1899) [*NELA 1899*]

23rd convention, 22–24 May 1900, Chicago (New York, 1900) [*NELA 1900*]

24th convention, 21–23 May 1901, Niagara Falls (New York, 1902) [*NELA 1901*]

25th convention, 20–22 May 1902, Cincinnati, Ohio (New York, 1902) [*NELA 1902*]

26th convention, 26–28 May 1903, Chicago (New York, 1903) [*NELA 1903*]

27th convention, 24–26 May 1904, Boston, Mass. (New York, 1904)
> *Vol. I. Papers, reports and discussions* [*NELA 1904-I*]
> *Vol. II. Question box and wrinkles* [*NELA 1904-II*]

28th convention, 6–11 June 1905, Denver-Colorado Springs, Colo. (New York, 1905)
> *Vol. I. Papers, reports and discussions* [*NELA 1905-I*]
> *Vol. II. Question box and wrinkles* [*NELA 1905-II*]

29th convention, 5–8 June 1906, Atlantic City, N. J. (New York, 1906)
> *Vol. I. Papers, reports and discussions* [*NELA 1906-I*]
> *Vol. II. Question box* [*NELA 1906-II*]
> *Vol. III. Executive sessions* [*NELA 1906-III*]

30th convention, 4–7 June 1907, Washington, D.C. (New York, 1907)
> *Vol. I. Papers, reports and discussions* [*NELA 1907-I*]
> *Vol. II. Commercial programs, question box* [*NELA 1907-II*]

31st convention, 19–22 May 1908, Chicago (n.p., n.d. [New York, 1908])
> *Vol. I. Papers, reports and discussions* [*NELA 1908-I*]

32nd convention, 1–4 June 1909, Atlantic City, N. J. (n.p., n.d. [New York, 1909])
> *Vol. I. General sessions: Papers, reports and discussions* [*NELA 1909-I*]
> *Vol. II. Technical and commercial sessions: Papers, reports and discussions* [*NELA 1909-II*]
> *Vol. III. Accounting sessions and executive and public policy committees: Papers, reports and discussions* [*NELA 1909-III*]

33rd convention, 23–27 May 1910, St. Louis, Mo. (n.p., n.d. [New York, 1910])
> *Vol. I. General sessions: Papers, reports and discussions* [*NELA 1910-I*]
> *Vol. II. Commercial sessions: Papers, reports and discussions* [*NELA 1910-II*]

34th convention, 29 May-2 June 1911, New York City (n.p., n.d. [New York, 1911])
> *Vol. I. General sessions and commercial sessions: Papers, reports and discussions* [*NELA 1911-I*]

Vol. II. Accounting sessions, technical sessions, power transmission sessions: Papers, reports and discussions [NELA 1911-II]

35th convention, 10–13 June 1912, Seattle, Wash. (n.p., n.d. [New York, 1912])

Vol. I. General sessions, executive sessions, public policy session, company sessions: Papers, reports and discussions [NELA 1912-I]

Vol. II. Commercial sessions: Papers, reports and discussions [NELA 1912-II]

Vol. III. Power transmission sessions; technical sessions: Papers, reports and discussions [NELA 1912-III]

Vol. IV. Accounting sessions: Papers, reports and discussions [NELA 1912-IV]

36th convention, 2–6 June 1913, Chicago (n.p., n.d. [New York, 1913])

Vol. I. General sessions, executive sessions, public policy sessions: Papers, reports and discussions [NELA 1913-I]

Hydro-electric and transmission sessions, technical sessions: Papers, reports and discussions [NELA 1913-II]

Commercial sessions [NELA 1913-III]

Accounting sessions: Papers, reports and discussions [NELA 1913-IV]

37th convention, 1–5 June 1914, Philadelphia (n.p., n.d. [New York, 1914])

General sessions, executive sessions, public policy sessions, Company Section sessions: Papers, reports and discussions [NELA 1914-I]

Hydro-electric and transmission sessions, technical sessions: Papers, reports and discussions [NELA 1914-II]

Commercial sessions: Papers, reports and discussions [NELA 1914-III]

Accounting sessions: Papers, reports and discussions [NELA 1914-IV]

38th convention, 7–11 June 1915, San Francisco (n.p., n.d. [New York, 1915])

General sessions, executive sessions, celebration exercises, public policy sessions: Papers, reports and discussions [NELA 1915-I]

Hydro-electric and transmission sessions, technical sessions: Papers, reports and discussions [NELA 1915-II]

Commercial sessions: Papers, reports and discussions [NELA 1915-III]

Accounting sessions: Papers, reports, and discussions [NELA 1915-IV]

39th convention, 22–26 May [1916], Chicago (n.p., n.d. [New York, 1916])

General sessions, executive sessions, company sessions: Papers, reports and discussions [NELA 1916-I]

Commercial sessions: Papers, reports and discussions [NELA 1916-II]

Technical and hydro-electric sessions: Papers, reports and discussions [NELA 1916-III]

Accounting section sessions: Papers, reports and discussions [NELA 1916-IV]

Electric Vehicle Section: Papers and reports [NELA 1916-V]

Proceedings of the war convention, 9–10 May 1917, New York City (n.p., n.d.)

General and executive sessions (Advance reports and papers prepared for presentation at the Fortieth Convention at Atlantic City, May 28th to June 1st, 1917, but received at the annual meeting in New York on May 9th and 10th, 1917, for publication in the Annual Proceedings) [NELA 1917-I]

Technical and Hydro-electric Section: Papers and reports (Advance . . . etc.) [NELA 1917-II]

Commercial Section: Papers and reports (Advance . . . etc.) [NELA 1917-III]

Accounting Section: Papers and reports (Advance . . . etc.) [NELA 1917-IV]

Electric Vehicle Section: Papers and reports (Advance . . . etc.) [NELA 1917-V]

Proceedings of the second war convention; Forty-first convention, held at Atlantic City, N. J., June 13–14, 1918 (n.p., n.d. [New York, 1918]) [NELA 1918]

42nd convention, 19–22 May 1919 (New York, n.d. [1919])
> *General and executive sessions: Papers, reports and discussions* [NELA 1919-I]
> *Technical Section sessions: Papers, reports and discussions* [NELA 1919-II]
> *Commercial Section sessions: Papers, reports and discussions* [NELA 1919-III]
> *Accounting Section sessions: Papers, reports and discussions* [NELA 1919-IV]
> *Electric Vehicle Section sessions: Papers, reports and discussions* [NELA 1919-V]

Forty-third convention; General sessions, public policy session, accounting sessions, commercial sessions, electric vehicle session, technical sessions: Papers, reports and discussions. Pasadena, Calif., May 18–22, 1920 (New York, n.d. [1920])
> [NELA 1920]

Proceedings of NELA; Forty-fourth convention, Chicago, Ill., May 31–June 3, 1921 (New York, n.d. [1921])
> *Vol. I* [NELA 1921-I]
> *Vol. II* [NELA 1921-II]

Proceedings of NELA; Forty-fifth convention, Atlantic City, N.J., May 15–19, 1922 (New York, n.d. [1922])
> *General volume (Pages 1 to 204 of the complete Proceedings)* [NELA 1922-I]
> *Accounting volume (Pages 205 to 338 of the complete Proceedings)* [NELA 1922-II]
> *Commercial volume (Pages 339 to 479 of the complete Proceedings)* [NELA 1922-III]
> *Technical volume (Pages 481 to 1251 of the complete Proceedings)* [NELA 1922-IV]

NELA; Proceedings, 46th convention, Hotel Commodore, New York City, June 4 to 8, 1923. Containing the Proceedings of the General and Executive, Public Policy, Customer Ownership, Accounting Section and Commercial Section sessions (New York, 1923)
> *Vol. 79* [NELA 1923-I]
> *Vol. 80* [NELA 1923-II]

NELA; Proceedings, Volume 81; 47th convention, Million Dollar Pier, Atlantic City, N.J., May 19 to 23, 1924; Containing the Transactions of the convention in full, in one volume (New York, 1924) [NELA 1924]

NELA; Proceedings, Volume 82; 48th convention, Exposition Auditorium, San Francisco, Calif., June 15 to 19, 1925. In one volume. (New York, 1925) [NELA 1925]

NELA; Proceedings, Volume 83; 49th convention, Million Dollar Pier, Atlantic City, N.J., May 17 to 21, 1926. In one volume. (New York, 1926) [NELA 1926]

NELA; Proceedings, Volume 84; 50th convention, Million Dollar Pier, Atlantic City, N.J., June 6 to 10, 1927. In one volume. (New York, 1927) [NELA 1927]

NELA; Proceedings, Volume 85; 51st convention, Million Dollar Pier, Atlantic City, N.J., June 4 to 8, 1928. In one volume. (New York, 1928) [NELA 1928]

NELA; Proceedings, Volume 86; 52nd convention, Atlantic City Auditorium and Convention Hall, Atlantic City, N.J., May 17 to 21, 1929. In one volume. (New York, 1929) [NELA 1929]

NELA; Proceedings, Volume 87; 53rd convention, Exposition Auditorium, Civic Center, San Francisco, Calif., June 16 to 20, 1930. In one volume. (New York, 1930) [NELA 1930]

NELA; Proceedings, Volume 88; 54th convention, Atlantic City Auditorium and Convention Hall, Atlantic City, N.J., June 8 to 12, 1931. In one volume. (New York, 1931) [NELA 1931]

NELA; Proceedings, Volume 89; 55th convention, Atlantic City Auditorium and Convention Hall, Atlantic City, N.J., June 5 to 10, 1932. In one volume. (New York, 1932) [NELA 1932]

Verband Deutscher Berufsfeuerwehren (see also Reutlinger):

Sechster Verbandstag am 21.–24. Juni 1906 in Aachen; Protokoll (n.p., n.d.) (Feuerwehrmuseum Fulda) [VDB, *Sechster Verbandstag . . . 1906*]

Siebenter Verbandstag am 13.–16. Juni 1907 in Stuttgart; Protokoll (n.p., n.d.) (Feuerwehrmuseum Fulda) [VDB, *Siebenter Verbandstag . . . 1907*]

Achter Verbandstag am 15.–17. Juni 1908 on Essen (Ruhr); Protokoll (n.p., n.d.) (Feuerwehrmuseum Fulda) [VDB, *Achter Verbandstag . . . 1908*]

Zehnter Verbandstag am 14. bis 16. Juni 1910 in Gross-Berlin (n.p., n.d.) (Feuerwehrmuseum Fulda) [VDB, *Zehnter Verbandstag . . . 1910*]

Elfter Verbandstag am 13. bis 15. Juni 1911 in Posen (Hamburg, n.d.) (Feuerwehrmuseum Fulda) [VDB, *Elfter Verbandstag . . . 1911*]

Zwölfter Verbandstag am 12. bis 14. Juni 1912 in Cöln a. Rhein (Hamburg, n.d.) [VDB, *Zwölfter Verbandstag . . . 1912*]

Dreizehnter Verbandstag am 4. bis 6. Juni 1913 in Stettin (Hamburg, n.d.) [VDB, *Dreizehnter Verbandstag . . . 1913*]

Verein Deutscher Berufsfeuerwehroffiziere (bisher Verband Deutscher Berufsfeuerwehren):

Bericht der 14. Tagung der "Vereinigung Deutscher Berufsfeuerwehroffiziere" am 9. bis 11. Juni 1914 in München (Hamburg, n.d.) [VDB, *14. Tagung . . . 1914*]

Verhandlungen des Internationalen Feuerwehrkongresses zu Berlin (Berlin, 1901) (Feuerwehrmuseum Berlin)

Articles, Brochures, and Monographs

50 Jahre Accumulatoren-Fabrik Aktiengesellschaft 1888–1938 (Berlin, 1928)

75 Jahre Berliner Feuerwehr 1851–1926 (Berlin, n.d.)

75 Jahre motorisierte Feuerwehr (n.p., n.d. [Stuttgart, 1964?])

75 Jahre Nutzfahrzeug-Entwicklung 1896–1971; Jubiläumsbericht der Daimler-Benz Aktiengesellschaft Stuttgart-Untertürkheim (n.p., n.d. [Stuttgart, 1971])

100 Jahre Berliner Feuerwehr 1851–1951 (Berlin, n.d.)

150 Jahre Gottfried Hagen (Köln-Kalk, n.d. [1977]) (brochure, Varta archives)

Dietmar Abt, *Die Erklärung der Technikgenese des Elektroautomobils* (Frankfurt am Main, 1998) (European university studies, series V, vol. 2295)

Michael C. C. Adams, *The great adventure: Male desire and the coming of World War I* (Bloomington, 1990)

W. Poynter Adams, *Electric and petrol-electric vehicles, being Part II of "Motor-car mechanism and management"* (London, 1908)

Akkumulatorenhandbuch der AFA (n.p., n.d.) (Varta archives)

Daniel Albert, "The psychotechnologist & the good driver: Granting admission to road society," *Automotive history review* 32 (Spring 1998), 38–42

Armen A. Alchian, "Uncertainty, evolution, and economic theory," *Journal of political economy* 58 (June 1950), 211–21

D. H. Aldcroft and M. J. Freeman, *Transport in the industrial revolution* (Manchester, 1983)

Allmers et al. (eds.), *Das deutsche Automobilwesen der Gegenwart* (Berlin, 1928)

F. R. Ankersmit, *De spiegel van het verleden; Exploraties I: geschiedtheorie* (Kampen, 1996)

F. R. Ankersmit, *De macht van representatie; Exploraties II: cultuurfilosofie & esthetica* (Kampen, 1996)

Suzanne Appelt, "Electric vehicles, a bibliography 1928–1966" (manuscript, Bonneville Power Administration, Portland, Ore., September 1966) (NAHC)

Suzanne Appelt, *Electric vehicles: A bibliography* (New York, 1970)

C. B. van Ardenne et al. (eds.), *Den Haag energiek: Hoofdstukken uit de geschiedenis van de energievoorziening in Den Haag* (The Hague, 1981)

M. Arkenbosch, G. Mom, and J. Nieuwland, *Het rijdend gedeelte* (Deventer, 1989) (part 4, vol. A of G. Mom [ed.], *De nieuwe Steinbuch: de automobiel*)

W. Brian Arthur, "Competing technologies, increasing returns, and lock-in by historical events," *The economic journal* 99 (March 1989), 116–31

William Aspray (ed.), *Technological competitiveness: Contemporary and historical perspectives on the electrical, electronics, and computer industries* (New York, 1993), 243–69

Henri Asselbergs (ed.), *Beijnes: een eeuw van arbeid* (n.p., n.d. [1938])

Autler Zucht-und Ruchlosigkeiten; ein Protest gegen die Schreckensherrschaft der Strasse; von einem Rechtsfreund (Berlin, 1909)

De automobiel in 's lands historie: 6–22 april 1903 (The Hague, 1904)

"Automobiles for physicians' use: Are they practical? are they desirable? are they economical? are they better than horses?" *Journal of the American Medical Association* 46, part 2, no. 14 (7 April 1906), 1172–1207

Automobiles of America (Detroit, 1961) (Automobile Manufacturers Association)

Philip S. Bagwell, *The transport revolution from 1770* (London, 1974)

Martijn Bakker, "Techniek als cultuurverschijnsel, een inleiding," in *Techniek als cultuurverschijnsel: Cursusboek* (Heerlen, 1996) (Open Universiteit, Leerstofgebied cultuurwetenschappen), 15–80

H. L. Barber, *Story of the automobile: Its history and development from 1760 to 1917; with an analysis of the standing and prospects of the automobile industry* (Chicago, 1917)

Jean-Pierre Bardou, Jean-Jacques Chanaron, Patrick Fridenson, and James M. Laux, *The automobile revolution: The impact of an industry* (Chapel Hill, 1982)

Dominique Barjot, "Advances in road construction technology in France," in Barker (ed.), *The economic and social effects of the spread of motor vehicles*, 291–312

Theo Barker, "A German centenary in 1986, a French in 1995 or the real beginnings about 1905?" in Barker (ed.), *The economic and social effects of the spread of motor vehicles*, 1–54

Theo Barker (ed.), *The economic and social effects of the spread of motor vehicles: An international centenary tribute* (Houndsmill, 1987)

T. C. Barker, "The delayed decline of the horse in the twentieth century," in Thompson (ed.), *Horses in European economic history*, 101–12

Theo Barker, "Towards an historical classification of urban transport development since the later eighteenth century," *Journal of transport history* 1, 3rd ser. (1980), 75–90.

T. C. Barker, "Passenger transport in nineteenth-century London," *Journal of transport history* 6 (1963–64), 166–74

T. C. Barker, "The spread of motor vehicles before 1914," in Charles P. Kindleberger and Guido di Tella (eds.), *Economics in the long view: Essays in honor of W. W. Rostow; vol. 2: Applications and cases, part I* (London, 1982), 149–67

T. C. Barker, "The international history of motor transport," *Journal of contemporary history* 20 (1985), 3–19

T. C. Barker, "Slow progress: Forty years of motoring research," *Journal of transport history* 14, 3rd ser. (1993), 142–65

Theo Barker and Dorian Gerhold, *The rise and rise of road transport, 1700–1990* (Houndsmills, 1993) (Studies in economic and social history)

T. C. Barker and Michael Robbins, *A history of London transport: Passenger travel and the development of the metropolis*, vol I: *The nineteenth century* (London, 1975)

T. C. Barker and Michael Robbins, *A history of London transport: Passenger travel and the development of the metropolis*, vol II: *The twentieth century to 1970* (London, 1974)

Paul Barrett, *The automobile and urban transit: The formation of public policy in Chicago, 1900–1930* (Philadelphia, 1983)

Otto Barsch, *Moderne Automobil-Strassenreinigungsmaschinen* (Berlin, 1919) (Autotechnische Bibliothek, Band 63)

George Basalla, *Geschiedenis van de technologie* (Utrecht, 1993)

L. Baudry de Saunier, *L'automobile théorique et pratique*, Tome I: *Le moteur: explications simples du fonctionnement des moteurs d'automobiles et de tous leurs organes* (Paris, n.d. [16e mille])

L. Baudry de Saunier, *L'automobile théorique et pratique*, Tome II: *Le chassis: explication simple du mécanisme complet d'une voiture automobile et de tous ses organes* (Paris, n.d. [6e mille])

L. Baudry de Saunier, Ch. Dolfuss, and E. de Geoffroy, *Histoire de la locomotion terrestre* (Paris, 1936)

Gerhard Bauer (ed.), *Berliner Strassenbahnen* (Berlin [DDR], 1987) (Strassenbahn-Archiv 5)

Günter Bayerl and Torsten Meyer, "Aufgaben einer Sozialgeschichte der Technik," *Blätter für Technikgeschichte*, 51./52. Heft (1989–90), 10–36

David Beasley, *The suppression of the automobile: Skulduggery at the crossroads* (New York, 1988)

W. Worby Beaumont, *Motor vehicles and motors: Their design construction and working by steam oil and electricity*, 2nd ed. (Westminster, 1902)

W. Worby Beaumont, *Motor vehicles and motors: Their design construction and working by steam oil and electricity*, vol. II (London, 1906)

W. Worby Beaumont, *Industrial electric vehicles and trucks* (London, 1920)

H. Beckmann, *Gegenwärtiger Stand der Technik stationärer und transportabeler Accumulatoren (Vortrag gehalten auf dem Internationalen Kongress für elektrotechnische Anwendungen Turin, 10.–17. September 1911)* (AFA brochure, Varta archives)

Wolfgang Behringer, "Der Fahrplan der Welt: Anmerkungen zu den Anfängen der europäischen Verkehrsrevolution," in Hans-Liudger Dienel and Helmuth Trischler (eds.), *Geschichte der Zukunft des Verkehrs: Verkehrskonzepte von der Frühen Neuzeit bis zum 21. Jahrhundert* (Frankfurt, 1997), 40–57

Hein Bekenkamp, Hanneke Boonstra et al., *Groningen toen* (Groningen, 1983)

René Bellu, *Toutes les Renault* (Paris, 1979)

Alain Beltran and Patrice A. Carré, *La fée et la servante: La société française face à l'électricité, XIXe–XXe siècle* (Paris, 1991)

Peter van den Berg, "Amsterdamse witkar na twee decennia definitief ter ziele," *De Volksrant* (18 January 1988)

Michael L. Berger, *The devil wagon in God's country: The automobile and social change in rural America, 1893–1929* (Hamden, Conn., 1979)

Michael L. Berger, "Women drivers!: The emergence of folklore and stereotypic opinions concerning feminine automotive behavior," *Women's studies international forum* 9, no. 3 (1986), 257–63

Die Berliner Feuerwehr 1901–1911: Als Nachtrag zur Geschichte des Korps aus Anlass des 60 jährigen Bestehens im amtlichen Auftrage bearbeitet (Berlin, 1911)

Jörg Jochen Berns, *Die Herkunft des Automobils aus Himmelstrionfo und Höllenmaschine* (Berlin, 1996)

Walter C. Bersey, *Electrically propelled carriages* (London, 1898)

Catherine Bertho-Lavenir, *La roue et le stylo: Comment nous sommes devenus touristes* (Paris, 1999)

Robert Besnier, *Les fiacres de Paris aux XVIIe et XVIIIe siècles* (Paris, 1972)

L. Betz, *Spezial-Lastautomobile, Band I: Kommunal-Automobile, Automobil-Kipper, Tank-Automobile, Sonderfahrzeuge* (Berlin, 1927)

Klaus Beyrer, *Die Postkutschenreise* (Tübingen, 1985)

Otto Julius Bierbaum, *Eine empfindsame Reise im Automobil: Von Berlin nach Sorrent und zurück an den Rhein; in Briefen an Freunde geschrieben* (Munich, 1979) (reprint of the first edition from 1903).

Wiebe Eco Bijker, *The social construction of technology* (Eijsden, 1990) (diss.)

Wiebe E. Bijker, *Democratisering van de technologische cultuur* (Eijsden, 1995) (Rijksuniversiteit Limburg, Maastricht, 24 March 1995) (inaugural lecture)

Wiebe E. Bijker, Thomas P. Hughes, and Trevor J. Pinch (eds.), *The social construction of technological systems: New directions in the sociology and history of technology* (Cambridge, Mass., 1987)

Bruce Bimber, "Karl Marx and the three faces of technological determinism," *Social studies of science* 20 (1990), 333–51

Charles W. Bishop, *La France et l'automobile: contribution française au développement économique et technique de l'automobilisme des origines à la deuxième guerre mondiale* (Paris, 1971)

Marc K. Blackburn, "A new form of transportation: The Quartermaster Corps and standardization of the United States Army's motor trucks 1907–1939" (diss., Temple University, 1992)

Marc K. Blackburn, *The United States Army and the motor truck: A case study in standardization* (Westport, Conn., 1996) (Contributions in military studies, no. 163)

J. Blok, *Haagse tram: Het Haagse trammaterieel van 1864 tot heden* (Rijswijk, n.d.)

A. Blondel and F. Paul-Dubois, *La traction électrique sur voies ferées: voie-matériel roulant-traction* (Paris, 1898)

Malcolm Bobbitt, *Taxi! The story of the "London" taxicab* (Dorchester, 1998)

H. de Boer, Th. Dobbelaar, and G. Mom, *De elektrische installatie* (Deventer, 1986) (part 8 of Mom [ed.], *De nieuwe Steinbuch: de automobiel*)

Horst Bomke and Wilfried Porsinger, "Strom macht mobil: Elektro-Strassenfahrzeuge und Elektrowasserfahrzeuge in Berlin," 2nd ed. (Berlin, 1986) (Bewag brochure)

Ariejan Bos, "De gebruikers geteld: het vroege Nederlandse automobilisme in cijfers en statistieken," in Bos et al., *Het paardloze voertuig*, 19–77

Ariejan Bos, Hans van Groningen, Gijs Mom (ed.), and Vincent van der Vinne, *Het paardloze voertuig: de auto in Nederland een eeuw geleden* (Deventer, 1996)

G. Bouchet, "La traction hippomobile dans les transports publics parisiens (1855–1914)," *Revue historique* 108me Année, 221 (1984), 125–34

Ghislaine Bouchet, *Le cheval à Paris de 1850 à 1914* (Genève, 1993) (diss.) (Mémoires et documents de l'Ecole des Chartes, 37)

Anne Boudou, "Les taxis parisiens de la fondation des Usines Renault aux 'Taxis de la Marne' 1898–1914" (unpubl. thesis, Université Paris X Nanterre, 1982)

B. Bowers, *A history of electric light and power* (Stevenage, 1982)

Geoffrey C. Bowker, "How things change: The history of sociotechnical sources," *Social studies of science* 26 (1996), 173–82

T. A. Boyd, "The self-starter," *Technology and culture* 9 (1968), 585–91

Klaus Brandhuber, *Die Insolvenz eines Familienkonzernes: Der wirtschaftliche Nieder-gang der BORGWARD-GRUPPE* (Cologne, 1988) (diss.)

Hans-Joachim Braun, "Gas oder Elektrizität? Zur Konkurrenz zweier Beleuchtungs-systeme, 1880–1914," *Technikgeschichte* 47, no. 1 (1980), 1–19

Hans-Joachim Braun, "Introduction" ("Symposium on 'failed innovations'"), *Social studies of science* 22 (1992), 213–30

Piers Brendon, *The motoring century: The story of the Royal Automobile Club* (London, 1997)

Grace R. Brigham, "Those elusive vehicles: A history of the Society of Automotive Historians," *Automotive history review* (Fall 1995), 3–10

Attilio Brilli, *Das rasende Leben: Die Anfänge des Reisens mit dem Automobil* (Berlin, 1999) (trans. of *La vita che corre* [Bologna, 1999])

A. C. Broeshart, *De geschiedenis van de brandweer in Nederland* (Rijswijk, 1986)

J. Graeme Bruce and Colin H. Curtis, *The London motor bus: Its origins and development* (London, 1973)

Hans Brunswig, *Feuerwehrfahrzeuge: Eine Darstellung ihrer Aufgaben, ihrer Entwick-lung und Ausführungsformen* (Hannover, 1957)

R. A. Buchanan, "Theory and narrative in the history of technology," *Technology and culture* (April 1991), 365–76

Günter Burkart, "Die kulturelle Durchsetzung des Automobilismus als Element der modernen Lebensweise," *Sowi* 25, no. 4 (1996), 250–56

G. Burkert, *Der technische Kraftfahrdienst bei der Deutschen Bundespost* (Goslar, 1963)

Konrad Buschmann, *Da ging die Post ab: Die Geschichte der Motorisierung der Post* (Aach, n.d.)

Richard Bussien, *Automobiltechnisches Handbuch* (Berlin, 1928)

Martin Caidin and Jay Barbree, *Bicycles in war* (New York, 1974)

Michel Callon, "The sociology of an actor-network: The case of the electric vehicle," in Michel Callon, John Law, and Arie Rip (eds.), *Mapping the dynamics of science and technology: Sociology of science in the real world* (Houndsmill, 1986), 19–34

Michel Callon, "Society in the making: The study of technology as a tool for sociologi-cal analysis," in Bijker et al., *The social construction of technological systems*, 83–103

Weert Canzler, "Das Automobil-Leitbild: Zur Konstruktion einer technischen Erfolgs-geschichte," in *Verkehrstechnik und individuelle Mobilität (Technikgeschichtliche Jahrestagung des VDI 1997, 13. und 14. Februar 1997, Düsseldorf, VDI-Haus)* (Düssel-dorf, 1997)

Weert Canzler and Andreas Knie, "Von der Automobilität zur Multimobilität: Die Krise des Automobils als Chance für eine neue Verkehrs- und Produktpolitik," in W. Fricke (ed.), *Jahrbuch Technik und Arbeit 1994* (Bonn, 1994), 171–81

Weert Canzler and Andreas Knie, *Das Ende des Automobils: Fakten und Trends zum Umbau der Autogesellschaft* (Heidelberg, 1994)

P. van Cappelle, *De electriciteit, hare voortbrenging en hare toepassing in de industrie en het maatschappelijk verkeer*, 4th ed. (Leiden, 1900)

P. van Cappelle, *De electriciteit, hare voortbrenging en hare toepassing in de industrie en het maatschappelijk verkeer* (Leiden, 1915)

W. Bernard Carlson, "Thomas Edison as a manager of R&D: The case of the alkaline storage battery, 1898–1915," *IEEE technology and society magazine* (December 1988), 4–12

Norman Miller Cary Jr., "The use of the motor vehicle in the United States Army, 1899–1939" (unpubl. diss., University of Georgia, 1980)

C. F. Caunter, *The history and development of cycles; as illustrated by the collection of cycles in the Science Museum* (London, 1955)

M. C. Chalumeau, "Die elektrischen Akkumulatoren-Fahrzeuge der stadt Lyon (Vortrag des . . . Chefingenieur der Stadt Lyon)" (undated typescript, Varta archives, file II D 0: "1909–1935")

Pierre Chany, *La fabuleuse histoire du cyclisme* (Paris, 1975)

Charles W. Cheape, *Moving the masses: Urban public transit in New York, Boston, and Philadelphia, 1880–1912* (Cambridge, Mass., 1980 [Harvard studies in business history, no. 31])

William R. Childs, *Trucking and the public interest: The emergence of federal regulation 1914–1940* (Knoxville, 1985)

Keith Chivers, "The supply of horses in Great Britain in the nineteenth century," in Thompson (ed), *Horses in European economic history*, 31–49

Marcel Clement, *Transport en economische ontwikkeling: Analyse van de modernisering van het transportsysteem in de provincie Groningen (1800–1914)* (Groningen, 1994) (diss.)

M. J. Colie, *Electric and hybrid vehicles* (Park Ridge, N.J., 1979)

E. J. T. Collins, "The farm horse economy of England and Wales in the early tractor age 1900–40," in Thompson (ed.), *Horses in European economic history*, 73–100

C. Colson, *Transports & tarifs* (Paris, 1890)

Mario Conte and Richard T. M. Smokers, "Overview of EV testing and demonstration fleets in Europe," *EVT '95: A Weva conference for electric vehicle research, development and operation; Conference Proceedings* (Paris, 13, 14, and 15 November 1995), vol. II, 433–42

Joseph Corn, *The winged gospel: America's romance with aviation, 1900–1950* (New York, 1983)

E. L. Cornwell, *Commercial road vehicles* (London, 1960)

Robin Cowan, "Nuclear power reactors: A study in technological lock-in," *Journal of economic history* 50, no. 3 (September 1990)

Donald T. Critchlow, "Studebaker Corporation," in May (ed.), *The automobile industry, 1896–1920*, 433–37

Donald T. Critchlow, *Studebaker, the life and death of an American corporation* (Bloomington, 1996)

Tarja Cronberg and Knut H. Sørensen (eds.), *Similar concerns, different styles? Technology studies in Western Europe (Proceedings of the COST A4 workshop in Ruvaslahti, Finland, 13 and 14 January 1994)* (Brussels, 1995)

Scott A. Cronk, *Building the E-motive industry: Essays and conversations about strategies for creating an electric vehicle industry* (Warrendale, 1995)

Donald Cross, "The development of traffic signs," *Industrial archaeology* 5 (1968) 266–72

Brian J. Cudahy, *Cash, tokens and transfers: A history of urban mass transit in North America* (New York, 1990)

H. C. Cushing Jr., *The electric vehicle hand-book*, 11th ed. (New York, 1923)

H. C. Cushing Jr. and Frank W. Smith, *The electric vehicle hand-book* (New York, 1913)

Daimler Feuerwehr-Fahrzeuge (n.p., n.d. [Stuttgart, 1914]) (brochure, DaimlerChrysler archives)

S. Daule, *Der Krieg gegen das Auto* (Leipzig, n.d.)

Paul A. David, *Technical choice, innovation and economic growth: Essays on American and British experience in the nineteenth century* (Cambridge, Mass., 1975)

Paul A. David and Julie Ann Bunn, "The economics of gateway technologies and network evolution: Lessons from electricity supply history," *Information economics and policy* 3 (1988), 3

Donald Finlay Davis, *Conspicuous production: Automobiles and elites in Detroit, 1899–1933* (Philadelphia, 1988)

H. P. W. Dekkers, *Historische belevenissen van een automobilist rond de eeuwwisseling* (brochure, Pionier Automobielenclub, 1979)

Alexander Demandt, *Ungeschehene Geschichte: Ein Traktat über die Frage: Was wäre geschehen, wenn . . . ?* (Göttingen, 1984)

Deuxième concours des poids lourds; Versailles, 1898; Rapport de la commission (Paris, 1899) (collection J.-L. Krieger, Paris)

Lothar Diehl, "Das Automobil in der wilhelminischen Gesellschaft: Alltagsgeschichtliche Aspekte einer technischen Innovation" (unpubl. master's thesis, Universität Tübingen, 1990)

Meinolf Dierkes and Andreas Knie, "Technikgenese: Zur Bedeutung von Organisationskulturen und Konstruktionstraditionen in der Entwicklung des Motorenbaus und der mechanischen Schreibtechniken: Vorstellung und Begründung eines Untersuchungsdesigns," in Burkart Lutz (ed.), *Technik in Alltag und Arbeit: Beiträge der Tagung des Verbunds Sozialwissenschaftliche Technikforschung (Bonn, 29./30.5.1989)* (Berlin, 1989), 203–18

James Rood Doolittle, *The romance of the automobile industry; being the story . . .* (New York, 1916)

Giovanni Dosi, "Technological paradigms and technological trajectories," *Research policy* 11 (1982), 147–62

Menno Duerksen, "Cars that go buzz in the night: A history of electrics," *Cars & parts* (August 1977), 9ff.

Perry Duis, *Challenging Chicago: Coping with everyday life, 1837–1920* (Urbana, 1998)

Richard van Dülmen (ed.), *Körper-Geschichten: Studien zur historischen Kulturforschung V* (Frankfurt am Main, 1996)

C. S. Dunbar, *Buses, trolleys & trams,* 3rd ed. (London, 1969)

Rudi Dünkelberg, "Elektroboote im Wandel der Zeit - eine Übersicht," *Elektrizitätswirtschaft* 91 (1992), Heft 10, 598–601

H. J. A. Duparc, *De Amsterdamse paardetrams* (Rotterdam, 1973)

Paul Dupuy, *La traction électrique: Tramways, locomotives et métropolitains électriques; tractions dans les mines, sur eau et sur route . . .* (Paris, 1897)

Heidrun Edelmann, *Vom Luxusgut zum Gebrauchsgegenstand: die Geschichte der Verbreitung von Personenkraftwagen in Deutschland* (Frankfurt am Main, 1989) (Schriftenreihe des Verbandes der Automobilindustrie, no. 60)

David Edgerton, "Tilting at paper tigers," *British journal for the history of science* 26 (1993), 67–75

W. J. C. Eikendal, *Constructie, werking, onderhoud en reparatie van den automobiel: handboek voor vakman en automobilist; eerste deel: De motor* (Amsterdam, n.d.)

The electric truck: A sketch indicating the development and present status of this modern successor of the horse (Pontiac, Mich., 1912) (brochure, General Motor Trucks Company, ENHS)

Electric vehicles and other alternatives to the internal combustion engine (Joint hearings before the Committee on Commerce and the Subcommittee on Air and Water Pollution of the Committee on Public Works; United States Senate, Ninetieth Congress, First Session . . . , March 14, 15, 16, 17, and April 10, 1967) (Washington, 1967)

Elektrofahrzeuge in kommunalen Betrieben: Tatsachen und Erfahrungen (Herausgegeben

von der Arbeitsgemeinschaft zur Förderung der Elektrowirtschaft (A. F. E.) Berlin)
(Berlin, n.d. [1938?])

Elektro-Strassenfahrzeuge: Mitteilungen über Forschungen zur Verbesserung der Verkehrsverhältnisse der Gemeinden (Forschung Stadtverkehr, Heft 28, Sonderheft) (Bonn, 1981)

Norbert Elias, *Über den Prozess der Zivilisation* (Basel, 1939), Band 2

Boelie Elzen, Johan Schot, and Remco Hoogma, "Strategies for influencing the car system," in Knut Holtan Sørensen (ed.), *The car and its environments: The past, present and future of the motorcar in Europe; Proceedings from the COST A4 Workshop in Trondheim, Norway, May 6–8 1993* (Brussels, 1994), 199–248

Clive Emsley, "'Mother, what *did* policemen do when there weren't any motors?' The law, the police and the regulation of motor traffic in England, 1900–1939," *The historical journal* 36, no. 2 (1993), 357–81

"Die Entwicklung des Kraftantriebes für Feuerwehrfahrzeuge mit Berücksichtigung der Möglichkeit der Anwendung modernster Kraftantriebe für die Zwecke der Feuerwehr" (typescript, Feuerwehrmuseum Berlin, document 325E)

Ralph C. Epstein, *The automobile industry: Its economic and commercial development* (Chicago, 1928)

A. Everitt, "Country carriers in the nineteenth century," *Journal of transport history*, New Series, 3 (1975–76), 179–202

Michael Evgénieff, *Droschkenbetrieb: Seine Organisation und wirtschaftspolitische Probleme* (Berlin, 1934) (diss.)

Jerry D. Falk, "The Woods electric," *Bulb horn* (April–June 1995), 14–17

D. Farman, *Les automobiles: Voitures, tramways et petits véhicules* (Paris, 1896)

J. A. Feith (ed.), *Groningse volksalmanak voor het jaar 1898: Jaarboekje voor geschiedenis, taal- en letterkunde der provincie Groningen* (Groningen, 1898)

Olaf von Fersen (ed.), *Ein Jahrhundert Automobiltechnik: Nutzfahrzeuge* (Düsseldorf, 1987)

Feuerschutz und Feuerrettungswesen beim Beginn des XX. Jahrhunderts: Berichtswerk über die Internationale Ausstellung für Feuerschutz und Feuerrettungswesen, Berlin 1901, bearbeitet im Auftrage des Königlich Preussischen Ministeriums des Innern (Berlin, 1902)

Claude S. Fischer, *America calling: A social history of the telephone to 1940* (Berkeley, 1992)

Reiner Flik, "Motorisierung und Automobilindustrie in Deutschland bis 1933: Der deutsche Personenkraftwagen und die 'Amerikanische Gefahr'," unpubl. "Habilitationsschrift," Katholische Universität Eichstatt, n.d. [1999]

James J. Flink, *America adopts the automobile, 1895–1910* (Cambridge, Mass., 1970)

James J. Flink, *The automobile age* (Cambridge, Mass., 1993)

Hans Fogelberg, "The electric car controversy: A social-constructivist interpretation of the California zero-emission vehicle mandate" (Göteborg, 1996) (unpubl. master's thesis, Chalmers University of Technology, Department of History of Technology and Industry)

C. Fontanon, *Mobilité de la population et transformations de l'espace urbain: le rôle des transports en commun dans la région parisienne 1855–1939*, vol. 1 (Paris, n.d.) (diss., Ecole des Hauts Etudes en Sciences sociales)

Henry Ford, *My life & work*, 7th ed. (London, 1925)

James Foreman-Peck, "'Technological lock-in' and the power source for the motor car," *Discussion papers in economic and social history* (May 1996)

James Foreman-Peck, "Death on the roads: Changing national responses to motor accidents," in Barker (ed.), *The economic and social effects of the spread of motor vehicles*, 264–90

James Foreman-Peck, "Was the electric car 'locked-out'?" (manuscript)

James Foreman-Peck and Masahiro Hayafuji, "Lock-in and Panglossian selection in technological choice: The power source for the motor car" (manuscript)

J. Freeke, *De kunst van het vervoer* (Den Haag, n.d. [1990])

C. L. Freeston, "Automobile clubs," in Harmsworth (ed.), *Motors and motor-driving*, 384–96

C. L. Freeston, "Races and trials," in Harmsworth (ed.), *Motors and motor-driving*, 402–8

C. L. Freeston, "Tyres," in Harmsworth (ed.), *Motors and motor-driving*, 224–42

Michael J. French, *The U.S. tire industry* (Boston, 1991)

M. J. Freeman and D. H. Aldcroft, *Transport in Victorian Britain* (Manchester, 1988)

Tom French, *Tyre technology* (Bristol, 1988)

Patrick Fridenson, *Histoire des usines Renault: I. Naissance de la grande entreprise 1898–1939* (Paris, 1972)

Hans Fründt, *Das Automobil und die Automobilindustrie in Deutschland* (Neustrelitz, 1911) (diss.)

J. Fuchs, *Die heerlijke auto's: De eerste halve eeuw autorijden in Nederland* (Amsterdam, 1970)

D. Gammrath and H. Jung, *Berliner Omnibusse* (Berlin, n.d.)

David Gartman, *Auto opium: A social history of American automobile design* (London, 1994)

Peter Gay, *The cultivation of hatred* (New York, 1993) (The Bourgeois experience: Victoria to Freud, vol. III)

Wolfgang H. Gebhardt, *Geschichte des deutschen Lkw-Baus; Band 1, 1896–1918* (Stuttgart, 1994)

Wolfgang Gebhardt, *Deutsche Omnibusse seit 1895* (Stuttgart, n.d. [1996])

Gedenkboek van het 25-jarig bestaan der Koninklijke Nederlandsche Automobiel Club, 1898–3 juli-1923 (Haarlem, 1923)

G. N. Georgano, *The American automobile: A centenary 1893–1993* (New York, 1992)

G. N. Georgano, *A history of the London taxicab* (Newton Abbot, 1972)

Nick Georgano, *Electric vehicles* (Princes Risborough, 1996) (Shire album 325)

Nick Georgano, *The London taxi* (n.p. [London], 1985) (Shire album 150)

G. N. Georgano (ed.), *The complete encyclopedia of motorcars 1885 to the present*, 2nd ed. (London, 1973)

G. N. Georgano, *Auto's uit de jaren 1886–1930* (Alphen aan den Rijn, 1986)

Manfred Gihl, *Rettungsfahrzeuge: Von der Krankenkutsche zum Notarztwagen* (Stuttgart, 1986)

Gorman Gilbert and Robert E. Samuels, *The taxicab: An urban transportation survivor* (Chapel Hill, 1982)

F. Girardault, *Les automobiles industrielles* (Paris, 1910)

Mark Girouard, *The return to Camelot: Chivalry and the English gentleman* (New Haven, 1981)

Charles N. Glaab and A. Theodore Brown, *A history of urban America* (New York, 1967)

G. Göbel, *Automobilmotoren: Kritische Betrachtung der Entwicklung der Automobil-Verbrennungsmotoren* (Vienna, 1905)

Stephen B. Goddard, *Getting there: The epic struggle between road and rail in the American century* (New York, 1994)

Gustav Goldbeck, "Entwicklungsstufen des Verbrennungsmotors," *Motortechnische Zeitschrift* 23, Heft 2 (February 1962), 76–80

C. Gordijn Jr., *Gedenkboek, uitgegeven ter gelegenheid van het vijftigjarig bestaan der Amsterdamsche beroepsbrandweer, 15 augustus 1874–15 augustus 1924* (n.p., n.d.)

H. de Graffigny, *La locomotion électrique* (Paris, n.d.)

H. de Graffigny, *Les moteurs légers, applicables à l'industrie, aux cycles et automobiles, à la navigation, à l'aéronautique, etc.* (Paris, 1899)

Gerald S. Graham, "The ascendancy of the sailing ship 1850–85," *Economic history review* 9 (August 1956), 74–88

Grand album illustré de l'industrie automobile pour l'année 1902 (Paris, n.d. [1902]) (Archives Départementales, Paris)

William Greenleaf, *Monopoly on wheels: Henry Ford and the Selden automobile patent* (Detroit, 1961)

J. A. Grégoire, *50 ans d'automobile: 2; la voiture électrique* (Paris, 1981)

K. Gries, "Die volkswirtschaftliche Bedeutung des Einsatzes von Elektrofahrzeugen im Stadtverkehr insbesondere unter Berücksichtigung der Berliner Verkehrsverhältnisse: Im Auftrag des RKW bearbeitet" (undated typescript [1951?], Varta archives)

Arnulf Grübler, *The rise and fall of infrastructures: Dynamics of evolution and technological change in transport* (Heidelberg, 1990)

Arnulf Grübler, *Technology and global change* (Cambridge, 1998)

Reiner Grundmann, "Gibt es eine Evolution von Technik? Überlegungen zum Automobil und zur Evolutionstheorie," *Technik und Gesellschaft* 7 (1994), 13–39

H. J. Habakkuk, *American and British technology in the nineteenth century: The search for labour-saving inventions* (Cambridge, 1962)

Rudolf Haller, "Deutsche Kommunalfahrzeuge: Eine Übersicht über das Bauprogramm 1951," *Übersee-Post* 3 (1951) (Varta archives, file II D 2: "1938–1954")

James R. Hansen, "Aviation history in the wider view," *Technology and culture* (July 1989), 643–56

Mikael Hård, *Machines are frozen spirit: The scientification of refrigeration and brewing in the 19th century—A Weberian interpretation* (Frankfurt am Main, 1994)

Mikael Hård, "Technology as practice: Local and global closure processes in diesel-engine design," *Social studies of science* 24 (1994), 549–85

M. Hård and A. Jamison, "Alternative cars: The contrasting stories of steam and diesel automotive engines," *Technology in society* 19, no. 2 (1997), 145–60

C. K. Harley, "On the persistence of old techniques: The case of North American wooden shipbuilding," *Journal of economic history* 33, no. 2 (1973), 372–89

Richard Harmond, "Progress and flight: An interpretation of the American cycle craze of the 1890s," *Journal of social history* 5 (Winter 1971–72), 235–57

Alfred C. Harmsworth (ed.), *Motors and motor-driving*, 2nd ed. (London, 1902)

Alfred C. Harmsworth, "The choice of a motor," in Harmsworth (ed.), *Motors and motor-driving*, 38–65

A. E. Harrison, "The origins and growth of the UK cycle industry to 1900," *Journal of transport history* (third series) 6, no. 1 (March 1985), 41–70

Paul N. Hasluck, *The automobile: A practical treatise on the construction of modern cars; steam, petrol, electric & petrol-electric*, 3 vols. (1909)

Barbara Haubner, *Nervenkitzel und Freizeitvergnügen: Automobilismus in Deutschland 1886–1914* (Göttingen, 1998)

Killingworth Hedges, *American electric street railways: Their construction and equipment*

with notes as to the cost of installation and of maintenance, also the advantages of
electric traction compared with other methods (London, 1894)

V. Heinz and V. Klement, *Z dejin automobilu* (Prague, 1931) (German translation: DaimlerChrysler archives)

Arnold Heller, *Motorwagen und Fahrzeugmaschinen für flüssigen Brennstoff: Ein Lehr-buch für den Selbstunterricht und für den Unterricht an technischen Lehranstalten* (Berlin, 1912) (reprint: Moers, 1985)

Hans Dieter Hellige, "Von der programmatischen zur empirischen Technikgenese-forschung: Ein technikhistorisches Analyseinstrumentarium für die prospektive Technikbewertung," *Technikgeschichte* 60, no. 3 (1993), 186–223

W. Hendlmeier, *Handbuch der deutschen Strassenbahngeschichte, Erster Band* (Munich, 1981)

W. Hendlmeier, *Von der Pferde-Eisenbahn zur Schnell-Strassenbahn* (Munich, 1968)

A. Herbst and J. Wu, "Some evidence of subsidization: The U.S. trucking industry, 1900–1920," *Journal of economic history* 33 (1973), 417–33

A. N. Hesselmans, "Elektriciteit," in Lintsen et al. (eds.), *Geschiedenis van de techniek in Nederland, vol. III*, 135–61

H. W. Heyman, "The electric vehicle (Address . . . at the 8th Annual Meeting of the Scottish Co-operative Transport Association . . . Glasgow . . . 11th March, 1952)" (Varta archives, file II D 2: "1938–1954")

H. W. Heyman, "The economic basis of battery electric road vehicle operation and manufacture" (paper no. 1206, 17 April 1952; source not indicated, probably *IMechE Proceedings;* Varta archives, file II D 2: "1938–1954")

Sigurd Hilkenbach and Wolfgang Kramer, *125 Jahre Strassenbahnen in Berlin* (Düssel-dorf, 1990)

Stanley M. Hills, *Battery-electric vehicles: Dealing with the construction and operation of all types of battery-operated electric vehicles and accessory equipment* (London, 1943)

Gardner D. Hiscox, *Horseless vehicles; automobiles, motorcycles operated by steam, hydro-carbon, electric and pneumatic motors: A practical treatise for automobilists, manu-facturers, capitalists, investors and everyone interested in the development, use and care of the automobile, including a special chapter on how to build an electric cab, with detail drawings* (New York, 1900)

Historical statistics of the United States, colonial times to 1970 (Bicentennial ed.) (Wash-ington, D.C., 1975)

Ulrico Hoepli, *Manuale dell'automobilista e guida pei meccanici conduttori d'automobili* (Milano, 1908)

Edmund Hoppe, *Die Akkumulatoren für Elektricität*, 2nd ed. (Berlin, 1892)

Karl H. Hörning, "Technik und Kultur: Ein verwickeltes Spiel der Praxis," in Jost Halfmann, Gotthard Bechmann, and Werner Rammert (eds.), *Technik und Gesellschaft, Jahrbuch 8* (Frankfurt, 1995), 131–51

Wolfgang Hornung-Arnegg, *Feuerwehrgeschichte: Brandschutz und Löschgerätetechnik von der Antike bis zur Gegenwart*, 4th ed. (Stuttgart, 1995)

Gerhard Horras, *Die Entwicklung des deutschen Automobilmarktes bis 1914* (Munich, 1982)

Horvat-Dembreville, "La voiture électrique sous l'occupation; première partie: les voitures 'de série,'" *Automobilia* (January 1998), 20–27

Gottfried Hösel, *Unser Abfall aller Zeiten: Eine Kulturgeschichte der Städtereinigung*, 2nd ed. (Munich, 1990)

David A. Hounshell, "Hughesian history of technology and Chandlerian business history: Parallels, departures, and critics," *History of technology* 12, no. 3 (1995), 205–24

David A. Hounshell, *From the American system to mass production 1800–1932: The development of manufacturing technology in the United States* (Baltimore, 1984)

E. S. Houwaart, "Medische statistiek," in Lintsen et al. (eds.), *Geschiedenis van de techniek in Nederland, vol. II*, 19–45

E. S. Houwaart, "Professionalisering en staatsvorming," in Lintsen et al. (eds.), *Geschiedenis van de techniek in Nederland, vol. II*, 81–92

[F.] Hubrig, "Das Elektrofahrzeug im Dienste der deutschen Reichspost, seine verkehrs- und volkswirtschaftliche Bedeutung" (Berlin, 1935) (Sonderabdruck aus dem *Archiv für Post und Telegraphie*, no. 5, Jahrgang 1935) (Varta archives, file II D 2: "1909–1935")

F. Hubrig, "Das Kraftfahrzeug im Dienst der Deutschen Reichspost," in Allmers et al. (eds.), *Das deutsche Automobilwesen der Gegenwart*, 115–18

Fritz Hubrig, "Über Erfahrungen im Einsatz von Elektrofahrzeugen im Postbetrieb" (Sonderdruck aus dem Jahrbuch *Schiene und Strasse 1952*) (Varta archives, file II D 2: "1938–1954")

Thomas P. Hughes, *Networks of power: Electrification in Western society, 1880–1930* (Baltimore, 1983)

Thomas Parke Hughes, *Elmer Sperry: Inventor and engineer* (Baltimore, 1971)

Thomas P. Hughes, *American genesis: A century of invention and technological enthusiasm 1870–1970* (New York, 1989)

Thomas P. Hughes, "The seamless web: Technology, science, etcetera, etcetera," *Social studies of science* 16 (1986), 281–92

Thomas P. Hughes, *Rescuing Prometheus* (New York, 1998)

Peter J. Hugill (review of Schiffer, *Taking charge*), *Technology and culture* 37, no. 2 (April 1996), 379–81

Josef Hüls, "Das deutsche Kraftdroschkengewerbe" (typescript, diss., Munich, 16 June 1931)

A. van Hulzen, *Utrecht en het verkeer, 1850–1910* (Baarn, 1987)

Wolfgang Huss and Wolf Schenk, *Omnibusgeschichte, Teil 1: Die Entwicklung bis 1924* (Munich, 1982)

Helmut Hütten, "100 Jahre Fahrzeugmotoren," *Motortechnische Zeitschrift* 47, Heft 4 (April 1986), 121–31 and "Teil 2": *Motortechnische Zeitschrift* 47, Heft 7/8 (July/August 1986), 313–16

Jacques Ickx, *Ainsi nacquit l'automobile, Tome I* (Lausanne, n.d.)

Gerhard Imohr, "Der Akku-Bus in der deutschen Bundesrepublik (Sonderdruck aus *Verkehr und technik*, . . . Heft 12/1953) (Varta archives, file II D 2: "1938–1954")

International dictionary on world automobile technicians (brochure, International Historical Commission of the FIA, Paris, October 1992)

Andrew C. Irvine, "The promotion and first twenty-two years history of a corporation in the electrical manufacturing industry" (unpubl. master's thesis, Temple University, Philadelpia, 1954)

Jan Jacobs, Peter Groote, and Jan-Egbert Sturm, "Waren investeringen in infrastructuur produktief in Nederland (1850–1913)?" *NEHA-Jaarboek voor economische, bedrijfs- en techniekgeschiedenis*, vol. 59 (Amsterdam, 1996), 238–57

Guillaume Jacquemyns, *Histoire contemporaine du Grand-Bruxelles* (Bruxelles, 1936)

Dieter Jarausch and Joachim Haase, *Die Stuttgarter Feuerwehr: Von den Anfängen der Brandbekämpfung und Brandverhütung bis zur Gegenwart; Chronik anlässlich des 100jährigen Jubiläums der Berufsfeuerwehr im Jahr 1991* (Stuttgart, 1991)

A. Jardillier, "Pascal's 'Carosses à cinq solz,'" *Journal of transport history* 7 (1965–66), 35–36

Hans Robert Jauss, *Literaturgeschichte als Provokation* (Frankfurt, 1970)

Sir Francis Jeune, "The charms of driving in motors," in Harmsworth (ed.), *Motors and motor-driving*, 341–45

Claude Johnson, *The early history of motoring* (London, n.d.)

Jubiläumsschrift 125 Jahre Berliner Feuerwehr (Berlin, n.d.)

M. Kallmann, *Wettbewerb und Prüfungsfahrten für elektrisch betriebene Fahrzeuge in Berlin, Frühjahr 1900: Bericht über die Ergebnisse* (Berlin, 1900)

H. P. Kaper and J. H. S. M. Veen, *De Rotterdamse paardetrams* (Rotterdam, 1974)

George Katona, *The powerful consumer: Psychological studies of the American economy* (New York, 1960)

George Katona, *Psychological analysis of economic behavior* (New York, 1951)

Rankin Kennedy, *The book of the motor car: A comprehensive and authoritative guide on the care, management, maintenance, and construction of the motor car and the motor cycle*, 3 vols. (London, n.d. [1913])

Stephen Kern, *The culture of time and space 1880–1918* (London, 1983)

Ketelhohn, *Das Elektrofahrzeug in der deutschen Kraftverkehrswirtschaft* (Heft 1 of: W. Schuster [ed.], *Wirtschaft und Technik*) (Berlin, 1940)

C. Kiesel, "Die 'DEW'-Elektrodroschke," *Elektrizitätswirtschaft: Mitteilungen der VDEW Nr. 439* (August 1927), 369–74

Beverly Rae Kimes and Henry Austin Clark Jr., *Standard catalog of American cars 1805–1942*, 2nd ed. (Iola, Wis., 1989)

Charles P. Kindleberger, *Manias, panics, and crashes: A history of financial crises* (London, 1978)

Charles B. King, *A golden anniversary 1895–1945: Personal side lights of America's first automobile race* (n.p. [New York], 1945) (published on his own)

Peter Kirchberg, "Das Wachstum der Produktivkräfte in der Geschichte des Kraftfahrzeugs, untersucht am Beispiel der Entwicklung der Technik des Kraftwagens in Deutschland von den Anfängen bis zur Weltwirtschaftskrise" (unpubl. Habilitationsschrift, Dresden, 1978)

David A. Kirsch, "Behind the numbers: Early quantitative data on the history of the American automobile industry," paper presented at the ICOHTEC Symposium, Prague, 24 August 2000

David Kirsch (review of Wakefield, *History of the electric automobile*), *Technology and culture* 36, no. 3 (July 1995), 710–12

David A. Kirsch, "The electric car and the burden of history: Studies in automotive systems rivalry in America, 1890–1996" (diss., Stanford University, 1996)

David Kirsch, "Technology, environment and public policy in perspective: Lessons from the history of the automobile," *Technical expertise and public decisions proceedings* (IEEE International Symposium on technology and society, Princeton, N.J., 21 June 1996), 67–75

David Kirsch, "Turning points in technology: Steam, gasoline and electric powered vehicles in America, 1890–1918," unpubl. lecture, SHOT conference, Charlottesville, Va., 22 October 1995

David A. Kirsch and Gijs P. A. Mom, "Visions of transportation: The EVC and the transition from service- to product-based mobility," *Business history review* 76 (Spring 2002), 75–110.

Edmund Klapper, *Die Entwicklung der deutschen Automobil-Industrie: Eine wirtschaftliche Monographie unter Berücksichtigung des Einflusses der Technik* (Berlin, 1910)

Ronald Kline and Trevor Pinch, "Users as agents of technological change: The social construction of the automobile in the rural United States," *Technology and culture* 37, no. 4 (October 1996), 763–95

G. Klinner and R. Welzel, "Sonderfahrzeuge für Städtereinigung," in Allmers et al. (eds.), *Das deutsche Automobilwesen der Gegenwart*, 124–26

Albert Kloss, *Elektrofahrzeuge: vom Windwagen zum Elektromobil* (Berlin, 1996)

Erwin Knaths, *Die Entwicklung des Berliner Droschkenfuhrwesens unter besonderer Berücksichtigung seiner Motorisierung: Eine verkehrsgeschichtliche Studie* (Marburg a.d. Lahn, 1929) (diss.)

Andreas Knie, *Diesel - Karriere einer Technik: Genese und Formierungsprozesse im Motorenbau* (Berlin, 1991)

Andreas Knie, *Wankel-Mut in der Autoindustrie: Anfang und Ende einer Antriebsalternative* (Berlin, 1994)

Andreas Knie, "Elektrofahrzeuge: Chancen oder Risiken? Rahmenbedingungen, Einsatzpotentiale, Nützungsprofile, technische Optionen und ökologische Bewertung von elektrisch betriebenen Strassenfahrzeugen" (Wissenschaftzentrum Berlin für Sozialforschung, March 1995)

Andreas Knie and Otto Berthold, "Das Ceteris paribus-Syndrom in der Mobilitätspolitik: Tatsächliche Nutzungsprofile von elektrischen Strassenfahrzeugen" (publication Wissenschaftszentrum Berlin für Sozialforschung and Technische Universität Berlin, February 1995)

Andreas Knie and Mikael Hård, "Die Dinge gegen den Strich bürsten: De-Konstruktionsübungen am Automobil," *Technikgeschichte* 60, no. 3 (1993), 224–42

Volkmar Köhler, "Deutsche Personenwagen-Fabrikate zwischen 1886 und 1965," *Tradition: Zeitschrift für Firmengeschichte und Unternehmerbiographie* (June 1966), 127–51

Paul A. C. Koistinen, *Mobilizing for modern war: The political economy of American warfare, 1865–1919* (Lawrence, 1997)

P. Kooij, *Groningen 1870–1914: Sociale verandering en economische ontwikkeling in een regionaal centrum* (n.p., n.d. [Groningen, 1986])

Wolfgang König, "Umbrüche und Umorientierungen - Kontinuität und Diskontinuität - Evolution und Revolution: Zur Theorie historischer zeitverläufe in der Wissenschafts- und Technikgechichte," in idem (ed.), *Umorientierungen: Wissenschaft, Technik und Gesellschaft im Wandel* (Frankfurt am Main, 1994), 9–31

Wolfgang König, "Technik, Macht und Markt: Eine Kritik der sozialwissenschaftlichen Technikgeneseforschung," *Technikgeschichte* 60, no. 3 (1993), 243–66

Steffen Koolmann, *Leitbilder der Technikentwicklung: Das Beispiel des Automobils* (Frankfurt, 1992) (diss.) (Campus Forschung, Band 687)

Thomas Köppen, "Elektromobilzeit: Die Jahrhundertwende und ihre ganz besonderen Automobile," *Kultur & Technik* 1 (1990), 50–55

Thomas Köppen, "Ferdinand Porsche, Ludwig Lohner und Emil Jellinek—frühe Innovatoren im Elektromobilbau: eine Falstudie über eine gescheiterte Innovation" (unpubl. master's thesis, TU Berlin, 1987)

Thomas Köppen, "Die Rolle der Firma Jakob Lohner & Co bei der Entwicklung von Hybridantrieben im Automobilbau," *Technikgeschichte* 55, Heft 2 (1988), 95–110

Thomas Köppen, "Die Unternehmensstrategien der städtischen Kutschenfabriken zu Beginn des 20.: Jahrhunderts am Beispiel der Wiener Hof-Wagenfabrik Jakob Lohner & Co.," *Zeitschrift für Unternehmensgeschichte* 38, Heft 3 (1993), 176–85

Arth. Korff-Petersen, "Gesundheitsgefährdung durch die Auspuffgase der Automobile," *Zeitschrift für Hygiene und Infektionskrankheiten* 69 (1911), 135–48

Robert J. Kothe, "Electric automobiles," in May (ed.), *The automobile industry, 1896–1920,* 172–73

G. Kraft, "Das gleislose Elektrofahrzeug im Kampf um seine Wiedereinführung" (Sonderdruck aus der Zeitschrift *Elektrizitätswirtschaft* [1955 Heft 11]) (Varta archives, file II D 2: "1938–1954")

G. Kraft, "Der neue Elektrowagen der Auto-Union bewährt sich" (Sonderdruck aus der Zeitschrift *Elektrizität* [1955, Heft 8]) (Varta archives, file II D 2: "1938–1954")

Eda Kranakis, "Technology assessment and the study of history," *Science, technology, & human values* 13, nos. 3 and 4 (Summer–Fall 1988), 290–307

Joachim Krause, "Das Fahrrad: Von der 'kindischen' Kombinatorik zur Montage," in Ruppert (ed.), *Fahrrad, Auto, Fernsehschrank,* 79–118

J. L. Krieger, "Un siècle de véhicules électriques en France (suite)," *La Berline* (newsletter of the "Amis du Musée de Compiègne") (February 1995), 3–6

L. Kriéger, M. Buchon, E. Lacroix, et al., *Le véhicule électrique utilitaire à accumulateurs: Conférences données à la Société des Ingénieurs de l'Automobile* (Paris, 1947)

Ulrich Kubisch, *Taxi: auf den Spuren des mobilen Gewerbes* (Berlin, 1984)

Ulrich Kubisch, *Borgward war nicht der Anfang: Hansa Lloyd Automobilbau in Bremen und Varel bis 1929* (Bremen, 1986)

Ulrich Kubisch, *Deutsche Automarken von A-Z* (Mainz, 199p3)

[Ulrich Kubisch], "Die Geschichte der Elektrodroschken: Strom contra Sprit," in Kubisch et al. (eds.), *Taxi,* 47–56

Ulrich Kubisch et al. (eds.), *Taxi: Das mobilste Gewerbe der Welt* (Berlin, n.d. [1993])

Klaus Kuhm, *Das eilige Jahrhundert: Einblicke in die automobile Gesellschaft* (Hamburg, 1995)

Klaus Kuhm, *Moderne und Asphalt: Die Automobilisierung als Prozess technologischer Integration und sozialer Vernetzung* (Pfaffenweiler, 1997)

K. Kühner, *Geschichtliches zum Fahrzeugantrieb* (Friedrichshafen, 1965)

H. C. Kuiler, *Verkeer en vervoer in Nederland* (Utrecht, 1949)

Yvette Kupélian, Jacques Kupélian, and Jacques Sirtaine, *De geschiedenis van de Belgische auto: Het fabelachtige verhaal van meer dan honderd automobielmerken* (Tielt, 1980)

Jul. Küster, "Zur Geschichte des Verbrennungsmotors," *Automobiltechnische Zeitschrift* Heft 10 (October 1938), 257–65

Tom LaMarre, "Detroit Electric: Society's town car," *Automobile quarterly* 27, no. 2 (second quarter 1989), 161–70

F. Van Z. Lane, *Motor truck transportation: The principles governing its success* (New York, 1921)

Philippe Laneyrie and Jacques Roux, "Transport traditionnel et innovation technique: L'exemple du taxi en France," *Culture technique* 19 (March 1989), 262–72

Todd R. La Porte, "The United States air traffic system: Increasing reliability in the midst of rapid growth," in Mayntz and Hughes (eds.), *The development of large technical systems,* 215–44

D. Larroque, "Apogée, déclin et relance du tramway en France," *Culture technique* 19 (March 1989), 54–62

Edward E. La Schum, *The electric motor truck: Selection of motor vehicle equipment, its operation and maintenance* (New York, 1924)

"The late W. C. Bersey," *The policy-holder: An insurance journal* (1950), 484–85 (archives Science Museum, London)

James M. Laux, *The European automobile industry* (New York, 1992)

James M. Laux, *In first gear: The French automobile industry to 1914* (Liverpool, 1976)

James M. Laux, "Trucks in the west during the First World War," *Journal of transport history* 6, no. 2 (September 1985), 64–70

Gérard Lavergne, *Manuel théorique et pratique de l'automobile sur route: Vapeur-pétrole-électricité* (Paris, 1900)

M. G. Lay, *Ways of the world: A history of the world's roads and of the vehicles that used them* (New Brunswick, N.J., 1992)

John Law, "Theory and narrative in the history of technology: Response," *Technology and culture* (April 1991), 377–84

W. J. M. Leideritz, *De paardetram in Nederland* (Alkmaar, n.d. [1970])

D. van Lente, H. W. Lintsen, M. S. C. Bakker, E. Homburg, J. W. Schot, and G. P. J. Verbong, "Techniek en modernisering," in Lintsen et al. (eds.), *Geschiedenis van de techniek in Nederland, vol. I*, 19–36

Dorothy Levitt, *The woman and the car: A chatty little handbook for all women who motor or who want to motor* (London, 1909)

David L. Lewis and Laurence Goldstein (eds.), *The automobile and American culture*, 2nd ed. (Ann Arbor, 1983)

John W. Lieb, *Electric vehicles: Report prepared for the Union Internationale des Producteurs et Distributeurs d'Energie Electrique, Paris Meeting, July 5th to 10th, 1928* (n.p., n.d.)

Jean Linamood, "A quick jolt of history: A century and a half of electric vehicles," *Car and driver* (June 1981), 51–52

Helmut Lindner, *Strom: Erzeugung, Verteilung und Anwendung der Elektrizität* (Reinbek bei Hamburg, 1985)

H. Lindner, "Mit dem Strom im Fluss. Kaiserliches Interesse an elektrisch angetriebenen Booten," *Kultur & Technik* 4 (1991), 16–17

Svank Lindqvist, "Telegraph diplomats: The United States' relations with France in 1848 and 1870," *Technology and culture* 40, no. 1 (January 1999), 1–25

Peter J. Ling, *America and the automobile: Technology, reform and social change* (Manchester, 1990)

H. W. Lintsen et al. (eds.), *Geschiedenis van de techniek in Nederland: De wording van een moderne samenleving 1800–1890, vol. I: Techniek en modernisering; landbouw en voeding* (Zutphen, 1992)

H. W. Lintsen et al. (eds.), *Geschiedenis van de techniek in Nederland: De wording van een moderne samenleving 1800–1890, vol. II: Gezondheid en openbare hygiëne, waterstaat en infrastructuur, papier, druk en communicatie* (Zutphen, 1993)

H. W. Lintsen et al. (eds.), *Geschiedenis van de techniek in Nederland: De wording van een moderne samenleving 1800–1890, vol. VI: Techniek en samenleving* (Zutphen, 1995)

P. W. Litchfield, *Thirty years of Goodyear 1898–1928: A statement to the stockholders of the Goodyear Tire & Rubber Company* (n.p., n.d. [1928]) (Goodyear archives, Folder 14. Finance. A-1037; Annual Report 1928)

B. Lizet, *Le cheval dans la vie quotidienne: Techniques et représentations du cheval de travail dans l'Europe industrielle* (Paris, 1982)

Lloyd-Wagen (n.p., n.d. [1908?]) (brochure of the Norddeutsche Automobil- & Motoren AG, archives Deutsches Museum, Munich)

Louis Lockert, *Les voitures électriques avec Supplément aux voitures à pétrole et Note sur les moteurs à acétylène et à alcool: Traité des véhicules automobiles sur route*, vol. 4 (Paris, 1897)

Alan P. Loeb, "Personal mobility in America: Its introduction and significance," paper presented at workshop, "The birth of the car—an international comparison", HTS-Autotechniek, Arnhem, The Netherlands, 13 September 1996

Alan P. Loeb, "Birth of the Kettering doctrine: Fordism, Sloanism and the discovery of tetraethyl lead" (unpubl. lecture, 41st Business History Conference, Fort Lauderdale, 19 March 1995)

Alan Loeb, "Lead in gasoline: The harmonious order" (part I of a manuscript for a book on the introduction of leaded gasoline in America; December 1994)

Alan P. Loeb, "Steam versus electric versus internal combustion: Choosing the vehicle technology at the start of the automotive age," lecture presented at the SAE Conference, Detroit, February 1997

Wilhelm Lohner, *Lohner-Automobile* (Graz, 1989)

Chris Lorenz, *De constructie van het verleden: Een inleiding in de theorie van de geschiedenis*, 4th ed. (Amsterdam, 1994)

Josef Löwy, *Das Elektromobil und seine Behandlung* (Leipzig, 1906)

G. Lucas, "Elektrokarren als Stromverbraucher," *Der Werbeleiter* (1933 Heft 10/11) (Sonderdruck) (Varta archives, file II D 2: "1909–1935")

Detlev Lüder, *Von der Sänfte zur Stadtbahn: Zur Geschichte des öffentlichen Nahverkehrs in Hannover* (Hannover, 1977)

J. R. Luurs, "De aanleg van verharde wegen in Drenthe, Groningen en Friesland, 1825–1925," *NEHA-jaarboek voor economische, bedrijfs- en techniekgeschiedenis*, vol. 59 (Amsterdam, 1996), 162–237

Sir John H. A. MacDonald, "Reminiscences," in Harmsworth (ed.), *Motors and motordriving*, 361–77

Erwin Maderholz, "Elektrofahrzeuge im Postdienst," *Archiv für deutsche Postgeschichte* (Heft 2/1981), 5–33

Vijay Mahajan and Robert A. Peterson, *Models for innovation diffusion* (Beverly Hills, 1985)

Susan Meikle Mandell, Stephen Peter Andrew, and Bernard Ross, *A historical survey of transit buses in the United States* (Warrendale, 1990) (SAE Special Publication, SP-842)

V. Christian Manz, "Ohne Stunk und Lärm: zur Geschichte des Elektroautos," *Oldtimer magazine* (July/August 1995), 35–50

"Market brief" (publication of the Electric Vehicle Association of America, March 1995)

C. W. Marshall, *Electric vehicles* (London, 1925)

Hiram P. Maxim, *Horseless carriage days* (New York, 1962) (republication of 1936)

George S. May (ed.), *The automobile industry, 1896–1920* (New York, 1990) (Encyclopedia of American business history and biography)

Hans W. Mayer, *Lloyd: Vom Elektromobil zur Arabella* (Stuttgart, 1989)

Renate Mayntz and Thomas P. Hughes (eds.), *The development of large technical systems* (Frankfurt am Main, 1988)

W. M. McBride, "Strategic determinism in technology selection: The electric battleship and U.S. naval-industrial relations," *Technology and culture* 2 (1992), 248–77

Monty McCord, *Police cars: A photographic history* (n.p., n.d.)

Jim McCraw, "Leader of the Stack," *Engine technology international* (2002), No. 1, 24–27

John P. McKay, "Comparative perspectives on transit in Europe and the United States, 1850–1914," in Tarr and Dupuy (eds.), *Technology and the rise of the networked city*, 3–23

John P. McKay, *Tramways and trolleys: The rise of the urban mass transport in Europe* (Princeton, 1976)

Clay McShane, *Down the asphalt path: The automobile and the American city* (New York, 1994)

Clay McShane, "Urban pathways: The street and highway, 1900–1940," in Tarr and Dupuy (eds.), *Technology and the rise of the networked city*, 67–85

Clay McShane and Joel Tarr, "The choice of horse over steam as a motive power for nineteenth century urban transportation in the U.S.," paper presented at the XXth International Congress of the History of Science, Liège, Belgium, 24 July 1997

R. J. Mecredy, "The petrol engine," in Harmsworth (ed.), *Motors and motor-driving*, 103–37

Anthonie W. M. Meijers, *Wat maakt een ingenieur?* (TU Delft, 6 March 1998) (inaugural lecture)

Martin V. Melosi, *Garbage in the cities: Refuse, reform, and the environment, 1880–1980* (College Station, Tex., 1981)

Martin V. Melosi, *Thomas A. Edison and the modernization of America* (n.p., 1990)

Christoph Maria Merki, "Das Rennen um Marktanteile: Eine Studie über das erste Jahrzehnt des französischen Automobilismus," *Zeitschrift für Unternehmensgeschichte* 43, no. 1 (1998), 69–91

Christoph Maria Merki, "Sociétés sportives et développement de l'automobilisme (1898–1930)," in Christophe Jaccoud, Laurent Tissot, and Yves Pedrazzini (eds.), *Sports en Suisse: Traditions, transitions et transformations* (Lausanne, 2000), 45–73

G. Mestayer, *Les véhicules électriques* (Paris, 1941)

Arnold Meyer, "Die Kosten des Strassentransport" (Separatdruck aus der Zeitschrift "Der Motorlastwagen," . . . Bern; Verband Schweizerischer Motorlastwagenbesitzer, Heft XVII) (Bern, 1952)

Guurt Meyloph and Willem Bosch, *Honderd jaar handel in paardekracht* (n.p., n.d.)

Félicien Michotte, *La circulation à Paris: Ses défauts - ses remèdes* (Paris, n.d.) (Archives de la Préfecture de Police, Paris, box D/b 500)

Ch. Milandre and R.-P. Bouquet, *Traité de la construction, de la conduite et de l'entretien des voitures automobiles* (4 vols., as of 1898)

Ch. Milandre and R.-P. Bouquet, *Traité de la construction, de la conduite et de l'entretien des voitures automobiles, publié sous la direction de Ch. Vigreux; Quatrième volume: Voitures automobiles électriques* (Paris, 1899) [Milandre and Bouquet, *Voitures automobiles électriques*]

Anne Millbrooke, "Technological systems compete at Otis: Hydraulic versus electric elevators," in Aspray (ed.), *Technological competitiveness*, 243–69

Minutes of the meeting of the Association of Electric Vehicle Manufacturers, held at the Hotel Belmont, 42d Street & Park Avenue, Borough of Manhattan, New York City, December 6, 1906, at 10 o'clock a.m. (n.p., n.d.) (ENHS)

Thomas J. Misa, "How machines make history, and how historians (and others) help them to do so," *Science, technology, & human values* 13, nos. 3 and 4 (Summer–Fall 1988), 308–31

Emily R. Mobley (ed.) (compiled by Delores E. Maximena), *Electric vehicles 1890–1966: A bibliography* (edition GMR-2425 by General Motors Research Laboratories Library, 13 May 1977)

Gijs Mom, "'Äffchen' und Blockbandtaxen," in Kubisch et al. (eds.), *Taxi*, 299–303

G. M. [Gijs Mom], "Bijbel van de autotechniek," *De auto* (February 1986), 16–19

Gijs Mom, "Competition and coexistence: Motorization of land transportation and the substitution of the horse," *Achse, Rad und Wagen* (forthcoming, in a German translation)

G. Mom, "Gasturbine als alternatieve voertuigaandrijving," *Polytechnisch tijdschrift, editie Werktuigbouw* (November 1991), 44–47

Gijs Mom, "'As reliable as a streetcar': European versus American experiences with early electric cars (1880–1925)," unpubl. lecture, SHOT conference, Charlottesville, Va., 22 October 1995

Gijs Mom, "Haver-en andere motoren: De Amsterdamse paardentaxi en het dilemma van de motorisering (1880–1925)," in Bos et al., *Het paardloze voertuig*, 165–267

Gijs Mom, "De auto: Van avonturenmachine naar gebruiksvoorwerp," in *Techniek als cultuurverschijnsel: casusboek* (Heerlen, 1996) (Open Universiteit, Leerstofgebied cultuurwetenschappen), 7–135

Gijs Mom (with the cooperation of Charley Werff and Ariejan Bos), *De auto: Van avonturenmachine naar gebruiksvoorwerp* (Deventer, 1997)

Gijs Mom, "Das Holzbrettchen in der schwarzen Kiste: Die Entwicklung des Elektromobilakkumulators bei und aus der Sicht der Accumulatoren-Fabrik AG (AFA) von 1902–1910," *Technikgeschichte* Bd. 63 (1996), No. 2, 119–51

Gijs Mom, "De moderne elektro-auto," in Mom and Van der Vinne, *De elektro-auto: een paard van Troje?* 11–110

Gijs Mom, "Wie Feuer und Wasser: Der Kampf um den Fahrzeugantrieb bei den deutschen Feuerwehr (1900–1940)," in Harry Niemann and Armin Hermann (eds.), *100 Jahre LKW: Geschichte und Zukunft des Nutzfahrzeuges* (Stuttgart, 1997), 263–320

Gijs Mom, "Das Auto als Prothese: Kultur und Technik bei der Bedienung des Automobils" (unpubl. presentation, workshop, "Strassenverkehr und Gesellschaft 1918–1933," Landesmuseum für Technik und Arbeit, Mannheim, 26 and 27 January 1995)

Gijs Mom, "Die Wunderbatterie oder Probleme bei der Definition des Elektrofahrzeugs am Beispiel der Brennstoffzelle" (unpubl. presentation, workshop, "Zur Karriere von Erfindungen," Landesmuseum für Technik und Arbeit, Mannheim, 29–30 June 1995)

Gijs Mom, "The electric vehicle: Male or female?" (unpubl. lecture, Intercontinental Mobile Sources/Clean Air Conference, Munich, 18–19 March 1996)

Gijs Mom, "De auto: Een zaak van ingenieurs?" (unpubl. presentation, Manifestatie "De kindertijd van de auto," Paneldiscussie "Autotechniek en sociale wetenschappen: een vruchtbare verbintenis?" HTS-Autotechniek, Arnhem, 14 September 1996)

Gijs Mom, "Waarom de elektro-auto opnieuw zal mislukken," in A. N. Cupédo (ed.), *Visie op mobiele techniek: ATC jubileum-uitgave* (edition on the occasion of the 50th anniversary of the Vereniging van Automobile Technicians ATC) (Doetinchem, 1996), 36–38

Gijs Mom, "De elektro-auto wordt niet begrepen," *Technisch weekblad* (20 September 1995)

Gijs Mom, "Inventing the miracle battery: Thomas Edison and the electric vehicle," *History of technology* 20 (1998), 18–45

Gijs Mom, "Main stream and its alternatives: The electric vehicle as a critical comment on the combustion engined car," *A future for the city: Electric Vehicle Symposium 15*, Brussels, 29 September to 3 October 1998 (Proceedings on CD-ROM)

Gijs Mom, "Competition and coexistence: Motorization of land transportation and the disappearance of the horse, 1850–1920," paper presented at the 25th Symposium of the International Committee for the History of Technology (ICOHTEC), 18–22 August 1998, Lisbon, Portugal

Gijs Mom, "Networks, systems and the European automobile: A plea for a mobility history programme" (Review essay for the first A^MES Workshop, Scenario 1: European Infrasystem; Torino, 2–4 November 2001)

G. Mom (ed.), *De nieuwe Steinbuch: De automobiel, handboek voor autobezitters, monteurs en technici* (Deventer, 1986 and following years)

Gijs P. A. Mom and David A. Kirsch, "Technologies in tension: Horses, electric trucks, and the motorization of American cities, 1900–1925," *Technology and culture* 42, no. 3 (July 2001), 489–518

G. Mom and H. Scheffers, *De aandrijflijn* (vol. 3A of G. Mom [ed.], *De nieuwe Steinbuch: De automobiel*) (Deventer, 1992)

G. Mom and H. Scheffers, *De complexe aandrijflijn* (vol. 3B of G. Mom [ed.], *De nieuwe Steinbuch: De automobiel*) (Deventer, 1993)

Gijs Mom and Peter Staal, "Autodiffusie in een klein vol land: Historiografie en verkenning van de massamotorisering in Nederland in international perspectief," in Yves Segers, Reginald Loyen, Guy Dejongh, and Erik Buyst (eds.), *Op weg naar een consumptiemaatschappij: Over het verbruik van voeding, kleding en luxegoederen in België en Nederland (19de–20ste eeuw)* (Amsterdam, 2002), 139–80

Gijs Mom and Vincent van der Vinne, "Geschiedenis van de elektrisch aangedreven auto: de eerste en de tweede generatie (1881–1914)," in idem, *De elektro-auto: een paard van Troje?* 111–93

Gijs Mom and Vincent van der Vinne, *De elektro-auto: een paard van Troje?* (Deventer, 1995)

Gijs Mom and Vincent van der Vinne, "De vroege elektro-auto in Nederland in het licht van de internationale strijd om een geschikte tractiewijze 1880–1920," *NEHA-Jaarboek voor economische, bedrijfs-en techniekgeschiedenis*, vol. 57 (Amsterdam, 1994), 370–416

Lord Montagu of Beaulieu and F. Wilson McComb, *Behind the wheel: The magic and manners of early motoring* (New York, 1977)

Lord Montagu of Beaulieu and Anthony Bird, *Steam cars 1770–1970* (London, 1971)

Henry Charles Moore, *Omnibuses and cabs: Their origin and history* (London, 1902)

Kurt Möser, "Amphibien, Landschiffe, Flugautos - utopische Fahrzeuge der Jahrhundertwende und die Durchsetzung des Benzinautomobils" (manuscript)

Kurt Möser, "Benz, Daimler, Maybach und das Strassenverkehr: Utopien und Realität der automobilen Gesellschaft" (Mannheim, December 1998) (LTA-Forschung; Reihe des Landesmuseum für Technik und Arbeit in Mannheim)

Kurt Möser, "'Knall auf Motor'—Die Liebesaffäre von Künstlern und Dichtern mit Motorfahrzeugen 1900–1930," in *Mannheims Motorradmeister: Franz Islinger gewinnt die Deutsche Motorradmeisterschaft 1926* (Ausstellungskatalog Mannheim, 1996)

Kurt Möser, "Kriegsgeschichte und Kriegsliteratur: Formen der Verarbeitung des Ersten Weltkrieges," *Militärgeschichtliche Mitteilungen* 2/86, 39–51

Kurt Möser, "Vom Fahren zum Verkehr: Zur Entwicklung des Verkehrssystems," *Ferrum* 62 (1990), 14–20

Kurt Möser, "World War One and the creation of desire for cars in Germany," in Strasser, McGovern, and Judt (eds.), *Getting and spending*

Eliane Mossé, "Sur quelques aspects de la croissance de l'industrie automobile en France depuis la fin du XIXe siècle," *Bulletin S. E. D. E. I. S.,* no. 691 (1 February 1958), supplément a

Motor cars and news of 1899 (Lloyd Clymer's historical scrapbook) (Los Angeles, 1955)

J. Mulder, *Tram en bus rond de Martini: De ontwikkeling van het openbaar vervoer in en om de stad Groningen* (Leiden, 1970)

Adolph Müller, *25 Jahre der Accumulatoren-Fabrik Aktiengesellschaft 1888–1913* (Berlin, 1913)

H. Müller, "Das Elektrofahrzeug als Grossabnehmer elektrischer Energie," *Elektrizitätswirtschaft* (June 1932) (Sonderdruck, Varta archives, file II D 2: "1909–1935")

Kenneth Murchison, *The dawn of motoring* (London, 1942)

Hans-Otto Neubauer, *Autos aus Berlin: Protos und NAG* (Stuttgart, 1983)

Ernst Neuburg, *Jahrbuch der Automobil- und Motorbootindustrie: Im Auftrage des Kaiserlichen Automobil-Clubs,* 4th ed. (Berlin, 1907), 114–19

St. John C. Nixon, *The antique automobile* (London, 1956)

David F. Noble, *The religion of technology: The divinity of man and the spirit of invention* (New York, 1997)

Jan van den Noort, *Licht op het GEB: Geschiedenis van het Gemeente-Energiebedrijf Rotterdam* (Rotterdam, 1993)

Jan Norbye, "Survey of the gasoline engine," *Automobile quarterly* 5, no. 1 (Summer 1966), 86–105

"Norddeutsche Automobil- und Motoren-Aktiengesellschaft Bremen-Hastedt," in Alexander Engel (ed.), *Historisch-biographische Blätter der Staat Bremen* (Berlin, 1900/1911), Bd. 3, 847–60, and 879–82 (Handelskammer Bremen)

S. V. Norton, *The motor truck as an aid to business profits* (Chicago, 1918)

David E. Nye, *Electrifying America: Social meanings of a new technology, 1880–1940* (Cambridge, Mass., 1990)

C. Tunstill Opperman, *Electric transport vehicles* (The Institution of Automobile Engineers, Proceedings, vol. XIII, paper no. 9, April 1919), 393–434

Fritz Ostermann, *Die wirtschaftliche Bedeutung der Benzinlastkraftwagen im Vergleich mit Pferdefuhrwerk, Zugmaschine, Elektromobil und Eisenbahn* (Cologne, 1927) (diss.)

Werner Oswald, *Mercedes-Benz Lastwagen und Omnibusse 1886–1986* (Stuttgart, 1986)

Werner Oswald and Manfred Gihl, *Kraftfahrzeuge der Feuerwehr und des Rettungsdienstes seit 1900* (Stuttgart, 1992)

Helmut Otto, "Die Herausbildung des Kraftfahrwesens im deutschen Heer bis 1914," *Militärgeschichte,* 3/1989, 227–36

Wim Oude Weernink, *Spyker, een Nederlands fabrikaat 1880–1926* (Den Haag, 1976)

R. J. Overy, "Cars, roads, and economic recovery in Germany, 1932–8," *The economic history review* 28 (1975), 466–83

Victor W. Pagé, *The modern gasoline automobile: Its design, construction, operation and maintenance,* 7th ed. (London, 1918)

Nicholas Papayanis, "The development of the Paris cab trade, 1855–1914," *Journal of transport history,* New Series (March 1987), 52–65

Nicholas Papayanis, *The coachmen of nineteenth-century Paris: Service workers and class consciousness* (Baton Rouge, 1993)

Nicholas Papayanis, *Horse-drawn cabs and omnibuses in Paris: The idea of circulation and the business of public transit* (Baton Rouge, 1996)

Bellamy Partridge, *Excuse my dust* (New York, 1943)

Harold C. Passer, *The electrical manufacturers, 1875–1900: A study in competition, entre-preneurship, technical change, and economic growth* (Cambridge, Mass., 1953)

Akos Paulinyi, "Wi(e)der eine neue Technikgeschichte (?)!" *Blätter für Technikgeschichte* 57–58 (1995–96), 39–48

Udo Paulitz, *Historische Feuerwehren: Fahrzeuge und Einsätze von 1900 bis 1970* (Stutt-gart, 1994)

Henry C. Pearson, *Rubber tires and all about them: Pneumatic, solid, cushion, combina-tion, for automobiles, omnibuses, cycles, and vehicles of every description* (New York, 1906)

Henry C. Pearson, *Pneumatic tires: Automobile, truck, airplane, motorcycle, bicycle: An encyclopedia of tire manufacture, history, processes, machinery, modern repair and re-building, patents, etc., etc., profusely illustrated* (New York, 1922)

Harold Pender and H. F. Thomson, *The economical transportation of merchandise in metropolitan districts* (n.p. [Boston], March 1912)

Harold Pender and H. F. Thomson, *Notes on the costs of motor trucking* (n.p. [Boston], n.d.) (Vehicle Research Bulletin, no. 2)

Harold Pender and H. F. Thomson, *Observations on horse and motor trucking* (n.p. [Boston], March 1913) (Vehicle Research Bulletin, no. 3)

L. Périssé, *Automobiles sur route* (Paris, n.d.)

Peter Peters, "Vergroeid met het gaspedaal: Nederland barst van de auto's," *Humanist* (February 1996), 23–28

Peter Peters, "De cultus van snelheid: Nederland barst van de auto's, deel 2," *Huma-nist* (March 1996), 27–32

Henri Petit, *La voiture électrique à accumulateurs* (Paris, 1943)

Trevor Pinch, "Gender and power relationships in the social construction of technol-ogy: Making sense of the rural car" (lecture, "Centre de recherche en histoire des sciences et des techniques," Paris, 6 December 1994) (see also: Kline and Pinch)

Harold L. Platt, *The Electric City: Energy and the growth of the Chicago area, 1880–1930* (Chicago, 1991)

H. G. Plust, "Das E-Fahrzeug aus der Sicht der gegenwärtigen und zukünftigen Energiesituation," *Elektrische Strassenfahrzeuge: Technik, Entwicklungsstand und Einsatzperspektive*, 2 (Lehrgang Nr. 2551/36.02, Fort- und Weiterbildungszen-trum, Technische Akademie Esslingen, Insitut des Kontaktstudiums an der Universität Stuttgart, 5–7 May 1975) (Varta-Kelkheim, EA/12)

Max Pöhler, "Das Elektrofahrzeug unter besonderer Berücksichtigung seiner Ver-wendungsmöglichkeit in städtischen Fuhrparksbetrieben (Vortrag . . . gehalten auf der Tagung des Verbandes Städtischer Fuhrparksbetriebe in Frankfurt a.M. am 18.8.1949)" (Varta archives, file II D 2: "1938–1954")

Max Pöhler, "Das Elektroauto in Vergangenheit und Gegenwart" (undated typescript [1967?], Varta archives)

P. H. Prasuhn, *Chronik der Strassenbahn* (Hannover, 1969)

Howard Lawrence Preston, *Dirt roads to Dixie: Accessibility and modernization in the South, 1885–1935* (Knoxville, 1991)

"Promoting the electric vehicle: Winning the female vote," *Automobile quarterly* 31, no. 1 (Fall 1992)

Rüdiger Rabenstein, *Radsport und Gesellschaft: ihre sozialgeschichtlichen Zusammen-hänge in der Zeit von 1867 bis 1914* (Hildesheim, 1991)

Robert Alan Raburn, "Motor freight and urban morphogenesis with reference to Cali-fornia and the West" (diss., University of California, Berkeley, 1988)

Joachim Radkau, *Technik in Deutschland: Vom 18. Jahrhundert bis zur Gegenwart* (Frankfurt am Main, 1989)

Joachim Radkau, "Technik im Temporausch der Jahrhundertwende," in Michael Salewski and Ilona Stölken-Fitschen (eds.), *Moderne Zeiten: Technik und Zeitgeist im 19. und 20. Jahrhundert* (Stuttgart, 1994), 61–76

Joachim Radkau, *Das Zeitalter der Nervosität: Deutschland zwischen Bismarck und Hitler* (Munich, 1998)

John B. Rae, *The American automobile industry* (Boston, 1984)

John B. Rae, *American automobile manufacturers: The first forty years* (Philadelphia, 1959)

John B. Rae, *The American automobile: A brief history* (Chicago, 1965)

John B. Rae, "Pope Manufacturing Company," in May (ed.), *The automobile industry,* 398–99

John B. Rae, "Electric Vehicle Company," in May (ed.), *The automobile industry,* 174–75

John B. Rae, "Albert Augustus Pope," in May (ed.), *The automobile industry,* 393–97

John B. Rae, "Hiram Percy Maxim," in May (ed.), *The automobile industry,* 327–29.

John B. Rae, "Why Michigan?" in Lewis and Goldstein (eds.), *The automobile and American culture,* 1–9

John B. Rae, "The Electric Vehicle Company: A monopoly that missed," *Business history review* 29, no. 4 (December 1955), 298–311

Kaushik Rajasheraka, "History of electric vehicles in General Motors," *IEEE Transactions on industry applications* (July/August 1994), 897–904

Werner Rammert, "Modelle der Technikgenese: Von der Macht und der Gemachtheit technischer Sachen in unserer Gesellschaft" (manuscript, also published in *Jahrbuch Arbeit und Technik 1994*)

Werner Rammert, "Technikgenese: Stand und Perspektiven der Sozialforschung zum Entstehungszusammenhang neuer Techniken," *Kölner Zeitschrift für Soziologie und Sozialpsychologie* 40 (1988), 747–61

Werner Rammert, "Entstehung und Entwicklung der Technik: Der Stand der Forschung zur Technikgenese in Deutschland," *Journal für Sozialforschung* 32 (1992), 177–208

A. S. Ramondt, *Handleiding voor den chauffeur-werktuigkundige* (Den Haag, n.d. [1903])

Michael Rauck, *Karl Freiherr Drais von Sauerbronn: Erfinder und Unternehmer (1785–1851)* (Wiesbaden, 1983)

Frédéric Régamey, *Vélocipédie et automobilisme* (Tours, 1898)

M. Reichel, *Der Automobil-Löschzug der Berufsfeuerwehr Hannover* (Berlin, 1903)

M. Reichel, *Denkschrift über die Vornahme von Versuchen mit Automobilfahrzeugen bei der Berliner Feuerwehr* (Berlin, 1906)

Branddirektor Reichel, *Bericht über das Ergebnis der mit Kraftfahrzeugen bei der Berliner Feuerwehr vorgenommenen Versuche* (26 December 1907, appendix with its own page numbers to: *Bericht Berlin 1907*) [Reichel, *Bericht über Versuche*]

Reichel, Bahrdt, and Rohnstock, *Bericht der Zentralstelle zwecks Sammlung der mit Feuerwehr-Automobilen gemachten Erfahrungen* (Berlin, 1907)

A. W. Reid, "The effective use of milk delivery transport," *Dairy industries* (May 1952), 422–28 (Varta archives, file II D 2: "1938–1954")

Siegfried Reinecke, *Mobile Zeiten: Eine Geschichte der Auto-Dichtung* (Bochum, 1986)

Georges Renoy, *Paris naguère: transports publics* (Zaltbommel, 1978)

D. W. Reutlinger, *Bericht über den vom 16. bis 19. Juni 1904 zu München abgehaltenen Vierten Verbandstag der Deutschen Berufsfeuerwehren* (Frankfurt a.M., 1904) (Sonderabdruck aus *Feuer und Wasser*) (Feuerwehrmuseum Fulda, document 1.5.3.2.-4/1A) [Reutlinger, *Bericht über den . . . Vierten Verbandstag*]

D. W. Reutlinger, *Bericht über den vom 21. bis 24. Juni 1905 zu Bremen abgehaltenen Fünften Verbandstag der Deutschen Berufsfeuerwehren* (Frankfurt, 1905) (Sonderabdruck aus *Feuer und Wasser*) [Reutlinger, *Bericht über den . . . Fünften Verbandstag*]

Georges Reverdy, *Les routes de France du XIXe siècle* (Paris, 1993)

Review of the research program of the Partnership for a New Generation of Vehicles; Second report (Washington, D.C., 1996)

A. Riedler, *Wissenschaftliche Automobil-Wertung: Berichte I–V des Laboratoriums für Kraftfahrzeuge an den Königlich Technischen Hochschule zu Berlin* (Berlin, 1911)

Alois Riedler, *Wissenschaftliche Automobil-Wertung: Berichte VI–X des Laboratorium für Kraftfahrzeuge an der Königlich Technischen Hochschule, Teil II* (Berlin, 1912)

A. Riedler, *The scientific determination of the merits of automobiles: Reports I–X of the Laboratory for Motor-Cars at the Royal Technical University, Berlin-Charlottenburg* (London, n.d.)

Gordon L. Rinschler and Tom Asmus, "Powerplant perspectives: part I," *Automotive engineering* (April 1995), 37–42; "part II" (May 1995), 37–41

Jean Robert, *Les tramways parisiens*, 3rd ed. (n.p., 1992)

Jean Robert, *Histoire des transports dans les villes de France* (Neuilly-sur-Seine, 1974) (published on his own) (collection H. M. Vellekoop)

Derek Roberts, *Cycling history: Myths and queries* (n.p. [Birmingham], 1991)

W. Rödiger, *Der elektrische Kraftwagen: Handbuch für Bau und Betrieb von Elektromobilen und Elektrokarren* (Berlin, 1927)

W. Rödiger, "Elektrofahrzeuge in der Kriegszeit," *Elektrizitätswirtschaft* (15 November 1939) (Sonderdruck, Varta archives, file II D 0: "Fahrzeugantrieb allgemein")

W. Rödiger, "Elektrische Fahrzeuge," in Allmers et al. (eds.), *Das deutsche Automobilwesen der Gegenwart*, 74–84

W. Rödiger, "Elektromobile und Elektrokarren: Verwendungszwecke und Betriebswirtschaftlichkeit," in Allmers et al. (eds.), *Das deutsche Automobilwesen der Gegenwart*, 131–34

Everett M. Rogers, *Diffusion of innovations*, 4th ed. (New York, 1995)

L. T. C. Rolt, *Horseless carriage: The motor-car in England* (London, 1950)

Günter Ropohl, *Eine Systemtheorie der Technik: Zur Grundlegung der Allgemeinen Technologie* (Munich, 1979)

Nathan Rosenberg, "On technological expectations," *The economic journal* 86 (September 1976), 523–35

Nathan Rosenberg, "Factors affecting the diffusion of technology," *Explorations in economic history* (1972), 3–33

Nathan Rosenberg, *Exploring the black box: Technology, economics, and history* (Cambridge, Mass., 1994)

Hugo Fischer von Röslerstamm, "Die Automobile," in *Berichte über die Weltausstellung in Paris 1900, Achter Band: Wasserbau, Schiffahrt, Ingenieurwesen, Automobile* (Vienna, 1901)

Claude Rouxel, "Frankreich auf der Suche nach dem idealen Taxi: Mit dem Taxi zur Front," in Kubisch et al. (eds.), *Taxi*, 263–84

Claude Rouxel, *La grande histoire des taxis français 1898–1988* (Pontoise, 1989)

Ken Ruddock, "Recharging an old idea: The hundred-year history of electric cars," *Automobile quarterly* 31, no. 1 (Fall 1992), 30–47

E. Rumpler (ed.), *Automobiltechnischer Kalender und Handbuch der Automobil-Industrie für 1907*, 4th ed. (Berlin, n.d.)

Wolfgang Ruppert (ed.), *Fahrrad, Auto, Fernsehschrank: zur Kulturgeschichte der All-tagsdinge* (Frankfurt, 1993)

Wolfgang Ruppert, "Das Auto: 'Herrschaft über Raum und Zeit'," in idem (ed.), *Fahrrad, Auto, Fernsehschrank*, 119–61

Stewart Russell, "The social construction of artefacts: A response to Pinch and Bijker," *Social studies of science* 16 (1986,) 331–46

J. Rutishauser, *Roues élastiques* (Paris, 1941)

Wolfgang Sachs, *Die Liebe zum Automobil: Ein Rückblick in die Geschichte unserer Wünsche* (Reinbek bei Hamburg, 1984)

Devendra Sahal, "Alternative conceptions of technology," *Research policy* 10 (1981), 2–24

Monique de Saint Martin, *L'espace de la noblesse* (Paris, 1993)

P. Paolo Saviotti and J. Stanley Metcalfe (eds.), *Evolutionary theories of economic and technological change: Present status and future prospects* (Chur, 1991)

G. Schaetzel, *Motor-Posten: Technik und Leistungsfähigkeit der heutigen Selbstfahrersys-teme und deren Verwendbarkeit für den öffentlichen Verkehr* (Munich, 1901)

Richard H. Schallenberg, *Bottled energy: Electrical engineering and the evolution of chemical energy storage* (Philadelphia, 1982)

Richard H. Schallenberg, "The anomalous storage battery: An American lag in early electrical engineering," *Technology and culture* 22 (October 1981), 725–52

Richard P. Scharchburg, *Carriages without horses* (Warrendale, 1993)

Martin Scharfe, "Die Nervosität des Automobilisten," in van Dülmen (ed.), *Körper-Geschichten*, 200–222

Martin Scharfe, "'Ungebundene Circulation der Individuen': Aspekte des Automobil-fahrens in der Frühzeit," *Zeitschrift für Volkskunde* 86 (1990), 216–43

Virginia Scharff, *Taking the wheel: Women and the coming of the motor age* (New York, 1991)

Virginia Scharff, "Gender, electricity, and automobility," in Wachs and Crawford (eds.), *The car and the city*, 75–85

Virginia J. Scharff, "Putting wheels on women's sphere," in Cheris Kramarae (ed.), *Technology and women's voices: Keeping in touch* (New York, 1988), 135–46

Virginia Scharff (review of Schiffer, *Taking charge*), *Isis* 86, no. 2 (1995), 351–52

E. Schatzberg, "The mechanization of urban transit in the United States: Electricity and its competitors," in Aspray (ed.), *Technological competitiveness*, 225–42

[J. B. Scherer], *1827–1952: Das Unternehmen der Firma Gottfried Hagen Köln-Kalk im Wandel der Zeiten* (Cologne, n.d. [1952])

Michael Schiffer, "Social theory and history in behavioral archaeology," in J. M. Skibo, W. H. Walker, and A. E. Nielsen (eds.), *Expanding archaeology* (Salt Lake City, 1995), 22–35

Michael Schiffer, "The electric automobile in the United States, 1895–1920" (manu-script)

Michael Brian Schiffer (with Tamara C. Butts and Kimberly K. Grimm), *Taking charge: The electric automobile in America* (Washington, D.C., 1994)

Friedrich Schildberger, "Die Entstehung der Automobilindustrie in England" (manu-script, DaimlerChrysler archives, Stuttgart)

Friedrich Schildberger, "Die Entstehung der Automobilindustrie in Frankreich" (manuscript, DaimlerChrysler archives, Stuttgart)

Friedrich Schildberger, "Die Entstehung der Automobilindustrie in den Vereinigten Staaten" (manuscript, DaimlerChrysler archives, Stuttgart)

Friedrich Schildberger, "Vom Deutzer Viertakt-Ottomotor zu den frühen Fahrzeug-motoren," *Automobiltechnische Zeitschrift* 3, 78 Heft (March 1976), 77–83

Wolfgang Schivelbusch, *Geschichte der Eisenbahnreise: zur Industrialisierung von Raum und Zeit im 19. Jahrhundert* (Munich, 1977)

[Wilhelm] Scholz, *Anlage zum Verhandlungsbericht über den 13. Verbandstag vom 4. bis 6. Juni 1913 in Stettin; statistische Zusammenstellungen der Automobilkommission* (Aachen, n.d. [1913]) [Scholz, *Anlage zum Verhandlungsbericht*]

Wilhelm Scholz, *Die Übergang zum Automobilbetriebe bei den Feuerwehren, seine Begründung, Durchführung und Ziele* (Aachen, 1914) (diss.)

Rainer Schönhammer, *In Bewegung: Zur Psychologie der Fortbewegung* (Munich, 1991)

J. W. Schot, "Innoveren in Nederland," in Lintsen et al. (eds.), *Geschiedenis van de techniek in Nederland, vol. VI*, 217–39

Johan W. Schot, "Constructive technology assessment and technology dynamics: The case of clean technologies," *Science, technology, & human values* 17, no. 1 (Winter 1992), 35–56

Johan Schot, "De inzet van Constructief Technology Assessment" (unpubl. presentation, lecture series, "Romantische kritiek op wetenschap en techniek," UT Twente, 17 January 1995)

Viktor Schützenhofer, "Lohner - vom Wagnergewerbe zur Grossindustrie," *Blätter für Technikgeschichte* (1950, Heft 12), 1–14.

Vanessa R. Schwartz, *Spectacular realities: Early mass culture in fin-de-siècle Paris* (Berkeley, 1998)

Bruno Schweder (ed.), *Forschen und Schaffen: Beiträge der AEG zur Entwicklung der Elektrotechnik bis zum Wiederaufbau nach dem zweiten Weltkrieg,* Band 2 (Berlin, 1965)

Philip Scranton, "Theory and narrative in the history of technology: Comment," *Technology and culture* (April 1991), 385–93

Philip Scranton, "Determinism and indeterminacy in the history of technology," *Technology and culture* (supplement to April 1995), S31–S52

H. C. Graf von Seherr-Thoss, *Die deutsche Automobilindustrie: Eine Dokumentation von 1886 bis 1979,* 2nd ed. (Stuttgart, 1979)

Craig R. Semsel, "More than an ocean apart: The street railway of Cleveland and Birmingham, 1880–1911," *Journal of transport history* (third series) 22, no. 1 (March 2001), 47–61

Gaston Sencier and A. Delasalle, *Les automobiles électriques* (Paris, 1901)

Hans Seper, *Österreichische Automobilgeschichte 1815 bis heute* (Vienna, 1986)

Hans Seper, "Daimler und Benz in Österreich" (manuscript, DaimlerChrysler archives, File "Beteiligungen Österreich")

Hans Seper, Helmut Krackowizer, and Alois Brusatti, *Österreichische Kraftfahrzeuge von Anbeginn bis heute,* 2nd ed. (Wels, 1984)

Sheldon R. Shacket, *The complete book of electric vehicles* (Chicago, 1979)

Michael Shnayerson, *The car that could: The inside story of GM's revolutionary electric vehicle* (New York, 1996)

Immo Sievers, *AutoCars: Die Beziehungen zwischen der englischen und der deutschen Automobilindustrie vor dem Ersten Weltkrieg* (Frankfurt am Main, 1995) (Europäische Hochschulschriften, Reihe III, Bd. 640) (diss.)

Herbert A. Simon, "A behavioral model of rational choice," *Quarterly journal of economics* 69 (1955), 99–118

Bayla Singer, "Automobiles and femininity," *Research in philosophy and technology* 13 (1993), 31–41

Chris Sinsabaugh, *Who, me? Forty years of automobile history* (Detroit, 1940)

Sergio Sismondo, "Some social constructions," *Social studies of science* 23 (1993), 515–53

J. W. Sluiter, *Beknopt overzicht van de Nederlandse spoor- en tramwegbedrijven* (Leiden, 1967)

Merritt Roe Smith and Leo Marx (eds.), *Does technology drive history? The dilemma of technological determinism* (Cambridge, Mass., 1994)

Pierre Souvestre, *Histoire de l'automobile* (Paris, 1907)

Daniel Sperling (with contributions from Mark A. Delucchi, Patricia M. Davis, and A. F. Burke), *Future drive: Electric vehicles and sustainable transportation* (Covelo, Calif., 1995)

"Spijkstaal 50 jaar: van ijzeren hond tot rooie haan," *Automobiel Management* (September 1988)

A. Stadie, "Der Personenkraftwagen: Verwendungszwecke und Betriebswirtschaftlichkeit," in Allmers et al. (eds.), *Das deutsche Automobilwesen der Gegenwart*, 94–102

Darwin H. Stapleton, "Walter C. Baker," in May (ed.), *The automobile industry*, 31–33

Darwin H. Stapleton, "Baker Motor Vehicle Company," in May (ed.), *The automobile industry*, 33–35

John M. Staudenmaier, S. J., "The politics of successful technologies," in Stephen H. Cutcliffe and Robert C. Post (eds.), *In context; history and the history of technology: Essays in honor of Melvin Kranzberg (Research in technology studies, vol. 1)* (Bethlehem, 1989), 150–71

Erwin Steinböck, *Lohner: Zu Land, zu Wasser und in der Luft* (Graz, 1984)

G. F. Steinbuch, *De automobiel: Handboek voor automobielbestuurders, monteurs en reparateurs*, 2 vols. (Deventer, 1922 and 1926)

J. St. Loe Strachey, "Roads: The return to the road," in Harmsworth (ed.), *Motors and motor-driving*, 346–55

Strassenbahnen mit Accumulatoren nach dem System der Accumulatoren-Fabrik Aktiengesellschaft Hagen i.W. (Hagen i. W., April 1897) (brochure, Varta archives)

Susan Strasser, Charles McGovern, and Matthias Judd (eds.), *Getting and spending: European and American consumer societies in the twentieth century* (Washington, D.C., 1998)

Ralf Stremmel, "Der Bestand 'VARTA Batterie AG' im Westfälischen Wirtschaftsarchiv, Dortmund," *Westfälische Forschungen*, 46/1996, 489–94

Ronald A. Stringer, "The Morrison electric: America's first automobile!?!," *Antique automobile* (January/February 1984), 32–35

Günter Strumpf, *Die Berliner Feuerwehr von den Anfängen bis zur Gegenwart* (Hanau, 1987)

Philip Sumner, "The evolution of the electric car," *Veteran and vintage magazine* 13, no. 1 (September 1968), 16–22

Anthony Sutcliffe, "Street transport in the second half of the nineteenth century: Mechanization delayed?" in Tarr and Dupuy (eds.), *Technology and the rise of the networked city*, 22–39

Anthony Sutcliffe, "Die Bedeutung der Innovation in der Mechanisierung städtischer Verkehrssysteme in Europa zwischen 1860 und 1914," in Horst Matzerath (ed.), *Stadt und Verkehr im Industriezeitalter* (Cologne, 1996), 231–41

Joel A. Tarr, *The search for the ultimate sink: Urban pollution in historical perspective* (Akron, 1996)

Joel A. Tarr and Gabriel Dupuy (eds.), *Technology and the rise of the networked city in Europe and America* (Philadelphia, 1988)

R. Thebis, *Elektrokarren: Allgemeinverständliche Darstellung des Aufbaus, des Betriebes und der Instandhaltung des Elektrokarrens* (Leipzig, 1927)

F. M. L. Thompson, *Victorian England: The horse-drawn society; an inaugural lecture* (London, 22 October 1970)

F. M. L. Thompson, "Horses and hay in Britain 1830–1918," in Thompson (ed.), *Horses in European economic history*, 50–72

F. M. L. Thompson, "Nineteenth-century horse sense," *The economic history review* 29 (1976), 60–79

F. M. L. Thompson (ed.), *Horses in European economic history: A preliminary canter* (Reading, 1983)

Harry F. Thomson, *Relative fields of horse, electric, and gasoline trucks* (n.p. [Boston], August 1914) (Vehicle Research Bulletin, no. 4)

H. F. Thomson, H. L. Manley and A. L. Pashek, *The delivery system of R. H. Macy & Co. of New York* (n.p. [Boston], September 1914) (Vehicle Research Bulletin, no. 6; reprint of *The commercial vehicle*, 15 December 1915)

Henry Thomson Bart., *The motor-car: An elementary handbook on its nature, use & management* (London, 1902)

Cecilia Ticki, *Shifting gears: Technology, literature, culture in modernist America* (Chapel Hill, 1987)

Jo Tollebeek and Tom Verschaffel, *De vreugden van Houssaye: Apologie van de historische interesse* (Amsterdam, 1992)

Eric Tompkins, *The history of the pneumatic tyre* (n.p. [Birmingham], 1981)

P. J. Troost, *Man en paard in oud-Rotterdam* (Rotterdam, 1985)

Jean Tuma, *Encyclopédie illustrée des transports* (Gründ, 1978)

André Turin, "La voiture mixte 'Kriéger'" (Extrait du *Journal technique et industriel*) (Paris, 1905) (collection J.-L. Krieger, Paris)

André Vant, *L'industrie du cycle dans la région stéphanoise* (Lyon, 1993)

"Vechten onder de motorkap," *Wetenschap, cultuur & samenleving* (September 1996), 7

A. J. Veenendaal Jr., "Spoorwegen," in Lintsen et al. (eds.), *Geschiedenis van de techniek in Nederland, vol. II*, 129–63

"Les véhicules industrielles," *La journée industrielle* (2 December 1924)

F. C. A. Veraart, "Geschiedenis van de fiets in Nederland 1870–1940: Van sportmiddel naar massavervoermiddel" (master's thesis, TU Eindhoven, 1995)

Verband deutscher Berufsfeuerwehren: see "Conference reports and annual reports"

G. J. Verburg, *Geschiedenis van de brandweerwagen* (Alkmaar, n.d.) (Alkenreeks)

L. J. Andrew Villalon and James M. Laux, "Steaming through New England with Locomobile," *Journal of transport history*, new series, 5, no. 2 (September 1979), 65–82

Simon P. Ville, *Transport and the development of the European economy, 1750–1918* (n.p., n.d.)

Walter G. Vincenti, "The retractable airplane landing gear and the Northrop 'anomaly': Variation-selection and the shaping of technology," *Technology and culture* (January 1994), 1–33

Walter G. Vincenti, "The scope for social impact in engineering outcomes: A diagrammatic aid to analysis," *Social studies of science* 21 (1991), 761–67

Walter G. Vincenti, *What engineers know and how they know it: Analytical studies from aeronautical history* (Baltimore, 1990)

Walter Vincenti, "The technical shaping of technology: Real-world constraints and technical logic in Edison's electrical lighting system," *Social studies of science* 25 (1995), 553–74

Vincent van der Vinne, *Spyker 1898–1926* (Amsterdam, 1998)

Vincent van der Vinne, "Automobielen in Nederland 1896–1940: Ondernemers, consumenten en overheid in een innovatieproces" (n.p., n.d.) (unpubl. master's thesis, University of Nijmegen)

Taylor Vinson, "Auto history at NAIAS," *SAH Journal: The newsletter of the Society of Automotive Historians* 167 (March/April 1997), 7

Erik van der Vleuten, *Electrifying Denmark: A symmetrical history of central and decentral electricity supply until 1970* (diss., University of Aarhus, 1998)

Erik van der Vleuten, "Autoproduction of electricity: Cases from Danish industry until 1960," *Polhem: Tidskrift för tecknikhistoria* 14 (1996), 118–54

Rudi Volti, "Why internal combustion?" *Invention & technology* (Fall 1990), 42–47

Martin Wachs and Margaret Crawford (eds.), *The car and the city: The automobile, the built environment, and daily urban life* (Ann Arbor, 1992)

E. J. Wade, *Secondary batteries: Their theory, construction and use*, 2nd ed. (London, 1908)

Richard Wager, *Golden wheels: The story of the automobiles made in Cleveland and Northeastern Ohio 1892–1932* (Cleveland, 1986)

Ernest H. Wakefield, *The consumer's electric car* (Ann Arbor, 1977)

Ernest Henry Wakefield, *History of the electric automobile: Battery-only powered cars* (Warrendale, 1994)

Ernest H. Wakefield, *History of the electric automobile: Hybrid electric vehicles* (Warrendale, 1998)

Martin Wallast, *Autobussen in Nederland: 90 jaar historie in woord en beeld* (Rijswijk, 1987)

Margaret Walsh, *Making connections: The long-distance bus industry in the USA* (Aldershot, 2000)

W. H. Ward, "The sailing ship effect," *Bulletin of the Institute of Physics and Physical Society* 18 (1967), 169

Philip Warren and Malcolm Linskey, *Taxicabs, a photographic history*, 2nd ed. (London, 1980)

Eugen Weber, *France, fin de siècle* (Cambridge, Mass., 1986)

Edna Robb Webster, *T. A. Willard, wizard of the storage battery: The biography of a famous inventor* (Sherman Oaks, Calif., 1976)

André Wegener Sleeswijk, "Hoe schrijven we de geschiedenis van de middeleeuwse techniek?" *Leidschrift* (June 1992), 23–41

Luth Westerkamp, *Das Elektro-Fahrzeug* (Berlin, 1928) (Autotechnische Bibliothek, Band 83)

Percival White, *Motor transportation of merchandise and passengers* (New York, 1923)

Adam Gowans Whyte, *Electricity in locomotion: An account of its mechanism, its achievements and its prospects* (Cambridge, 1911)

Gerhard Wilke (ed.), *Denkschrift Elektrospeicherfahrzeuge (Im Auftrag der Deutschen Forschungsgemeinschaft, Teil 11/1969)* (Wiesbaden, 1970)

R. Thomas Wilson, *Baker Raulang: The first hundred years, eighteen fifty three nineteen fifty three* (Cleveland, 1953)

R. Winckler, "'Strom statt Benzin': Akkumulator-Fahrzeuge und ihre Bedeutung als Stromverbraucher (Vortrag bei dem Verband der Elektrizitätswerke Rheinlands und Westfalens . . . am 17. Mai 1934)" (typescript, Varta archives, file II D 2: "1909–1935")

Langdon Winner, "Upon opening the black box and finding it empty: Social constructivism and the philosophy of technology," *Science, technology, & human values* 18, no. 3 (Summer 1993), 362–78

Langdon Winner, *Autonomous technology: Technics-out-of-control as a theme in political thought* (Cambridge, Mass., 1977)

Robert Wohl, *A passion for wings: Aviation and the Western imagination 1908–1918* (New Haven, 1994)

Roland Wolf, *Le véhicule électrique gagne le coeur de la ville* (n.p., n.d. [Paris, 1995])

C. E. Woods, *The electric automobile, its construction, care and operation* (Chicago, 1900)

Genevieve Wren, "Andrew Lawrence Riker," in May (ed.), *The automobile industry,* 407–8

Genevieve Wren, "Pedro G. Salom," in May (ed.), *The automobile industry,* 408–9

James Wren, "George H. Day," in May (ed.), *The automobile industry,* 120–22

Karl S. Zahm, "Owen & Entz: an electrifying combination: The story of the Owen-Magnetic," *Bulb horn* (January–March 1991), 27–31 and (April–June 1991), 34–39

Angela Zatsch, *Staatsmacht und Motorisierung am Morgen des Automobilzeitalters* (Konstanz, 1993) (diss.)

Max R. Zechlin, "Feuerlöschautomobile," in *Handbuch der Automobilindustrie 1909,* 11–22 (collection Hans-Otto Neubauer)

Index

Page numbers in *italics* refer to illustrations, graphs, and tables.

Dla Adama od wychowanicyni Hanny Siuber.

Szczecin, 19 luty 2015 r.

ZIEMIA
Poznaj jej sekrety i tajemnice

BŁĘKITNA PLANETA
w Układzie Słonecznym

Ziemia to trzecia pod względem oddalenia od Słońca planeta
Układu Słonecznego. Jest też największą spośród planet
posiadających skalną powierzchnię, choć jednocześnie
jest karzełkiem w porównaniu z gazowymi olbrzymami,
takimi jak na przykład Jowisz. Wyjątkową cechą
Ziemi jest to, że rozwinęło się na niej życie.

MERKURY   **WENUS**   **MARS**

ZIEMIA

- Średnia odległość
  od Słońca: 149,6 mln km
- Masa: $5,975 \times 10^{24}$ kg
- Okres obiegu wokół Słońca:
  365 dni 5 h 48 min 46 s
- Czas obrotu wokół własnej
  osi: 23 h 56 min 4 s
- Księżyce: 1

JOWISZ

Obiegając Słońce, Ziemia każdego roku
pokonuje odległość prawie 940 milionów kilometrów!

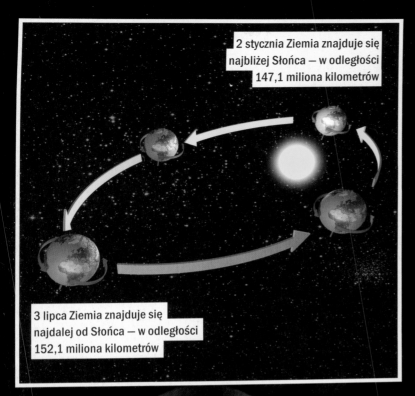

2 stycznia Ziemia znajduje się najbliżej Słońca — w odległości 147,1 miliona kilometrów

3 lipca Ziemia znajduje się najdalej od Słońca — w odległości 152,1 miliona kilometrów

RUCH ZIEMI WOKÓŁ SŁOŃCA I OSI

Ziemia obiega Słońce z prędkością ponad 108 tysięcy kilometrów na godzinę. Aby zatoczyć pełne koło, potrzebuje 365 dni 5 godzin 48 minut i 46 sekund. Okres ten nazywamy ROKIEM. Ziemia obraca się też wokół własnej osi z prędkością, która na równiku wynosi 1674,4 kilometry na godzinę. Aby wykonać pełen obrót, planeta potrzebuje 23 godzin 56 minut i 4 sekund. Czas ten nazywamy DNIEM. Efektem ruchu obrotowego jest obserwowana z Ziemi wędrówka Słońca i gwiazd po niebie, dzięki czemu mówimy o PORACH DNIA. Ziemia obraca się wokół osi, która nie jest prostopadła do płaszczyzny obiegu. To nachylenie wynosi 23,5 stopnia. Efektem ruchu obiegowego i nachylenia osi Ziemi są PORY ROKU.

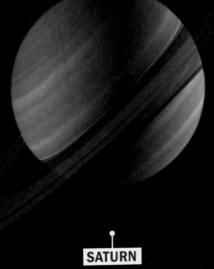

SATURN

URAN

NEPTUN

DROGA MLECZNA

Słońce wraz z całym Układem Słonecznym obiega centrum naszej galaktyki — Drogi Mlecznej — z prędkością 792 tysięcy kilometrów na godzinę. Galaktyka z kolei pędzi w przestrzeni kosmicznej z zawrotną prędkością ponad 3,5 miliona kilometrów na godzinę! Z taką prędkością podróżujemy przez Kosmos nawet wtedy, gdy wygodnie śpimy w łóżkach...

←Słońce

ZIEMIA
Jej powstanie i ewolucja

Ziemia ma około 4,6 miliarda lat, podobnie jak inne planety
Układu Słonecznego i samo Słońce. Powstała jako skupisko pyłów
i gazów krążących wokół protosłońca. Zanim uformowała się planeta,
jaką znamy, przez miliony lat Ziemia rosła, zbierając kosmiczną materię,
a bombardowana przez asteroidy rozgrzewała się, tworząc
kulę stopionych skał i metali. Później jej powierzchnia ostygła na tyle,
że mogła się na niej zgromadzić woda.

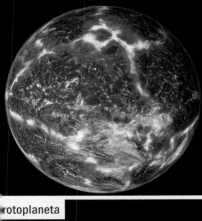

Protoplaneta

Wielkie bombardowanie

Uderzenie obiektu wielkości Marsa

POWSTANIE ZIEMI

Powstawanie brył materii — PROTOPLANET — w dysku wokół protosłońca. Jedną z takich protopanet była Ziemia.

Wielkie bombardowanie i przetopienie materii. Około 4,5 miliarda lat temu w planetę uderzył obiekt o rozmiarach Marsa. Oba ciała połączyły się, a z wyrzuconej w przestrzeń materii powstał Księżyc.

• Początkowo na powierzchni Ziemi nie istniały oceany, a atmosfera była pozbawiona tlenu. Dopiero po ochłodzeniu skorupy możliwe było skroplenie wody z atmosfery i powstanie wód.

Przez większość swojej historii krajobraz Ziemi prawdopodobnie przypominał marsjański

HISTORIA ZIEMI

4,6 Ga*	powstanie Ziemi
4,2 Ga	najstarsze znane skały
4,1 Ga	ustalenie skorupy
4,0 Ga	wielkie bombardowanie meteorytami zakończone ostatecznie 3,5 Ga, schłodzenie powierzchni do temperatury poniżej 100 stopni Celsjusza; powstanie oceanów
3,8 Ga	powstanie życia
3,7 Ga	najstarsze znane skały osadowe
3,5 Ga	utworzenie jądra Ziemi
3,5 Ga	powstanie organizmów fotosyntetyzujących (bakterii) — tlen w atmosferze
3,0 Ga	wielkie wichury i ogromne pływy
2,7 Ga	najstarsze organizmy jądrowe
1,2 Ga	najstarsze organizmy wielokomórkowe
0,8 Ga	Ziemia-śnieżka
0,6 Ga	skład atmosfery zbliżony do dzisiejszej
0,6 Ga	„eksplozja" życia
0,4 Ga	zawartość tlenu w atmosferze podobna do dzisiejszej
Dziś	

---

* Ga to skrót od angielskiego *gigaannum*, czyli „miliard lat". 4,6 Ga oznacza więc 4,6 miliarda lat temu.

BUDOWA PLANETY
i ruchy płyt litosfery

Przez wiele setek lat ludzie zastanawiali się, co znajduje się
pod ich stopami. Dzięki wierceniom udało się sprawdzić,
co jest na głębokości 13 kilometrów, lecz to niewiele w porównaniu
z promieniem Ziemi, który wynosi ponad 6 tysięcy kilometrów.
Gdyby porównać Ziemię do jabłka, to nie udało się przewiercić nawet
przez jego skórkę. Fizycy potrafią jednak określić jak zbudowana jest Ziemi
dzięki badaniom przebiegu fal uderzeniowych w jej wnętrzu —
ten dział nauki nazywa się sejsmologią.

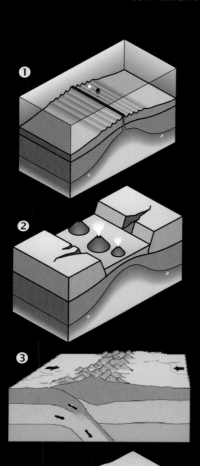

TYPY KONTAKTÓW PŁYT

1. Atlantyk jest młodym oceanem, który wciąż się
rozszerza. Dokładnie w połowie jego szerokości
znajduje się grzbiet oceaniczny, który jest
ogromnym pęknięciem skorupy oceanicznej.
W pęknięcie to wlewa się roztopiona skała,
rozpychająca dno na boki.

2. Ruchy konwekcyjne pod kontynentami
mogą być tak bardzo silne, że doprowadzają
do rozerwania lądu. Powstaje wtedy rów, do
którego wlewa się ocean.

3. Kiedy zderzą się dwa kontynenty, dochodzi
do wypiętrzenia gór. Góry to sprasowane
i nawarstwione osady z dna morza, które niegdyś
dzieliło dwa lady. To dlatego na ich szczytach
można znaleźć skamieniałości morskich zwierząt.
Przykładem takich gór są Himalaje.

4. Zetknięcie grubej i sztywnej skorupy
kontynentalnej z cienką skorupą oceaniczną
doprowadza do tego, że ta ostatnia zatapia się
pod lądem. Zagięcie obu części powoduje
powstanie na dnie bardzo głębokich rowów.
Przykładem jest Rów Mariański o głębokości
11 kilometrów. Jest to strefa częstych trzęsień

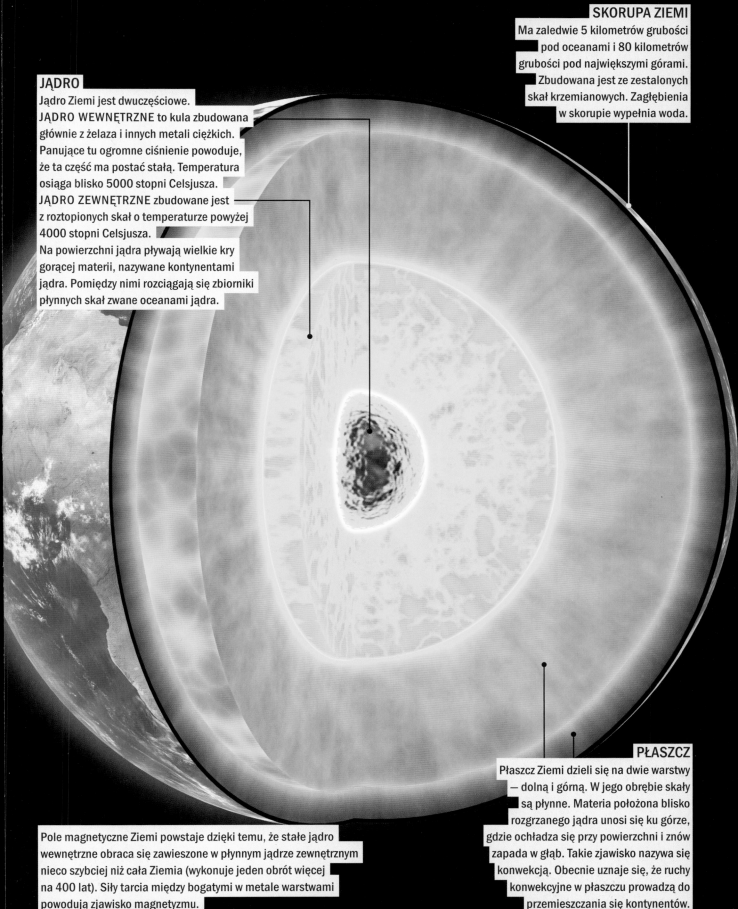

SKORUPA ZIEMI

Ma zaledwie 5 kilometrów grubości pod oceanami i 80 kilometrów grubości pod największymi górami. Zbudowana jest ze zestalonych skał krzemianowych. Zagłębienia w skorupie wypełnia woda.

JĄDRO

Jądro Ziemi jest dwuczęściowe.
JĄDRO WEWNĘTRZNE to kula zbudowana głównie z żelaza i innych metali ciężkich. Panujące tu ogromne ciśnienie powoduje, że ta część ma postać stałą. Temperatura osiąga blisko 5000 stopni Celsjusza.
JĄDRO ZEWNĘTRZNE zbudowane jest z roztopionych skał o temperaturze powyżej 4000 stopni Celsjusza.
Na powierzchni jądra pływają wielkie kry gorącej materii, nazywane kontynentami jądra. Pomiędzy nimi rozciągają się zbiorniki płynnych skał zwane oceanami jądra.

PŁASZCZ

Płaszcz Ziemi dzieli się na dwie warstwy — dolną i górną. W jego obrębie skały są płynne. Materia położona blisko rozgrzanego jądra unosi się ku górze, gdzie ochładza się przy powierzchni i znów zapada w głąb. Takie zjawisko nazywa się konwekcją. Obecnie uznaje się, że ruchy konwekcyjne w płaszczu prowadzą do przemieszczania się kontynentów.

Pole magnetyczne Ziemi powstaje dzięki temu, że stałe jądro wewnętrzne obraca się zawieszone w płynnym jądrze zewnętrznym nieco szybciej niż cała Ziemia (wykonuje jeden obrót więcej na 400 lat). Siły tarcia między bogatymi w metale warstwami powodują zjawisko magnetyzmu.

LĄDY I OCEANY
Wędrówka kontynentów

Łatwo sobie wyobrazić, że Ziemia nie zawsze wyglądała tak, jak dziś.
Również w przyszłości naszą planetę czekają zmiany układu
lądów i oceanów. Co kilkaset milionów lat wszystkie lądy zderzają się,
tworząc jeden superkontynent otoczony wszechoceanem. Ostatnim z takich
superkontynentów była Pangea, która powstała około 300 milionów lat temu
i zaczęła się ponownie rozpadać po 50 milionach lat.

Park Narodowy Thingvellir na Islandii — miejsce styku płyt tektonicznych eurazjatyckiej i północnoamerykańskiej;
powierzchnia ziemi poprzecinana jest licznymi szczelinami

PŁYTY TEKTONICZNE I DRYF KONTYNENTALNY

Powierzchnia Ziemi składa się z płyt litosfery, które stale się przemieszczają. Tym samym zmieniają się
kształty kontynentów i oceanów. Wędrówka (dryf) kontynentów odbywa się od przynajmniej
2,5 miliarda lat. Świadczą o tym najstarsze skały w Kanadzie, których wiek geolodzy oceniają
na 2 miliardy lat. Choć życie człowieka jest za krótkie, by dostrzec ogrom zmian, można zaobserwować
na przykład, że Ameryka Północna oddala się od Europy o około 1,5 centymetra każdego roku.

DRYF KONTYNENTALNY W OSTATNICH 300 MILIONACH LAT

300 MILIONÓW LAT TEMU

Na Ziemi istniał jeden superkontynent — Pangea, który był otoczony przez superocean — Panthalassę.

200 MILIONÓW LAT TEMU

Superkontynent rozpadł się na dwie masy lądowe — Laurazję i Gondwanę. Pomiędzy nie wlało się morze Tetyda. Powstał północny Ocean Atlantycki.

135 MILIONÓW LAT TEMU

Laurazja i Gondwana uległy dalszemu podziałowi. Powstały kontynenty, które zaczęły się od siebie oddalać.

65 MILIONÓW LAT TEMU

Znane nam obecnie rozmieszczenie mas lądowych. Największym istniejącym kontynentem jest Eurazja.

ZIEMIA W PRZYSZŁOŚCI

Sądzi się, że za jakieś 50 milionów lat Europa połączy się z Afryką. Tym samym zniknie Morze Śródziemne. Natomiast za 250 milionów lat wszystkie lądy ponownie utworzą superkontynent.

WULKANY
Potęga magmy z wnętrza Ziemi

Choć powierzchnia naszej planety wygląda na spokojną, w jej wnętrzu wiele się dzieje. Ogromne ciśnienie i wysokie temperatury powodują, że materia miesza się i przemieszcza. Objawem tego bogatego życia wewnętrznego są wulkany. Ich obserwacja daje pojęcie o potędze drzemiącej wewnątrz naszej planety.

RODZAJE WULKANÓW

Najpowszechniejszy podział wulkanów wynika z obserwacji ich działalności. Dzielimy je na CZYNNE (np. Wezuwiusz, Etna); DRZEMIĄCE (np. hawajski Mauna Kea) i WYGASŁE (np. stożki wulkaniczne w Polsce).

Ze względu na miejsce erupcji wulkany dzielimy na powierzchniowe, podmorskie i podlodowcowe. Jeszcze inny podział bierze pod uwagę kształt otworu wulkanicznego. Rozróżniamy tu wulkany centralne, szczelinowe i mieszane.

ŚMIERĆ WULKANU

Najczęściej wulkany przestają działać na skutek wystygnięcia komory magmowej. Czasem jednak zdarza się, że koniec wulkanu jest dramatyczny i ulega on całkowitemu zniszczeniu. W wyniku zapadnięcia górnej części wzniesienia na skutek opróżnienia zbiornika magmowego powstaje KALDERA. Wulkanami kalderowymi są na przykład Wezuwiusz i Kilimandżaro.

Kaldery często zostają wypełnione wodą tworzącą jeziora (na zdjęciu jezioro Quilotoa w Ekwadorze).

CHMURA POPIOŁÓW, GAZÓW I PARY
Pióropusz materiałów wyrzucanych przez wulkan może wznosić się na wysokość ponad 80 kilometrów, a popioły mogą unosić się w powietrzu nawet przez kilka lat

ERUPCJA LAWY
Magma wypływająca na powierzchnię nosi nazwę lawy

KRATER

POTOK LAWY
Lawa spływa po stokach wulkanu, paląc wszystko na swojej drodze

STOŻEK WULKANICZNY
Jest zbudowany z warstw lawy i popiołów

BŁOTNE LAWINY
Wulkany to często wysokie wzniesienia, na których szczytach gromadzi się lód i śnieg. Ciepło wydobywające się z wnętrza góry może spowodować ich topnienie i katastrofalne w skutkach spływy błota, tak zwane lahary.

KOMIN
Kanał, którym magma wydostaje się na powierzchnię krateru

OGNISKO MAGMOWE
Komora wypełniona magmą o temperaturze 1100 stopni Celsjusza

TRZĘSIENIA I TSUNAMI
Niszczycielskie kataklizmy

Trzęsienia ziemi to bodaj najbardziej niszczycielskie kataklizmy znane na naszej planecie. Dochodzi do nich co trzy sekundy w jakiejś części globu. Większość z nich nie przynosi szkody, ale co jakiś czas dochodzi do silnych wstrząsów powodujących ogromne zniszczenia i niosących śmierć tysiącom ludzi. Wielkie fale morskie, zwane tsunami, powodują, że skutki trzęsienia ziemi mogą być odczuwane tysiące kilometrów od epicentrum.

JAK POWSTAJE TRZĘSIENIE ZIEMI

Sztywna skorupa ziemska podlega ciągłym naprężeniom: jest ściskana, rozciągana, wyginania, podgrzewana i chłodzona. Na skutek tego pęka, a pęknięcia te nazywa się USKOKAMI. Umożliwiają one przemieszczanie się skał względem siebie w dowolnym kierunku. Między blokami skalnymi występują jednak ogromne siły związane z tarciem sąsiednich bloków o siebie.

Jeśli tarcie nie pozwala na wzajemne przesunięcie bloków, wzdłuż linii uskoku gromadzi się naprężenie. Kiedy siła naprężeń staje się większa niż siła tarcia, następuje gwałtowne przemieszczenie bloków względem siebie i chwilowe rozładowanie napięć. To właśnie ten moment odczuwany jako trzęsienie ziemi.

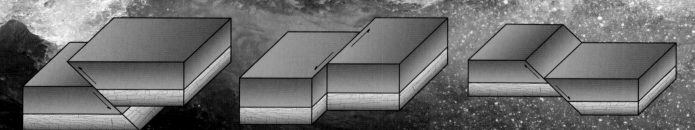

Rodzaje uskoków (od lewej): uskok odwrócony, przesuwczy i normalny

POWSTAWANIE TSUNAMI

Tsunami jest najczęściej wywołane przez trzęsienie ziemi. Gwałtowne przemieszczenie się dużej ilości wody na dnie oceanicznym powoduje lokalne wypiętrzenie fal, które rozchodzą się koliście we wszystkich kierunkach. Na głębokim otwartym oceanie nie stanowi to zagrożenia. Taka fala porusza się wprawdzie z prędkością dochodzącą do 900 kilometrów na godzinę, ale jest raczej niewysoka (zazwyczaj do 1 metra wysokości, często zaledwie pół metra) i przemieszcza się na duże odległości właściwie niezauważalnie.

Kiedy fala zbliża się do lądu, co prawda zwalnia do prędkości 50 kilometrów na godzinę, ale rośnie do wysokości 10—20 metrów, tworząc pionową ścianę wody, która uderza w brzeg z ogromną siłą. Woda wdziera się na ląd (nawet do kilometra w głąb), niszcząc wszystko, co napotka na swojej drodze.

TSUNAMI

To słowo pochodzenia japońskiego. Oznacza „fala portowa", bo dopiero docierając do brzegu, ujawnia swoją niszczycielską moc. Japonia doświadcza tsunami mniej więcej co 7 lat.

MINERAŁY
Podstawowy składnik skorupy

Minerały są podstawowym składnikiem skorupy ziemskiej, ale też innych planet skalistych czy asteroid. O ich barwie, twardości i innych cechach decyduje skład chemiczny. Obecnie znamy niemal 5 tysięcy różnych minerałów. Ponieważ skorupa Ziemi składa się w trzech czwartych z krzemu i tlenu, to właśnie te pierwiastki tworzą najwięcej minerałów.

MINERAŁ CZY NIE MINERAŁ

Ciekawymi substancjami są rodzima RTĘĆ (na zdjęciu) i woda. Substancje te w temperaturze pokojowej są płynami, ale w temperaturach poniżej zera uzyskują strukturę krystaliczną (woda w temperaturze poniżej 0 stopni Celsjusza, rtęć w temperaturze poniżej –39 stopni Celsjusza). Wtedy stają się minerałami.

NAJWIĘKSZE Z NAJWIĘKSZYCH

Największe znane kryształy tworzy minerał o nazwie BERYL (na zdjęciu beryl w granicie). Gigantyczny okaz znaleziony na Madagaskarze miał długość 18 metrów, szerokość 3,5 metra i wagę ponad 350 ton! Największe nagromadzenie olbrzymich kryształów SELENITU (odmiany popularnego gipsu) znaleziono w 2000 roku w Kryształowej Jaskini w Naica. Znajduje się tu całe mnóstwo kryształów, spośród których rekordowe rozmiary ma okaz o długości 11 metrów.

HALIT

Choć niewiele osób kojarzy nazwę halit, minerał ten odegrał jedną z najdonioślejszych ról w historii ludzkości. Sól, bo o niej mowa, wydobywano w kopalniach już dwa tysiące lat przed naszą erą, toczono o nią wojny, używano jako pieniędzy. Nie tylko sprawiała, że jedzenia było smaczniejsze, ale także zapobiegała jego psuciu.

DIAMENT I BRYLANT

Diament to rzadki minerał tworzący przezroczyste kryształy. To że jest rzadki, bardzo twardy i wyjątkowo ozdobny sprawia, iż jest poszukiwanym oraz cennym minerałem. Oszlifowany diament nosi nazwę brylantu i jest wykorzystywany w przemyśle jubilerskim.

GRAFIT

Bliskim krewnym diamentu jest grafit, wykorzystywany choćby do produkcji ołówków. Podobnie jak jego brat zbudowany jest wyłącznie z atomów węgla. W graficie atomy wiążą się ze sobą, tworząc warstwy, które łatwo oddzielić, choćby przez pocieranie o papier, w diamencie natomiast łączą się w sztywne sieci, dając minerał o największej znanej twardości.

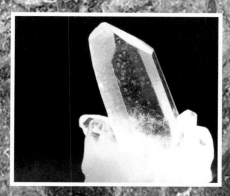

KWARC

Jeśli mają dość miejsca i czasu, minerały krystalizują w regularne formy. Kiedy jednak ciekła skała ulega schładzaniu, kryształy nie zdążą się w pełni wykształcić. Im szybsze stygnięcie, tym mniejsze kryształy. W wolno stygnącym pod powierzchnią Ziemi granicie bez problemu można dostrzec kryształy kwarcu, trudno natomiast w ryolicie, który jest wynikiem szybkiego krzepnięcia lawy na powierzchni Ziemi.

RUBIN I SZAFIR

Korund jest bardzo twardym bezbarwnym minerałem. Kolorów dodają mu domieszki innych substancji, czyniąc z niego poszukiwany kamień ozdobny. Korund zanieczyszczony chromem ma barwę czerwoną i zwie się rubinem, a zanieczyszczony żelazem i tytanem to niebieskawy szafir.

BURSZTYN I PERŁA

Nie wszyscy o tym wiedzą, ale minerałami nie są substancje pochodzenia roślinnego lub zwierzęcego. Nie można zatem nazwać minerałem ani bursztynu (skamieniałej żywicy drzew), ani perły (wytworu płaszcza małży).

SKAŁY
magmowe, osadowe i metamorficzne

Naturalne nagromadzenia minerałów nazywa się skałami. Skały tworzyły
się już u zarania historii Ziemi, kiedy powstawała pierwsza skorupa,
i tworzą się do dziś. Znamy wiele ich rodzajów o najróżniejszym wyglądzie
i właściwościach. Jednak nie są one wieczne – z czasem ulegają zniszczeniu,
by utworzyć inne rodzaje skał.

Powstawanie i przemiany skał to najczęściej procesy bardzo powolne, bo trwające nieraz
miliony, a nawet dziesiątki milionów lat. Zdarza się jednak, że skały tworzą się bardzo
gwałtownie, na przykład podczas wybuchu wulkanu i wypływu lawy. Rodzaje skał
oraz ich przemiany poznamy na przykładzie popularnego granitu.

SKAŁY MAGMOWE

Powstają z roztopionej gorącej magmy, głęboko pod powierzchnią ziemi. Powolne stygnięcie pozwala na wykształcenie się dużych kryształów.

SKAŁY OSADOWE

Powstają z produktów niszczenia rozmaitych skał (w tym również innych osadowych). Proces niszczenia skał nazywamy wietrzeniem.

SKAŁY METAMORFICZNE

Powstają na drodze przeobrażenia (metamorfozy) innych rodzajów skał wskutek działania wysokiej temperatury, wysokiego ciśnienia lub obu tych czynników jednocześnie.

GRANIT — JEGO POWSTAWANIE I PRZEMIANY

- Kiedy magma szybko zastyga blisko powierzchni lub wręcz na niej, powstaje drobnokrystaliczna skała o składzie identycznym jak granit, zwana RYOLITEM.

- GRANIT to najczęściej jasna skała, składająca się przede wszystkim z trzech minerałów: przezroczystego kwarcu, białego lub różowego skalenia potasowego, a także z ciemnych kryształów łyszczyku. Granit tworzy się podczas powolnego stygnięcia ogromnych zbiorników magmy głęboko pod ziemią. Powstałe w ten sposób masywy są czasem wynoszone i tworzą góry. Granit jest twardą oraz wytrzymałą skałą, dlatego chętnie wydobywa się go jako materiał budowlany i ozdobny.

- PEGMATYT to skała powstała po zakrzepnięciu głównej masy magmy z jej resztek. Powolne krzepnięcie pozwoliło jej na wytworzenie ogromnych nieraz kryształów. Mimo tego składem skała ta odpowiada granitowi.

- Wietrzenie granitów powoduje ich rozpad na drobne okruchy skalne. Najmniej odpornym na niszczenie minerałem granitu jest łyszczyk, dlatego jako pierwszy zostaje silnie rozdrobniony, a potem wywiany lub wypłukany z osadu. Pozostałe drobiny kwarcu i skalenia tworzą luźny PIASEK ARKOZOWY.

- Po długim czasie ziarna piasku arkozowego mogą ulec spojeniu. Powstaje wtedy skała osadowa zwana ARKOZĄ.

- Najbardziej odpornym na wietrzenie minerałem granitu jest kwarc i często zdarza się, że tylko on pozostanie w osadzie. Taką luźną skałę nazywamy PIASKIEM.

- Kiedy ziarna piasku zlepią się w litą masę, powstaje skała zwana PIASKOWCEM.

- Duże nagromadzenie osadów (np. w morzach czy oceanach) może powodować ich zapadanie się pod ciężarem wciąż tworzących się nowych warstw. Gdy piaskowiec znajdzie się dość głęboko, działają na niego ciśnienie i wysoka temperatura z wnętrza Ziemi. Ziarna ulegają wtedy częściowemu przetopieniu. Tak powstaje bardzo twarda skała zwana KWARCYTEM.

- Granit poddany działaniu wysokiego ciśnienia i temperatury we wnętrzu Ziemi ulega częściowemu przetopieniu. Tak powstaje GNEJS, który składem niczym nie różni się od granitu, z którego powstał. Gnejs, podobnie jak granit, może na skutek wynoszenia znaleźć się na powierzchni, a wtedy ulegnie wietrzeniu.

- Wszystkie rodzaje skał, jeśli tylko znajdą się wystarczająco głęboko pod powierzchnią ziemi, ulegają stopieniu i wchodzą w skład magmy, z której z kolei powstają SKAŁY MAGMOWE. Z magm tworzą się skały o różnym składzie mineralnym. Każda z nich ma swoją własną nazwę. Jedną z tych skał jest granit.

WODA
Jej obieg w przyrodzie

Bez występowania wody w płynnej postaci niemożliwe byłoby powstanie
i rozwój życia na Ziemi. Woda bezustannie krąży w przyrodzie,
zasilając chmury, rzeki i oceany, nawadniając gleby oraz
wypełniając komórki istot żywych.

OBIEG WODY W PRZYRODZIE

Krążenie wody w przyrodzie zachodzi dzięki energii cieplnej Słońca oraz sile ciężkości, które sprawiają, że woda wciąż się przemieszcza między atmosferą, hydrosferą a litosferą. Na cykl składa się: parowanie z powierzchni Ziemi, przemieszczanie się pary wodnej w atmosferze, kondensacja pary wodnej i opady atmosferyczne, wsiąkanie wody opadowej w głąb powierzchni Ziemi, spływ powierzchniowy oraz odpływ podziemny.

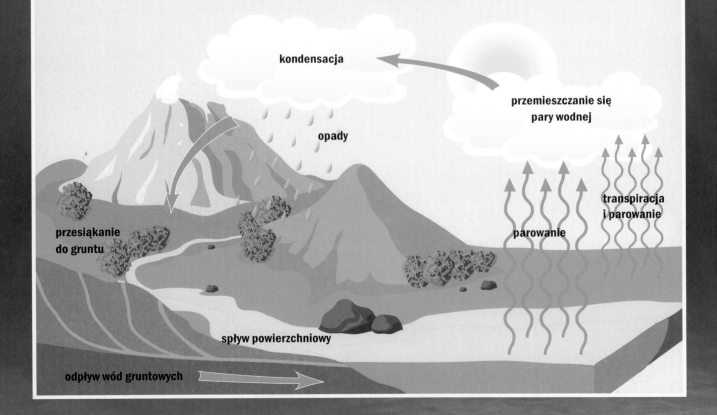

kondensacja

przemieszczanie się
pary wodnej

opady

transpiracja
i parowanie

przesiąkanie
do gruntu

parowanie

spływ powierzchniowy

odpływ wód gruntowych

WODA W LICZBACH

- Woda zajmuje ponad 70% powierzchni naszej planety.

- Ponad 97% jej objętości to wody słone, a zaledwie niecałe 3% — wody słodkie.

- 75% zasobów wody słodkiej zmagazynowanych jest w lodowcach, a 25% pod powierzchnią. Zalewie 1% wody słodkiej (a więc 0,05% objętości całej wody na Ziemi) to woda w jeziorach, rzekach i atmosferze.

- Ponadto duże ilości wody występują w głębszych warstwach Ziemi. Nie potrafimy jednak oszacować jej ilości.

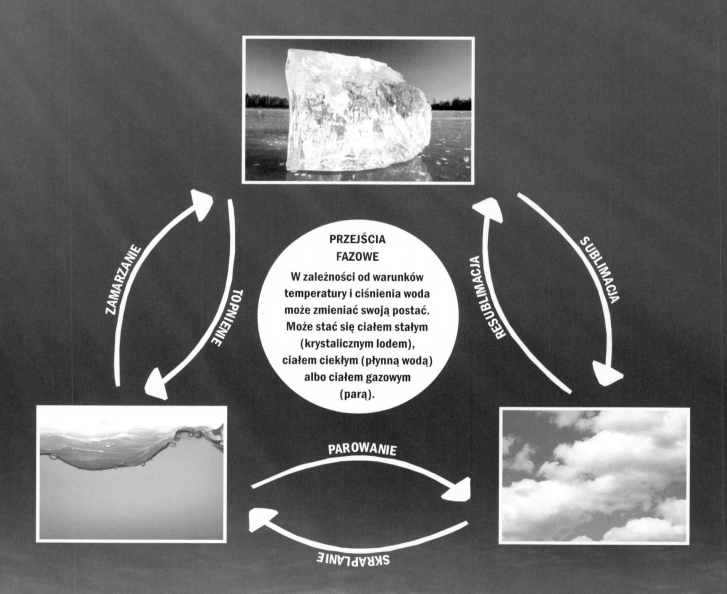

PRZEJŚCIA FAZOWE

W zależności od warunków temperatury i ciśnienia woda może zmieniać swoją postać. Może stać się ciałem stałym (krystalicznym lodem), ciałem ciekłym (płynną wodą) albo ciałem gazowym (parą).

ZAMARZANIE

TOPNIENIE

RESUBLIMACJA

SUBLIMACJA

PAROWANIE

SKRAPLANIE

ATMOSFERA
Ochronna powłoka gazowa

Ziemska atmosfera to złożona głównie z azotu gruba powłoka gazowa otaczająca cały glob. Dzięki niej temperatura na Ziemi jest w miarę wyrównana, a do jej powierzchni nie docierają szkodliwe promienie słoneczne, mogące zagrozić życiu. Ogrzanie lub ochładzanie powietrza, zmiany jego ciśnienia, ruch obrotowy Ziemi oraz inne czynniki powodują, że powietrze w atmosferze nieustannie się miesza. Stale zmieniająca się atmosfera dostarcza nam widowiskowych zjawisk.

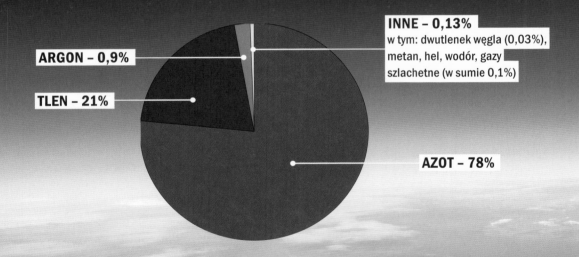

ARGON – 0,9%

INNE – 0,13%
w tym: dwutlenek węgla (0,03%), metan, hel, wodór, gazy szlachetne (w sumie 0,1%)

TLEN – 21%

AZOT – 78%

PROCENTOWY SKŁAD POWIETRZA PRZY POWIERZCHNI (BEZ PARY WODNEJ)

Zawartość poszczególnych gazów w atmosferze zmienia się wraz z wysokością, choć do 25 kilometrów pozostaje w zasadzie bez zmian. Na wysokości 180 do 500 kilometrów tlen staje się dominującym gazem, a powyżej 750 kilometrów najwięcej jest helu. Z kolei zawartość pary wodnej podlega bardzo wyraźnym zmianom. Najwięcej jest jej tuż przy powierzchni, na wysokości 1,5 kilometra jest jej już o połowę mniej. Powyżej 12 kilometrów w zasadzie nie występuje.

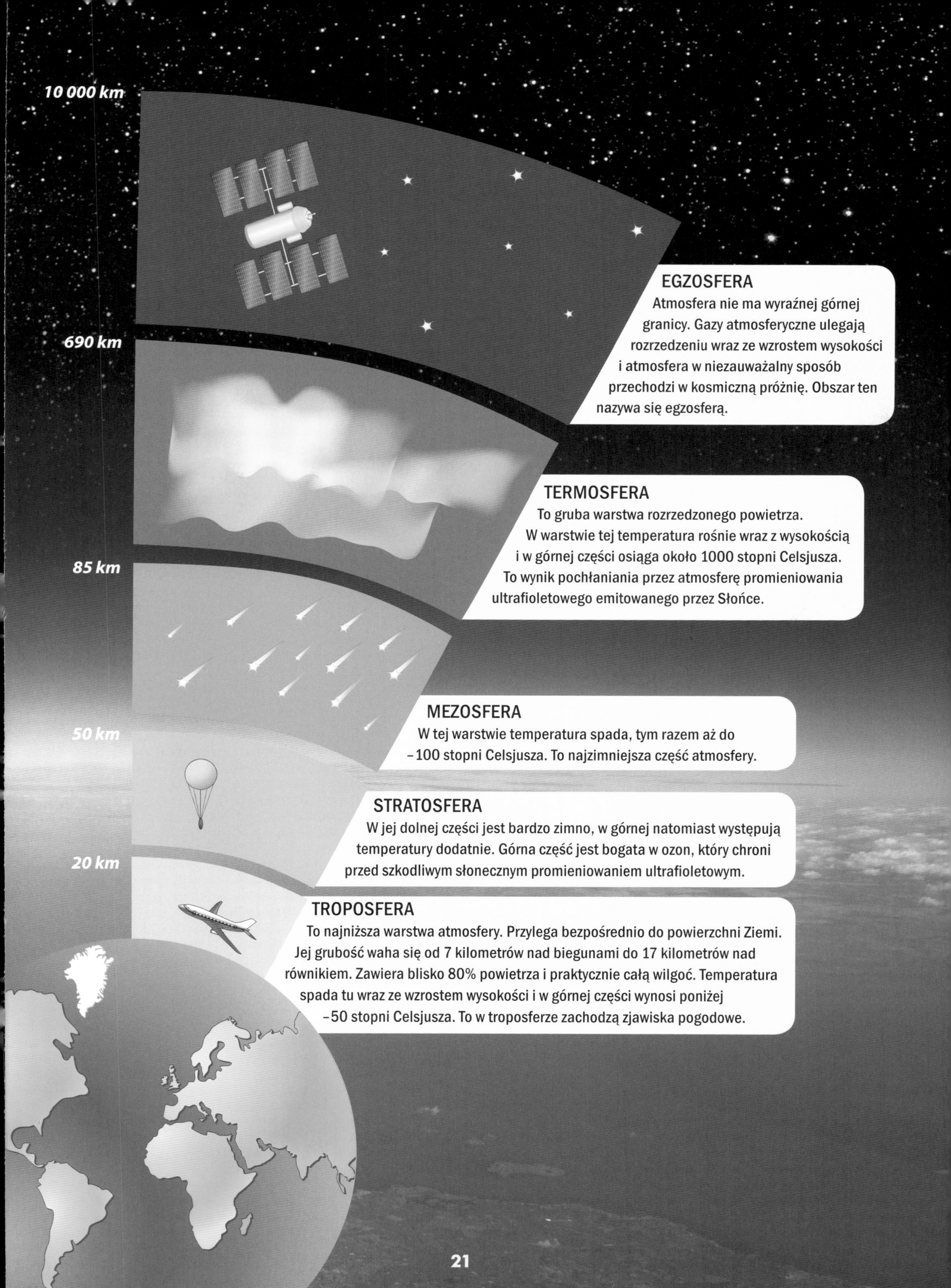

10 000 km

690 km

85 km

50 km

20 km

EGZOSFERA

Atmosfera nie ma wyraźnej górnej granicy. Gazy atmosferyczne ulegają rozrzedzeniu wraz ze wzrostem wysokości i atmosfera w niezauważalny sposób przechodzi w kosmiczną próżnię. Obszar ten nazywa się egzosferą.

TERMOSFERA

To gruba warstwa rozrzedzonego powietrza. W warstwie tej temperatura rośnie wraz z wysokością i w górnej części osiąga około 1000 stopni Celsjusza. To wynik pochłaniania przez atmosferę promieniowania ultrafioletowego emitowanego przez Słońce.

MEZOSFERA

W tej warstwie temperatura spada, tym razem aż do – 100 stopni Celsjusza. To najzimniejsza część atmosfery.

STRATOSFERA

W jej dolnej części jest bardzo zimno, w górnej natomiast występują temperatury dodatnie. Górna część jest bogata w ozon, który chroni przed szkodliwym słonecznym promieniowaniem ultrafioletowym.

TROPOSFERA

To najniższa warstwa atmosfery. Przylega bezpośrednio do powierzchni Ziemi. Jej grubość waha się od 7 kilometrów nad biegunami do 17 kilometrów nad równikiem. Zawiera blisko 80% powietrza i praktycznie całą wilgoć. Temperatura spada tu wraz ze wzrostem wysokości i w górnej części wynosi poniżej – 50 stopni Celsjusza. To w troposferze zachodzą zjawiska pogodowe.

21

EFEKT CIEPLARNIANY
Zmiany w klimacie Ziemi

Efekt cieplarniany powstaje wtedy, gdy w atmosferze kumuluje się
duża objętość takich gazów jak dwutlenek węgla czy metan. Rozgrzana
przez promienie Słońca powierzchnia Ziemi wytwarza promieniowanie
podczerwone, które odbija się od warstwy tych gazów i nie może uciec
w przestrzeń kosmiczną. Atmosfera coraz bardziej się nagrzewa,
nie mogąc pozbyć się nadmiaru ciepła. Powoduje to poważne zmiany
w klimacie oraz topnienie lodowców i lądolodów.

**SKUTKI
OCIEPLENIA**

Ocieplenie klimatu może
przynieść katastrofalne skutki.
Naukowcy przewidują, że dojdzie
do zaburzenia produkcji żywności,
zniszczone zostaną lasy, zatopione
obszary przybrzeżne, a miliony
osób będzie musiało
migrować.

Słońce

Przez dwutlenek węgla nagromadzony w stratosferze
ciepło odbija się od warstwy gazów i wraca na Ziemię

ZORZA POLARNA
Niezwykły spektakl natury

Zorza polarna to bardzo widowiskowe zjawisko świecenia górnych warstw atmosfery na skutek bombardowania jej elektronami wyrzucanymi przez Słońce. Zorze obserwuje się w okolicach biegunowych i podbiegunowych.

Świetlne smugi, draperie, łuki, kurtyny i girlandy mogą się mienić różnymi kolorami, w zależności od tego, jakie cząsteczki gazu lub atomu zostały wzbudzone — tlen świeci na zielono i czerwono, azot na purpurowo, a wodór i hel na niebiesko

KLIMAT
i zjawiska atmosferyczne

Klimat to warunki pogodowe charakterystyczne dla danego obszaru.
Są one zależne przede wszystkim od położenia geograficznego,
czyli wysokości nad poziomem morza i szerokości geograficznej.
Ta ostatnia określa wielkość otrzymywanego promieniowania słonecznego,
natomiast wraz z wysokością zmienia się temperatura, ciśnienie
i wilgotność powietrza.

STREFY KLIMATYCZNE I TYPY KLIMATU

Czynniki wpływające na klimat mają tendencję do strefowości, stąd i klimaty układają się strefowo, w przybliżeniu równoleżnikowo. Wyróżniamy strefę:
- równikową
- zwrotnikową
- podzwrotnikową
- umiarkowaną (chłodną i ciepłą)
- okołobiegunową.

W każdej z tych stref istnieje kilka typów klimatu, zależnych między innymi od wzniesienia terenu i odległości od oceanu. Przykładowo są to klimaty: górski, morski, kontynentalny, suchy czy wilgotny. Klimat Polski jest zazwyczaj określany jako umiarkowany, przejściowy między morskim a kontynentalnym.

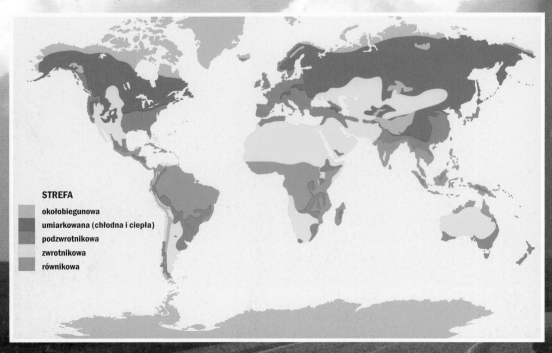

STREFA
- okołobiegunowa
- umiarkowana (chłodna i ciepła)
- podzwrotnikowa
- zwrotnikowa
- równikowa

ZJAWISKA ATMOSFERYCZNE I POWODOWANE PRZEZ NIE KLĘSKI ŻYWIOŁOWE

Czasami siły natury przybierają groźne i niszczycielskie oblicze. BURZA potrafi przynieść ze sobą gwałtowny i obfity deszcz wywołujący powódź, gradobicie czy przerażające, gwałtowne TORNADO. Może także wzniecić wielkie pożary. Bywa, że niż atmosferyczny niesie sztormowe wiatry czy długotrwałe deszcze prowadzące do katastrofalnych powodzi. W obszarach okołozwrotnikowych niż może skutkować CYKLONEM, który w jednych rejonach świata nazywany jest huraganem, a w innych — tajfunem. Wyż natomiast potrafi przynieść SUSZĘ, a na obszarach o mocno zanieczyszczonym powietrzu spowodować klęskę SMOGU.

Na zdjęciach klęski żywiołowe (od góry i lewej) — burza, tropikalny cyklon, szusza, pożar

25

RZEŹBA ZIEMI
Ukształtowanie powierzchni

Choć góry czy oceany wydają się niezniszczalne i wieczne, powierzchnia skorupy ziemskiej wciąż się zmienia. Morza zostają sprasowane przez zbliżające się do siebie masy kontynentów i wypiętrzone w wysokie góry. Góry rozpadają się na pył, który gromadzi się w morzach, by w przyszłości stać się materiałem na nowe góry... Te ciągłe zmiany zawdzięcza Ziemia dwóm podstawowym zjawiskom: wynoszeniu i wietrzeniu.

HIPSOGRAFICZNA KRZYWA ZIEMI

Strukturę wysokościową i głębokościową na Ziemi przedstawia krzywa hipsograficzna. Pokazuje ona ukształtowanie terenu i pozwala na obliczenie objętości lądów i mórz. Na Ziemi istnieje kilka pięter wysokościowych: rowy oceaniczne, dna basenów oceanicznych, stok kontynentalny, szelfy, niziny i wyżyny i góry.

Różnica między najwyższym i najniższym punktem wynosi ponad 20 kilometrów. Elementami pionowego ukształtowania lądu są góry, wyżyny, niziny, depresje, szelf i stok kontynentalny (dwa ostatnie znajdują się pod powierzchnią wody, ale zaliczane są do obszarów lądowych), a dna oceanu — głębie oceaniczne, grzbiety oceaniczne i rowy oceaniczne.

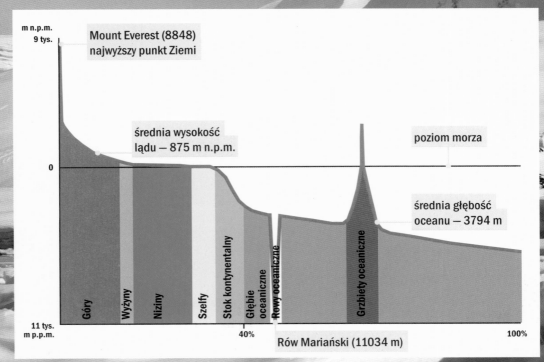

m n.p.m.
9 tys.

Mount Everest (8848) najwyższy punkt Ziemi

średnia wysokość lądu — 875 m n.p.m.

poziom morza

0

średnia głębość oceanu — 3794 m

Góry · Wyżyny · Niziny · Szelfy · Stok kontynentalny · Głębie oceaniczne · Rowy oceaniczne · Grzbiety oceaniczne

11 tys.
m p.p.m.

40%

100%

Rów Mariański (11034 m)

Góry powstają zazwyczaj w procesie wypiętrzenia lub wskutek długotrwałych procesów wulkanicznych. Ostateczny kształt nadają im procesy rzeźbotwórcze — na przykład wietrzenie czy erozja. Na zdjęciu himalajski Mount Everest, najwyższy szczyt świata (8848 m n.p.m.).

GEOMORFOLOGIA

Rzeźba powierzchni Ziemi to wszystkie formy na powierzchni naszej planety, które zostały ukształtowane w wyniku procesów geologicznych. Badaniem, opisem ich rozwoju i określaniem ich wieku zajmuje się geomorfologia.

Wietrzenie i erozja bez ustanku zachodzą na powierzchni Ziemi i w jej przypowierzchniowej strefie. Procesy te prowadzą do rozpadu skał. Zanim jednak do tego dojdzie, skały przybierają fantazyjne kształty. Na zdjęciu (od lewej): Park Narodowy Arches (Stany Zjednoczone) i wapienne twory w stanie Wiktoria w Australii.

ZAMIESZKAŁA PLANETA
Powstanie i ewolucja życia na Ziemi

Ziemia to jedyna planeta, na której rozwinęły się skomplikowane,
a nawet inteligentne formy życia. Jest więc wyjątkowym miejscem
w Układzie Słonecznym. Nie znamy podobnych miejsc we wszechświecie.
Życie na Ziemi powstało prawdopodobnie 3,8 miliarda lat temu. To zaledwie
niecałe 800 milionów lat po utworzeniu się planety i niemal natychmiast
po zakrzepnięciu skorupy i utworzeniu się oceanów.

JAK POWSTAŁO ŻYCIE?

Są dwie teorie powstania życia na Ziemi. Jedna mówi o tym, że życie narodziło się w ziemskich oceanach na skutek przemian chemicznych zawartych w nich cząsteczek. Inna zakłada, że prymitywne formy życia przybyły na Ziemię wraz z upadającymi meteorytami, życie zaś powstało na jakiejś nieznanej nam, odległej planecie.

Życie na Ziemi mogło powstać i może nadal trwać dzięki obecności wody w stanie płynnym. Gdyby nasza planeta była bliżej Słońca, cała woda znajdowałaby się w atmosferze w postaci chmur, a gdyby była dalej — woda zamieniłaby się w lód.

Ta wyjątkowa sytuacja nie będzie jednak trwać wiecznie. Za jakiś miliard lat życie na Ziemi stanie się utrudnione, a może wręcz niemożliwe. Słońce będzie świecić kilkakrotnie razy jaśniej. Ostateczny kres życia nastąpi za około cztery miliardy lat, kiedy Słońce spuchnie i stanie się gwiazdą zwaną czerwonym karłem. Dojdzie wtedy do takiego rozgrzania powierzchni Ziemi, że nie tylko woda wyparuje, ale ulegnie przetopieniu ziemska skorupa.

HISTORIA ŻYCIA NA ZIEMI

1,2 GA NAJSTARSZE ORGANIZMY WIELOKOMÓRKOWE
Przez większość czasu życie rozwijało się w oceanach. Na lądach nie było to możliwe z powodu silnego promieniowania ultrafioletowego, które niszczyło organizmy. Woda natomiast stanowiła naturalny filtr UV. Dopiero produkowanie tlenu przez organizmy żywe doprowadziło do powstania ochronnej warstwy ozonowej.

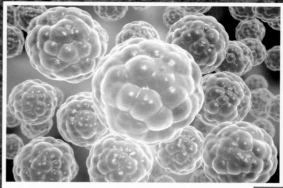
Piersze organizmy wielokomórkowe — wizja artysty

0,57 GA PIERWSZE ZWIERZĘTA TKANKOWE
Żyjące w tym okresie zwierzęta bezkręgowe wytwarzały pancerze, skorupki i inne elementy twarde, dzięki którym wiele z nich zachowało się jako skamieniałości.

0,47 GA PIERWSZE ROŚLINY NACZYNIOWE
Miały niewielkie rozmiary i prostą budowę; zasiedlały zbiorniki słodkowodne, a później obszary lądowe.

0,43 GA NAJSTARSZE RYBY
Były to tak zwane ryby pancerne; ich ciało pokrywały płytki kostne.

Trylobit — kopalny stawonóg

0,38 GA PIERWSZE OWADY
Najwcześniejsze owady były bezskrzydłe, później pojawiły się ważki, koniki polne i… karaluchy.

0,37 GA PIERWSZE KRĘGOWCE LĄDOWE
Pierwszym etapem rozwoju kręgowców lądowych były płazy, które jednak szybko straciły na znaczeniu na rzecz gadów.

0,25 GA NAJWIĘKSZE WYMIERANIE W HISTORII
Wyginęło wtedy około 96% organizmów.

Ważka kopalna

0,23 GA PIERWSZE DINOZAURY
Pierwsze „straszne jaszczury" były wielkości psa i poruszały się szybko na dwóch nogach.

0,22 GA PIERWSZE SSAKI
Były niewielkie; dopóki nie wymarły dinozaury, nie odgrywały większej roli w środowisku.

0,15 GA PIERWSZE PTAKI
Wywodziły się od dinozaurów.

0,06 GA WYMARCIE DINOZAURÓW
Wraz z dinozaurami z powierzchni Ziemi zniknęło niemal dwie trzecie organizmów.

0,03 GA EWOLUCJA CZŁOWIEKOWATYCH

Dinozaur, czyli „starszny jaszczur" — wizja artysty

KSIĘŻYC
i jego wpływ na Ziemię

Księżyc jest najbliższym Ziemi ciałem niebieskim. Jego skład wskazuje jednoznacznie, że jest jej „dzieckiem" — powstał z wyniku oderwania dużych mas skorupy i płaszcza ziemskiego w jakiejś dawnej katastrofie. Choć znajduje się setki tysięcy kilometrów od Ziemi, nie sposób nie zauważyć jego wpływu na naszą planetę.

DWIE STRONY KSIĘŻYCA

Na skutek powiązania ruchu obrotowego Ziemi z ruchem obiegowym Księżyca dla ziemskiego obserwatora widoczna jest zawsze tylko jedna strona Księżyca. W oczy rzucają się na niej gładkie, ciemne miejsca. Przez długi czas uważano, że to morza wypełnione wodą. Wierzono więc, że na Księżycu istnieje życie. W rzeczywistości plamy to zastygła lawa, która rozlała się po powierzchni naszego satelity jakieś 3—4 miliardy lat temu. Strona przeciwna, której z Ziemi nie jesteśmy w stanie dojrzeć, jest gęsto usiana kraterami uderzeniowymi.

Mimo że masa naszego satelity to tylko ułamek masy Ziemi, przyciąganie Księżyca powoduje przemieszczanie się ogromnych objętości wody w morzach i oceanach. Zjawisko to nazywamy pływami. Kiedy Słońce i Księżyc są w jednej linii (czyli podczas nowiu i pełni Księżyca), ich oddziaływanie się sumuje i powstają największe pływy, tak zwane syzygijne. Kiedy natomiast nie są w jednej linii (czyli w pierwszej lub trzeciej kwadrze Księżyca), pływy są mniejsze.

PRZYPŁYWY I ODPŁYWY

Pływy, czyli regularne, dobowe przypływy i odpływy morza były znane ludziom od dawna. Jednak dopiero w XVII wieku zajęto się nimi naukowo. W 1687 roku zjawisko opisał matematycznie Isaac Newton.

FAZY KSIĘŻYCA

Słońce oświetla widziany z Ziemi Księżyc pod różnym kątem. Kiedy znajduje się przed Księżycem, widzimy całą jego tarczę i mówimy wtedy o pełni. Kiedy Słońce znajduje się za Księżycem, cała tarcza pogrąża się w cieniu i takie zjawisko nazywamy nowiem.

SPIS TREŚCI

ZDJĘCIA I ILUSTRACJE

g — góra, ś — środek, d — dół, l — lewa, p — prawa

© Copyright by Damidos sp. z o.o. 2013

Tekst: Mariusz Lubka

Redakcja: Iwona Baturo | baturo.pl

Korekta: Agnieszka Szmuc

Projekt graficzny i skład: Iwona Baturo | baturo.pl

Projekt okładki: Magdalena Muszyńska, Izabela Surdykowska-Jurek | Czartart

Wyłączny dystrybutor:

TROY-DYSTRYBUCJA sp. z o.o.

tel./faks: 32 258 95 79

e-mail: troy@troy.net.pl

Zapraszamy na: www.troy.net.pl

Znajdź nas na:

www.facebook.com/TROY.DYSTRYBUCJA